Mechanical Wave Vibrations

Mechanical Wave Vibrations

Analysis and Control

Chunhui Mei
University of Michigan-Dearborn
Dearborn, MI, USA

This edition first published 2023

Registered Offices
John Wiley & Sons, Inc., 111 River Street, Hoboken, NJ 07030, USA
John Wiley & Sons Ltd, The Atrium, Southern Gate, Chichester, West Sussex, PO19 8SQ, UK

For details of our global editorial offices, customer services, and more information about Wiley products visit us at www.wiley.com.

Wiley also publishes its books in a variety of electronic formats and by print-on-demand. Some content that appears in standard print versions of this book may not be available in other formats.

Library of Congress Cataloging-in-Publication Data
Names: Mei, Chunhui, author.
Title: Mechanical wave vibrations : analysis and control / Chunhui Mei.
Description: Chichester, West Sussex, UK : John Wiley & Sons, 2023. | Includes bibliographical references and index. | Summary: "In this book titled Mechanical Wave Vibrations, vibrations in solid structures are viewed as waves that propagate along uniform waveguides, and are reflected and transmitted incident upon discontinuities similar to light and sound waves. The wave vibration description is particularly useful for structures consisting of onedimensional structural elements where a finite number of waves with given directions of propagation exist. In conventional textbooks on mechanical vibrations vibration problems in distributed structures are solved as boundary value problems. The coverage is typically limited to the analysis of a single type of vibration in a simple beam element due to the complexity in boundaries imposed by builtup structures."-- Provided by publisher.
Identifiers: LCCN 2022060202 (print) | LCCN 2022060203 (ebook) | ISBN 9781119135043 (hardback) | ISBN 9781119135067 (pdf) | ISBN 9781119135050 (epub) | ISBN 9781119135074 (ebook)
Subjects: LCSH: Vibration. | Waves.
Classification: LCC QC136 .M45 2023 (print) | LCC QC136 (ebook) | DDC 531/.32--dc23/eng20230429
LC record available at https://lccn.loc.gov/2022060202
LC ebook record available at https://lccn.loc.gov/2022060203

Cover Image: Courtesy of the Author
Cover Design: Wiley

Set in 9.5/12.5pt STIXTwoText by Integra Software Services Pvt. Ltd, Pondicherry, India
Printed and bound by CPI Group (UK) Ltd, Croydon, CR0 4YY

C9781119135043_030723

To my mother Yaoguang Zeng who still takes care of me the same way she did before I left home for college, and my late father Fuchu Mei who was always proud whatever my pursuit was.

Contents

Preface

In this book titled *Mechanical Wave Vibrations*, vibrations in solid structures are viewed as waves that propagate along uniform waveguides and are reflected and transmitted incident upon discontinuities, similar to light and sound waves. The wave vibration description is particularly useful for structures consisting of one-dimensional structural elements where a finite number of waves with given directions of propagation exist.

In conventional textbooks on mechanical vibrations, vibration problems in distributed structures are solved as boundary value problems. The coverage is typically limited to the analysis of a single type of vibration in a simple beam element because of the complexity in boundaries of each structural element imposed by built-up structures.

From the wave standpoint, however, a structure, regardless of its complexity, consists of only two components, namely, structural elements and structural joints. Vibrations propagate along uniform structural elements, and are reflected and/or transmitted at structural discontinuities such as joints and boundaries. Assembling these propagation, reflection, and transmission relationships provides a concise and systematic approach for vibration analysis of a complex structure.

Unlike the conventional modal vibration analysis approach that has been taught in standard vibration courses for decades, the wave vibration analysis approach is seldom taught to students and the related knowledge is limited to the research community through journal or conference publications.

This textbook is written with both undergraduate and graduate students in mind. The author hopes to see this wave-based vibration analysis approach incorporated into the engineering curriculum to allow engineering students a better understanding of mechanical vibrations and to equip them with additional tools for solving practical vibration problems. In addition, researchers and educators in the vibration and control field will find this book helpful.

This textbook is written in such a way that there is no prior knowledge on vibrations needed, although it requires knowledge on mechanics of materials and dynamics at an undergraduate level. As a result, courses based on this textbook can be offered prior to, concurrent with, or after any conventional courses on mechanical vibrations. Students are expected to either be familiar with or be willing to learn MATLAB technical computing language. Sample MATALB scripts for numerical simulations are provided at the end of most chapters.

This book is organized as follows. Chapter 1 is devoted to the coverage of sign conventions and the derivation of equations of motion using the Newtonian approach. Sign conventions, which are often a source of error for engineering analysis, play important roles not only in the derivation of governing equations of motion for bending, longitudinal, and torsional vibrations, but also in wave vibration analysis.

In Chapters 2 and 3, longitudinal and bending vibrations in beams are studied, both based on elementary vibration theories. Fundamental concepts related to wave vibration analysis are introduced, such as the propagation of vibration waves along uniform structural elements (the waveguides) and the reflection of vibration waves at classical and non-classical boundaries (the discontinuities). Free and forced longitudinal and bending vibrations are analyzed from the wave vibration standpoint. Natural frequencies, modeshapes, as well as steady state frequency responses are obtained and compared with experimental results.

Chapter 4 studies both longitudinal and bending waves in beams on a Winkler elastic foundation. The concepts of cut-off frequency and wave mode transition are introduced. The analysis is presented in non-dimensional form, a different form than the previous chapters.

Chapter 5 studies vibration waves in composite beams, in which the concept of coupled waves caused by material coupling in a composite beam is introduced.

In Chapter 6, coupled vibration motions along the radial and tangential directions in a thin curved beam are analyzed based on Love's vibration theory. Cut-off frequencies, wave mode transitions, and dispersion relationships are studied. Wave reflections at classical and non-classical boundaries are derived. Natural frequencies, modeshapes, as well as steady state frequency responses are obtained from the wave vibration standpoint.

Chapters 7 and 8 cover out-of-plane and in-plane vibrations in rectangular plates with at least one pair of opposite edges simply supported, which is required for closed form solutions to exist in plates.

Chapters 9 and 10 advance the coverage of bending and longitudinal vibrations of Chapters 3 and 2 by taking into account effects that are neglected in the elementary theories. For example, the effect of rotary inertia and transverse shear deformation for bending vibration neglected in the Euler–Bernoulli bending vibration theory, are included in part or in full by the Rayleigh, Shear, and Timoshenko bending vibration theories. Free and forced longitudinal and bending vibrations are analyzed with comparison to experimental results.

In Chapters 11 and 12, vibrations in built-up planar and space frames are studied. An angle joint, in general, introduces wave mode conversion. As a result, multiple wave types co-exist in built-up frame structures. For example, in a planar frame that undergoes in-plane vibrations, in-plane bending and longitudinal waves co-exist. In built-up space frames, in- and out-of-plane bending, longitudinal, and torsional vibrations co-exist. Solving vibration problems of such complexity has proven to be challenging by the conventional modal analysis approach, however, the wave-based vibration analysis approach is seen to offer a concise assembly approach for systematically analyzing complex vibrations in built-up planar as well as space frames.

The final two chapters of this book, Chapters 13 and 14, are devoted to vibration control from the wave standpoint, either by adding discontinuities to the path of wave propagation for the purpose of altering vibration characteristics of a structure or by minimizing the transmitted and/or reflected vibration energy in a structure.

It is recommended to cover Chapters 1, 2, 3, 4, 5, 9, 11, 13, and 14 in an introductory course on Mechanical Wave Vibration at undergraduate level. The remaining chapters can be selected and added at an instructor's discretion for a similar course at graduate level.

Acknowledgement

First and foremost, I wish to thank my Ph.D. advisor and lifelong role model, Brian Mace. Brian brought me into the field of mechanical wave vibrations and trustfully handed me this book project. He has always been there whenever I needed guidance and encouragement.

I am grateful to my academic brother Neil Harland. As the first reader of this book manuscript, Neil has provided much valuable and constructive feedback. His time and effort are greatly appreciated.

Most importantly, I wish to thank my mom Yaoguang, my husband Chundao, my sons Yonglu and Yongwei, and my daughter-in-law Yujiang, for their love, support, and patience.

Last but not least, I would like to acknowledge the Mechanical Engineering Department and the College of Engineering and Computer Science at the University of Michigan-Dearborn for jointly purchasing a laptop computer for this book project.

About the Companion Website

This book is accompanied by a companion website:

www.wiley.com/go/Mei/MechanicalWaveVibrations

This website includes

- Instructor Solutions Manual
- Instructor PPT Slides

1

Sign Conventions and Equations of Motion Derivations

Sign conventions and coordinate systems play important roles in wave vibration analysis and in the derivation of governing equations of motion for bending, longitudinal, and torsional vibrations.

In this book, Cartesian coordinate system is adopted. For a planar structure, the x- and y-axis of a two-dimensional Cartesian coordinate system are chosen to be in the plane of the structure. The x-axis is always chosen to be along the longitudinal axis of a member. The axial and shear force are parallel to the x- and y-axis, respectively. Angle φ is defined by the right hand rule rotation from the x-axis to the y-axis.

For an in depth understanding of sign conventions, which are often a source of error for engineering analysis, the governing equations of motion are derived using the Newtonian approach following various sign conventions.

1.1 Derivation of the Bending Equations of Motion by Various Sign Conventions

Figure 1.1 shows the positive sign directions for internal resistant shear force V and bending moment M of four possible sets of conventions. In the figure, subscripts L and R denote parameters related to the left and right side of the cut section, respectively. The set shown in Figure 1.1a is a convention that has been popularly adopted by many authors in textbooks and research papers, while the remaining sets presented in Figures 1.1b to 1.1d are less often adopted.

The best way to interpret a sign convention is to look at how the internal resistant forces and moments deform or rotate the corresponding element. In Sets (a) and (b) shown in Figure 1.1, the shear force is positive when it rotates its element along the positive direction of angle φ. The convention for the bending moment is defined differently. In Set (a) the bending moment is positive when it bends its element concave towards the positive y-axis direction; however, in Set (b) the positive bending moment is when it bends its element convex towards the positive y-axis direction. In Sets (c) and (d) in Figure 1.1, the shear force is positive when it rotates its element along the negative direction of angle φ. The convention for the bending moment is defined differently, in Set (c) the bending moment is positive when it bends its element concave towards the positive y-axis direction; while in Set (d) the positive bending moment is when it bends its element convex towards the positive y-axis direction.

This deformation and rotation based interpretation holds regardless of the orientation of the beam; one only needs to be consistent with the choice of the coordinate system and the definition of positive sign directions.

Consider now, as shown in Figure 1.2, a beam of length L that is subjected to an external distributed transverse load of $f(x,t)$ per unit length. The x-axis is chosen to be along the neutral axis of the beam, t is the time, and $y(x,t)$ is the transverse deflection of the beam.

In the absence of axial loading, the bending equations of motion of the beam derived using the four sets of sign conventions shown in Figure 1.1 are presented below. Figures 1.3a to 1.3d are the free body diagrams of a differential element of the beam according to the four sets of sign conventions of Figures 1.1a to 1.1d, respectively. The bending moments and shear forces on both sides of the differential element are with positive sign directions by the corresponding sign conventions.

Mechanical Wave Vibrations: Analysis and Control, First Edition. Chunhui Mei.
© 2023 Chunhui Mei. Published 2023 by John Wiley & Sons Ltd.
Companion Website: www.wiley.com/go/Mei/MechanicalWaveVibrations

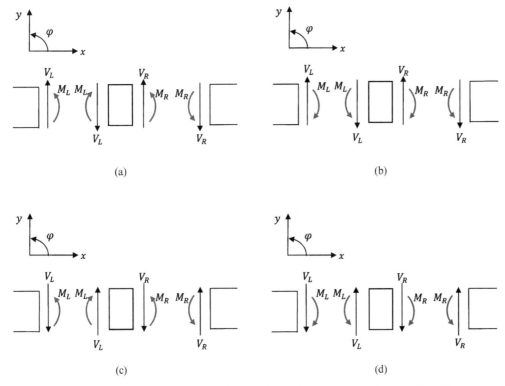

Figure 1.1 Definitions of positive sign directions for internal resistant shear force and bending moment by various sign conventions: (a) sign convention 1, (b) sign convention 2, (c) sign convention 3, and (d) sign convention 4.

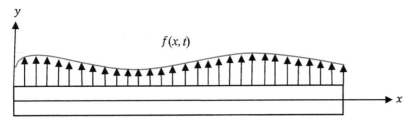

Figure 1.2 A beam in bending vibration.

1.1.1 According to Euler–Bernoulli Bending Vibration Theory

The bending equations of motion in the Euler–Bernoulli (or thin beam) theory are derived based on the following three assumptions. First, the neutral axis does not experience any change in length. Second, all cross sections remain planar and perpendicular to the neutral axis. Third, deformation at the cross section within its own plane is negligibly small. In other words, the rotation of cross sections of the beam is neglected compared to the translation, and the angular distortion due to shear is neglected compared to the bending deformation.

The concept of curvature of a beam is central to the understanding of beam bending. Mathematically, the radius of curvature R of a curve $y(x,t)$ can be found using the following formula

$$R = \frac{\left[1 + \left(\frac{dy(x,t)}{dx}\right)^2\right]^{3/2}}{\frac{d^2y(x,t)}{dx^2}} \tag{1.1}$$

For a beam element in a practical engineering structure that undergoes bending vibration, the transverse deflection of the centerline $y(x,t)$ normally forms a shallow curve because of limitations set forth by engineering design codes on allowable deflection of engineering structures. Consequently, the slope of the deformation curve $\frac{dy(x,t)}{dx}$ is normally very small, and its square is negligible when compared to unity. Therefore, the radius of curvature as defined above can be approximated by

$$R \approx \frac{1}{\dfrac{d^2 y(x,t)}{dx^2}}$$

(1.2)

By definition, the neutral axis, which lies on the *x*-axis, does not experience any change in length. Consequently, the lengths of the neutral axis of the differential element remain the same amount of *dx* before and after deformation, as shown in Figure 1.4.

In the absence of axial loading, the longitudinal strain in the beam is produced only from bending and by definition of strain,

$$\varepsilon = \frac{ds - dx}{dx}$$

(1.3)

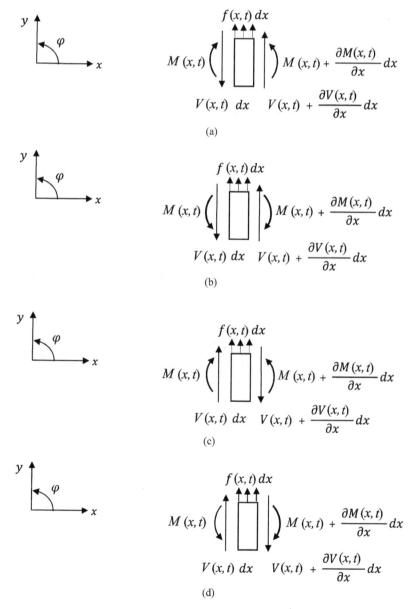

(a)

(b)

(c)

(d)

Figure 1.3 Free body diagram of a beam element in bending vibration by the various sign conventions defined in Figure 1.1: (a) sign convention 1, (b) sign convention 2, (c) sign convention 3, and (d) sign convention 4.

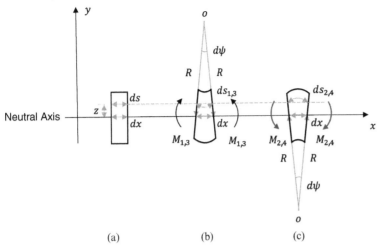

Figure 1.4 Strain and radius of curvature: (a) before deformation, (b) after concave bending deformation, and (c) after convex bending deformation.

There are two types of bending deformations with reference to the positive y-axis, concave and convex, because of internal bending moments M_1 or M_3 and M_2 or M_4, respectively. The normal strains $\varepsilon_{1,3}$ and $\varepsilon_{2,4}$ on the differential element that is a distance z above the neutral axis correspond to the concave and convex deformations shown in Figures 1.4b and 1.4c are

$$\varepsilon_{1,3} = \frac{ds_{1,3} - dx}{dx} = \frac{(R - z)d\psi - Rd\psi}{Rd\psi} = -\frac{z}{R} \tag{1.4a}$$

$$\varepsilon_{2,4} = \frac{ds_{2,4} - dx}{dx} = \frac{(R + z)d\psi - Rd\psi}{Rd\psi} = \frac{z}{R} \tag{1.4b}$$

where R is the radius of curvature of the transverse deflection of the centerline $y(x,t)$, and $\psi(x,t)$ is the angle of rotation of the cross section due to bending. Subscripts 1, 2, 3, and 4 denote parameters related to sign conventions 1, 2, 3, and 4 defined in Figure 1.1.

For concave deformation, strains $\varepsilon_{1,3}$ are negative above the neutral axis (where z is positive) because of compressive normal stress in the region caused by the internal resistant bending moment at the given direction. Strains $\varepsilon_{1,3}$ are positive below the neutral axis (where z is negative) because of tensile normal stress in the region caused by the internal resistant bending moment at the given direction. This explains the negative sign in Eq. (1.4a). For convex deformation, the region above the neutral axis (where z is positive) is subject to positive strain, and below the neutral axis (where z is negative) is subject to negative strain; hence, strains $\varepsilon_{2,4}$ carry the same sign as z, as reflected in Eq. (1.4b).

For homogeneous materials behaving in a linear elastic manner, the stress σ and strain ε are related by the Young's modulus E,

$$\sigma = E\varepsilon \tag{1.5}$$

From Eqs. (1.4a), (1.4b), and (1.5),

$$\sigma_{1,3} = E\varepsilon_{1,3} = -E\frac{z}{R} \tag{1.6a}$$

$$\sigma_{2,4} = E\varepsilon_{2,4} = E\frac{z}{R} \tag{1.6b}$$

Balancing the internal normal stress σ and the internal bending moment M requires

$$M = -\int_A z(\sigma dA) \tag{1.7}$$

Substituting Eqs. (1.6a) and (1.6b) into Eq. (1.7) gives

$$M_{1,3} = -\int_A z\left(-E\frac{z}{R}dA\right) = \frac{E}{R}\int_A z^2 dA = \frac{E}{R}I(x) \tag{1.8a}$$

$$M_{2,4} = -\int_A z\left(E\frac{z}{R}dA\right) = -\frac{E}{R}\int_A z^2 dA = -\frac{E}{R}I(x) \tag{1.8b}$$

In Eqs. (1.7), (1.8a), and (1.8b), A is the area of the cross section and $I(x) = \int_A z^2 dA$ is the area moment of inertia about the centroidal axis that is normal to the plane of bending.

Substituting Eq. (1.2) into Eqs. (1.8a) and (1.8b) gives

$$M_{1,3} = EI(x)\frac{d^2 y(x,t)}{dx^2} \tag{1.9a}$$

$$M_{2,4} = -EI(x)\frac{d^2 y(x,t)}{dx^2} \tag{1.9b}$$

From the free body diagrams of Figure 1.3, the force equations along the y-axis direction obtained by sign conventions 1, 2, 3, and 4 are

$$\left[V(x,t) + \frac{\partial V(x,t)}{\partial x}dx\right] - V(x,t) + f(x,t)dx = \rho A(x)dx\frac{\partial^2 y(x,t)}{\partial t^2}, \qquad 0 < x < L \tag{1.10a}$$

$$\left[V(x,t) + \frac{\partial V(x,t)}{\partial x}dx\right] - V(x,t) + f(x,t)dx = \rho A(x)dx\frac{\partial^2 y(x,t)}{\partial t^2}, \qquad 0 < x < L \tag{1.10b}$$

$$-\left[V(x,t) + \frac{\partial V(x,t)}{\partial x}dx\right] + V(x,t) + f(x,t)dx = \rho A(x)dx\frac{\partial^2 y(x,t)}{\partial t^2}, \qquad 0 < x < L \tag{1.10c}$$

$$-\left[V(x,t) + \frac{\partial V(x,t)}{\partial x}dx\right] + V(x,t) + f(x,t)dx = \rho A(x)dx\frac{\partial^2 y(x,t)}{\partial t^2}, \qquad 0 < x < L \tag{1.10d}$$

where ρ is the volume mass density of the beam. Note that Eqs. (1.10a) and (1.10b) are identical, so are Eqs. (1.10c) and (1.10d).

Simplifying the above equations gives

$$\frac{\partial V(x,t)}{\partial x} + f(x,t) = \rho A(x)\frac{\partial^2 y(x,t)}{\partial t^2}, \qquad 0 < x < L \tag{1.11a}$$

$$\frac{\partial V(x,t)}{\partial x} + f(x,t) = \rho A(x)\frac{\partial^2 y(x,t)}{\partial t^2}, \qquad 0 < x < L \tag{1.11b}$$

$$-\frac{\partial V(x,t)}{\partial x} + f(x,t) = \rho A(x)\frac{\partial^2 y(x,t)}{\partial t^2}, \qquad 0 < x < L \tag{1.11c}$$

$$-\frac{\partial V(x,t)}{\partial x} + f(x,t) = \rho A(x)\frac{\partial^2 y(x,t)}{\partial t^2}, \quad 0 < x < L \tag{1.11d}$$

From the free body diagrams of Figure 1.3, the moment equations by sign conventions 1, 2, 3, and 4, under the assumption that the rotary inertia is negligibly small, are

$$\left[M(x,t)+\frac{\partial M(x,t)}{\partial x}dx\right]-M(x,t)+\left[V(x,t)+\frac{\partial V(x,t)}{\partial x}dx\right]dx + f(x,t)dx\frac{dx}{2} = 0, \qquad 0 < x < L \tag{1.12a}$$

$$-\left[M(x,t)+\frac{\partial M(x,t)}{\partial x}dx\right]+M(x,t)+\left[V(x,t)+\frac{\partial V(x,t)}{\partial x}dx\right]dx + f(x,t)dx\frac{dx}{2} = 0, \quad 0 < x < L \tag{1.12b}$$

$$\left[M(x,t)+\frac{\partial M(x,t)}{\partial x}dx\right]-M(x,t)-\left[V(x,t)+\frac{\partial V(x,t)}{\partial x}dx\right]dx + f(x,t)dx\frac{dx}{2} = 0, \qquad 0 < x < L \tag{1.12c}$$

$$-\left[M(x,t)+\frac{\partial M(x,t)}{\partial x}dx\right]+M(x,t)-\left[V(x,t)+\frac{\partial V(x,t)}{\partial x}dx\right]dx + f(x,t)dx\frac{dx}{2} = 0, \quad 0 < x < L \tag{1.12d}$$

Neglecting second-order terms in dx and canceling appropriate terms, Eqs. (1.12a-d) reduce to

$$\frac{\partial M(x,t)}{\partial x}+V(x,t)=0, \qquad 0 < x < L \tag{1.13a}$$

$$-\frac{\partial M(x,t)}{\partial x}+V(x,t)=0, \quad 0 < x < L \tag{1.13b}$$

$$\frac{\partial M(x,t)}{\partial x}-V(x,t)=0, \qquad 0 < x < L \tag{1.13c}$$

$$-\frac{\partial M(x,t)}{\partial x}-V(x,t)=0, \qquad 0 < x < L \tag{1.13d}$$

From Eqs. (1.9a), (1.9b), and (1.13a-d),

$$V_{1,2}(x,t)=-\frac{\partial}{\partial x}\left[EI(x)\frac{\partial^2 y(x,t)}{\partial x^2}\right], \quad 0 < x < L \tag{1.14a}$$

$$V_{3,4}(x,t)=\frac{\partial}{\partial x}\left[EI(x)\frac{\partial^2 y(x,t)}{\partial x^2}\right], \quad 0 < x < L \tag{1.14b}$$

From Eqs. (1.14a), (1.14b), and (1.11a-d), a unique fourth order partial differential governing equation of a beam in bending, regardless of the sign conventions, is obtained,

$$\frac{\partial^2}{\partial x^2}\left[EI(x)\frac{\partial^2 y(x,t)}{\partial x^2}\right]+\rho A(x)\frac{\partial^2 y(x,t)}{\partial t^2}=f(x,t), \quad 0 < x < L \tag{1.15}$$

It is important to note that although the equations of motion are not sign convention dependent, the relationships between moments/forces and deflections are. Their relationships in the Euler–Bernoulli bending vibration theory, corresponding to the various sign conventions illustrated in Figure 1.1, are as given in Eqs. (1.9a), (1.9b), (1.14a), and (1.14b).

1.1.2 According to Timoshenko Bending Vibration Theory

The Timoshenko bending vibration theory is applied in the derivation of the governing equations of motion of a beam in bending vibration. It is still assumed that the neutral axis does not experience any change in length and the cross section remains a plane after deformation. However, the restrictions imposed by the Euler–Bernoulli bending vibration theory on the rotation of the differential element and angular distortion due to shear are eliminated. In other words, the rotation of the differential element does not have to be negligible compared to the translation, and the angular distortion due to shear does not have to be small in relation to the bending deformation. Consequently, the analysis based on the Timoshenko bending vibration theory provides greater accuracy at higher frequencies than that of the Euler–Bernoulli bending vibration theory.

The normal strains $\varepsilon_{1,3}$ and $\varepsilon_{2,4}$ on the differential element at a distance z above the neutral axis corresponding to the concave and convex deformation shown in Figures 1.4b and 1.4c are

$$\varepsilon_{1,3} = \frac{ds_{1,3} - dx}{dx} = \frac{(R-z)d\psi - Rd\psi}{dx} = -z\frac{d\psi}{dx} \tag{1.16a}$$

$$\varepsilon_{2,4} = \frac{ds_{2,4} - dx}{dx} = \frac{(R+z)d\psi - Rd\psi}{dx} = z\frac{d\psi}{dx} \tag{1.16b}$$

where R is the radius of curvature of the transverse deflection of the centerline $y(x,t)$ and $\psi(x,t)$ is the angle of rotation of the cross section due to bending. Subscripts 1, 2, 3, and 4 denote that the parameters are related to sign conventions 1, 2, 3, and 4, as defined in Figure 1.1, respectively.

From Eqs. (1.5), (1.16a), and (1.16b),

$$\sigma_{1,3} = E\varepsilon_{1,3} = -Ez\frac{d\psi}{dx} \tag{1.17a}$$

$$\sigma_{2,4} = E\varepsilon_{2,4} = Ez\frac{d\psi}{dx} \tag{1.17b}$$

Substituting Eqs. (1.17a) and (1.17b) into Eq. (1.7) gives

$$M_{1,3} = -\int_A z\left(-Ez\frac{d\psi}{dx}dA\right) = E\frac{d\psi}{dx}\int_A z^2 dA = EI(x)\frac{d\psi}{dx} \tag{1.18a}$$

$$M_{2,4} = -\int_A z\left(Ez\frac{d\psi}{dx}dA\right) = -E\frac{d\psi}{dx}\int_A z^2 dA = -EI(x)\frac{d\psi}{dx} \tag{1.18b}$$

where A is the area of the cross section and $I(x) = \int_A z^2 dA$ is the area moment of inertia about the centroidal axis that is normal to the plane of bending.

The bending moment causes the cross section to rotate an angle $\psi(x,t)$. The net shear force acting on the section also contributes a shear deformation $\alpha(x,t)$. As a result, the total deformation $\dfrac{\partial y(x,t)}{\partial x}$ of the neutral line of the differential element of a beam comprises two components, namely bending deformation $\psi(x,t)$ and shear deformation $\alpha(x,t)$. Figure 1.5 shows the deformations corresponding to the four sets of sign conventions. They are related by the following algebraic summation,

$$\frac{\partial y(x,t)}{\partial x} = \psi(x,t) + \alpha(x,t) \tag{1.19}$$

Equation (1.19) gives

$$\alpha(x,t) = \frac{\partial y(x,t)}{\partial x} - \psi(x,t) \tag{1.20}$$

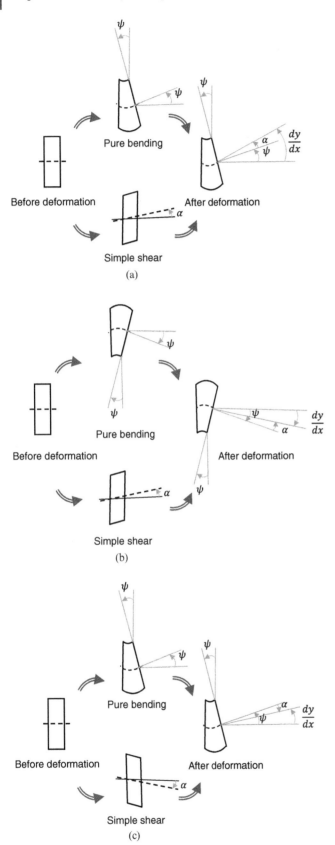

Figure 1.5 Deformation of a beam in bending subject to positive moment and shear force by the various sign conventions defined in Figure 1.1: (a) sign convention 1, (b) sign convention 2, (c) sign convention 3, and (d) sign convention 4.

$\alpha(x,t)$ is a positive value corresponding to sign conventions 1 and 2, and it is negative for sign conventions 3 and 4.

Recall that in the Euler–Bernoulli bending vibration theory, the shear deformation $\alpha(x,t)$ is assumed negligibly small, from Eq. (1.20), one has $\psi(x,t) = \dfrac{\partial y(x,t)}{\partial x}$.

For homogeneous materials behaving in a linear elastic manner, the shear stress τ and shear strain α are related by Hooke's law,

$$\tau = G\alpha \tag{1.21}$$

where G is the shear modulus.

The internal resistant shear forces corresponding to the positive bending moment and shear force defined in Figure 1.1 are

$$V_{1,2} = |\tau| A\kappa = GA\kappa \left[\frac{\partial y(x,t)}{\partial x} - \psi(x,t) \right] \tag{1.22a}$$

$$V_{3,4} = |\tau| A\kappa = GA\kappa \left[\psi(x,t) - \frac{\partial y(x,t)}{\partial x} \right] \tag{1.22b}$$

where κ is the shear coefficient and has been tabulated in (Cowper 1966).

From the free body diagrams of Figure 1.3, the force equations along the y-axis direction obtained by sign conventions 1, 2, 3, and 4 are

$$\left[V(x,t) + \frac{\partial V(x,t)}{\partial x} dx \right] - V(x,t) + f(x,t)dx$$

$$= \rho A(x)dx \frac{\partial^2 y(x,t)}{\partial t^2}, \quad 0 < x < L \tag{1.23a}$$

$$\left[V(x,t) + \frac{\partial V(x,t)}{\partial x} dx \right] - V(x,t) + f(x,t)dx$$

$$= \rho A(x)dx \frac{\partial^2 y(x,t)}{\partial t^2}, \quad 0 < x < L \tag{1.23b}$$

$$-\left[V(x,t) + \frac{\partial V(x,t)}{\partial x} dx \right] + V(x,t) + f(x,t)dx$$

$$= \rho A(x)dx \frac{\partial^2 y(x,t)}{\partial t^2}, \quad 0 < x < L \tag{1.23c}$$

$$-\left[V(x,t) + \frac{\partial V(x,t)}{\partial x} dx \right] + V(x,t) + f(x,t)dx$$

$$= \rho A(x)dx \frac{\partial^2 y(x,t)}{\partial t^2}, \quad 0 < x < L \tag{1.23d}$$

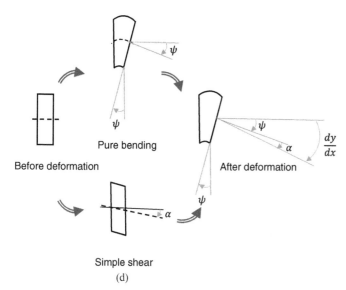

Before deformation

Pure bending

After deformation

ψ

ψ

α

$\frac{dy}{dx}$

ψ

α

Simple shear

(d)

Figure 1.5 (Cont'd)

Note that Eqs. (1.23a) and (1.23b) are the same, so are Eqs. (1.23c) and (1.23d). Simplifying the above equations gives

$$\frac{\partial V(x,t)}{\partial x} + f(x,t) = \rho A(x)\frac{\partial^2 y(x,t)}{\partial t^2}, \qquad 0 < x < L \quad (1.24a)$$

$$\frac{\partial V(x,t)}{\partial x} + f(x,t) = \rho A(x)\frac{\partial^2 y(x,t)}{\partial t^2}, \qquad 0 < x < L \quad (1.24b)$$

$$-\frac{\partial V(x,t)}{\partial x} + f(x,t) = \rho A(x)\frac{\partial^2 y(x,t)}{\partial t^2}, \qquad 0 < x < L \quad (1.24c)$$

$$-\frac{\partial V(x,t)}{\partial x} + f(x,t) = \rho A(x)\frac{\partial^2 y(x,t)}{\partial t^2}, \qquad 0 < x < L \quad (1.24d)$$

From Eqs. (1.22a), (1.22b), and (1.24a-d), a unique force-based equation of motion, regardless of the sign conventions, is obtained,

$$GA\kappa\left[\frac{\partial \psi(x,t)}{\partial x} - \frac{\partial^2 y(x,t)}{\partial x^2}\right] + \rho A(x)\frac{\partial^2 y(x,t)}{\partial t^2} = f(x,t) \tag{1.25}$$

From the free body diagrams of Figure 1.3, the moment equations by sign conventions 1, 2, 3, and 4 are

$$\left[M(x,t) + \frac{\partial M(x,t)}{\partial x}dx\right] - M(x,t) + \left[V(x,t) + \frac{\partial V(x,t)}{\partial x}dx\right]dx + f(x,t)dx\frac{dx}{2} = \rho I(x)dx\frac{\partial^2 \psi(x,t)}{\partial t^2}, \quad 0 < x < L \quad (1.26a)$$

$$-\left[M(x,t) + \frac{\partial M(x,t)}{\partial x}dx\right] + M(x,t) + \left[V(x,t) + \frac{\partial V(x,t)}{\partial x}dx\right]dx + f(x,t)dx\frac{dx}{2} = \rho I(x)dx\frac{\partial^2 \psi(x,t)}{\partial t^2}, \quad 0 < x < L \quad (1.26b)$$

$$\left[M(x,t) + \frac{\partial M(x,t)}{\partial x}dx\right] - M(x,t) - \left[V(x,t) + \frac{\partial V(x,t)}{\partial x}dx\right]dx + f(x,t)dx\frac{dx}{2} = \rho I(x)dx\frac{\partial^2 \psi(x,t)}{\partial t^2}, \quad 0 < x < L \quad (1.26c)$$

$$-\left[M(x,t) + \frac{\partial M(x,t)}{\partial x}dx\right] + M(x,t) - \left[V(x,t) + \frac{\partial V(x,t)}{\partial x}dx\right]dx + f(x,t)dx\frac{dx}{2} = \rho I(x)dx\frac{\partial^2 \psi(x,t)}{\partial t^2}, \quad 0 < x < L \quad (1.26d)$$

Neglecting second-order terms in dx and canceling appropriate terms, Eqs. (1.26a-d) reduce to

$$\frac{\partial M(x,t)}{\partial x} + V(x,t) = \rho I(x)\frac{\partial^2 \psi(x,t)}{\partial t^2}, \qquad 0 < x < L \tag{1.27a}$$

$$-\frac{\partial M(x,t)}{\partial x} + V(x,t) = \rho I(x)\frac{\partial^2 \psi(x,t)}{\partial t^2}, \qquad 0 < x < L \tag{1.27b}$$

$$\frac{\partial M(x,t)}{\partial x} - V(x,t) = \rho I(x)\frac{\partial^2 \psi(x,t)}{\partial t^2}, \qquad 0 < x < L \tag{1.27c}$$

$$-\frac{\partial M(x,t)}{\partial x} - V(x,t) = \rho I(x)\frac{\partial^2 \psi(x,t)}{\partial t^2}, \qquad 0 < x < L \tag{1.27d}$$

From Eqs. (1.18a), (1.1.8b), (1.22a), (1.22b), and (1.27a-d), a unique moment-based equation of motion, regardless of the sign conventions, is obtained,

$$EI(x)\frac{\partial^2 \psi(x,t)}{\partial x^2}+GA\kappa\left[\frac{\partial y(x,t)}{\partial x}-\psi(x,t)\right]-\rho I(x)\frac{\partial^2 \psi(x,t)}{\partial t^2}=0 \qquad (1.28)$$

It is important to note that although the equations of motion are not sign convention dependent, the relationships between moments/forces and deflections are. Their relationships by the Timoshenko bending vibration theory, corresponding to the various sign conventions illustrated in Figure 1.1, are as given in Eqs. (1.18a), (1.18b), (1.22a), and (1.22b).

1.2 Derivation of the Elementary Longitudinal Equation of Motion by Various Sign Conventions

Figure 1.6 shows positive sign directions of internal resistant axial force F for two types of sign conventions. In the figure, subscripts "L" and "R" denote parameters related to the left and right side of the cut section, respectively.

In Figure 1.6a, the internal resistant axial force $F(x,t)$ is defined positive when it stretches the element, which is a convention that has been popularly adopted by many authors in textbooks and research papers. In Figure 1.6b, the internal resistant axial force $F(x,t)$ is defined positive when it compresses the element, which is less often adopted.

Consider a beam that is subjected to an external distributed axial load of $p(x,t)$ per unit length, as shown in Figure 1.7, where the x-axis is chosen to be along the centerline of the beam and t is the time.

First, consider the sign convention of Figure 1.6a. From the free body diagram of Figure 1.8a,

$$\sum F_x = p(x,t)dx + F(x,t)+\frac{\partial F(x,t)}{\partial x}dx - F(x,t) \qquad (1.29a)$$

Simplifying Eq. (1.29a),

$$\sum F_x = p(x,t)dx + \frac{\partial F(x,t)}{\partial x}dx \qquad (1.29b)$$

Applying Newton's second law,

$$p(x,t)dx + \frac{\partial F(x,t)}{\partial x}dx = \rho(x)A(x)dx\frac{\partial^2 u(x,t)}{\partial t^2} \qquad (1.30)$$

where $u(x,t)$ is the longitudinal deflection of the beam, $\rho(x)$ is the volume mass density, and $A(x)$ is the cross-sectional area of the beam.

Simplifying Eq. (1.30) gives

$$p(x,t) + \frac{\partial F(x,t)}{\partial x} = \rho(x)A(x)\frac{\partial^2 u(x,t)}{\partial t^2} \qquad (1.31)$$

From the mechanics of materials and the sign convention of Figure 1.6a (Hibbeler 2017), the internal resistant force $F(x,t)$ is related to the axial deflection $u(x,t)$ by

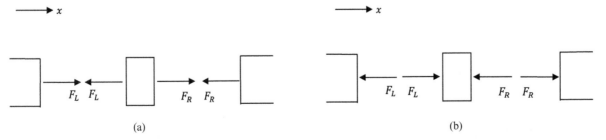

(a)   (b)

Figure 1.6 Definitions of positive sign directions for internal resistant axial force by various sign conventions: (a) sign convention 1 and (b) sign convention 2.

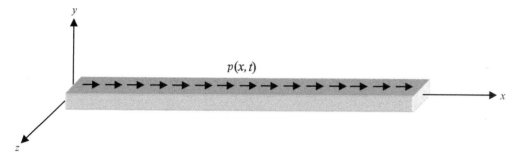

Figure 1.7 A beam in longitudinal vibration.

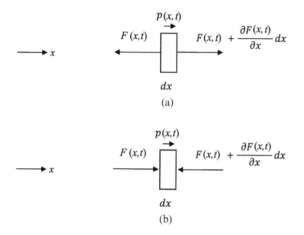

Figure 1.8 Free body diagram of a beam element in longitudinal vibration by the various sign conventions defined in Figure 1.6: (a) sign convention 1 and (b) sign convention 2.

$$F(x,t) = EA(x)\frac{\partial u(x,t)}{\partial x} \tag{1.32}$$

Substituting Eq. (1.32) into Eq. (1.31) gives the equation of motion for longitudinal vibration,

$$\rho(x)A(x)\frac{\partial^2 u(x,t)}{\partial t^2} - \frac{\partial}{\partial x}\left[EA(x)\frac{\partial u(x,t)}{\partial x}\right] = p(x,t) \tag{1.33}$$

Next, consider the sign convention of Figure 1.6b. From the free body diagram of Figure 1.8b,

$$\sum F_x = -F(x,t) - \frac{\partial F(x,t)}{\partial x}dx + F(x,t) + p(x,t)dx \tag{1.34a}$$

Simplifying Eq. (1.34a),

$$\sum F_x = -\frac{\partial F(x,t)}{\partial x}dx + p(x,t)dx \tag{1.34b}$$

Applying Newton's second law,

$$-\frac{\partial F(x,t)}{\partial x}dx + p(x,t)dx = \rho(x)A(x)dx\frac{\partial^2 u(x,t)}{\partial t^2} \tag{1.35}$$

Simplifying Eq. (1.35) gives

$$-\frac{\partial F(x,t)}{\partial x} + p(x,t) = \rho(x)A(x)\frac{\partial^2 u(x,t)}{\partial t^2} \tag{1.36}$$

From the mechanics of materials and the sign convention of Figure 1.6b, the internal resistant force $F(x,t)$ is related to the axial deflection $u(x,t)$ by

$$F(x,t) = -EA(x)\frac{\partial u(x,t)}{\partial x} \tag{1.37}$$

Substituting Eq. (1.37) into Eq. (1.36) gives the same equation of motion for longitudinal vibration obtained in Eq. (1.33).

It is important to note that although the equations of motion are not sign convention dependent, the relationships between forces and deflections by sign conventions 1 and 2 are described in Eqs. (1.32) and (1.37), respectively.

1.3 Derivation of the Elementary Torsional Equation of Motion by Various Sign Conventions

Figure 1.9 shows positive sign directions of internal resistant torque T for two types of sign conventions. In the figure, subscripts L and R denote parameters related to the left and right side of the cut section, respectively. The double arrow notation by the right hand rule is adopted for torque and torsional deflection, where the double arrow shows the direction the thumb points.

In Figure 1.9a, the internal resistant torque $T(x,t)$ is defined positive when the thumb directs outwards from the beam element by following the right hand rule, which is a convention that has been popularly adopted by many authors in textbooks and research papers. In Figure 1.9b, the internal resistant torque $T(x,t)$ is defined positive when the thumb directs inwards to the beam element by following the right hand rule, which is less often adopted.

Consider a shaft that is subjected to an external distributed torque load of $q(x,t)$ per unit length, as shown in Figure 1.10, where the x-axis is chosen to be along the centerline of the shaft and t is the time.

First, consider the sign convention of Figure 1.9a. From the free body diagram of Figure 1.11a,

$$\sum T_x = q(x,t)dx + T(x,t) + \frac{\partial T(x,t)}{\partial x}dx - T(x,t) \tag{1.38a}$$

Simplifying Eq. (1.38a),

$$\sum T_x = q(x,t)dx + \frac{\partial T(x,t)}{\partial x}dx \tag{1.38b}$$

Applying Newton's second law for rotational motion,

$$q(x,t)dx + \frac{\partial T(x,t)}{\partial x}dx = \rho(x)J(x)dx\frac{\partial^2\theta(x,t)}{\partial t^2} \tag{1.39}$$

where $\theta(x,t)$ is the torsional deflection, $\rho(x)$ is the volume mass density of the shaft, and $J(x)$ is the polar moment of inertia of the cross-sectional area.

Simplifying Eq. (1.39) gives

$$q(x,t) + \frac{\partial T(x,t)}{\partial x} = \rho(x)J(x)\frac{\partial^2\theta(x,t)}{\partial t^2} \tag{1.40}$$

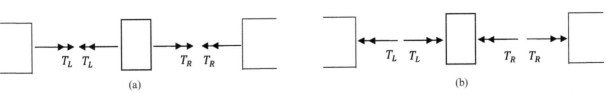

Figure 1.9 Definitions of positive sign directions for internal resistant torque by various sign conventions: (a) sign convention 1 and (b) sign convention 2.

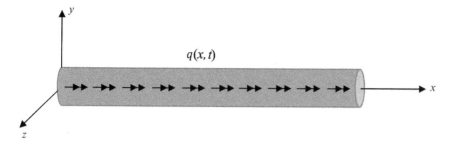

Figure 1.10 A shaft in torsional vibration.

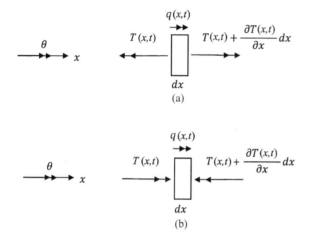

Figure 1.11 Free body diagram of a beam element in torsional vibration by the various sign conventions defined in Figure 1.9: (a) sign convention 1 and (b) sign convention 2.

From the mechanics of materials and the sign convention of Figure 1.9a (Hibbeler 2017), the internal resistant torque $T(x,t)$ is related to the torsional deflection $\theta(x,t)$ by

$$T(x,t) = GJ(x)\frac{\partial\theta(x,t)}{\partial x} \tag{1.41}$$

where $GJ(x)$ is the torsional rigidity of a beam whose cross section is rotationally symmetric. For a rotationally asymmetric cross section, an adjustment coefficient is normally needed for obtaining an equivalent torsional rigidity.

Substituting Eq. (1.41) into Eq. (1.40) gives the equation of motion for torsional vibration of a beam whose cross section is rotationally symmetric,

$$\rho(x)J(x)\frac{\partial^2\theta(x,t)}{\partial t^2} - \frac{\partial}{\partial x}\left[GJ(x)\frac{\partial\theta(x,t)}{\partial x}\right] = q(x,t) \tag{1.42}$$

Next, consider the sign convention of Figure 1.9b. From the free body diagram of Figure 1.11b,

$$\sum T_x = -T(x,t) - \frac{\partial T(x,t)}{\partial x}dx + T(x,t) + q(x,t)dx \tag{1.43a}$$

Simplifying Eq. (1.43a),

$$\sum T_x = \frac{-\partial T(x,t)}{\partial x}dx + q(x,t)dx \tag{1.43b}$$

Applying Newton's second law for rotational motion,

$$-\frac{\partial T(x,t)}{\partial x}dx + q(x,t)dx = \rho(x)J(x)dx\frac{\partial^2\theta(x,t)}{\partial t^2} \tag{1.44}$$

Simplifying Eq. (1.44) gives

$$-\frac{\partial T(x,t)}{\partial x} + q(x,t) = \rho(x)J(x)\frac{\partial^2\theta(x,t)}{\partial t^2} \tag{1.45}$$

For a beam whose cross section is rotationally symmetric, from the mechanics of materials and the sign convention of Figure 1.9b, the internal resistant torque $T(x,t)$ is related to the torsional deflection $\theta(x,t)$ by

$$T(x,t) = -GJ(x)\frac{\partial\theta(x,t)}{\partial x} \tag{1.46}$$

Substituting Eq. (1.46) into Eq. (1.45) gives the same equation of motion for torsional vibration as Eq. (1.42). The relationships between torques and deflections by sign conventions 1 and 2 are described in Eqs. (1.41) and (1.46), respectively.

In summary, the equations of motion are not sign convention dependent; however, the relationships between moments/forces/torques and deflections are.

References

Cowper G. R. The Shear Coefficient in Timoshenko's Beam Theory, *Journal of Applied Mechanics*, 33, 335–340 (1966).

Hibbeler R. C. *Statics and Mechanics of Materials*, Pearson, Prentice Hall (2017).

2

Longitudinal Waves in Beams

Similar to the propagation of acoustical waves and electro-magnetic waves through gaseous, liquid, and solid media, mechanical vibrations can be described as waves that propagate in a solid medium (Graff 1975; Cremer et. al. 1987; and Doyle 1989). From the wave standpoint, vibrations in a distributed or continuous structure can be viewed as waves that propagate along uniform waveguides and are reflected and transmitted at structural discontinuities (Mace 1984). The propagation relationships of waves are governed by the equations of motion of a beam for free vibration, and the reflection and transmission relationships are determined by the equilibrium and continuity at a structural discontinuity. Assembling these propagation, reflection, and transmission relationships provides a concise and systematic approach for vibration analysis of a distributed or continuous structure.

In this chapter, fundamental concepts related to longitudinal waves are introduced, such as the propagation coefficient of longitudinal vibration waves along a uniform beam (the waveguide) and the reflection coefficient of longitudinal vibration waves at either a classical or non-classical boundary (the discontinuity). Natural frequencies, modeshapes, as well as steady state frequency responses, are obtained, with comparison to experimental results. MATLAB scripts for numerical simulations are provided.

2.1 The Governing Equation and the Propagation Relationships

The governing equation of longitudinal vibrations based on the elementary theory is

$$\rho(x)A(x)\frac{\partial^2 u(x,t)}{\partial t^2} - \frac{\partial}{\partial x}\left[EA(x)\frac{\partial u(x,t)}{\partial x}\right] = p(x,t) \tag{2.1}$$

where x is the position along the beam axis, t is the time, $A(x)$ is the cross-sectional area, $u(x,t)$ is the longitudinal deflection of the centerline of the beam, $p(x,t)$ is the externally applied longitudinal force per unit length, E is the Young's modulus, and $\rho(x)$ is the volume mass density of the beam.

For a uniform beam subjected to no external force, the differential equation for free longitudinal vibration is

$$\rho A\frac{\partial^2 u(x,t)}{\partial t^2} - EA\frac{\partial^2 u(x,t)}{\partial x^2} = 0 \tag{2.2}$$

Assuming time harmonic motion and using separation of variables, the solution to Eq. (2.2) can be written in the form $u(x,t) = u_0 e^{-ikx}e^{i\omega t}$, where u_0 is the amplitude of the longitudinal deflection of the centerline of the beam, ω is the circular frequency, k is the wavenumber, and i is the imaginary unit. Substituting this expression for $u(x,t)$ into Eq. (2.2) gives the dispersion equation from which the longitudinal wavenumber is obtained,

$$k = \pm\sqrt{\frac{\rho}{E}}\,\omega \tag{2.3}$$

Mechanical Wave Vibrations: Analysis and Control, First Edition. Chunhui Mei.
© 2023 Chunhui Mei. Published 2023 by John Wiley & Sons Ltd.
Companion Website: www.wiley.com/go/Mei/MechanicalWaveVibrations

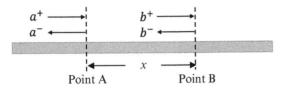

Figure 2.1 Wave propagation relationships.

which is a function of circular frequency ω. The $\pm$ sign indicates that longitudinal waves in the beam travel in both the positive and negative directions. It can be seen from Eq. (2.3) that longitudinal waves are non-dispersive because the phase velocity $\dfrac{\omega}{k}$ is not frequency dependent.

With time dependence $e^{i\omega t}$ suppressed, the solution to Eq. (2.2) can be written as

$$u(x) = a^+ e^{-ik_l x} + a^- e^{ik_l x} \tag{2.4}$$

where the longitudinal wavenumber $k_l = |k|$ and subscript l stands for longitudinal waves. Superscripts $+$ and $-$ in wave amplitude denote positive- and negative-going propagating longitudinal waves.

Consider two points A and B on a uniform beam that are a distance x apart, as shown in Figure 2.1. Waves propagate from one point to the other, with the propagation relationships determined by the appropriate wavenumber. The positive- and negative-going waves at Points A and B are related by

$$b^+ = f(x)a^+, \ a^- = f(x)b^- \tag{2.5}$$

where $f(x) = e^{-ik_l x}$ is the longitudinal propagation coefficient for a distance x.

2.2 Wave Reflection at Classical and Non-Classical Boundaries

Figure 2.2 shows the sign convention adopted in this chapter.

The reflection coefficients at classical fixed and free boundaries, as well as at non-classical boundaries with spring, damper, and mass attachments, are derived.

Figure 2.3 illustrates a boundary with an elastic spring. An incident wave is reflected at the boundary, the incident wave a^+ and reflected wave a^- are related by reflection coefficient r,

$$a^- = ra^+ \tag{2.6}$$

where r is determined by boundary condition.

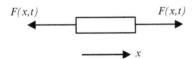

Figure 2.2 Definition of positive directions of internal resistant forces along the longitudinal direction.

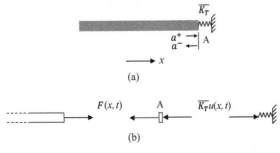

Figure 2.3 (a) Boundary with a spring attachment and (b) free body diagram.

According to the Elementary longitudinal vibration theory, the internal resistant longitudinal force and longitudinal deflection are related by

$$F(x,t) = EA\frac{\partial u(x,t)}{\partial x} \tag{2.7}$$

a) Free boundary

At a free boundary, the equilibrium condition is

$$F(x,t) = 0 \tag{2.8}$$

From Eqs. (2.7) and (2.8),

$$EA\frac{\partial u(x,t)}{\partial x} = 0 \tag{2.9}$$

From Eq. (2.4),

$$\frac{\partial u(x)}{\partial x} = -ik_l a^+ e^{-ik_l x} + ik_l a^- e^{ik_l x} \tag{2.10}$$

Choosing the origin at the boundary, from Eqs. (2.9) and (2.10),

$$-ik_l a^+ + ik_l a^- = 0 \tag{2.11}$$

From Eqs. (2.6) and (2.11), the reflection coefficient at a free boundary is

$$r = 1 \tag{2.12}$$

b) Fixed boundary

The boundary condition at a fixed boundary is

$$u(x,t) = 0 \tag{2.13}$$

From Eqs. (2.4) and (2.13),

$$a^+ e^{-ik_l x} + a^- e^{ik_l x} = 0 \tag{2.14}$$

Choosing the origin at the boundary, from Eq. (2.14),

$$a^+ + a^- = 0 \tag{2.15}$$

From Eqs. (2.6) and (2.15), the reflection coefficient at a fixed boundary is

$$r = -1 \tag{2.16}$$

c) Boundary with a spring attachment

From the free body diagram of Figure 2.3,

$$-F(x,t) - \overline{K_T}u(x,t) = 0 \tag{2.17}$$

where $\overline{K_T}$ is the stiffness of the spring and $\overline{K_T}u(x,t)$ is the resistant force from the spring by Hooke's law.
From Eqs. (2.4), (2.7), and (2.17),

$$-EA(-ik_l a^+ e^{-ik_l x} + ik_l a^- e^{ik_l x}) - \overline{K_T}(a^+ e^{-ik_l x} + a^- e^{ik_l x}) = 0 \tag{2.18}$$

Choosing the origin at the spring attached boundary, from Eq. (2.18),

$$EA(ik_l a^+ - ik_l a^-) - \overline{K_T}(a^+ + a^-) = 0 \tag{2.19}$$

Combining like terms in Eq. (2.19) gives

$$\left(\overline{K_T} + ik_l EA\right)a^- = \left(-\overline{K_T} + ik_l EA\right)a^+ \tag{2.20}$$

From Eqs. (2.6) and (2.20), the reflection coefficient at a spring attached boundary is

$$r = \frac{-\overline{K_T} + ik_l EA}{\overline{K_T} + ik_l EA} \tag{2.21a}$$

or equivalently is

$$r = \frac{-1 + ik_l EA / \overline{K_T}}{1 + ik_l EA / \overline{K_T}} \tag{2.21b}$$

Under the special situation where $\overline{K_T}$ approaches zero or infinity, the spring attached end is physically equivalent to a free or a fixed end. Substituting $\overline{K_T} = 0$ into Eq. (2.21a) gives $r = 1$, which is the reflection coefficient at a free boundary. Substituting $\overline{K_T} = \infty$ into Eq. (2.21b) gives $r = -1$, which is the reflection coefficient at a fixed boundary.

d) Boundary with a viscous damper attachment

Figure 2.4 shows a boundary with a viscous damper attachment and its free body diagram. From the free body diagram,

$$-F(x,t) - \overline{C_T}\frac{\partial u(x,t)}{\partial t} = 0 \tag{2.22}$$

where $\overline{C_T}$ is the viscous damping constant and $\overline{C_T}\dfrac{\partial u(x,t)}{\partial t}$ is the resistant force from the viscous damper.

Because $\dfrac{\partial u(x,t)}{\partial t} = i\omega u(x,t)$ for time harmonic motion, Eq. (2.22) can be written as

$$-F(x,t) - i\omega\overline{C_T}u(x,t) = 0 \tag{2.23}$$

From Eqs. (2.4), (2.7), and (2.23),

$$-EA(-ik_l a^+ e^{-ik_l x} + ik_l a^- e^{ik_l x}) - i\omega\overline{C_T}(a^+ e^{-ik_l x} + a^- e^{ik_l x}) = 0 \tag{2.24}$$

Choosing the origin at the viscous damper attached boundary, Eq. (2.24) becomes

$$-EA(-ik_l a^+ + ik_l a^-) - i\omega\overline{C_T}(a^+ + a^-) = 0 \tag{2.25}$$

Combining like terms in Eq. (2.25) gives

$$(i\omega\overline{C_T} + ik_l EA)a^- = (-i\omega\overline{C_T} + ik_l EA)a^+ \tag{2.26}$$

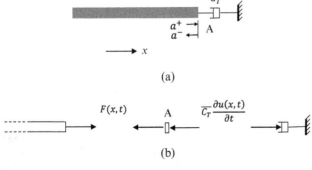

(a)

(b)

Figure 2.4 (a) Boundary with a viscous damper attachment and (b) free body diagram.

From Eqs. (2.6) and (2.26),

$$r = \frac{-\omega\overline{C_T} + k_l EA}{\omega\overline{C_T} + k_l EA} = \frac{-1 + k_l EA / \left(\omega\overline{C_T}\right)}{1 + k_l EA / \left(\omega\overline{C_T}\right)} \tag{2.27}$$

It is not difficult to see from Eq. (2.17) and Eq. (2.23) that when a boundary comprises a combined spring and viscous damper attachment, one may treat it as a spring attachment of complex stiffness $\overline{K_T} + i\omega\overline{C_T}$. In other words, the viscous damping effect can be taken into account by adding a frequency dependent imaginary term to the spring stiffness, that is, by introducing a dynamic spring stiffness.

e) Boundary with a mass attachment

Figure 2.5 shows a boundary with a lumped mass attachment and its free body diagram. Applying Newton's second law to the lumped mass and from Eq. (2.7),

$$-EA\frac{\partial u(x,t)}{\partial x} = m\frac{\partial u_m^2(x,t)}{\partial t^2} \tag{2.28}$$

where m and u_m are the mass and displacement of the lumped rigid body attachment, respectively.

From Eqs. (2.4), (2.28), and the relationship that $\dfrac{\partial^2 u(x,t)}{\partial t^2} = -\omega^2 u(x,t)$ for time harmonic motion,

$$-EA(-ik_l a^+ e^{-ik_l x} + ik_l a^- e^{ik_l x}) = -m\omega^2(a^+ e^{-ik_l x} + a^- e^{ik_l x}) \tag{2.29}$$

Note that because the mass block is modeled as a rigid body, the displacement of the attached mass block u_m is equal to the longitudinal deflection of the beam at the point where the mass block is attached.

Choosing the origin at the boundary, Eq. (2.29) becomes

$$EA(-ik_l a^+ + ik_l a^-) = m\omega^2(a^+ + a^-) \tag{2.30}$$

Combining like terms in Eq. (2.30) gives

$$(-m\omega^2 + ik_l EA)a^- = (m\omega^2 + ik_l EA)a^+ \tag{2.31}$$

From Eqs. (2.6) and (2.31), the reflection coefficient at a mass attached boundary is

$$r = \frac{m\omega^2 + ik_l EA}{-m\omega^2 + ik_l EA} \tag{2.32a}$$

or equivalently is

$$r = \frac{1 + ik_l EA / (m\omega^2)}{-1 + ik_l EA / (m\omega^2)} \tag{2.32b}$$

Under the special situation where m approaches zero or infinity, the mass attached end is physically equivalent to a free or a fixed end. Substituting $m = 0$ into Eq. (2.32a) gives $r = 1$, which is the reflection coefficient of a free boundary. Substituting $m = \infty$ into Eq. (2.32b) gives $r = -1$, which is the reflection coefficient of a fixed boundary.

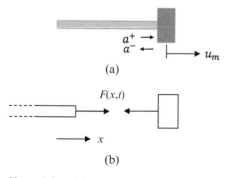

(a)

(b)

Figure 2.5 (a) Boundary with a mass attachment and (b) free body diagram.

Table 2.1 Reflection coefficients at various boundaries.

Boundary Condition	Free	Fixed	Spring	Viscous Damper	Mass
Reflection Coefficient	1	−1	$\dfrac{-1+ik_lEA/\overline{K_T}}{1+ik_lEA/K_T}$	$\dfrac{-1+k_lEA/\left(\omega\overline{C_T}\right)}{1+k_lEA/\left(\omega\overline{C_T}\right)}$	$\dfrac{1+ik_lEA/\left(m\omega^2\right)}{-1+ik_lEA/\left(m\omega^2\right)}$

The reflection coefficients at various boundaries obtained above are summarized in Table 2.1. Note that the reflection coefficients at spring, viscous damper, and mass attached boundaries are frequency dependent.

2.3 Free Vibration Analysis in Finite Beams – Natural Frequencies and Modeshapes

Free vibration analysis of a longitudinally vibrating uniform beam from the wave standpoint involves a concise and systematic assembling process.

Figure 2.6 illustrates a uniform beam of length L. The following reflection and propagation relationships exist:

$$a^+ = r_A\,a^-;\ b^- = r_B\,b^+;\ b^+ = f(L)\,a^+;\ a^- = f(L)\,b^- \tag{2.33}$$

where $f(L) = e^{-ik_lL}$ is the longitudinal propagation coefficient for a distance L, and r_A and r_B are the reflection coefficients at Boundaries A and B, respectively.

The above relationships can be written in matrix form,

$$\begin{bmatrix} -1 & r_A & 0 & 0 \\ 0 & 0 & r_B & -1 \\ f(L) & 0 & -1 & 0 \\ 0 & -1 & 0 & f(L) \end{bmatrix}\begin{bmatrix} a^+ \\ a^- \\ b^+ \\ b^- \end{bmatrix} = 0 \tag{2.34}$$

Equation (2.34) can be written in the form of

$$\mathbf{A_0z_0} = \mathbf{0} \tag{2.35}$$

where $\mathbf{A_0}$ is a 4 by 4 square coefficient matrix and $\mathbf{z_0}$ is a wave vector containing four wave components of Eq. (2.34).

Setting the determinant of the coefficient matrix $\mathbf{A_0}$ to zero gives the characteristic equation. MATLAB symbolic scripts for characteristic expressions of various boundaries can be found in Section 2.6.

Natural frequencies of a structure are roots of its characteristic equation, which correspond to the non-trivial solution of Eq. (2.34), or equivalently Eq. (2.35). Three factors determine the natural frequencies of a structure, namely, material, dimension, and boundary conditions.

Natural frequency expressions or equations corresponding to some example combination of boundary conditions listed in Table 2.1 are discussed as follows.

a) Free-Free Boundary

Substituting the reflection coefficient at a free boundary from Table 2.1 into the coefficient matrix $\mathbf{A_0}$ of Eq. (2.35) and setting the determinant of $\mathbf{A_0}$ to zero, as shown in the MATLAB scripts in Section 2.6, gives the following characteristic equation,

Figure 2.6 Wave propagation and reflection in a uniform beam.

$$\frac{1}{e^{i2k_l L}} - 1 = 0 \tag{2.36}$$

Rearranging Eq. (2.36) and noting that $k_l = \omega\sqrt{\rho/E}$,

$$e^{i2L\omega\sqrt{\rho/E}} = 1 \tag{2.37}$$

Equating phase angles on the left and right side of Eq. (2.37),

$$2L\omega_n\sqrt{\rho/E} = 2n\pi, \text{ where } n = 0, 1, 2, \ldots \tag{2.38}$$

The natural frequency ω_n is solved from Eq. (2.38),

$$\omega_n = \frac{n\pi}{L}\sqrt{E/\rho}, \text{ where } n = 0, 1, 2, \ldots \tag{2.39}$$

Note that $n = 0$ corresponds to the rigid body mode of the beam with a free-free boundary.

Introducing non-dimensional frequency $\Omega = L\omega\sqrt{\rho/E}$, from Eq. (2.39) the non-dimensional natural frequency corresponding to a free-free boundary is

$$\Omega_n = n\pi, \text{ where } n = 0, 1, 2, \ldots \tag{2.40}$$

b) Fixed-Free Boundary

Substituting the reflection coefficients for fixed and free boundaries from Table 2.1 into the coefficient matrix $\mathbf{A}_0$ of Eq. (2.35) and setting the determinant of $\mathbf{A}_0$ to zero, as shown in the MATLAB scripts in Section 2.6, gives the following characteristic equation,

$$-\frac{1}{e^{i2k_l L}} - 1 = 0 \tag{2.41}$$

Rearranging Eq. (2.41) and noting that $k_l = \omega\sqrt{\rho/E}$,

$$e^{i2L\omega\sqrt{\rho/E}} = -1 \tag{2.42}$$

Equating phase angles on the left and right side of Eq. (2.42) gives

$$2L\omega_n\sqrt{\rho/E} = -\pi + 2n\pi, \text{ where } n = 1, 2, \ldots \tag{2.43}$$

The natural frequency ω_n is solved from Eq. (2.43),

$$\omega_n = \frac{(2n-1)\pi}{2L}\sqrt{E/\rho} \tag{2.44}$$

In terms of non-dimensional frequency Ω_n, from Eq. (2.44), the natural frequency for a fixed-free boundary is

$$\Omega_n = \frac{(2n-1)\pi}{2}, \text{ where } n = 1, 2, \ldots \tag{2.45}$$

c) Fixed-Fixed Boundary

Substituting the reflection coefficient at a fixed boundary from Table 2.1 into Eq. (2.35) and setting the corresponding characteristic expression to zero, as shown in the MATLAB scripts in Section 2.6, gives an identical characteristic equation to Eq. (2.36), which was obtained for a beam with a free-free boundary. From this characteristic equation, the natural frequency for a fixed-fixed beam is

$$\omega_n = \frac{n\pi}{L}\sqrt{E/\rho}, \text{ where } n = 1, 2, \ldots \tag{2.46}$$

or in terms of non-dimensional frequencies $\Omega = L\omega\sqrt{\rho/E}$ is

$$\Omega_n = n\pi, \text{ where } n = 1, 2, \dots \tag{2.47}$$

Although the natural frequencies are given by the same expressions as those of a beam with a free-free boundary, integer n starts from 1 instead of 0. Recall that for a beam with a free-free boundary, $n = 0$ corresponds to the rigid body mode of the beam.

d) Fixed-Spring Boundary

Substituting the reflection coefficients for fixed and spring boundaries from Table 2.1 into the coefficient matrix $\mathbf{A}_0$ of Eq. (2.35) and setting the determinant of $\mathbf{A}_0$ to zero, as shown in the MATLAB scripts in Section 2.6, gives the following characteristic equation,

$$\frac{\overline{K_T}}{e^{i2k_lL}} - \overline{K_T} - iAEk_l\frac{1}{e^{i2k_lL}} - iAEk_l = 0 \tag{2.48}$$

Multiplying both sides of Eq. (2.48) by L, the characteristic equation in terms of non-dimensional frequency $\Omega = L\omega\sqrt{\rho/E} = k_lL$ is

$$\frac{\overline{K_T}L}{e^{i2\Omega}} - \overline{K_T}L - iAE\Omega\frac{1}{e^{i2\Omega}} - iAE\Omega = 0 \tag{2.49}$$

Solving for $e^{i2\Omega}$ from Eq. (2.49) gives

$$e^{i2\Omega} = \frac{\overline{K_T}L - iAE\Omega}{\overline{K_T}L + iAE\Omega} \tag{2.50}$$

The non-dimensional natural frequencies Ω_n are obtained by solving the above nonlinear equation, usually numerically.

e) Fixed-Mass Boundary

Substituting the reflection coefficients for fixed and lumped mass boundaries from Table 2.1 into the coefficient matrix $\mathbf{A}_0$ of Eq. (2.35) and setting the determinant of $\mathbf{A}_0$ to zero, as shown in the MATLAB scripts in Section 2.6, gives the following characteristic equation,

$$\frac{iAEk_l}{e^{i2k_lL}} + \frac{\omega^2 m}{e^{i2k_lL}} - \omega^2 m + iAEk_l = 0 \tag{2.51}$$

Multiplying both sides by L, the characteristic equation in terms of non-dimensional frequency $\Omega = L\omega\sqrt{\rho/E} = k_lL$ is

$$\frac{iAE\Omega}{e^{i2\Omega}} + \frac{\omega^2 mL}{e^{i2\Omega}} - \omega^2 mL + iAE\Omega = 0 \tag{2.52}$$

Solving for $e^{i2\Omega}$ gives

$$e^{i2\Omega} = \frac{\omega^2 mL + iAE\Omega}{\omega^2 mL - iAE\Omega} \tag{2.53}$$

The non-dimensional natural frequencies Ω_n can be obtained by solving the above nonlinear equation numerically.

f) Fixed-Damper Boundary

Substituting the reflection coefficients for fixed and viscous damper boundaries from Table 2.1 into the coefficient matrix $\mathbf{A}_0$ of Eq. (2.35) and setting the determinant of $\mathbf{A}_0$ to zero, as shown in the MATLAB scripts in Section 2.6, gives the following characteristic equation,

$$-\frac{AEk_l}{e^{i2k_lL}} + \frac{\omega\overline{C_T}}{e^{i2k_lL}} - \omega\overline{C_T} - AEk_l = 0 \tag{2.54}$$

Multiplying both sides of Eq. (2.54) by L, the characteristic equation in terms of non-dimensional frequency $\Omega = L\omega\sqrt{\rho/E} = k_l L$ is

$$-\frac{AE\Omega}{e^{i2\Omega}} + \frac{\omega\overline{C_T}L}{e^{i2\Omega}} - \omega\overline{C_T}L - AE\Omega = 0 \tag{2.55}$$

Solving for $e^{i2\Omega}$ from Eq. (2.55) gives

$$e^{i2\Omega} = \frac{\omega\overline{C_T}L - AE\Omega}{\omega\overline{C_T}L + AE\Omega} \tag{2.56}$$

The non-dimensional natural frequency Ω_n can be obtained by solving the above nonlinear equation numerically.

Table 2.2 summarizes the non-dimensional natural frequency expressions or equations of a beam with the example combination of boundary conditions studied above.

When a distributed structure is excited at one of its natural frequencies, the motion of the structure exhibits a certain shape, which is called the modeshape. A natural frequency and its corresponding modeshape are jointly called a mode of vibration.

The modeshape at a given natural frequency can be found by eliminating any one row of the coefficient matrix $\mathbf{A}_0$ of Eq. (2.35), and using the remaining rows to solve the wave components in terms of an arbitrarily chosen wave component. For example, one may eliminate the first row and solve wave components a^-, b^+, and b^- in terms of wave component a^+, which gives

$$\begin{bmatrix} 0 & r_B & -1 \\ 0 & -1 & 0 \\ -1 & 0 & f(L) \end{bmatrix} \begin{bmatrix} a^- \\ b^+ \\ b^- \end{bmatrix} = \begin{bmatrix} 0 \\ -f(L) \\ 0 \end{bmatrix} a^+ \tag{2.57}$$

From which,

$$\begin{bmatrix} a^- \\ b^+ \\ b^- \end{bmatrix} = \begin{bmatrix} 0 & r_B & -1 \\ 0 & -1 & 0 \\ -1 & 0 & f(L) \end{bmatrix}^{-1} \begin{bmatrix} 0 \\ -f(L) \\ 0 \end{bmatrix} a^+ \tag{2.58}$$

Table 2.2 Non-dimensional natural frequency expressions/equations for various boundaries.

Boundary Conditions	Natural Frequencies
Free-Free	$\Omega_n = n\pi,\ n = 0, 1, 2, \ldots$
Fixed-Free	$\Omega_n = \dfrac{(2n-1)\pi}{2},\ n = 1, 2, \ldots$
Fixed-Fixed	$\Omega_n = n\pi,\ n = 1, 2, \ldots$
Fixed-Spring	$e^{i2\Omega} = \dfrac{\overline{K_T}L - iAE\Omega}{\overline{K_T}L + iAE\Omega}$
Fixed-Mass	$e^{i2\Omega} = \dfrac{\omega^2 mL + iAE\Omega}{\omega^2 mL - iAE\Omega}$
Fixed-Damper	$e^{i2\Omega} = \dfrac{\omega\overline{C_T}L - AE\Omega}{\omega\overline{C_T}L + AE\Omega}$

With the wave components obtained from Eq. (2.58), the deflection of any point on the uniform beam shown in Figure 2.6 can be found. For example, the response at a point that is x distance away from Boundary A can be obtained from waves at Boundary A as

$$u(x) = f(x)a^+ + f(-x)a^- \tag{2.59}$$

Or equivalently, the response can be obtained from waves at Boundary B, which are a distance $L-x$ away, as

$$u(x) = f(-(L-x))b^+ + f(L-x)b^- \tag{2.60}$$

Plotting the response at all points along the structure at a chosen natural frequency gives the modeshape corresponding to that natural frequency.

While the natural frequency values are determined uniquely, the modeshape amplitudes are not. Because the wave components are found in terms of a reference wave component, such as a^+ in the above example. A normalization process is often needed to provide uniqueness in the modeshapes. In this book, each modeshape is normalized to have the maximum amplitude being unity.

2.4 Force Generated Waves and Forced Vibration Analysis of Finite Beams

An externally applied force has the effect of injecting waves into a continuous structure. Figure 2.7 illustrates a point axial force $\bar{F}$ and waves $a^\pm$ and $b^\pm$ at the point where the axial force is applied. These waves are injected by and related to the externally applied force.

The longitudinal deflections on the left and right side of the point where the force is applied are the same. Denoting physical parameters on the left and right side of the point where the force is applied using subscripts − and +, respectively,

$$u_- = u_+ \tag{2.61}$$

As shown in the free body diagram of Figure 2.7, the externally applied axial force is related to the internal resistant forces on the left and right side of the point where the external force is applied,

$$\bar{F} = F_- - F_+ \tag{2.62}$$

The relationships described in Eqs. (2.61) and (2.62) are called continuity and equilibrium equations, respectively.

With the origin chosen at the point where the external force is applied, from Eq. (2.4) and Figure 2.7a,

$$u_- = a^+ + a^-, \text{ and } u_+ = b^+ + b^- \tag{2.63}$$

From Eqs. (2.4) and (2.7) and Figure 2.7,

$$F_- = -ik_l EAa^+ + ik_l EAa^-, \text{ and } F_+ = -ik_l EAb^+ + ik_l EAb^- \tag{2.64}$$

The relationships between the externally applied force and its generated waves are obtained from Eqs. (2.61) to (2.64),

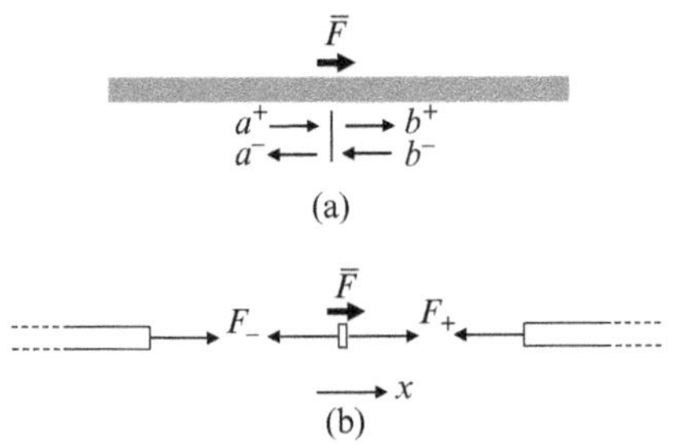

Figure 2.7 (a) Waves generated by an external axial force applied on the span and (b) free body diagram.

$$b^+ - a^+ = -q, \text{ and } b^- - a^- = q \qquad (2.65)$$

where

$$q = i\frac{\bar{F}}{2EAk_l} \qquad (2.66)$$

With the availability of force generated waves, propagation relationships, and boundary reflections, forced vibration in a beam can be obtained.

Figure 2.8 illustrates a beam with an external excitation force applied at an arbitrary point G on the beam. The following wave relationships exist:

- At Boundaries C and D, from Eq. (2.6),

$$c^+ = r_C \, c^-, \, d^- = r_D \, d^+ \qquad (2.67)$$

  where r_C and r_D are the reflection coefficients at Boundaries C and D, respectively.

- At Point G where the external force is applied, from Eq. (2.65),

$$g_2^+ - g_1^+ = -q, \, g_2^- - g_1^- = q \qquad (2.68)$$

  where q is obtained in Eq. (2.66).

- The propagation relationships, from Eq. (2.5), are

$$g_1^+ = f(L_1) \, c^+, \, c^- = f(L_1) \, g_1^-, \, g_2^- = f(L_2) \, d^-, \, d^+ = f(L_2) \, g_2^+ \qquad (2.69)$$

  where the longitudinal propagation coefficient for a distance x is $f(x) = e^{-ik_l x}$.

Assembling the wave relationships from Eqs. (2.67) to (2.69) into matrix form,

$$\begin{bmatrix} -1 & r_C & 0 & 0 & 0 & 0 & 0 & 0 \\ 0 & 0 & r_D & -1 & 0 & 0 & 0 & 0 \\ 0 & 0 & 0 & 0 & -1 & 0 & 1 & 0 \\ 0 & 0 & 0 & 0 & 0 & -1 & 0 & 1 \\ f(L_1) & 0 & 0 & 0 & -1 & 0 & 0 & 0 \\ 0 & -1 & 0 & 0 & 0 & f(L_1) & 0 & 0 \\ 0 & 0 & 0 & f(L_2) & 0 & 0 & 0 & -1 \\ 0 & 0 & -1 & 0 & 0 & 0 & f(L_2) & 0 \end{bmatrix} \begin{bmatrix} c^+ \\ c^- \\ d^+ \\ d^- \\ g_1^+ \\ g_1^- \\ g_2^+ \\ g_2^- \end{bmatrix} = \begin{bmatrix} 0 \\ 0 \\ -q \\ q \\ 0 \\ 0 \\ 0 \\ 0 \end{bmatrix} \qquad (2.70)$$

which can be written in the form of

$$\mathbf{A}_f \mathbf{z}_f = \mathbf{F} \qquad (2.71)$$

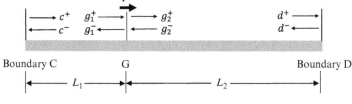

Figure 2.8 Waves in a uniform beam with an external axial force applied on the span.

where $\mathbf{A}_f$ is an 8 by 8 square coefficient matrix, $\mathbf{z}_f$ is a wave vector containing eight wave components, and $\mathbf{F}$ is a vector related to the externally applied force of Eq. (2.70).

Solving for $\mathbf{z}_f$ from Eq. (2.71),

$$\mathbf{z}_f = \mathbf{A}_f^{-1}\mathbf{F} \tag{2.72}$$

Equation (2.72) provides the responses of all wave components, from which the deflection of any point on the beam can be obtained. For example, in Figure 2.8, the deflection of a point located between Boundary C and the excitation point G that is a distance x_1 from Point G is

$$u_1 = f(x_1)g_1^- + f(-x_1)g_1^+ \tag{2.73}$$

or equivalently is

$$u_1 = f(L_1 - x_1)c^+ + f(-(L_1 - x_1))c^- \tag{2.74}$$

Similarly, the deflection of a point located between Boundary D and the excitation Point G in Figure 2.8 that is a distance x_2 from Point G is

$$u_2 = f(x_2)g_2^+ + f(-x_2)g_2^- \tag{2.75}$$

or equivalently is

$$u_2 = f(L_2 - x_2)d^- + f(-(L_2 - x_2))d^+ \tag{2.76}$$

An external force may be applied at a free end of a beam, as shown in Figure 2.9. This situation is considered below.

The free body diagram of Figure 2.9 shows that when an external force is applied at a free end of a beam, the externally applied axial force is related to the internal resistant force by the following equilibrium equation

$$\bar{F} = F \tag{2.77}$$

With the origin chosen at the free end where the external force is applied, from Figure 2.9 and Eqs. (2.4) and (2.7),

$$F = -ik_l EAa^+ + ik_l EAa^- \tag{2.78}$$

From Eqs. (2.77) and (2.78), the relationship between an externally applied end force and its generated waves is

$$a^+ - a^- = 2q \tag{2.79}$$

where q is given by Eq. (2.66).

Note that when there is no external force, that is $\bar{F} = 0$, the external force applied end becomes a free end. From Eqs. (2.66) and (2.79), $a^- = a^+$. This indicates that the reflection coefficient $r = 1$, which is the same as what is obtained in Eq. (2.12) for a free boundary, as expected.

With the availability of force generated waves, propagation relationships, and boundary reflections, forced vibration in a beam caused by an external force applied at a free end can be obtained.

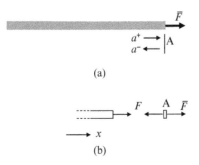

Figure 2.9 (a) Waves generated by an end force on a beam and (b) free body diagram.

Figure 2.10 illustrates a beam with an external excitation point force applied at End D. The following wave relationships exist:

- At Boundary C

$$c^+ = r_C \, c^-$$

(2.80)

where r_C is the reflection coefficient at Boundary C.

- At the external force applied at end D, from Eq. (2.79),

$$d^+ - d^- = 2q$$

(2.81)

where q is given by Eq. (2.66).

- The propagation relationships are

$$d^+ = f(L) \, c^+, \; c^- = f(L) \, d^-$$

(2.82)

where the longitudinal propagation coefficient for a distance L is $f(L) = e^{-ik_l L}$.
Assembling the wave relationships from Eqs. (2.80) to (2.82) into matrix form

$$\begin{bmatrix} -1 & r_C & 0 & 0 \\ 0 & 0 & 1 & -1 \\ f(L) & 0 & -1 & 0 \\ 0 & -1 & 0 & f(L) \end{bmatrix} \begin{bmatrix} c^+ \\ c^- \\ d^+ \\ d^- \end{bmatrix} = \begin{bmatrix} 0 \\ 2q \\ 0 \\ 0 \end{bmatrix}$$

(2.83)

Equation (2.83) can be written in the form of Eq. (2.71), and the wave components can be solved by Eq. (2.72). In which $\mathbf{A}_f$ is a 4 by 4 square coefficient matrix, $\mathbf{z}_f$ is a wave vector containing four wave components, and $\mathbf{F}$ is a vector related to externally applied force of Eq. (2.83).

With the availability of responses of wave components, the deflection of any point on the beam can be obtained. For example, the deflection of a point located a distance x from Boundary C in Figure 2.10 is

$$u = f(x)c^+ + f(-x)c^-$$

(2.84)

or equivalently is

$$u = f(L-x)d^- + f(-(L-x))d^+$$

(2.85)

2.5 Numerical Examples and Experimental Studies

Vibrations in a uniform steel beam are analyzed using the wave vibration approach and tested experimentally.

The material and geometrical properties of the steel beam are: the Young's modulus is 198.87 GN/m^2 and mass density is 7664.5 kg/m^3. The cross section of the beam is a rectangular shape whose dimensions are 12.7 mm by 25.4 mm (or 0.5 in by 1.0 in). The length of the beam is 914.4 mm (or 3.0 ft). In the following numerical and experimental examples, both boundaries of the beam are free.

Figure 2.10 Waves in a uniform beam with an external axial force applied at a free end.

In free wave vibration analysis of the uniform beam, the natural frequencies are obtained from Eq. (2.35) by substituting the propagation and boundary reflection coefficients into coefficient matrix $\mathbf{A}_0$ and solving for the characteristic polynomial of matrix $\mathbf{A}_0$. The magnitude of the characteristic polynomial is shown in Figure 2.11. The roots of the characteristic polynomial are the natural frequencies of the beam. The theoretical values of cyclic natural frequencies calculated using the frequency expression for free-free boundary from Table 2.2 are marked on in Figure 2.11 using asterisks. The natural frequencies obtained from the wave-based free vibration analysis are identical to the theoretical values.

The modeshapes are obtained following the procedures described in Eqs. (2.57–2.60) and are shown in Figure 2.12. Note that the straight line represents the modeshape of the rigid body mode of the beam with free-free boundary.

What is of particular interest on the modeshape plots are the nodal points, which are where the amplitudes of a modeshape become zero. On the one hand, when the beam is excited at a non-nodal position with a frequency that is identical to a natural frequency, the nodal points of the modeshape corresponding to this natural frequency are stationary. On the other hand, if the excitation force is applied at one of the nodal points of a particular natural frequency, this specific mode will not be excited, resulting in zero contribution to the dynamic response from this mode. Marked on Figure 2.12 are two example nodal points, which are discussed in forced vibration analysis that follows.

MATLAB scripts for free vibration analysis can be found in Section 2.6.

For forced wave vibration analysis of the uniform beam with an external excitation force applied at a point on the span, the analysis follows the procedures described in Eqs. (2.67–2.76).

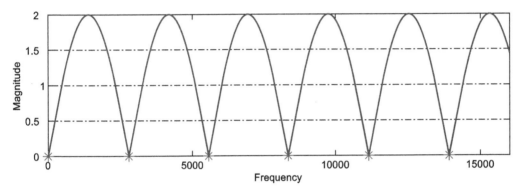

Figure 2.11 Magnitude of the characteristic polynomial.

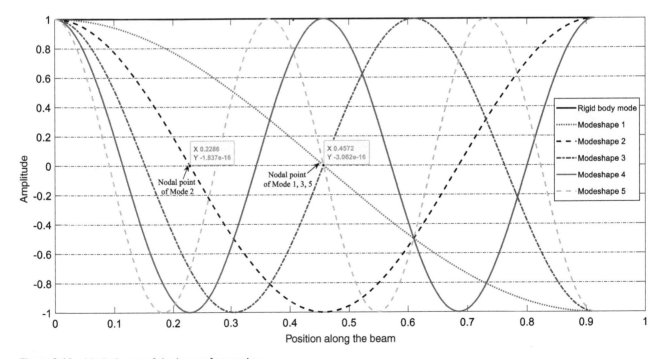

Figure 2.12 Modeshapes of the lowest few modes.

To obtain forced responses, the external excitation and response observation locations need to be determined. Table 2.3 lists these points, where all distances are measured from Boundary C of Figure 2.8. Because a point on the structure can either be a nodal point or a non-nodal point of a certain mode, the selection covers both scenarios.

Figures 2.13 and 2.14 show the receptance frequency responses (displacement/force), with the external excitation force applied at a non-nodal point (Test 1) and a nodal point (Test 2) as shown in Table 2.3. In both tests, responses are observed at the same three locations listed in Table 2.3, covering both non-nodal and nodal points.

The frequency range for forced vibration analysis is from 0 to 16,000 *Hz*, which is kept the same as that in free vibration analysis. In this frequency range, there are five non-rigid body modes contributing to the longitudinal dynamic responses.

As expected, it is seen from the receptance frequency responses in Figures 2.13 and 2.14 that resonant responses always occur at natural frequencies, and the resonant peaks line up regardless of the external excitation and response observation locations. Examining the response curves in both figures, however, one finds that not all resonant peaks occur at all natural frequencies. This is because the occurrence of resonant peaks in frequency responses is closely related to the location of the external excitation as well as the location of the response observation, relative to a nodal point, as discussed in the modeshape curves in free vibration analysis.

Table 2.4 summarizes the occurrence of resonant peaks in the frequency responses of Figures 2.13 and 2.14. Locations of external force and response observation relative to a nodal point are available from Table 2.3. Information provided in these two tables will allow one to understand the reasons behind the appearance and disappearance of resonant peaks in the frequency responses.

Recall that in Test 1, the external excitation force is applied at a non-nodal point. Among the three observation locations, Location 1 is at a non-nodal point; Location 2 is at a nodal point of odd modes 1, 3, and 5; and Location 3 is a nodal point of Mode 2. This explains why at Location 1, all resonant peaks appear; at Location 2, the resonant peaks corresponding to odd modes 1, 3, and 5 disappear; and at Location 3, the resonant peak corresponding to Mode 2 disappears.

In Test 2, with the external excitation force applied point purposely chosen at a nodal point of odd modes 1, 3, and 5, resonant peaks corresponding to these three modes all disappear, regardless of the observation location. At Location 3, an additional resonant peak disappears, which corresponds to Mode 2. This is because this observation point is at a nodal point of Mode 2.

The above studies point out the importance of nodal points. On the one hand, a nodal point needs to be avoided in identifying natural frequencies of a structure because of the disappearance of resonant peaks that results in miscounting of

Table 2.3 Locations of external excitation and response observation.

External force applied point	Test 1	490.2 *mm* (a non-nodal point of the lowest five modes)
	Test 2	457.2 *mm* (a nodal point of odd modes 1, 3, 5, ...)
Response observed points	Location 1	599.4 *mm* (a non-nodal point of the lowest five modes)
	Location 2	457.2 *mm* (a nodal point of odd modes 1, 3, 5, ...)
	Location 3	228.6 *mm* (a nodal point of Mode 2)

Table 2.4 Occurrence of resonant peaks.

Occurrence of resonant peaks		Mode 0	Mode 1	Mode 2	Mode 3	Mode 4	Mode 5
Test 1 (Figure 2.13)	Location 1	✓	✓	✓	✓	✓	✓
	Location 2	✓	✗	✓	✗	✓	✗
	Location 3	✓	✓	✗	✓	✓	✓
Test 2 (Figure 2.14)	Location 1	✓	✗	✓	✗	✓	✗
	Location 2	✓	✗	✓	✗	✓	✗
	Location 3	✓	✗	✗	✗	✓	✗

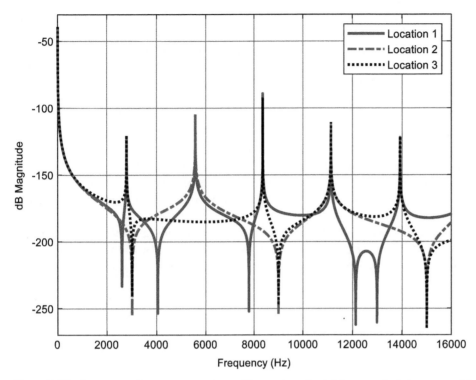

Figure 2.13 Receptance frequency responses with an external axial force applied at a non-nodal point.

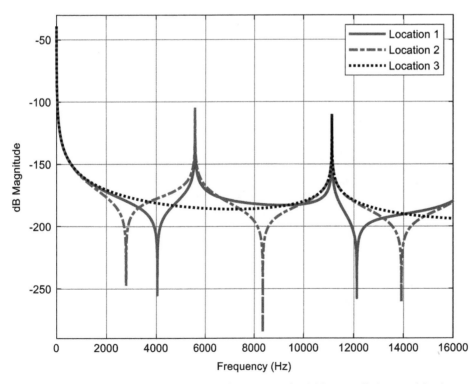

Figure 2.14 Receptance frequency responses with an external axial force applied at a nodal point.

natural frequencies. One may avoid or reduce the related problem by exciting a structure and observing the responses at a variety of locations. On the other hand, a nodal point is an ideal location for mounting equipment on a structure from the vibration isolation standpoint because of its stationary nature.

MATLAB scripts for forced vibration analysis can be found in Section 2.6.

Vibration tests are conducted on the beam, and the experimental setup is shown in Figure 2.15. To have free-free boundary conditions, the metal beam is suspended by two thin wires. The test is conducted with an impact force applied at one end of the beam using a Brüel & Kjær Type 8202 impact hammer. Vibrations are picked up by a Brüel & Kjær Type 4397 accelerometer, which is mounted at the other end of the beam. Frequency responses are obtained using the PULSE unit from Brüel & Kjær. The frequency span and frequency resolution are 6400 *Hz* and 1 *Hz*, respectively.

Because the external excitation force is applied at a free end, the analysis follows the procedures described in Eqs. (2.80) to (2.85). The analytical results and experimental results are overlaid for comparison.

Figure 2.15 Experimental setup for vibration test on a steel beam.

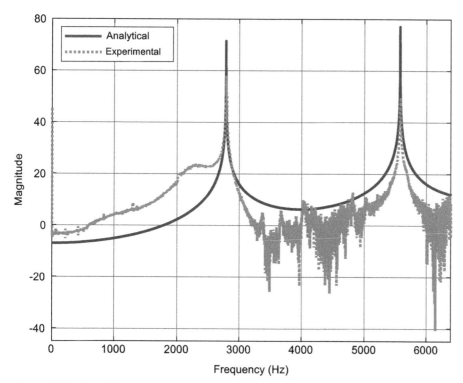

Figure 2.16 Overlaid experimental and analytical inertance frequency responses.

Figure 2.16 shows the overlaid Inertance Frequency Responses (Acceleration/Force) of the beam. The experimental and analytical results agree well. The discrepancy in the magnitudes is due to structural damping being neglected in the analytical study.

The equations of motion for longitudinally vibrating beams, torsionally vibrating shafts, and transversely vibrating string are only different in format (Meirovitch 2001). Consequently, the analyses of torsionally vibrating shafts and transversely vibrating string follow the same procedure as that of longitudinally vibrating beams described in this chapter.

2.6 MATLAB Scripts

Script 1: Symbolic scripts for characteristic expressions of various boundaries

```
clc, close all, clear all

% omega is for circular frequency, kl= omega*sqrt(p/E)

syms kl p E A KT M omega c L rA rB

r_free=1;
r_fixed=-1;
r_spring=(-1+i*kl*E*A/KT)/(1+i*kl*E*A/KT);
r_damper=(-1+kl*E*A/(omega*c))/(1+kl*E*A/(omega*c));
r_mass=(1+i*kl*E*A/(M*omega^2))/(-1+i*kl*E*A/(M*omega ^2));

fL=exp(-i*kl*L)

%The characteristic matrix
CE_Matrix=[-1,rA,0,0;0,0,rB,-1;fL,0,-1,0;0,-1,0,fL]

CEMatrix_FreeFree=subs(CE_Matrix,{rA, rB, fL},{r_free, r_free, fL})
CEMatrix_FixedFree=subs(CE_Matrix,{rA, rB, fL},{r_free, r_fixed, fL})
CEMatrix_FixedFixed=subs(CE_Matrix,{rA, rB, fL},{r_fixed, r_fixed, fL})
CEMatrix_FixedSpring=subs(CE_Matrix,{rA, rB, fL},{r_fixed, r_spring, fL})
CEMatrix_FixedMass=subs(CE_Matrix,{rA, rB, fL},{r_fixed, r_mass, fL})
CEMatrix_FixedDamper=subs(CE_Matrix,{rA, rB, fL},{r_fixed, r_damper, fL})

CE_FreeFree=simplify(det(CEMatrix_FreeFree))
CE_FixedFree=simplify(det(CEMatrix_FixedFree))
CE_FixedFixed=simplify(det(CEMatrix_FixedFixed))
CE_FixedSpring=simplify(det(CEMatrix_FixedSpring))
CE_FixedMass=simplify(det(CEMatrix_FixedMass))
CE_FixedDamper=simplify(det(CEMatrix_FixedDamper))
```

Script 2: Scripts for obtaining natural frequencies from free vibration analysis

```
clc, close all, clear all

%Material and geometrical parameters of the steel beam
E=198.87*10^9;      %Young's modulus
p=7664.5;           %Volume mass density
L=36.0*2.54/100;    %Beam length
b=2.54/100;         %Width
h=0.5*2.54/100;     %Thickness
```

```
rfixed=-1;          % Reflection at a fixed end
rfree=1;            % Reflection at a free end

rA=rfree;           % Assuming boundary conditions at A
rB=rfree;           % Assuming boundary conditions at B

f1=0;               % Starting frequency in Hz
f2=16000;           % Ending frequency in Hz
stepsize=0.1;
freq=f1:stepsize:f2;

for n=1:length(freq)
w=2*pi*freq(n);
kl=sqrt(p/E)*w;     % Wavenumber

fL=exp(-i*kl*L);    % Propagation relationship

%Forming the characteristic matrix, wave components: a+,a-,b+,b-
BigA=[-1,rA,0,0;0,0,rB,-1;fL,0,-1,0;0,-1,0,fL];
result(n)=det(BigA);     % The determinant at each frequency
end

%Plotting out the magnitude response of the characteristic polynomial
figure(1),plot(freq,abs(result),'b','LineWidth',2), grid on, hold on
xlabel('Frequency (Hz)')
ylabel('Magnitude')
title('Magnitude of the characteristic polynomial')

%Overlaying theoretical values of natural frequency to figure 1
for n=1:6
wn(n)=(n-1)*pi*sqrt(E/p)/L;  %For free-free beam
fn(n)=wn(n)/2/pi;
end

figure(1), plot(fn, zeros(1,length(fn)),'*r','MarkerSize',8)
xlabel('Frequency')
ylabel('Magnitude')
```

Script 3: Scripts for obtaining modeshapes from free vibration analysis

```
clc, close all, clear all

%Material and geometrical parameters of the steel beam
E=198.87*10^9;      %Young's modulus
p=7664.5;           %Volume mass density
L=36.0*2.54/100;    %Beam length
b=2.54/100;         %Width
h=0.5*2.54/100;     %Thickness

rfixed=-1;          % Reflection at a fixed end
rfrec=1;            % Reflection at a free end
```

```
rA=rfree;         % Assuming boundary conditions at A
rB=rfree;         % Assuming boundary conditions at B

Points=input('Enter number of points for calculating modeshapes(100): ')

X=0:L/Points:L;

%Theoretical natural frequency values for free-free beam
for n=1:6
for n0=1:length(X)
wn(n)=(n-1)*pi*sqrt(E/p)/L;
fn(n)=wn(n)/2/pi;

kl=sqrt(p/E)*wn(n);       % Wavenumber

fL=exp(-i*kl*L);       % Propagation relationship

%Wave components: a+,a-,b+,b-
BigA=[-1,rA,0,0;0,0,rB,-1;fL,0,-1,0;0,-1,0,fL];

%Solving wave components a-,b+,b- in terms of a+ for modeshapes
BigB=BigA(2:4,2:4);
BigC= -BigA(2:4,1);
WaveComponents=inv(BigB)* BigC;
aplus=1;       %Set a+ to 1
aminus= WaveComponents(1)* aplus;
bplus= WaveComponents(2)* aplus;
bminus= WaveComponents(3)* aplus;

x= X(n0);
fx1= exp(-i*kl*x);
fx2= exp(-i*kl*(-x));

u(n,n0)= aplus* fx1+ aminus* fx2 ;
end

%Plot out normalized modeshapes
umax(n)=max(real(u(n,:)));  %Modeshape normalization
figure(1), plot(X,real(u(n,:))/umax(n),'LineWidth',2),hold on
grid on
end
legend('Rigid body mode', 'Modeshape 1', 'Modeshape 2', 'Modeshape 3', 'Modeshape 4', 'Modeshape 5')
xlabel('Position along the beam')
ylabel('Amplitude')
title('Normalized modeshapes')
```

Script 4: Scripts for forced vibration analysis

```
clc; clear all; close all;
```

```
rfixed = -1;
rfree = 1;

rc=rfree;
rd=rfree;

%Steel beam
E=198.87*10^9;
p=7664.5;

%Beam dimension
L=36.0*2.54/100;
b=2.54/100;
h=0.5*2.54/100;

EA=E*b*h;

%L1=L/2            %External force applied at a nodal point (of all odd modes)
L1=19.3*2.54/100   %External force applied at a non-nodal point
L2=L-L1;
xd=L1;

x1sensor=23.6*2.54/100   %Sensor at a non-nodal point
x2sensor=L/2             %Sensor at a nodal point of odd modes: Mode 1, 3 5, ...
x3sensor=9.0*2.54/100    %Sensor at a nodal point of Mode 2

I = 1/12*b*h^3;

EA=E*b*h;
EI=E*I;

f1 = 0; % Starting frequency in Hz
f2 = 16000; % Ending frequency in Hz
stepsize=1;
freq = f1:stepsize:f2;

for n = 1:length(freq)

w = 2*pi*freq(n);
k = sqrt(p/E)*w;

fL = exp(-i*k*L);
fL1 = exp(-i*k*L1);
fL2 = exp(-i*k*L2);

fx1FromG2=exp(-i*k*(x1sensor-L1));
fx2FromC=exp(-i*k*x2sensor);
fx3FromC=exp(-i*k*x3sensor);

%Forming the characteristic matrix

%Force generated waves
```

```
F=1;
fforce=(F/(2*EA*k))*i;

%Building up the characteristic matrix

A=zeros(8,8);

%Boundary reflections
RefC=[-1 rc];
RefD=[rd -1];

ExternalForce=[0 -1 0 1;-1 0 1 0];

%c+ c- g1+ g1- g2+ g2- (force G) d+ d-

A(1,1:2)=RefC;
A(2,7:8)=RefD;
A(3:4,3:6)=ExternalForce;

A(5:6,1:8)=[fL1 0 -1 0 0 0 0 0;0 -1 0 fL1 0 0 0 0];

A(7:8,1:8)=[0 0 0 0 fL2 0 -1 0;0 0 0 0 0 -1 0 fL2];

%(g2-)-(g1-)=fforce)
%(g2+)-(g1+)=-fforce)
B=[0;0;fforce;-fforce;0;0;0;0];

ForcedResponse=A\B;

%c+ c- g1+ g1- g2+ g2- (force G) d+ d-

cplus=ForcedResponse(1);
cminus=ForcedResponse(2);

g1plus=ForcedResponse(3);
g1minus=ForcedResponse(4);
g2plus=ForcedResponse(5);
g2minus=ForcedResponse(6);

dplus=ForcedResponse(7);
dminus=ForcedResponse(8);

Y1sensor(n)=fx1FromG2*g2plus+inv(fx1FromG2)*g2minus;
Y2sensor(n)=fx2FromC*cplus+inv(fx2FromC)*cminus;
Y3sensor(n)=fx3FromC*cplus+inv(fx3FromC)*cminus;

YboundaryC(n)=cplus+cminus;
YboundaryD(n)=dplus+dminus;
YforcepointG(n)=g1plus+g1minus;

result(n)=det(A);

end
```

```
figure(1);
plot(freq,(abs(result)),'b','LineWidth',2);
xlabel('Frequency (Hz)')
ylabel('Magnitude')
title('Magnitude of the characteristic polynomial')
grid on
hold on;

%Theoretical natural frequency values for free-free beam
for n=1:6
wn(n)=(n-1)*pi*sqrt(E/p)/L;    % Circular natural frequency (rad/s)
fn(n)=wn(n)/2/pi;              % Cyclic natural frequency (Hz)
end
figure(1), plot(fn, zeros(1,length(fn)),'*r','MarkerSize',8)

figure(2);
plot(freq,20*log10(abs(result)),'b','LineWidth',2);
xlabel('Frequency (Hz)')
ylabel('dB Magnitude')
grid on
hold on;
title('dB magnitude response of the characteristic polynomial')

figure(10), plot(freq,20*log10(abs(Y1sensor)),'b','LineWidth',2); hold on
plot(freq,20*log10(abs(Y2sensor)),'-.r','LineWidth',2); hold on
plot(freq,20*log10(abs(Y3sensor)),':k','LineWidth',2); hold on
legend('Location 1','Location 2','Location 3')
grid on
xlabel('Frequency (Hz)')
ylabel('dB Magnitude')
title('Forced responses at the three sensor locations')

figure(20), plot(freq,20*log10(abs(YboundaryC)),'b','LineWidth',2); hold on
plot(freq,20*log10(abs(YboundaryD)),'-.r','LineWidth',2); hold on
plot(freq,20*log10(abs(YforcepointG)),':k','LineWidth',2); hold on
legend('Boundary C','Boundary D','Force applied point G')
grid on
xlabel('Frequency (Hz)')
ylabel('dB Magnitude')
title('Forced responses at the two boundaries and the force applied point')
```

References

Cremer L., Heckl M., Ungar E. E. *Structure-Borne Sound*, Springer-Verlag, Berlin (1987).

Doyle J. F. *Wave Propagation in Structures*, Spring-Verlag, New York (1989).

Graff K. F. *Wave Motion in Elastic Solids*, Ohio State University Press, Columbus, Ohio (1975).

Mace B. R. Wave Reflection and Transmission in Beams, *Journal of Sound and Vibration*, 97(2), 237–246 (1984).

Meirovitch L. *Fundamentals of Vibrations*, McGraw Hill, New York (2001).

Homework Project

Wave Analysis of Free and Forced Longitudinal Vibrations in a Uniform Beam

The figure below shows a uniform steel beam that has one end free and the other end attached to a spring whose stiffness is $\overline{K_T} = 10^8 \ N/m$.

The material and geometrical properties of the steel beam are: the Young's modulus is $E = 198.87 \ GN/m^2$ and mass density is $7664.5 \ kg/m^3$. The cross section of the beam is a rectangular shape whose dimensions are $12.7 \ mm$ by $25.4 \ mm$ (or 0.5 *in* by 1.0 *in*). The length of the beam is $914.4 \ mm$ (or 3.0 *ft*).

Complete the following tasks:

i) Perform free longitudinal vibration analysis of this uniform steel beam using the wave analysis approach introduced in this chapter, plot the magnitude of the characteristic polynomial, and read the lowest five natural frequencies from the plot.

ii) Plot the modeshapes of the beam corresponding to the lowest five natural frequencies that are found in (i).

iii) Obtain the receptance frequency response of the beam to include the lowest five modes by applying an external excitation force at 500 *mm* and observing the response at 600 *mm*, both measured from the free end.

iv) Examining the receptance frequency response from (iii), do you find any resonance peak of the lowest five modes missing? Why?

v) Let the spring have a large stiffness value $\overline{K_T} = 10^{12} \ N/m$, overlay the magnitude of the characteristic polynomials of this beam of free-spring boundaries, and the same beam with free-fixed boundaries. How are their natural frequencies related? Why?

vi) Let the spring have a small stiffness value $\overline{K_T} = 10^4 \ N/m$, overlay the magnitude of the characteristic polynomials of this beam of free-spring boundaries, and the same beam with free-free boundaries, how are their natural frequencies related? Why?

3

Bending Waves in Beams

Bending vibrations in a uniform beam are analyzed from the wave standpoint, where bending vibrations are described as waves that propagate along uniform waveguides and are reflected and transmitted at structural discontinuities. Compared to longitudinal vibrations studied in Chapter 2, bending vibrations are more complex. This is reflected in the higher fourth-order partial differential governing equations for bending vibrations. As a result, the propagation and reflection scalar coefficients in longitudinal vibrations become propagation and reflection matrices for bending vibrations (Mace 1984). The wave assembling approach, however, remains the same.

The propagation matrix for bending waves in a uniform beam is obtained from its governing equations of motion for free vibration. The reflection and transmission matrices are determined by the equilibrium and continuity at a structural discontinuity. As an introductory chapter to bending vibrations, the study is based on the Euler–Bernoulli bending vibration theory. Natural frequencies, modeshapes, as well as steady state frequency responses are obtained, with comparison to experimental results. MATLAB scripts for numerical simulations are provided.

3.1 The Governing Equation and the Propagation Relationships

According to the Euler–Bernoulli bending vibration theory, the governing equation of bending vibrations, which are sometimes called flexural vibrations, in a beam is

$$\frac{\partial^2}{\partial x^2}\left[EI(x)\frac{\partial^2 y(x,t)}{\partial x^2}\right]+\rho(x)A(x)\frac{\partial^2 y(x,t)}{\partial t^2}=f(x,t) \tag{3.1}$$

where x is the position along the beam axis, t is the time, $A(x)$ is the cross-sectional area, $I(x)$ is the area moment of inertia of the cross section, E is the Young's modulus, and $\rho(x)$ is the volume mass density of the beam. $y(x,t)$ is the bending deflection of the centerline of the beam, and $f(x,t)$ is the externally applied transverse force per unit length.

For a uniform beam subjected to no external force, the differential equation for free bending vibration is

$$EI\frac{\partial^4 y(x,t)}{\partial x^4}+\rho A\frac{\partial^2 y(x,t)}{\partial t^2}=0 \tag{3.2}$$

Assuming time harmonic motion and using separation of variables, the solution to Eq. (3.2) can be written in the form $y(x,t)=y_0 e^{-ikx}e^{i\omega t}$, where y_0 is the amplitude of the bending deflection of the centerline of the beam, ω is the circular frequency, k is the wavenumber, and i is the imaginary unit. Substituting this expression for $y(x,t)$ into Eq. (3.2) gives the dispersion equation, from which the bending wavenumber is obtained,

$$k=\pm\sqrt[4]{\rho A\omega^2 / EI}, \text{ and } k=\pm i\sqrt[4]{\rho A\omega^2 / EI} \tag{3.3}$$

It can be seen from Eq. (3.3) that bending waves are dispersive because the phase velocity $\dfrac{\omega}{k}$ is frequency dependent.

Mechanical Wave Vibrations: Analysis and Control, First Edition. Chunhui Mei.
© 2023 Chunhui Mei. Published 2023 by John Wiley & Sons Ltd.
Companion Website: www.wiley.com/go/Mei/MechanicalWaveVibrations

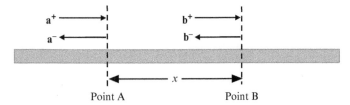

Figure 3.1 Wave propagation relationships.

With time dependence $e^{i\omega t}$ suppressed, the solutions to Eq. (3.2) can be written as

$$y = a_1^+ e^{-ik_1 x} + a_2^+ e^{-k_2 x} + a_1^- e^{ik_1 x} + a_2^- e^{k_2 x} \tag{3.4}$$

where $k_1 = k_2 = k_b$, $k_b = |k| = \sqrt[4]{\rho A \omega^2 / EI}$, and subscript b stands for bending waves. Superscripts $+$ and $-$ and subscripts 1 and 2 in wave amplitude denote positive- and negative-going propagating and decaying bending waves, respectively.

Consider two points A and B on a uniform beam that are a distance x apart, as shown in Figure 3.1. Waves propagate from one point to the other, with propagation relationships determined by the appropriate wavenumber. The positive- and negative-going waves at Points A and B are related

$$\mathbf{b}^+ = \mathbf{f}(x)\mathbf{a}^+, \ \mathbf{a}^- = \mathbf{f}(x)\mathbf{b}^- \tag{3.5a}$$

where

$$\mathbf{a}^+ = \begin{bmatrix} a_1^+ \\ a_2^+ \end{bmatrix}, \mathbf{a}^- = \begin{bmatrix} a_1^- \\ a_2^- \end{bmatrix}, \mathbf{b}^+ = \begin{bmatrix} b_1^+ \\ b_2^+ \end{bmatrix}, \mathbf{b}^- = \begin{bmatrix} b_1^- \\ b_2^- \end{bmatrix} \tag{3.5b}$$

are the wave vectors and

$$\mathbf{f}(x) = \begin{bmatrix} e^{-ik_1 x} & 0 \\ 0 & e^{-k_2 x} \end{bmatrix} \tag{3.5c}$$

is the bending propagation matrix for a distance x.

3.2 Wave Reflection at Classical and Non-Classical Boundaries

Figure 3.2 shows the sign convention adopted in the analysis that follows.

By the sign convention defined in Figure 3.2 and according to the Euler–Bernoulli bending vibration theory, the internal resistant shear force $V(x,t)$ and bending moment $M(x,t)$ are related to the transverse deflection $y(x,t)$ and bending slope $\psi(x,t)$,

$$V(x,t) = -EI\frac{\partial^3 y(x,t)}{\partial x^3}, \ M(x,t) = EI\frac{\partial \psi(x,t)}{\partial x} = EI\frac{\partial^2 y(x,t)}{\partial x^2} \tag{3.6}$$

because $\psi(x,t) = \dfrac{\partial y(x,t)}{\partial x}$ according to the Euler–Bernoulli bending vibration theory.

The reflection matrices at classical free, clamped, and pinned (also called simply-supported) boundaries, as well as at non-classical boundaries with spring and mass attachments, are derived.

Figure 3.3 illustrates a boundary with translational and rotational constraints. Incident waves are reflected at the boundary, and the incident waves $\mathbf{a}^+$ and reflected waves $\mathbf{a}^-$ are related by reflection matrix $\mathbf{r}$,

$$\mathbf{a}^- = \mathbf{r}\mathbf{a}^+ \tag{3.7}$$

where $\mathbf{r}$ is determined by boundary conditions.

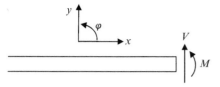

Figure 3.2 Definition of positive internal resistant shear force and bending moment.

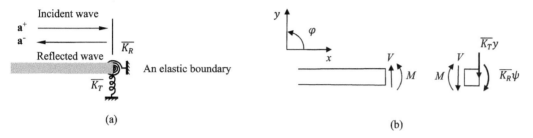

Figure 3.3 (a) Boundary with spring attachments and (b) free body diagram.

a) Free Boundary

At a free boundary, the equilibrium conditions are

$$M(x,t)=0,\ V(x,t)=0 \tag{3.8}$$

From Eqs. (3.4), (3.6), and (3.8),

$$EI\left(\left[(-ik_1)^2 a_1^+ e^{-ik_1 x}+(-k_2)^2 a_2^+ e^{-k_2 x}+(ik_1)^2 a_1^- e^{ik_1 x}+(k_2)^2 a_2^- e^{k_2 x}\right]\right)=0 \tag{3.9a}$$

$$-EI\left(\left[(-ik_1)^3 a_1^+ e^{-ik_1 x}+(-k_2)^3 a_2^+ e^{-k_2 x}+(ik_1)^3 a_1^- e^{ik_1 x}+(k_2)^3 a_2^- e^{k_2 x}\right]\right)=0 \tag{3.9b}$$

Choosing the origin at the boundary, noting that by the Euler–Bernoulli bending vibration theory $k_1=k_2$, and putting Eqs. (3.9a) and (3.9b) into matrix form,

$$\begin{bmatrix} -1 & 1 \\ i & -1 \end{bmatrix}\begin{bmatrix} a_1^+ \\ a_2^+ \end{bmatrix}-\begin{bmatrix} 1 & -1 \\ i & -1 \end{bmatrix}\begin{bmatrix} a_1^- \\ a_2^- \end{bmatrix}=0 \tag{3.10}$$

Solving the reflected waves $\mathbf{a}^-=\begin{bmatrix} a_1^- \\ a_2^- \end{bmatrix}$ in terms of the incident waves $\mathbf{a}^+=\begin{bmatrix} a_1^+ \\ a_2^+ \end{bmatrix}$,

$$\begin{bmatrix} a_1^- \\ a_2^- \end{bmatrix}=\begin{bmatrix} 1 & -1 \\ i & -1 \end{bmatrix}^{-1}\begin{bmatrix} -1 & 1 \\ i & -1 \end{bmatrix}\begin{bmatrix} a_1^+ \\ a_2^+ \end{bmatrix}=\begin{bmatrix} -i & 1+i \\ 1-i & i \end{bmatrix}\begin{bmatrix} a_1^+ \\ a_2^+ \end{bmatrix} \tag{3.11}$$

From Eqs. (3.5b), (3.7), and (3.11), the reflection matrix at a free boundary is

$$\mathbf{r}=\begin{bmatrix} -i & 1+i \\ 1-i & i \end{bmatrix} \tag{3.12}$$

b) Clamped boundary

The boundary conditions at a clamped boundary are

$$y(x,t)=0,\ \psi(x,t)=\frac{\partial y(x,t)}{\partial x}=0 \tag{3.13}$$

From Eqs. (3.4) and (3.13),

$$a_1^+ e^{-ik_1 x} + a_2^+ e^{-k_2 x} + a_1^- e^{ik_1 x} + a_2^- e^{k_2 x} = 0 \tag{3.14a}$$

$$-ik_1 a_1^+ e^{-ik_1 x} - k_2 a_2^+ e^{-k_2 x} + ik_1 a_1^- e^{ik_1 x} + k_2 a_2^- e^{k_2 x} = 0 \tag{3.14b}$$

Choosing the origin at the boundary, noting that by the Euler–Bernoulli bending vibration theory $k_1 = k_2$, and putting Eqs. (3.14a) and (3.14b) into matrix form,

$$\begin{bmatrix} 1 & 1 \\ -i & -1 \end{bmatrix} \begin{bmatrix} a_1^+ \\ a_2^+ \end{bmatrix} + \begin{bmatrix} 1 & 1 \\ i & 1 \end{bmatrix} \begin{bmatrix} a_1^- \\ a_2^- \end{bmatrix} = 0 \tag{3.15}$$

From Eqs. (3.5b), (3.7), and (3.15), the reflection matrix at a clamped boundary is

$$\mathbf{r} = \begin{bmatrix} -i & -1-i \\ -1+i & i \end{bmatrix} \tag{3.16}$$

c) Pinned Boundary

The boundary conditions at a pinned boundary are

$$y(x,t) = 0, \; M(x,t) = 0 \tag{3.17}$$

From Eqs. (3.4), (3.6), and (3.17),

$$a_1^+ e^{-ik_1 x} + a_2^+ e^{-k_2 x} + a_1^- e^{ik_1 x} + a_2^- e^{k_2 x} = 0 \tag{3.18a}$$

$$EI\left(\left[(-ik_1)^2 a_1^+ e^{-ik_1 x} + (-k_2)^2 a_2^+ e^{-k_2 x} + (ik_1)^2 a_1^- e^{ik_1 x} + (k_2)^2 a_2^- e^{k_2 x}\right]\right) = 0 \tag{3.18b}$$

Choosing the origin at the boundary, noting that by the Euler–Bernoulli bending vibration theory $k_1 = k_2$, and putting Eqs. (3.18a) and (3.18b) into matrix form,

$$\begin{bmatrix} 1 & 1 \\ -i & -1 \end{bmatrix} \begin{bmatrix} a_1^+ \\ a_2^+ \end{bmatrix} + \begin{bmatrix} 1 & 1 \\ i & 1 \end{bmatrix} \begin{bmatrix} a_1^- \\ a_2^- \end{bmatrix} = 0 \tag{3.19}$$

From Eqs. (3.5b), (3.7), and (3.19), the reflection matrix at a pinned boundary is

$$\mathbf{r} = \begin{bmatrix} -1 & 0 \\ 0 & -1 \end{bmatrix} \tag{3.20}$$

d) Boundary with Spring Attachments

Figure 3.3 shows a boundary with spring attachments. The translational and rotational stiffness are $\overline{K_T}$ and $\overline{K_R}$, respectively.

Note that the viscous damping effect, if it exists, can be taken into account by adding a frequency dependent imaginary term to the spring stiffness, that is, by introducing a dynamic spring stiffness. Denoting viscous damping constants for translational and rotational motion by $\overline{C_T}$ and $\overline{C_R}$, one may add $i\omega\overline{C_T}$ and $i\omega\overline{C_R}$ terms to the translational and rotational stiffness $\overline{K_T}$ and $\overline{K_R}$, respectively. In other words, a complex spring stiffness represents a viscous damping effect.

From the free body diagram of Figure 3.3,

$$-V(x,t) - \overline{K_T} y(x,t) = 0, \; -M(x,t) - \overline{K_R}\frac{\partial y(x,t)}{\partial x} = 0 \tag{3.21}$$

From Eqs. (3.4), (3.6), and (3.21), with the origin chosen at the spring attached boundary,

$$EI(ik_1^3 a_1^+ - k_2^3 a_2^+ - ik_1^3 a_1^- + k_2^3 a_2^-) - \overline{K_T}(a_1^+ + a_2^+ + a_1^- + a_2^-) = 0 \tag{3.22a}$$

$$EI(k_1^2 a_1^+ - k_2^2 a_2^+ + k_1^2 a_1^- - k_2^2 a_2^-) - \overline{K_R}(-ik_1 a_1^+ - k_2 a_2^+ + ik_1 a_1^- + k_2 a_2^-) = 0 \tag{3.22b}$$

Combining like terms in Eqs. (3.22a) and (3.22b) gives

$$(iEIk_1^3 - \overline{K_T})a_1^+ + (-EIk_2^3 - \overline{K_T})a_2^+ + (-iEIk_1^3 - \overline{K_T})a_1^- + (EIk_2^3 - \overline{K_T})a_2^- = 0 \tag{3.23a}$$

$$(EIk_1^2 + i\overline{K_R}k_1)a_1^+ + (-EIk_2^2 + \overline{K_R}k_2)a_2^+ + (EIk_1^2 - i\overline{K_R}k_1)a_1^- + (-EIk_2^2 - \overline{K_R}k_2)a_2^- = 0 \tag{3.23b}$$

Equations (3.23a) and (3.23b) can be placed into matrix form,

$$\begin{bmatrix} iEIk_1^3 - \overline{K_T} & -EIk_2^3 - \overline{K_T} \\ EIk_1^2 + i\overline{K_R}k_1 & -EIk_2^2 + \overline{K_R}k_2 \end{bmatrix}\begin{bmatrix} a_1^+ \\ a_2^+ \end{bmatrix} + \begin{bmatrix} -iEIk_1^3 - \overline{K_T} & EIk_2^3 - \overline{K_T} \\ EIk_1^2 - i\overline{K_R}k_1 & -EIk_2^2 - \overline{K_R}k_2 \end{bmatrix}\begin{bmatrix} a_1^- \\ a_2^- \end{bmatrix} = 0 \tag{3.24}$$

From Eqs. (3.5b), (3.7), and (3.24), the reflection matrix at a spring attached boundary is

$$\mathbf{r} = \begin{bmatrix} iEIk_1^3 + \overline{K_T} & -EIk_2^3 + \overline{K_T} \\ -EIk_1^2 + i\overline{K_R}k_1 & EIk_2^2 + \overline{K_R}k_2 \end{bmatrix}^{-1}\begin{bmatrix} iEIk_1^3 - \overline{K_T} & -EIk_2^3 - \overline{K_T} \\ EIk_1^2 + i\overline{K_R}k_1 & -EIk_2^2 + \overline{K_R}k_2 \end{bmatrix} \tag{3.25}$$

Under the special situation where $\overline{K_T}$ and $\overline{K_R}$ approach zero or infinity, the spring attached end is physically equivalent to one of the classical boundaries.

- Free boundary when $\overline{K_T} = 0$ and $\overline{K_R} = 0$

Substituting the corresponding values for $\overline{K_T}$ and $\overline{K_R}$ into Eq. (3.25) and noting that $k_1 = k_2$ gives the same reflection matrix at a free boundary obtained in Eq. (3.12).

- Clamped boundary when $\overline{K_T} = \infty$ and $\overline{K_R} = \infty$

Dividing both sides of Eq. (3.23a) by $\overline{K_T}$, dividing both sides of Eq. (3.23b) by $\overline{K_R}$, and following a similar procedure described above, a reflection matrix that is equivalent to Eq. (3.25) is obtained,

$$\mathbf{r} = \begin{bmatrix} i\dfrac{EI}{K_T}k_1^3 + 1 & -\dfrac{EI}{K_T}k_2^3 + 1 \\ -\dfrac{EI}{K_R}k_1^2 + ik_1 & \dfrac{EI}{K_R}k_2^2 + k_2 \end{bmatrix}^{-1}\begin{bmatrix} i\dfrac{EI}{K_T}k_1^3 - 1 & -\dfrac{EI}{K_T}k_2^3 - 1 \\ \dfrac{EI}{K_R}k_1^2 + iK_Rk_1 & -\dfrac{EI}{K_R}k_2^2 + k_2 \end{bmatrix} \tag{3.26}$$

Substituting $\overline{K_T} = \infty$ and $\overline{K_R} = \infty$ into Eq. (3.26) and noting that $k_1 = k_2$ gives the same reflection matrix at a clamped boundary obtained in Eq. (3.16).

- Pinned boundary when $\overline{K_T} = \infty$ and $\overline{K_R} = 0$

Dividing both sides of Eq. (3.23a) by $\overline{K_T}$, combining it with Eq. (3.23b), and following a similar procedure described above, a reflection matrix that is equivalent to both Eqs. (3.25) and (3.26) is obtained,

$$\mathbf{r} = \begin{bmatrix} i\dfrac{EI}{K_T}k_1^3 + 1 & -\dfrac{EI}{K_T}k_2^3 + 1 \\ -EIk_1^2 + i\overline{K_R}k_1 & EIk_2^2 + \overline{K_R}k_2 \end{bmatrix}^{-1}\begin{bmatrix} i\dfrac{EI}{K_T}k_1^3 - 1 & -\dfrac{EI}{K_T}k_2^3 - 1 \\ EIk_1^2 + i\overline{K_R}k_1 & -EIk_2^2 + \overline{K_R}k_2 \end{bmatrix} \tag{3.27}$$

Substituting $\overline{K_T} = \infty$ and $\overline{K_R} = 0$ into Eq. (3.27) and noting that $k_1 = k_2$ gives the same reflection matrix at a pinned boundary obtained in Eq. (3.20).

e) Boundary with a Mass Attachment

Figure 3.4 shows a boundary with a lumped mass attachment and its free body diagram. Applying Newton's second law for translational and rotational motion to the lumped mass gives

$$-V = m\frac{\partial^2 y_m(x,t)}{\partial t^2}, \quad -M + V\frac{h}{2} = J_m\frac{\partial^2 \psi_m(x,t)}{\partial t^2} \tag{3.28}$$

where m is the mass, h is the width of the attached rigid body mass block, and J_m is the mass moment of inertia of the mass block about an axis normal to the x–y plane and passing its center of mass. y_m and ψ_m are the transverse displacement and rotation of the mass block, respectively, which are related to the transverse deflection and bending slope of the beam at the point where the mass block is attached,

$$y_m = y + \frac{h}{2}\psi_m, \text{ and } \psi_m = \psi = \frac{\partial y}{\partial x} \tag{3.29}$$

where the mass block is modeled as a rigid body.

From Eqs. (3.4), (3.6), (3.28), and (3.29), with the mass attached point chosen as the origin,

$$\left(iEIk_1^3 + m\omega^2\left(1 - ik_1\frac{h}{2}\right)\right)a_1^+ + \left(-EIk_2^3 + m\omega^2\left(1 - k_2\frac{h}{2}\right)\right)a_2^+$$
$$+ \left(-iEIk_1^3 + m\omega^2\left(1 + ik_1\frac{h}{2}\right)\right)a_1^- + \left(EIk_2^3 + m\omega^2\left(1 + k_2\frac{h}{2}\right)\right)a_2^- = 0 \tag{3.30a}$$

$$\left(-iEIk_1^3\frac{h}{2} + EIk_1^2 - ik_1J_m\omega^2\right)a_1^+ + \left(EIk_2^3\frac{h}{2} - EIk_2^2 - k_2J_m\omega^2\right)a_2^+$$
$$+ \left(iEIk_1^3\frac{h}{2} + EIk_1^2 + ik_1J_m\omega^2\right)a_1^- + \left(-EIk_2^3\frac{h}{2} - EIk_2^2 + k_2J_m\omega^2\right)a_2^- = 0 \tag{3.30b}$$

Equations (3.30a) and (3.30b) can be placed into matrix form,

$$\begin{bmatrix} iEIk_1^3 + m\omega^2\left(1 - ik_1\frac{h}{2}\right) & -EIk_2^3 + m\omega^2\left(1 - k_2\frac{h}{2}\right) \\ -iEIk_1^3\frac{h}{2} + EIk_1^2 - ik_1J_m\omega^2 & EIk_2^3\frac{h}{2} - EIk_2^2 - k_2J_m\omega^2 \end{bmatrix}\begin{bmatrix} a_1^+ \\ a_2^+ \end{bmatrix}$$
$$+ \begin{bmatrix} -iEIk_1^3 + m\omega^2\left(1 + ik_1\frac{h}{2}\right) & EIk_2^3 + m\omega^2\left(1 + k_2\frac{h}{2}\right) \\ iEIk_1^3\frac{h}{2} + EIk_1^2 + ik_1J_m\omega^2 & -EIk_2^3\frac{h}{2} - EIk_2^2 + k_2J_m\omega^2 \end{bmatrix}\begin{bmatrix} a_1^- \\ a_2^- \end{bmatrix} = 0 \tag{3.31}$$

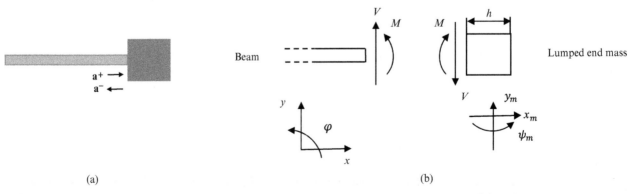

(a)   (b)

Figure 3.4 (a) Boundary with a mass attachment and (b) free body diagram.

From Eqs. (3.5b), (3.7), and (3.31), the reflection matrix at a lumped mass attached boundary is

$$
\mathbf{r} = \left[\begin{array}{cc} iEIk_1^3 - m\omega^2\left(1 + ik_1\dfrac{h}{2}\right) & -EIk_2^3 - m\omega^2\left(1 + k_2\dfrac{h}{2}\right) \\[2mm] -iEIk_1^3\dfrac{h}{2} - EIk_1^2 - ik_1 J_m\omega^2 & EIk_2^3\dfrac{h}{2} + EIk_2^2 - k_2 J_m\omega^2 \end{array} \right]^{-1}
\left[\begin{array}{cc} iEIk_1^3 + m\omega^2\left(1 - ik_1\dfrac{h}{2}\right) & -EIk_2^3 + m\omega^2\left(1 - k_2\dfrac{h}{2}\right) \\[2mm] -iEIk_1^3\dfrac{h}{2} + EIk_1^2 - ik_1 J_m\omega^2 & EIk_2^3\dfrac{h}{2} - EIk_2^2 - k_2 J_m\omega^2 \end{array} \right]
\tag{3.32}
$$

Under the special situation where $h = 0$, and m and J_m approach either zero or infinity, the mass attached end is physically equivalent to a free, clamped, or pinned end.

- Free boundary when $m = 0$ and $J_m = 0$

Substituting $m = 0$, $J_m = 0$, and $h = 0$ into Eq. (3.32), the same reflection matrix for a free boundary as Eq. (3.12) is obtained.

- Clamped boundary when $m = \infty$ and $J_m = \infty$

First dividing the first and the second row of Eq. (3.32) by m and J_m, respectively, then substituting $m = \infty$, $J_m = \infty$, and $h = 0$ into the matrix equation, the same reflection matrix for a clamped boundary as Eq. (3.16) is obtained.

- Pinned boundary when $m = \infty$ and $J_m = 0$

First dividing the first row of Eq. (3.32) by m, then substituting $m = \infty$, $J_m = 0$, and $h = 0$ into the matrix equation, the same reflection matrix for a pinned boundary as Eq. (3.20) is obtained.

The reflection matrices at the various boundaries are summarized in Table 3.1. MATLAB symbolic scripts for obtaining the reflection matrix at a rigid mass block attached end can be found in Section 3.6.

Table 3.1 Reflection matrices at various boundaries.

Boundary Condition	Reflection Matrix
Free	$\begin{bmatrix} -i & i+1 \\ 1-i & i \end{bmatrix}$
Clamped	$\begin{bmatrix} -i & -(i+1) \\ -(1-i) & i \end{bmatrix}$
Pinned	$\begin{bmatrix} -1 & 0 \\ 0 & -1 \end{bmatrix}$
Spring	$\begin{bmatrix} iEIk_1^3 + \overline{K_T} & -EIk_2^3 + \overline{K_T} \\ -EIk_1^2 + i\overline{K_R}k_1 & EIk_2^2 + \overline{K_R}k_2 \end{bmatrix}^{-1} \begin{bmatrix} iEIk_1^3 - \overline{K_T} & -EIk_2^3 - \overline{K_T} \\ EIk_1^2 + i\overline{K_R}k_1 & -EIk_2^2 + \overline{K_R}k_2 \end{bmatrix}$
Mass	$\left[\begin{array}{cc} iEIk_1^3 - m\omega^2\left(1 + ik_1\dfrac{h}{2}\right) & -EIk_2^3 - m\omega^2\left(1 + k_2\dfrac{h}{2}\right) \\[2mm] -iEIk_1^3\dfrac{h}{2} - EIk_1^2 - ik_1 J_m\omega^2 & EIk_2^3\dfrac{h}{2} + EIk_2^2 - k_2 J_m\omega^2 \end{array} \right]^{-1} \left[\begin{array}{cc} iEIk_1^3 + m\omega^2\left(1 - ik_1\dfrac{h}{2}\right) & -EIk_2^3 + m\omega^2\left(1 - k_2\dfrac{h}{2}\right) \\[2mm] -iEIk_1^3\dfrac{h}{2} + EIk_1^2 - ik_1 J_m\omega^2 & EIk_2^3\dfrac{h}{2} - EIk_2^2 - k_2 J_m\omega^2 \end{array} \right]$

3.3 Free Vibration Analysis in Finite Beams – Natural Frequencies and Modeshapes

Free vibration analysis of a flexurally vibrating uniform beam from the wave standpoint involves a concise and systematic assembling process, similar to that of an axially vibrating uniform beam. The main difference is that the propagation and reflection coefficients in axial vibrations now become propagation and reflection matrices because a set of bending waves consists of two components: a propagating wave and a decaying wave.

Figure 3.5 illustrates a uniform beam of length L. The following reflection and propagation relationships exist:

$$\mathbf{a}^+ = \mathbf{r}_A\,\mathbf{a}^-;\ \mathbf{b}^- = \mathbf{r}_B\,\mathbf{b}^+;\ \mathbf{b}^+ = \mathbf{f}(L)\,\mathbf{a}^+;\ \mathbf{a}^- = \mathbf{f}(L)\,\mathbf{b}^- \tag{3.33}$$

where $\mathbf{f}(L)$ is the bending propagation matrix of Eq. (3.5c) for a distance L and $\mathbf{r}_A$ and $\mathbf{r}_B$ are the reflection matrices at Boundaries A and B, respectively.

The above relationships can be written in matrix form,

$$\begin{bmatrix} -\mathbf{I} & \mathbf{r}_A & \mathbf{0} & \mathbf{0} \\ \mathbf{0} & \mathbf{0} & \mathbf{r}_B & -\mathbf{I} \\ \mathbf{f}(L) & \mathbf{0} & -\mathbf{I} & \mathbf{0} \\ \mathbf{0} & -\mathbf{I} & \mathbf{0} & \mathbf{f}(L) \end{bmatrix} \begin{bmatrix} \mathbf{a}^+ \\ \mathbf{a}^- \\ \mathbf{b}^+ \\ \mathbf{b}^- \end{bmatrix} = \mathbf{0} \tag{3.34}$$

where $\mathbf{I}$ and $\mathbf{0}$ denote an identity matrix and a zero matrix, respectively.

Equation (3.34) can be written in the form of

$$\mathbf{A}_0\mathbf{z}_0 = \mathbf{0} \tag{3.35}$$

where $\mathbf{A}_0$ is an 8 by 8 square coefficient matrix and $\mathbf{z}_0$ is a wave vector containing eight wave components of Eq. (3.34).

Setting the determinant of the coefficient matrix $\mathbf{A}_0$ to zero gives the characteristic equation. MATLAB symbolic scripts for obtaining the characteristic expressions are available in Section 3.6.

Natural frequencies of a structure are roots of its characteristic equation, which correspond to the non-trivial solution of Eq. (3.34), or equivalently, Eq. (3.35). Natural frequencies of a beam are determined by three factors, namely, material, dimension, and boundary conditions.

Characteristic equations corresponding to some example classical boundary conditions, namely, the free, clamped, and pinned boundaries, are obtained in the following study, where non-dimensional frequency $\Omega = k_b L$ is introduced for brevity. The natural frequencies can be solved either numerically or analytically from the corresponding characteristic equations.

a) Free-Free Boundary

Substituting the reflection matrix at a free boundary from Table 3.1 into the coefficient matrix $\mathbf{A}_0$ of Eq. (3.35) and setting the determinant of $\mathbf{A}_0$ to zero, as shown in the MATLAB scripts, gives the following complex characteristic equation

$$e^{-2(1+i)\Omega}(e^{2\Omega} + e^{i2\Omega} - 4e^{(1+i)\Omega} + e^{2(1+i)\Omega} + 1) = 0 \tag{3.36}$$

Figure 3.5 Wave propagation and reflection in a uniform beam.

Because the magnitude of the first factor is nonzero, Eq. (3.36) can only be satisfied when

$$e^{2\Omega} + e^{i2\Omega} - 4e^{(1+i)\Omega} + e^{2(1+i)\Omega} + 1 = 0 \tag{3.37}$$

that is, when the magnitude of the second factor equals zero.

Although Eq. (3.37) cannot be solved analytically, numerical solutions can be obtained using MATLAB. The characteristic equation and natural frequencies are listed in Table 3.2.

b) Clamped-Free Boundary

Substituting the reflection matrices for clamped and free boundaries from Table 3.1 into the coefficient matrix $\mathbf{A}_0$ of Eq. (3.35) and setting the determinant of $\mathbf{A}_0$ to zero, as shown in the MATLAB scripts, gives the following characteristic equation

$$e^{-2(1+i)\Omega}(e^{2\Omega} + e^{i2\Omega} + 4e^{(1+i)\Omega} + e^{2(1+i)\Omega} + 1) = 0 \tag{3.38}$$

Because the magnitude of the first factor is nonzero, Eq. (3.38) can only be satisfied when

$$e^{2\Omega} + e^{i2\Omega} + 4e^{(1+i)\Omega} + e^{2(1+i)\Omega} + 1 = 0 \tag{3.39}$$

that is, when the magnitude of the second factor equals zero.

Equation (3.39) cannot be solved analytically; however, numerical solutions can be obtained using MATLAB. The characteristic equation and natural frequencies are listed in Table 3.2.

Table 3.2 Characteristic expressions and natural frequencies for classical boundaries.

Boundary Conditions	Characteristic Equations	Non-dimensional Natural Frequencies Ω_n (Inman 2022)
Free-Free	$e^{2\Omega} + e^{i2\Omega} - 4e^{(1+i)\Omega} + e^{2(1+i)\Omega} + 1 = 0$	0 (rigid body mode)
		4.73004074
		7.85320462
		10.9956078
		14.1371655
		17.2787597
		$\dfrac{(2n+1)\pi}{2}$, for $n > 5$
Clamped-Clamped	$e^{2\Omega} + e^{i2\Omega} - 4e^{(1+i)\Omega} + e^{2(1+i)\Omega} + 1 = 0$ (identical to that of a free-free boundary)	Same as above, excluding 0 value
Clamped-Free	$e^{2\Omega} + e^{i2\Omega} + 4e^{(1+i)\Omega} + e^{2(1+i)\Omega} + 1 = 0$	1.87510407
		4.69409113
		7.85475744
		10.99554073
		14.13716839
		$\dfrac{(2n-1)\pi}{2}$, for $n > 5$
Clamped-Pinned	$ie^{2\Omega} - ie^{i2\Omega} - e^{2(1+i)\Omega} + 1 = 0$	3.92660231
		7.06858275
		10.21017612
		13.35176878
		16.49336143
		$\dfrac{(4n+1)\pi}{4}$, for $n > 5$
Pinned-Pinned	$e^{i2\Omega} - 1 = 0$	$n\pi$ where $n = 1, 2, \ldots$

c) Clamped-Clamped Boundary

Substituting the reflection matrix at a clamped boundary from Table 3.1 into the coefficient matrix $\mathbf{A}_0$ of Eq. (3.35) and setting the determinant of $\mathbf{A}_0$ to zero, as shown in the MATLAB scripts, gives the following complex characteristic equation

$$e^{-2(1+i)\Omega}(e^{2\Omega} + e^{i2\Omega} - 4e^{(1+i)\Omega} + e^{2(1+i)\Omega} + 1) = 0 \qquad (3.40)$$

One may have noticed that Eq. (3.40) is identical to Eq. (3.36) for a beam with a free-free boundary. The only difference is that there exists a rigid body mode in Eq. (3.36), which corresponds to a natural frequency value of zero.

Because the magnitude of the first factor is nonzero, Eq. (3.40) can only be satisfied when

$$e^{2\Omega} + e^{i2\Omega} - 4e^{(1+i)\Omega} + e^{2(1+i)\Omega} + 1 = 0 \qquad (3.41)$$

that is, when the magnitude of the second factor equals zero.

Equation (3.41) cannot be solved analytically. Numerical solutions are obtained using MATLAB. The characteristic equation and natural frequencies are listed in Table 3.2.

d) Clamped-Pinned Boundary

Substituting the reflection matrices for clamped and pinned boundaries from Table 3.1 into the coefficient matrix $\mathbf{A}_0$ of Eq. (3.35) and setting the determinant of $\mathbf{A}_0$ to zero, as shown in the MATLAB scripts, gives the following characteristic equation

$$e^{-2(1+i)\Omega}(ie^{2\Omega} - ie^{i2\Omega} - e^{2(1+i)\Omega} + 1) = 0 \qquad (3.42)$$

Because the magnitude of the first factor is nonzero, Eq. (3.42) can only be satisfied when

$$ie^{2\Omega} - ie^{i2\Omega} - e^{2(1+i)\Omega} + 1 = 0 \qquad (3.43)$$

that is, when the magnitude of the second factor equals zero.

Because Eq. (3.43) cannot be solved analytically, numerical solutions are obtained using MATLAB. The characteristic equation and natural frequencies are listed in Table 3.2.

e) Pinned-Pinned Boundary

Substituting the reflection matrix at a pinned boundary from Table 3.1 into the coefficient matrix $\mathbf{A}_0$ of Eq. (3.35) and setting the determinant of $\mathbf{A}_0$ to zero, as shown in the MATLAB scripts, gives the following characteristic equation

$$e^{-2(1+i)\Omega}(e^{2\Omega} - 1)(e^{i2\Omega} - 1) = 0 \qquad (3.44)$$

Because the magnitude of neither the first nor the second factor is zero for nonzero frequency, Eq. (3.44) can only be satisfied when

$$e^{i2\Omega} - 1 = 0 \qquad (3.45)$$

Equation (3.45) can be solved analytically by rearranging it as

$$e^{i2\Omega} = 1 \qquad (3.46)$$

Equating the phase angles on the left and right sides of Eq. (3.46) and excluding zero frequency value,

$$2\Omega = 2n\pi, \text{ where } n = 1, 2, \dots \qquad (3.47)$$

The non-dimensional natural frequency Ω_n is solved from Eq. (3.47) as

$$\Omega_n = n\pi, \text{ where } n = 1, 2, \dots \qquad (3.48)$$

The characteristic equation and natural frequencies are listed in Table 3.2.

When a distributed structure is excited at one of its natural frequencies, the motion of the structure exhibits a certain shape, namely, the modeshape. The modeshape at a given natural frequency can be found by eliminating any one row of the coefficient matrix $\mathbf{A}_0$ of Eq. (3.35) and using the remaining rows to solve the wave components in terms of an arbitrarily chosen wave component. In expanded form, Eq. (3.35) can be written as

$$
\begin{bmatrix}
-1 & 0 & (\mathbf{r}_A)_{11} & (\mathbf{r}_A)_{12} & 0 & 0 & 0 & 0 \\
0 & -1 & (\mathbf{r}_A)_{21} & (\mathbf{r}_A)_{22} & 0 & 0 & 0 & 0 \\
0 & 0 & 0 & 0 & (\mathbf{r}_B)_{11} & (\mathbf{r}_B)_{12} & -1 & 0 \\
0 & 0 & 0 & 0 & (\mathbf{r}_B)_{21} & (\mathbf{r}_B)_{22} & 0 & -1 \\
\mathbf{f}(L)_{11} & \mathbf{f}(L)_{12} & 0 & 0 & -1 & 0 & 0 & 0 \\
\mathbf{f}(L)_{21} & \mathbf{f}(L)_{22} & 0 & 0 & 0 & -1 & 0 & 0 \\
0 & 0 & -1 & 0 & 0 & 0 & \mathbf{f}(L)_{11} & \mathbf{f}(L)_{12} \\
0 & 0 & 0 & -1 & 0 & 0 & \mathbf{f}(L)_{21} & \mathbf{f}(L)_{22}
\end{bmatrix}
\begin{bmatrix}
a_1^+ \\ a_2^+ \\ a_1^- \\ a_2^- \\ b_1^+ \\ b_2^+ \\ b_1^- \\ b_2^-
\end{bmatrix}
=
\begin{bmatrix}
0 \\ 0 \\ 0 \\ 0 \\ 0 \\ 0 \\ 0 \\ 0
\end{bmatrix}
\tag{3.49}
$$

where the numerical subscripts in matrices $\mathbf{r}_A$, $\mathbf{r}_B$, and $\mathbf{f}(L)$ denote the location of the corresponding element in the 2 by 2 reflection and propagation matrices, with the leading digit for the row number and the other digit for the column number.

One may eliminate, for example, the first row and solve the remaining wave components in terms of a chosen wave component, such as a_1^+, which gives

$$
\begin{bmatrix}
-1 & (\mathbf{r}_A)_{21} & (\mathbf{r}_A)_{22} & 0 & 0 & 0 & 0 \\
0 & 0 & 0 & (\mathbf{r}_B)_{11} & (\mathbf{r}_B)_{12} & -1 & 0 \\
0 & 0 & 0 & (\mathbf{r}_B)_{21} & (\mathbf{r}_B)_{22} & 0 & -1 \\
\mathbf{f}(L)_{12} & 0 & 0 & -1 & 0 & 0 & 0 \\
\mathbf{f}(L)_{22} & 0 & 0 & 0 & -1 & 0 & 0 \\
0 & -1 & 0 & 0 & 0 & \mathbf{f}(L)_{11} & \mathbf{f}(L)_{12} \\
0 & 0 & -1 & 0 & 0 & \mathbf{f}(L)_{21} & \mathbf{f}(L)_{22}
\end{bmatrix}
\begin{bmatrix}
a_2^+ \\ a_1^- \\ a_2^- \\ b_1^+ \\ b_2^+ \\ b_1^- \\ b_2^-
\end{bmatrix}
=-
\begin{bmatrix}
0 \\ 0 \\ 0 \\ \mathbf{f}(L)_{11} \\ \mathbf{f}(L)_{21} \\ 0 \\ 0
\end{bmatrix}
a_1^+
\tag{3.50}
$$

From which

$$
\begin{bmatrix}
a_2^+ \\ a_1^- \\ a_2^- \\ b_1^+ \\ b_2^+ \\ b_1^- \\ b_2^-
\end{bmatrix}
=
\begin{bmatrix}
-1 & (\mathbf{r}_A)_{21} & (\mathbf{r}_A)_{22} & 0 & 0 & 0 & 0 \\
0 & 0 & 0 & (\mathbf{r}_B)_{11} & (\mathbf{r}_B)_{12} & -1 & 0 \\
0 & 0 & 0 & (\mathbf{r}_B)_{21} & (\mathbf{r}_B)_{22} & 0 & -1 \\
\mathbf{f}(L)_{12} & 0 & 0 & -1 & 0 & 0 & 0 \\
\mathbf{f}(L)_{22} & 0 & 0 & 0 & -1 & 0 & 0 \\
0 & -1 & 0 & 0 & 0 & \mathbf{f}(L)_{11} & \mathbf{f}(L)_{12} \\
0 & 0 & -1 & 0 & 0 & \mathbf{f}(L)_{21} & \mathbf{f}(L)_{22}
\end{bmatrix}^{-1}
\begin{bmatrix}
0 \\ 0 \\ 0 \\ -\mathbf{f}(L)_{11} \\ -\mathbf{f}(L)_{21} \\ 0 \\ 0
\end{bmatrix}
a_1^+
\tag{3.51}
$$

With the wave components obtained from Eq. (3.51), the deflection of any point on the uniform beam shown in Figure 3.5 can be found. For example, the response at a point that is x distance away from Boundary A can be obtained from waves at Boundary A,

$$
y(x) = \begin{bmatrix} 1 & 1 \end{bmatrix} \mathbf{f}(x)\mathbf{a}^+ + \begin{bmatrix} 1 & 1 \end{bmatrix} \mathbf{f}(-x)\mathbf{a}^-
\tag{3.52}
$$

Or equivalently, the response can be obtained from waves at Boundary B, which are a distance $(L-x)$ away,

$$
y(x) = \begin{bmatrix} 1 & 1 \end{bmatrix} \mathbf{f}(-(L-x))\mathbf{b}^+ + \begin{bmatrix} 1 & 1 \end{bmatrix} \mathbf{f}(L-x)\mathbf{b}^-
\tag{3.53}
$$

Plotting the response at all points along the structure at a chosen natural frequency gives the modeshape corresponding to that natural frequency.

While the natural frequency values are determined uniquely, the modeshape amplitudes are not. Because the wave components are solved in terms of an arbitrary reference wave component, which is a_1^+ in the above example. A normalization process is often needed to provide uniqueness in the modeshapes. In this book, each modeshape is normalized to have the maximum amplitude being unity.

3.4 Force Generated Waves and Forced Vibration Analysis of Finite Beams

An externally applied force or moment has the effect of injecting waves into a continuous structure. Figure 3.6 illustrates a point transverse force $\bar{Q}$, a bending moment $\bar{M}$, and waves $\mathbf{a}^{\pm}$ and $\mathbf{b}^{\pm}$ at the point where force and moment are applied. These bending waves are injected by and related to the externally applied force and moment.

Because the bending deflections on the left and right side of the point where the force and moment are applied are the same, denoting physical parameters on the left and right side of the point where the force and moment are applied using subscripts − and +, respectively,

$$y_- = y_+, \; \psi_- = \psi_+ \tag{3.54}$$

From the free body diagram of Figure 3.6, the externally applied transverse force and bending moment are related to the internal resistant forces and moments on the left and right side of the point where the external force and moment are applied,

$$\bar{Q} = V_- - V_+, \; \bar{M} = M_- - M_+ \tag{3.55}$$

The relationships described in Eqs. (3.54) and (3.55) are called the continuity and equilibrium equations, respectively.

From Eq. (3.4) and Figure 3.6a,

$$y_- = a_1^+ e^{-ik_1 x} + a_2^+ e^{-k_2 x} + a_1^- e^{ik_1 x} + a_2^- e^{k_2 x} \tag{3.56a}$$

$$y_+ = b_1^+ e^{-ik_1 x} + b_2^+ e^{-k_2 x} + b_1^- e^{ik_1 x} + b_2^- e^{k_2 x} \tag{3.56b}$$

According to the Euler–Bernoulli bending vibration theory $\psi(x,t) = \dfrac{\partial y(x,t)}{\partial x}$, from Eqs. (3.56a) and (3.56b),

$$\psi_- = \frac{\partial y_-}{\partial x} = -ik_1 a_1^+ e^{-ik_1 x} - k_2 a_2^+ e^{-k_2 x} + ik_1 a_1^- e^{ik_1 x} + k_2 a_2^- e^{k_2 x} \tag{3.57a}$$

$$\psi_+ = \frac{\partial y_+}{\partial x} = -ik_1 b_1^+ e^{-ik_1 x} - k_2 b_2^+ e^{-k_2 x} + ik_1 b_1^- e^{ik_1 x} + k_2 b_2^- e^{k_2 x} \tag{3.57b}$$

With the origin chosen at the point where the external force and moment are applied, from Eqs. (3.54), (3.56a), (3.56b), (3.57a), and (3.57b), the continuity conditions become

$$a_1^+ + a_2^+ + a_1^- + a_2^- = b_1^+ + b_2^+ + b_1^- + b_2^- \tag{3.58a}$$

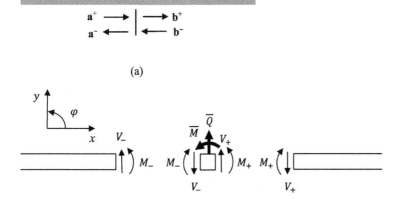

(a)

(b)

Figure 3.6 (a) Waves generated by externally applied force and moment applied on the span and (b) free body diagram.

$$-ik_1 a_1^+ - k_2 a_2^+ + ik_1 a_1^- + k_2 a_2^- = -ik_1 b_1^+ - k_2 b_2^+ + ik_1 b_1^- + k_2 b_2^- \tag{3.58b}$$

Noting that $k_1 = k_2$ by the Euler–Bernoulli bending vibration theory and putting Eqs. (3.58a) and (3.58b) into matrix form,

$$\begin{bmatrix} 1 & 1 \\ -i & -1 \end{bmatrix}(\mathbf{a}^+ - \mathbf{b}^+) + \begin{bmatrix} 1 & 1 \\ i & 1 \end{bmatrix}(\mathbf{a}^- - \mathbf{b}^-) = 0 \tag{3.59}$$

From Eq. (3.6) and Figure 3.6,

$$V_- = -EI \frac{\partial^3 y_-}{\partial x^3}, \quad M_- = EI \frac{\partial \psi_-}{\partial x} = EI \frac{\partial^2 y_-}{\partial x^2} \tag{3.60a}$$

$$V_+ = -EI \frac{\partial^3 y_+}{\partial x^3}, \quad M_+ = EI \frac{\partial \psi_+}{\partial x} = EI \frac{\partial^2 y_+}{\partial x^2} \tag{3.60b}$$

From Eqs. (3.55), (3.60a) and (3.60b),

$$\bar{Q} = -EI \left(\frac{\partial^3 y_-}{\partial x^3} - \frac{\partial^3 y_+}{\partial x^3} \right), \quad \bar{M} = EI \left(\frac{\partial^2 y_-}{\partial x^2} - \frac{\partial^2 y_+}{\partial x^2} \right) \tag{3.61}$$

With the origin chosen at the point where the external force and moment are applied, from Eqs. (3.56a), (3.56b), and (3.61), the equilibrium conditions become

$$\bar{Q} = -EI \left(\left[(-ik_1)^3 a_1^+ + (-k_2)^3 a_2^+ + (ik_1)^3 a_1^- + (k_2)^3 a_2^- \right] - \left[(-ik_1)^3 b_1^+ + (-k_2)^3 b_2^+ + (ik_1)^3 b_1^- + (k_2)^3 b_2^- \right] \right) \tag{3.62a}$$

$$\bar{M} = EI \left(\left[(-ik_1)^2 a_1^+ + (-k_2)^2 a_2^+ + (ik_1)^2 a_1^- + (k_2)^2 a_2^- \right] - \left[(-ik_1)^2 b_1^+ + (-k_2)^2 b_2^+ + (ik)^2 b_1^- + (k_2)^2 b_2^- \right] \right) \tag{3.62b}$$

Noting that $k_1 = k_2 = k_b$ by the Euler–Bernoulli bending vibration theory and putting Eqs. (3.62a) and (3.62b) into matrix form,

$$\begin{bmatrix} i & -1 \\ -1 & 1 \end{bmatrix}(\mathbf{a}^+ - \mathbf{b}^+) + \begin{bmatrix} -i & 1 \\ -1 & 1 \end{bmatrix}(\mathbf{a}^- - \mathbf{b}^-) = \begin{bmatrix} -\dfrac{\bar{Q}}{EIk_b^3} \\[2mm] \dfrac{\bar{M}}{EIk_b^2} \end{bmatrix} \tag{3.63}$$

From Eqs. (3.59) and (3.63), the relationships between the injected bending waves and the externally applied transverse force and bending moment are

$$\mathbf{b}^+ - \mathbf{a}^+ = \mathbf{q} + \mathbf{m}, \quad \mathbf{a}^- - \mathbf{b}^- = \mathbf{q} - \mathbf{m} \tag{3.64a}$$

where the load vectors are

$$\mathbf{q} = \begin{bmatrix} -i \\ -1 \end{bmatrix} \frac{\bar{Q}}{4EIk_b^3}, \quad \mathbf{m} = \begin{bmatrix} 1 \\ -1 \end{bmatrix} \frac{\bar{M}}{4EIk_b^2} \tag{3.64b}$$

With the availability of force generated waves, propagation relationships, and boundary reflections, forced vibration in a beam can be obtained.

Figure 3.7 illustrates a beam with an external transverse force and bending moment applied on the beam at an arbitrary point G that is L_1 and L_2 away from Boundaries C and D, respectively. The following wave relationships exist:

- At Boundaries C and D, from Eq. (3.7),

$$\mathbf{c}^+ = \mathbf{r}_C \, \mathbf{c}^-, \, \mathbf{d}^- = \mathbf{r}_D \, \mathbf{d}^+ \tag{3.65}$$

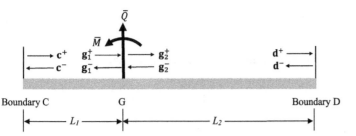

Figure 3.7 Waves in a uniform beam with external transverse force and bending moment applied on the span.

where $\mathbf{r}_C$ and $\mathbf{r}_D$ are the reflection matrices at Boundaries C and D, respectively.

- At Point G where the external force and moment are applied, from Eq. (3.64a),

$$\mathbf{g}_2^+ - \mathbf{g}_1^+ = \mathbf{q} + \mathbf{m}, \ \mathbf{g}_1^- - \mathbf{g}_2^- = \mathbf{q} - \mathbf{m} \tag{3.66}$$

where $\mathbf{q}$ and $\mathbf{m}$ are given in Eq. (3.64b).

- The propagation relationships, from Eq. (3.5a), are

$$\mathbf{g}_1^+ = \mathbf{f}(L_1)\,\mathbf{c}^+, \ \mathbf{c}^- = \mathbf{f}(L_1)\,\mathbf{g}_1^-, \ \mathbf{g}_2^- = \mathbf{f}(L_2)\,\mathbf{d}^-, \ \mathbf{d}^+ = \mathbf{f}(L_2)\,\mathbf{g}_2^+ \tag{3.67}$$

where the bending propagation matrix $\mathbf{f}(x)$ for a distance x is given by Eq. (3.5c).
Assembling the wave relationships from Eqs. (3.65) to (3.67) into matrix form,

$$
\begin{bmatrix}
-\mathbf{I} & \mathbf{r}_C & \mathbf{0} & \mathbf{0} & \mathbf{0} & \mathbf{0} & \mathbf{0} & \mathbf{0} \\
\mathbf{0} & \mathbf{0} & \mathbf{r}_D & -\mathbf{I} & \mathbf{0} & \mathbf{0} & \mathbf{0} & \mathbf{0} \\
\mathbf{0} & \mathbf{0} & \mathbf{0} & \mathbf{0} & -\mathbf{I} & \mathbf{0} & \mathbf{I} & \mathbf{0} \\
\mathbf{0} & \mathbf{0} & \mathbf{0} & \mathbf{0} & \mathbf{0} & \mathbf{I} & \mathbf{0} & -\mathbf{I} \\
\mathbf{f}(L_1) & \mathbf{0} & \mathbf{0} & \mathbf{0} & -\mathbf{I} & \mathbf{0} & \mathbf{0} & \mathbf{0} \\
\mathbf{0} & -\mathbf{I} & \mathbf{0} & \mathbf{0} & \mathbf{0} & \mathbf{f}(L_1) & \mathbf{0} & \mathbf{0} \\
\mathbf{0} & \mathbf{0} & \mathbf{0} & \mathbf{f}(L_2) & \mathbf{0} & \mathbf{0} & \mathbf{0} & -\mathbf{I} \\
\mathbf{0} & \mathbf{0} & -\mathbf{I} & \mathbf{0} & \mathbf{0} & \mathbf{0} & \mathbf{f}(L_2) & \mathbf{0}
\end{bmatrix}
\begin{bmatrix}
\mathbf{c}^+ \\ \mathbf{c}^- \\ \mathbf{d}^+ \\ \mathbf{d}^- \\ \mathbf{g}_1^+ \\ \mathbf{g}_1^- \\ \mathbf{g}_2^+ \\ \mathbf{g}_2^-
\end{bmatrix}
=
\begin{bmatrix}
\mathbf{0} \\ \mathbf{0} \\ \mathbf{q}+\mathbf{m} \\ \mathbf{q}-\mathbf{m} \\ \mathbf{0} \\ \mathbf{0} \\ \mathbf{0} \\ \mathbf{0}
\end{bmatrix}
\tag{3.68}
$$

where $\mathbf{I}$ and $\mathbf{0}$ denote an identity matrix and a zero matrix, respectively.

Equation (3.68) can be expressed in the form of

$$\mathbf{A}_f \mathbf{z}_f = \mathbf{F} \tag{3.69}$$

where $\mathbf{A}_f$ is a 16 by 16 square coefficient matrix, $\mathbf{z}_f$ is a wave vector containing 16 wave components, and $\mathbf{F}$ is the load vector of Eq. (3.68).

Solving for $\mathbf{z}_f$ from Eq. (3.69),

$$\mathbf{z}_f = \mathbf{A}_f^{-1}\mathbf{F} \tag{3.70}$$

Equation (3.70) provides the responses of all wave components, from which the deflection of any point on the beam can be obtained. For example, in Figure 3.7, the deflection of a point located between Boundary C and the excitation point G that is a distance x_1 from Point G is

$$y_1 = \begin{bmatrix} 1 & 1 \end{bmatrix} \mathbf{f}(x_1)\mathbf{g}_1^- + \begin{bmatrix} 1 & 1 \end{bmatrix} \mathbf{f}(-x_1)\mathbf{g}_1^+ \tag{3.71}$$

or equivalently is

$$y_1 = \begin{bmatrix} 1 & 1 \end{bmatrix} \mathbf{f}(L_1 - x_1)\mathbf{c}^+ + \begin{bmatrix} 1 & 1 \end{bmatrix} \mathbf{f}(-(L_1 - x_1))\mathbf{c}^- \tag{3.72}$$

Similarly, the deflection of a point located between Boundary D and the excitation point G in Figure 3.7 that is a distance x_2 from Point G is

$$y_2 = \begin{bmatrix} 1 & 1 \end{bmatrix} \mathbf{f}(x_2)\mathbf{g}_2^+ + \begin{bmatrix} 1 & 1 \end{bmatrix} \mathbf{f}(-x_2)\mathbf{g}_2^- \tag{3.73}$$

or equivalently is

$$y_2 = \begin{bmatrix} 1 & 1 \end{bmatrix} \mathbf{f}(L_2 - x_2)\mathbf{d}^- + \begin{bmatrix} 1 & 1 \end{bmatrix} \mathbf{f}(-(L_2 - x_2))\mathbf{d}^+ \tag{3.74}$$

An external force/moment may be applied at a free end of a beam, as shown in Figure 3.8. From the free body diagram of Figure 3.8, the externally applied force and moment are related to the internal resistant force and moment by the following equilibrium equations

$$\bar{Q} = V, \ \bar{M} = M \tag{3.75}$$

From Eq. (3.4) and Figure 3.8,

$$y = a_1^+ e^{-ik_1 x} + a_2^+ e^{-k_2 x} + a_1^- e^{ik_1 x} + a_2^- e^{k_2 x} \tag{3.76}$$

According to the Euler–Bernoulli bending vibration theory $\psi(x,t) = \dfrac{\partial y(x,t)}{\partial x}$, from Eq. (3.76),

$$\psi = \frac{\partial y}{\partial x} = -ik_1 a_1^+ e^{-ik_1 x} - k_2 a_2^+ e^{-k_2 x} + ik_1 a_1^- e^{ik_1 x} + k_2 a_2^- e^{k_2 x} \tag{3.77}$$

From Eqs. (3.6) and (3.75),

$$\bar{Q} = -EI\frac{\partial^3 y}{\partial x^3}, \ \bar{M} = EI\frac{\partial \psi}{\partial x} = EI\frac{\partial^2 y}{\partial x^2} \tag{3.78}$$

With the origin chosen at the point where the external force and moment are applied, from Eqs. (3.76) to (3.78), the equilibrium conditions become

$$\bar{Q} = -EI\left[\left[(-ik_1)^3 a_1^+ + (-k_2)^3 a_2^+ + (ik_1)^3 a_1^- + (k_2)^3 a_2^- \right] \right] \tag{3.79a}$$

$$\bar{M} = EI\left[\left[(-ik_1)^2 a_1^+ + (-k_2)^2 a_2^+ + (ik_1)^2 a_1^- + (k_2)^2 a_2^- \right] \right] \tag{3.79b}$$

Noting that $k_1 = k_2 = k_b$ by the Euler–Bernoulli bending vibration theory and putting Eqs. (3.79a) and (3.79b) into matrix form,

$$\begin{bmatrix} i & -1 \\ -1 & 1 \end{bmatrix} \mathbf{a}^+ + \begin{bmatrix} i & 1 \\ -1 & 1 \end{bmatrix} \mathbf{a}^- = \begin{bmatrix} -\dfrac{\bar{Q}}{EIk_b^3} \\ \dfrac{\bar{M}}{EIk_b^2} \end{bmatrix} \tag{3.80}$$

(a)   (b)

Figure 3.8 (a) Waves generated by an end force and an end moment on a beam and (b) free body diagram.

In the special case of $\bar{Q} = 0$ and $\bar{M} = 0$, which represents a free boundary, Eq. (3.80) becomes

$$\begin{bmatrix} i & -1 \\ -1 & 1 \end{bmatrix} \mathbf{a}^+ + \begin{bmatrix} -i & 1 \\ -1 & 1 \end{bmatrix} \mathbf{a}^- = 0 \tag{3.81}$$

From Eqs. (3.7) and (3.81), the reflection matrix is

$$\mathbf{r} = \begin{bmatrix} -i & 1+i \\ 1-i & i \end{bmatrix} \tag{3.82}$$

which, as expected, is identical to the reflection matrix of a free boundary obtained in Eq. (3.12).

With the availability of force generated waves, propagation relationships, and boundary reflections, forced vibration in a beam with external excitations applied at a free end can be obtained.

Figure 3.9 illustrates a beam with an external transverse force and bending moment applied at End D. The following wave relationships exist:

- At Boundary C, from Eq. (3.7),

$$\mathbf{c}^+ = \mathbf{r}_C \, \mathbf{c}^- \tag{3.83}$$

  where $\mathbf{r}_C$ is the reflection matrix at Boundary C.

- At external force and moment applied end D, from Eq. (3.80),

$$\begin{bmatrix} i & -1 \\ -1 & 1 \end{bmatrix} \mathbf{d}^+ + \begin{bmatrix} -i & 1 \\ -1 & 1 \end{bmatrix} \mathbf{d}^- = \begin{bmatrix} -\dfrac{\bar{Q}}{EIk_b^3} \\ \dfrac{\bar{M}}{EIk_b^2} \end{bmatrix} \tag{3.84}$$

- The propagation relationships, according to Eq. (3.5a), are

$$\mathbf{d}^+ = \mathbf{f}(L) \, \mathbf{c}^+, \; \mathbf{c}^- = \mathbf{f}(L) \, \mathbf{d}^- \tag{3.85}$$

  where the bending propagation matrix $\mathbf{f}(x)$ for a distance x is given by Eq. (3.5c).

  Assembling the wave relationships from Eqs. (3.83) to (3.85) into matrix form,

$$\begin{bmatrix} -\mathbf{I} & \mathbf{r}_C & \mathbf{0} & \mathbf{0} \\ \mathbf{0} & \mathbf{0} & \mathbf{q}_{c1} & \mathbf{q}_{c2} \\ \mathbf{f}(L) & \mathbf{0} & -\mathbf{I} & \mathbf{0} \\ \mathbf{0} & -\mathbf{I} & \mathbf{0} & \mathbf{f}(L) \end{bmatrix} \begin{bmatrix} \mathbf{c}^+ \\ \mathbf{c}^- \\ \mathbf{d}^+ \\ \mathbf{d}^- \end{bmatrix} = \begin{bmatrix} \mathbf{0} \\ \mathbf{q}_0 \\ \mathbf{0} \\ \mathbf{0} \end{bmatrix} \tag{3.86a}$$

where

$$\mathbf{q}_{c1} = \begin{bmatrix} i & -1 \\ -1 & 1 \end{bmatrix}, \; \mathbf{q}_{c2} = \begin{bmatrix} -i & 1 \\ -1 & 1 \end{bmatrix}, \; \mathbf{q}_0 = \begin{bmatrix} -\dfrac{\bar{Q}}{EIk_b^3} \\ \dfrac{\bar{M}}{EIk_b^2} \end{bmatrix} \tag{3.86b}$$

and $\mathbf{I}$ and $\mathbf{0}$ denote an identity matrix and a zero matrix, respectively.

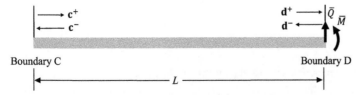

Figure 3.9 Waves in a uniform beam with external force and moment applied at a free end.

Equation (3.86a) can be written in the form of Eq. (3.69), and the wave components can be solved from Eq. (3.70). In which $\mathbf{A}_f$ is an 8 by 8 square coefficient matrix, $\mathbf{z}_f$ is a wave vector containing eight wave components, and $\mathbf{F}$ is a vector related to externally applied force and moment of Eq. (3.86a).

With the availability of responses of wave components, the deflection of any point on the beam can be obtained. For example, the deflection of a point located a distance x from Boundary C in Figure 3.9 is

$$y = \begin{bmatrix} 1 & 1 \end{bmatrix} \mathbf{f}(x)\mathbf{c}^+ + \begin{bmatrix} 1 & 1 \end{bmatrix} \mathbf{f}(-x)\mathbf{c}^- \tag{3.87}$$

or equivalently is

$$y = \begin{bmatrix} 1 & 1 \end{bmatrix} \mathbf{f}(L-x)\mathbf{d}^- + \begin{bmatrix} 1 & 1 \end{bmatrix} \mathbf{f}(-(L-x))\mathbf{d}^+ \tag{3.88}$$

3.5 Numerical Examples and Experimental Studies

Vibrations in a uniform steel beam are analyzed using the wave vibration approach and tested experimentally.

The material and geometrical properties of the steel beam are: the Young's modulus and mass density are $198.87\ GN/m^2$ and $7664.5\ kg/m^3$, respectively. The cross section of the beam is a rectangular shape whose width and thickness are 25.4 mm and $12.7\ mm$ (or $1.0\ in$ and $0.5\ in$), respectively. The length of the beam is $914.4\ mm$ (or $3.0\ ft$). In the following numerical and experimental examples, both boundaries of the beam are free.

In free wave vibration analysis of the uniform beam, the natural frequencies are obtained from Eq. (3.35) by substituting the propagation and boundary reflection matrices into the coefficient matrix $\mathbf{A}_0$ and solving for the characteristic polynomial of matrix $\mathbf{A}_0$. The magnitude of the characteristic polynomial is shown in Figure 3.10. The roots of the characteristic polynomial are the natural frequencies of the beam. The theoretical values of natural frequencies calculated using the frequency expression for free-free boundary from Table 3.2 are marked on the same graph using asterisks. The natural frequencies obtained from the wave-based free vibration analysis are identical to the theoretical values.

Figure 3.11 shows the modeshapes of the lowest five non-rigid body modes, obtained following the procedures described in Eqs. (3.49) to (3.53). What is of particular interest on the modeshape plots are the nodal points, which are where the amplitudes of a modeshape become zero. On the one hand, when a beam is excited at a non-nodal position with a frequency that is identical to a natural frequency, the nodal points of the modeshape corresponding to this natural frequency are stationary. On the other hand, if the excitation force is applied at one of the nodal points of a particular natural frequency, this specific mode will not be excited, resulting in zero contribution to the dynamic response from this mode. Marked on Figure 3.11 are two example nodal points, which are discussed in forced vibration analysis that follows.

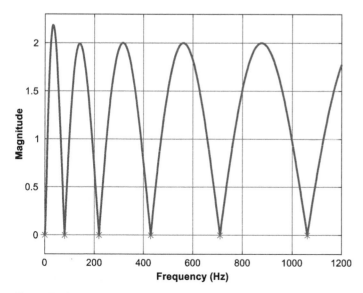

Figure 3.10 Magnitude of the characteristic polynomial.

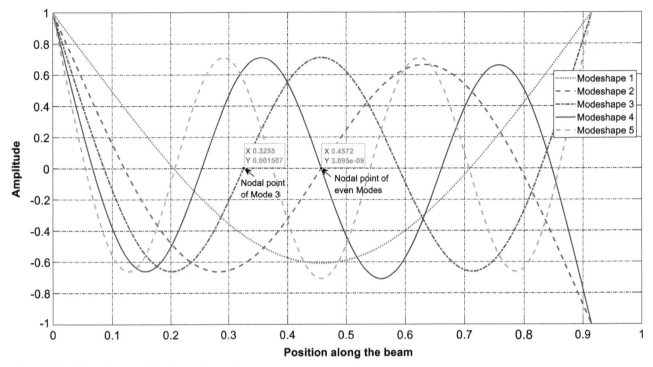

Figure 3.11 Modeshapes of the lowest few modes.

Table 3.3 Locations of external excitation and response observation.

External force and moment applied point	Test 1	500.0 *mm* (a non-nodal point of the lowest five modes)
	Test 2	457.2 *mm* (a nodal point of even modes 2, 4, 6 ...)
Responses observed points	Location 1	600.0 *mm* (a non-nodal point of the lowest five modes)
	Location 2	457.2 *mm* (a nodal point of even modes 2, 4, 6 ...)
	Location 3	325.5 *mm* (a nodal point of mode 3)

MATLAB scripts for free vibration analysis are available in Section 3.6.

For forced wave vibration analysis of a uniform beam with external excitation force and moment applied at a point on the span, the analysis follows the procedures described in Eqs. (3.65) to (3.74).

To obtain forced responses, external excitation and response observation locations need to be determined. Table 3.3 lists these points, all distances are measured from Boundary C of Figure 3.7. Because a point on the structure can either be a nodal point or a non-nodal point of a certain mode, the selection covers both scenarios.

Figures 3.12 and 3.13 show the receptance frequency responses (displacement/force), with the external excitation force applied at a non-nodal point (Test 1) and a nodal point (Test 2), as shown in Table 3.3, respectively. In both tests, responses are observed at the same three locations listed in Table 3.3, covering both non-nodal and nodal points.

The frequency range for forced vibration analysis is from 0 to 1200 *Hz*, which is kept the same as that in free vibration analysis. In this frequency range, there are five non-rigid body modes contributing to the bending dynamic responses. MATLAB scripts for forced vibration analysis are available in Section 3.6.

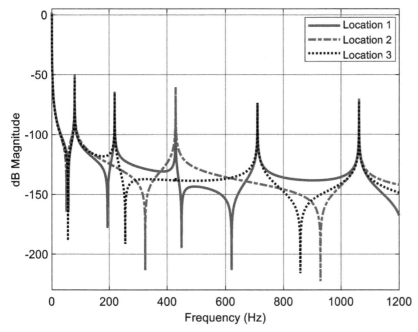

Figure 3.12 Receptance frequency responses with an external force applied at a non-nodal point.

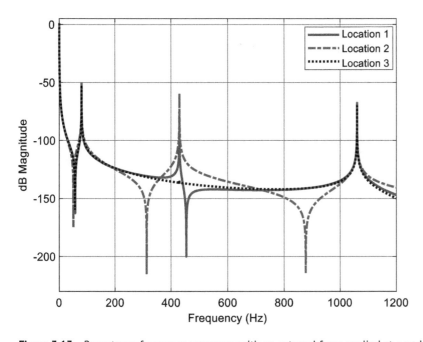

Figure 3.13 Receptance frequency responses with an external force applied at a nodal point.

As expected, it is seen from the receptance frequency responses in Figures 3.12 and 3.13 that resonant responses always occur at natural frequencies, and the resonant peaks line up regardless of the force application and the response observation locations. Examining the response curves in both figures, however, one finds that not all resonant peaks occur at all natural frequencies. This is because the occurrence of resonant peaks in frequency responses is closely related to the location of external excitation as well as the location of response observation relative to a nodal point, as discussed in mode-shape curves in free vibration analysis.

Table 3.4 summarizes the occurrence of resonant peaks in the frequency responses of Figures 3.12 and 3.13. Locations of external excitation and response observation relative to a nodal point are shown in Table 3.3. Information provided in these two tables will allow one to understand the reasons behind the appearance and disappearance of resonant peaks in the frequency responses.

Recall that in Test 1, the external excitation force is applied at a non-nodal point. Among the three observation locations, Location 1 is at a non-nodal point, Location 2 is at a nodal point of nonzero even modes, and Location 3 is a nodal point of Mode 3. This explains why at Location 1, all resonant peaks appear; at Location 2, the resonant peaks corresponding to even Modes 2 and 4 disappear; and at Location 3, the resonant peak corresponding to Mode 3 disappears.

In Test 2, with the point where the external excitation force is applied purposely chosen at a nodal point of nonzero even modes, it is seen that resonant peaks corresponding to Modes 2 and 4 disappear, regardless of the observation locations. At Location 3, an additional resonant peak disappears, which corresponds to Mode 3. This is because the observation point is at a nodal point of Mode 3.

The above studies point out the importance of nodal points. On the one hand, a nodal point needs to be avoided in identifying natural frequencies of a structure because of the disappearance of resonant peaks that results in miscounting of natural frequencies. One may avoid or reduce the related problem by exciting a structure and observing the responses at a variety of locations. On the other hand, a nodal point is an ideal location for mounting equipment on a structure from a vibration isolation standpoint because of its stationary nature.

Vibration tests are conducted on the beam, and the experimental set up is the same as that of Figure 2.15. To have free-free boundary conditions, the metal beam is suspended by two thin wires. The test is conducted with an impact force applied on the beam using a Brüel & Kjær Type 8202 impact hammer along the direction normal to the plane formed by the beam and the wires. Vibrations are picked up by a Brüel & Kjær Type 4397 accelerometer. The impact force is applied at 489.0 *mm* measured from a free end, and the dynamic responses are taken at 593.9 *mm*, measured from the same free end. Frequency responses are obtained using a PULSE unit from Brüel & Kjær. The frequency span and frequency resolution are 6400 *Hz* and 1 *Hz*, respectively.

Because the external excitation force is applied on the span of the beam, the analysis follows the procedures described in Eqs. (3.65) to (3.74). The analytical results and experimental results are overlaid for comparison.

Figure 3.14 shows the overlaid inertance frequency responses (Acceleration/Force) of the beam. The experimental and analytical results agree well in the lower frequency range. As frequency increases, the analytical study overpredicts the natural frequencies. This is because the analysis is based on the Euler–Bernoulli bending vibration theory. It calls for advanced theory for modeling bending vibrations at higher frequencies. The discrepancy in the magnitudes is because of structural damping being neglected in the analytical study.

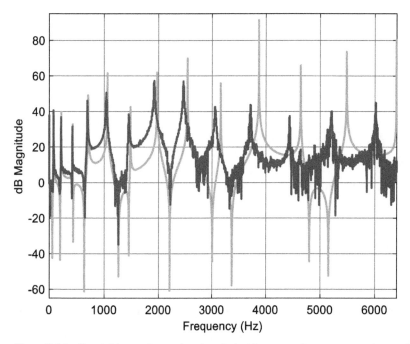

Figure 3.14 Overlaid experimental and analytical inertance frequency responses with width and thickness of the cross section being 25.4 *mm* and 12.7 *mm*, respectively.

Table 3.4 Occurrence of resonant peaks.

Occurrence of resonant peaks		Mode 0	Mode 1	Mode 2	Mode 3	Mode 4	Mode 5
Test 1 (Figure 3.12)	Location 1	✓	✓	✓	✓	✓	✓
	Location 2	✓	✓	✗	✓	✗	✓
	Location 3	✓	✓	✓	✗	✓	✓
Test 2 (Figure 3.13)	Location 1	✓	✓	✗	✓	✗	✓
	Location 2	✓	✓	✗	✓	✗	✓
	Location 3	✓	✓	✗	✗	✗	✓

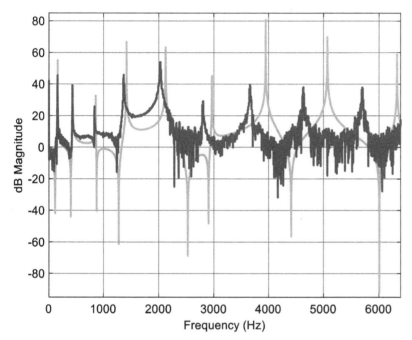

Figure 3.15 Overlaid experimental and analytical inertance frequency responses with width and thickness of the cross section being 12.7 *mm* and 25.4 *mm*, respectively.

Experiments have also been conducted by rotating the beam 90° about its longitudinal axis, so that the width and thickness are now 12.7 *mm* and 25.4 *mm* (or 0.5 *in* and 1.0 *in*), respectively. The overlaid inertance frequency responses (acceleration/force) from experimental and analytical studies are presented in Figure 3.15. Again, the experimental and analytical results agree well in the lower frequency range. Discrepancies occur as frequency increases. The analytical study based on the Euler–Bernoulli bending vibration theory again overpredicts the natural frequencies. It confirms the need for advanced bending vibration theory for modeling vibrations at higher frequencies. This will be addressed in Chapter 9 of this book.

3.6 MATLAB Scripts

<u>Script 1:</u> Symbolic scripts for reflection matrix at a rigid mass block attached end

```
clc, close all, clear all

syms k1 k2 k h m EI Jm omega a1plus a2plus a1minus a2minus x real

w=a1plus*exp(-i*k1*x)+ a2plus*exp(-k2*x)+ a1minus*exp(i*k1*x)+ a2minus*exp(k2*x)
```

```
w1=diff(w,x)
w2=diff(w,x,2)
w3=diff(w,x,3)

phi=w1
phi1= diff(phi,'x')
phim=phi
ym=w+(h/2)*phim
V=-EI*w3
M=EI*phi1

Expression1x=-V-m*(-omega^2)*ym
Expression2x=-M+V*(h/2)-Jm*(-omega^2)*phim

Expression1=subs(Expression1x,{x},{0})
Expression2=subs(Expression2x,{x},{0})

a1plusCoeff1=subs(Expression1,{a1plus a2plus a1minus a2minus},{1 0 0 0})
a2plusCoeff1=subs(Expression1,{a1plus a2plus a1minus a2minus},{0 1 0 0})
a1minusCoeff1=subs(Expression1,{a1plus a2plus a1minus a2minus},{0 0 1 0})
a2minusCoeff1=subs(Expression1,{a1plus a2plus a1minus a2minus},{0 0 0 1})

a1plusCoeff2=subs(Expression2,{a1plus a2plus a1minus a2minus},{1 0 0 0})
a2plusCoeff2=subs(Expression2,{a1plus a2plus a1minus a2minus},{0 1 0 0})
a1minusCoeff2=subs(Expression2,{a1plus a2plus a1minus a2minus},{0 0 1 0})
a2minusCoeff2=subs(Expression2,{a1plus a2plus a1minus a2minus},{0 0 0 1})

AminusCoeff=[a1minusCoeff1, a2minusCoeff1;a1minusCoeff2,a2minusCoeff2]
AplusCoeff=[a1plusCoeff1, a2plusCoeff1;a1plusCoeff2,a2plusCoeff2]
rEndMass=simplify(-inv(AminusCoeff)*(AplusCoeff))
```

Script 2: Symbolic scripts for characteristic expressions of various classical boundaries

```
clc, close all, clear all

syms k k1 k2 EI L rA rB I O f

r_clamped=[-i,-(1+i);-(1-i),i];

r_simp=[-1,0;0,-1];

r_free=[-i,1+i;1-i,i];

fL=diag([exp(-i*k*L),exp(-k*L)]);

%The Characteristic Matrices

CEMatrix=[-I,rA,O,O;O,O,rB,-I;f,O,-I,O;O,-I,O,f]

CEMatrixFreeFree=subs(CEMatrix,{rA, rB, f, I, O},{r_free, r_free, fL, eye(2), zeros(2)})
CEMatrixFreeClamped=subs(CEMatrix,{rA, rB, f, I, O},{r_free, r_clamped, fL, eye(2), zeros(2)})
CEMatrixClampedClamped=subs(CEMatrix,{rA, rB, f, I, O},{r_clamped, r_clamped, fL, eye(2), zeros(2)})
CEMatrixSimpleSimple=subs(CEMatrix,{rA, rB, f, I, O},{r_simp, r_simp, fL, eye(2), zeros(2)})
CEMatrixClampedSimple=subs(CEMatrix,{rA, rB, f, I, O},{r_clamped, r_simp, fL, eye(2), zeros(2)})
```

%The Characteristic Expressions

CEExpressionFreeFree=simplify(det(CEMatrixFreeFree))
CEExpressionFreeClamped=simplify(det(CEMatrixFreeClamped))
CEExpressionClampedClamped=simplify(det(CEMatrixClampedClamped))
CEExpressionSimpleSimple=simplify(det(CEMatrixSimpleSimple))
CEExpressionClampedSimple=simplify(det(CEMatrixClampedSimple))

<u>Script 3</u>: Scripts for obtaining natural frequencies from free vibration analysis

```
clc; clear all; close all;

rclamped = [ -i -(1+i); -(1-i) i];
rpinned = [-1 0; 0 -1];
rfree = [ -i (1+i); (1-i) i];

rc=rfree;
rd=rfree;

%Material and geometrical parameters of the steel beam
E=198.87*10^9;
p=7664.5;
L=36.0*2.54/100;    %Beam length
b=2.54/100;         %Width
h=0.5*2.54/100;     %Thickness

A = b*h;
I = 1/12*b*h^3;
EA=E*A;
EI=E*I;

f1 = 0; % Starting frequency in Hz
f2 = 1200; % Ending frequency in Hz
stepsize = 0.1;
freq = f1:stepsize:f2;

for n = 1:length(freq)

w = 2*pi*freq(n);

k = sqrt(sqrt(p*A*w^2/EI));
k1 = k;
k2 = k;

fL = diag([exp(-i*k1*L) exp(-k2*L)]);

% Forming the characteristic matrix

BigA=zeros(8,8);

%Boundary reflections
RefC=[-eye(2) rc];
RefD=[rd -eye(2)];
```

```
%Propagations along L
PropL=[fL zeros(2) -eye(2) zeros(2); zeros(2) -eye(2) zeros(2) fL];

% Wave component sequence c1+ c2+ c1- c2-  d1+ d2+ d1- d2-

BigA(1:2,1:4)=RefC;
BigA(3:4,5:8)=RefD;
BigA(5:8,1:8)=PropL;

result(n)=det(BigA);

end

figure(1);
plot(freq,abs(result),'b','LineWidth',2);
xlabel('Frequency (Hz)')
ylabel('Magnitude')
axis([0 1200 -0.2 2.3])
grid on
hold on;

figure(2);
plot(freq,20*log10(abs(result)),'b','LineWidth',2);
xlabel('Frequency (Hz)')
ylabel('dB Magnitude')
grid on
hold on;
```

Script 4: Scripts for obtaining modeshapes from free vibration analysis

```
clc; clear all; close all;

rclamped = [ -i -(1+i); -(1-i) i];
rpinned = [-1 0; 0 -1];
rfree = [ -i (1+i); (1-i) i];

rc=rfree;
rd=rfree;

%Material and geometrical parameters of the steel beam
E=198.87*10^9;
p=7664.5;
L=36.0*2.54/100;   %Beam length
b=2.54/100;        %Width
h=0.5*2.54/100;    %Thickness

A = b*h;
I = 1/12*b*h^3;
EA=E*A;
EI=E*I;

%Theoretical natural frequency values of the lowest five modes of the free-free beam
%(excluding the rigid body mode)
```

```matlab
wn =[4.73004074 7.85320462 10.9956078 14.1371655 17.2787597].^2/L^2*sqrt(E*I/p/A);

point=100;

for n=1:length(wn)

w = wn(n);

%Bending wavenumber of Euler–Bernoulli beam
k = sqrt(sqrt(p*A*w^2/EI));
k1 = k;
k2 = k;

fL = diag([exp(-i*k1*L) exp(-k2*L)]);

BigA=zeros(8,8);

%Boundary reflections
RefC=[-eye(2) rc];
RefD=[rd -eye(2)];

%Propagations along L
PropL=[fL zeros(2) -eye(2) zeros(2); zeros(2) -eye(2) zeros(2) fL];

BigA(1:2,1:4)=RefC;
BigA(3:4,5:8)=RefD;
BigA(5:8,1:8)=PropL;

%Component sequence in vector
% c1+ c2+ c1- c2- d1+ d2+ d1- d2-

F=BigA(2:8,2:8);
G=-BigA(2:8,1);          %in terms of c1+
cp1=1;
invFG=(inv(F)*G)*cp1;

cp2=invFG(1);
cm1=invFG(2);
cm2=invFG(3);
dp1=invFG(4);
dp2=invFG(5);
dm1=invFG(6);
dm2=invFG(7);

X=0:L/point:L;

for nn=1:(point+1)

  x=X(nn);
  Y(n,nn)=cp1*exp(-i*k1*x)+cp2*exp(-k2*x)+cm1*exp(i*k1*x)+cm2*exp(k2*x);

end
end
```

```
for n=1:length(wn)
figure(1)
plot(X,real(Y(n,:))/(max(abs(real((Y(n,:)))))),'LineWidth',2) %Modeshape normalization

hold on
grid on
end
legend('Modeshape 1', 'Modeshape 2', 'Modeshape 3', 'Modeshape 4', 'Modeshape 5')
xlabel('Position along the beam')
ylabel('Amplitude')
title('Normalized modeshapes')
```

Script 5 MATLAB scripts for forced vibration analysis

```
clc; clear all; close all

rclamped = [ -i -(1+i);-(1-i) i];
rpinned = [-1 0;0 -1];
rfree = [ -i (1+i);(1-i) i];

rc=rfree;
rd=rfree;

%Beam length, width, and thickness
L=36.0*2.54/100;   %Length
b=2.54/100          %Width
h=0.5*2.54/100       %Thickness

E=198.87*10^9;
p=7664.5

A = b*h;
I = 1/12*b*h^3;

EA=E*A;
EI=E*I;

%Disturbance force applied point (Two choices, select one, comment out the other)
%L1=0.4572 %A nodal point for even mode, except for Mode 0, the rigid body mode
L1=0.5    %A non-nodal point for the lowest 5 non-rigid body modes

xd=L1;
L2=L-L1;

%Sensors to view responses
x3sensor=0.3255;  %Nodal point of mode 3
x2sensor=0.4572;  %A nodal point for even mode, except for Mode 0, the rigid body mode
x1sensor=0.6;      %A non-nodal point for the lowest 5 non-rigid body modes

f1 = 0; % Starting frequency in Hz
f2 = 1200; % Ending frequency in Hz
stepsize=0.1

freq = f1:stepsize:f2;
```

```
for n = 1:length(freq)

w = 2*pi*freq(n);

%Bending wavenumber of Euler–Bernoulli beam
k = sqrt(sqrt(p*A*w^2/EI));
k1 = k;
k2 = k;

fL1 = diag([exp(-i*k1*L1) exp(-k2*L1)]);
fL2 = diag([exp(-i*k1*L2) exp(-k2*L2)]);

fx1FromG2=diag([exp(-i*k1*(x1sensor-L1)) exp(-k1*(x1sensor-L1))]);
fx2FromG2=diag([exp(-i*k1*abs(x2sensor-L1)) exp(-k1*abs(x2sensor-L1))]);
fx3FromC=diag([exp(-i*k1*x3sensor) exp(-k1*x3sensor)]);

%Forming the characteristic matrix

%Force generated waves
%(g12-)-(g11-)=-qforce+mforce
%(g12+)-(g11+)=qforce+mforce
ShearForce=1;
Moment=0;

%By the Euler–Bernoulli bending vibration theory
qforce=(ShearForce/(4*EI*k1^3))*[-i;-1];
mforce=(Moment/(4*EI*k1^2))*[1;-1];

%Building up the characteristic matrix

BigA=zeros(16,16);

%Boundary reflections
RefC=[-eye(2) rc];
RefD=[rd -eye(2)];

BigB=[zeros(2,1);zeros(2,1);-
qforce+mforce;qforce+mforce;zeros(2,1);zeros(2,1);zeros(2,1);zeros(2,1)];

ExternalForce1=[zeros(2) -eye(2) zeros(2) eye(2)];
ExternalForce2=[-eye(2) zeros(2) eye(2) zeros(2)];

%Sequence of wave components: c+ c- g1+ g1- g2+ g2- (force G) d+ d- (2 by 2 matrix form)
%In expanded form: c1+ c2+ c1- c2-  g11+ g12+ g11- g12- g21+ g22+ g21- g22- d1+ d2+ d1- d2-

%Propagations along L1 and L2
PropL1=[fL1 zeros(2) -eye(2) zeros(2); zeros(2) -eye(2) zeros(2) fL1];
PropL2=[fL2 zeros(2) -eye(2) zeros(2); zeros(2) -eye(2) zeros(2) fL2];

BigA(1:2,1:4)=RefC;
BigA(3:4,13:16)=RefD;
BigA(5:6,5:12)=ExternalForce1;
BigA(7:8,5:12)=ExternalForce2;
```

```
BigA(9:12,1:8)=PropL1;
BigA(13:16,9:16)=PropL2;

ForcedResponse=BigA\BigB;

cplus=ForcedResponse(1:2);
cminus=ForcedResponse(3:4);

g1plus=ForcedResponse(5:6);
g1minus=ForcedResponse(7:8);
g2plus=ForcedResponse(9:10);
g2minus=ForcedResponse(11:12);

dplus=ForcedResponse(13:14);
dminus=ForcedResponse(15:16);

Y1sensor(n)=[1 1]*fx1FromG2*g2plus+[1 1]*inv(fx1FromG2)*g2minus;
Y2sensor(n)=[1 1]*inv(fx2FromG2)*g1plus+[1 1]*fx2FromG2*g1minus;
Y3sensor(n)=[1 1]*fx3FromC*cplus+[1 1]*inv(fx3FromC)*cminus;

result(n)=det(BigA);

end

figure(1);
plot(freq,(abs(result)),'b','LineWidth',2);
xlabel('Frequency (Hz)')
ylabel('Magnitude')
grid on
hold on;

figure(2);
plot(freq,20*log10(abs(result)),'b','LineWidth',2);
xlabel('Frequency (Hz)')
ylabel('dB Magnitude')
grid on
hold on;

figure(3), plot(freq,20*log10(abs(Y1sensor)),'b','LineWidth',2); hold on
plot(freq,20*log10(abs(Y2sensor)),'-.r','LineWidth',2); hold on
plot(freq,20*log10(abs(Y3sensor)),':k','LineWidth',2); hold on
legend('Location 1','Location 2','Location 3')
grid on
xlabel('Frequency (Hz)')
ylabel('dB Magnitude')
axis([0 1200 -230 5])
```

References

Inman D. J. *Engineering Vibration*, 5th Edition, Pearson, New Jersey (2022).

Mace B. R. Wave Reflection and Transmission in Beams, *Journal of Sound and Vibration*, 97(2), 237–246 (1984).

Homework Project

Wave Analysis of Free and Forced In-Plane Bending Vibrations in a Uniform Cantilever Beam with a Rigid Body Mass Attached at Its Free End

Consider in-plane bending vibration, that is, bending vibrations in the x–y plane, of a steel beam with an end mass, as shown in the figure below. The uniform steel beam that has one end clamped and the other end attached to a rigid mass block whose mass is $m = 1.7\ kg$, width is $h = 80\ mm$, and mass moment of inertia about an axis normal to x–y plane and passing its center of mass is $J_m = 0.001\ kgm^2$.

The material and geometrical properties of the steel beam are: the Young's modulus is $E = 198.87\ GN/m^2$ and mass density is $\rho = 7664.5\ kg/m^3$. The cross section of the beam is a rectangular shape whose thickness is $t = 12.7\ mm$ and depth is $d = 25.4\ mm$. The length of the beam is $L = 914.4\ mm$.

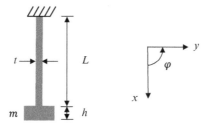

Complete the following tasks:

i) Perform free in-plane bending vibration analysis of this uniform steel beam using the wave analysis approach introduced in this chapter, plot the magnitude of the characteristic polynomial, read the lowest five natural frequencies from the plot.

ii) Plot the modeshapes of the beam corresponding to the lowest five natural frequencies idnentified in (i).

iii) Obtain the receptance frequency response of the beam to include the lowest five modes by applying an external excitation force at 400 mm and observing the response at 700 mm, both measured from the clamped end.

iv) Examining the receptance frequency response from (iii). Do you find any resonance peak of the lowest five modes missing? Why?

v) For the three special situations in the table below, overlay the magnitude of the characteristic polynomials of this beam of clamped-mass boundaries under each special situation tabulated below, and the same beam with the corresponding classical boundaries, that is,

- Overlay the magnitude of the characteristic polynomial of the beam with clamped-mass boundaries under Situation 1 in the table with that of the same beam of clamped-free boundaries.
- Overlay the magnitude of the characteristic polynomial of the beam with clamped-mass boundaries under Situation 2 in the table with that of the same beam of clamped-clamped boundaries.
- Overlay the magnitude of the characteristic polynomial of the beam with clamped-mass boundaries under Situation 3 in the table with that of the same beam of clamped-pinned boundaries.

In each overlaid magnitudes of the characteristic polynomials, how are their natural frequencies related? Are they as expected? Why?

Special situations	m (kg)	Jm (kgm²)	h (mm)	Equivalency check
Situation 1	0	0	0	free boundary
Situation 2	1.7×10^3	0.001×10^6	0	clamped boundary
Situation 3	1.7×10^3	0	0	pinned boundary

4

Waves in Beams on a Winkler Elastic Foundation

Longitudinal and bending vibration waves in beams resting on a Winkler elastic foundation are studied. A Winkler elastic foundation is the simplest model for soil–structure interactions by assuming linear and local relationships between the elastic force from the foundation and the deflection of a beam. The concepts of cut-off frequency and wave mode transition are introduced. The analysis is presented in non-dimensional form, a different form compared to those of the previous chapters.

4.1 Longitudinal Waves in Beams

4.1.1 The Governing Equation and the Propagation Relationships

Figure 4.1 shows an axially vibrating uniform beam lying on a Winkler elastic foundation. The governing equation of motion for free vibration is (De Rosa and Maurizi 2000)

$$\rho A \frac{\partial^2 u(x,t)}{\partial t^2} - EA \frac{\partial^2 u(x,t)}{\partial x^2} + k_w u(x,t) = 0 \tag{4.1}$$

where x is the position along the beam axis, t is the time, A is the cross-sectional area, $u(x,t)$ is the longitudinal deflection of the centerline of the beam, E is the Young's modulus, ρ is the volume mass density of the beam, and k_w is the axial stiffness of the Winkler elastic foundation.

The internal resistant axial force $F(x,t)$ and axial deflection $u(x,t)$, according to the sign convention defined in Figure 4.2, are related by

$$F(x,t) = EA \frac{\partial u(x,t)}{\partial x} \tag{4.2}$$

Assuming time harmonic motion and using separation of variables $u(x,t) = U(x)e^{i\omega t}$, where ω is the circular frequency and i is the imaginary unit, the differential equation for free longitudinal motion becomes

$$\rho A \omega^2 U(x) + EA \frac{\partial^2 U(x)}{\partial x^2} - k_w U(x) = 0 \tag{4.3}$$

In non-dimensional form, Eq. (4.3) is

$$\frac{\partial^2 U(\xi)}{\partial \xi^2} - \left(K_w - \Omega^2\right) U(\xi) = 0 \tag{4.4}$$

where the non-dimensional position ξ along the beam axis, the non-dimensional frequency Ω, and the non-dimensional axial stiffness of the Winkler elastic foundation K_w are

Mechanical Wave Vibrations: Analysis and Control, First Edition. Chunhui Mei.
© 2023 Chunhui Mei. Published 2023 by John Wiley & Sons Ltd.
Companion Website: www.wiley.com/go/Mei/MechanicalWaveVibrations

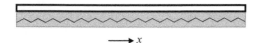

Figure 4.1 An axially vibrating beam on a Winkler elastic foundation.

Figure 4.2 Definition of positive internal resistant axial force.

$$\xi = \frac{x}{L}, \ \Omega^2 = \frac{\rho A}{EA}\omega^2 L^2, \ K_w = \frac{k_w L^2}{EA} \tag{4.5}$$

in which L is the length of the beam.

The solution to Eq. (4.4) can be written in the form of $U(\xi) = u_0 e^{-ik\xi}$, where u_0 is the amplitude of the longitudinal deflection of the centerline of the beam, k is the wavenumber, and i is the imaginary unit. Substituting this expression for $U(\xi)$ into Eq. (4.4) gives the dispersion equation, from which the longitudinal wavenumber is obtained,

$$k^2 = \Omega^2 - K_w \tag{4.6}$$

Equation (4.6) shows that there exists a cut-off frequency $\Omega_c = \sqrt{K_w}$,

$$k_{1,2} = \pm k_l, \quad \text{for } \Omega > \Omega_c \tag{4.7a}$$

$$k_{1,2} = \pm ik_l, \quad \text{for } \Omega < \Omega_c \tag{4.7b}$$

where $k_l = \sqrt{\left|\Omega^2 - K_w\right|}$ and subscript l stands for longitudinal waves. Equation (4.7a) corresponds to a pair of positive- and negative-going propagating waves and Eq. (4.7b) corresponds to a pair of positive- and negative-going decaying waves. This indicates that a wave mode transition occurs at the cut-off frequency.

The solution to Eq. (4.4), the non-dimensional homogeneous axial vibration equation, in terms of the two longitudinal wave components can be written as

$$u(\xi) = a^+ e^{-ik\xi} + a^- e^{ik\xi} \tag{4.8}$$

where $k = k_l$ for $\Omega > \Omega_c$ and $k = ik_l$ for $\Omega < \Omega_c$. Superscripts $+$ and $-$ in wave amplitude denote positive- and negative-going longitudinal wave, respectively.

For two points A and B on a uniform beam that are a distance ξ apart, as shown in Figure 4.3, the propagation relationships between them are determined by their distance and the wavenumber. The positive- and negative-going waves at Points A and B are related by

$$a^- = f(\xi)b^-, \ b^+ = f(\xi)a^+ \tag{4.9a}$$

where

$$f(\xi) = e^{-ik\xi} \tag{4.9b}$$

is the longitudinal propagation coefficient corresponding to a distance ξ.

4.1.2 Wave Reflection at Boundaries

Since $\dfrac{\partial u(x,t)}{\partial x} = \dfrac{\partial u(\xi,t)}{\partial \xi}\dfrac{\partial \xi}{\partial x} = \dfrac{1}{L}\dfrac{\partial u(\xi,t)}{\partial \xi}$, Equation (4.2) can be written in terms of non-dimensional displacement ξ,

$$F(\xi,t) = \frac{EA}{L}u(\xi,t) \tag{4.10}$$

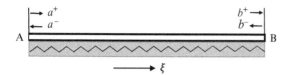

Figure 4.3 Wave propagation relationships.

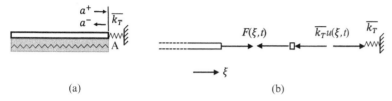

Figure 4.4 (a) A beam with an elastic boundary and (b) free body diagram.

As shown in Figure 4.4a, an incident wave is reflected at a boundary. The incident wave a^+ and reflected wave a^- are related by reflection coefficient r,

$$a^- = ra^+ \tag{4.11}$$

where r is determined by boundary condition.

From the free body diagram of Figure 4.4,

$$-F(\xi,t) - \overline{k_T}u(\xi,t) = 0 \tag{4.12}$$

where $\overline{K_T}$ is the stiffness of the spring and $\overline{k_T}u(\xi,t)$ is the resistant force from the spring by Hooke's law.

In non-dimensional form, from Eqs. (4.8), (4.10), and (4.12),

$$-\frac{EA}{L}\left(-ika^+e^{-ik\xi} + ika^-e^{ik\xi}\right) - \overline{k_T}\left(a^+e^{-ik\xi} + a^-e^{ik\xi}\right) = 0 \tag{4.13}$$

Choosing the origin at the boundary and collecting like terms,

$$(-\overline{k_T}L + ikEA)a^+ = (\overline{k_T}L + ikEA)a^- \tag{4.14}$$

From Eqs. (4.11) and (4.14), the reflection coefficient r at an elastic boundary is

$$r = \frac{-\overline{k_T}L + ikEA}{\overline{k_T}L + ikEA} \tag{4.15a}$$

or equivalently is

$$r = \frac{-1 + ikX_T}{1 + ikX_T} \tag{4.15b}$$

where $X_T = \dfrac{EA}{\overline{k_T}L}$ is the non-dimensional flexibility coefficient.

Under the special situation where $\overline{k_T}$ approaches zero or infinity, the spring attached end is physically equivalent to a free or a fixed end, respectively. From Eqs. (4.15a) and (4.15b), the reflection coefficients are $r = 1$ and $r = -1$, which correspond to the reflection coefficients of a free and a fixed boundary, respectively.

4.1.3 Free Wave Vibration Analysis

Free vibration analysis of an axially vibrating uniform beam on a Winkler elastic foundation follows a similar procedure as that of an axially vibrating beam described in Chapter 2.

From the wave standpoint, vibrations of a uniform beam on a Winkler elastic foundation shown in Figure 4.3 can be viewed as waves that propagate along the beam, and are reflected at the boundaries. In a non-dimensional coordinate system, the propagation and reflection relationships are

$$b^+ = f(1)\, a^+; \; a^- = f(1)\, b^-; \; a^+ = r_A\, a^-; \; b^- = r_B\, b^+ \tag{4.16}$$

where r_A and r_B are the reflection coefficients at Boundaries A and B, respectively. $f(\xi)$ is the longitudinal propagation coefficient for a distance ξ given in Eq. (4.9b). The total length of a beam in dimensionless unit, from Eq. (4.5), is $\xi = 1$.

The above relationships can be written in matrix form,

$$\begin{bmatrix} f(1) & 0 & -1 & 0 \\ 0 & -1 & 0 & f(1) \\ -1 & r_A & 0 & 0 \\ 0 & 0 & r_B & -1 \end{bmatrix} \begin{bmatrix} a^+ \\ a^- \\ b^+ \\ b^- \end{bmatrix} = 0 \tag{4.17}$$

Natural frequencies of a structure are roots of its characteristic equation, which correspond to the non-trivial solution of Eq. (4.17). For the non-trivial solution, the determinant of the coefficient matrix is zero, which allows the natural frequencies of the beam to be solved.

The modeshape at a given natural frequency can be found by eliminating any one row of the coefficient matrix of Eq. (4.17), and using the remaining rows to solve the wave components in terms of any chosen wave component. The response of any point located on the structure can be obtained. For example, the response at a point that is a distance ξ away from Boundary A is

$$u(\xi) = f(\xi)a^+ + f(-\xi)a^- \tag{4.18a}$$

or equivalently is

$$u(\xi) = f(1-\xi)b^- + f(-(1-\xi))b^+ \tag{4.18b}$$

Plotting the response at all points along the structure at a chosen natural frequency gives the modeshape corresponding to that natural frequency, as described in Chapters 2 and 3.

4.1.4 Force Generated Waves and Forced Vibration Analysis of Finite Beams

An externally applied force has the effect of injecting waves into a continuous structure. Figure 4.5 illustrates a point axial force $\bar{F}$ and waves $a^\pm$ and $b^\pm$ at the point where the axial force is applied. These waves are injected by and related to the externally applied force.

First, from the continuity at the point where the force is applied, the longitudinal deflections on the left and right side of the point where the force is applied are equal. Denoting physical parameters on the left and right side of the point where the force is applied using subscripts − and +, respectively,

$$u_- = u_+ \tag{4.19}$$

Second, consider the equilibrium at the point where the force is applied. From the free body diagram of Figure 4.5, the externally applied axial force is related to the internal resistant forces on the left and right side of the point where the external force is applied by

$$\bar{F} = F_- - F_+ \tag{4.20}$$

With the origin chosen at the point where the external force is applied, from Eq. (4.8) and Figure 4.5a,

$$u_- = a^+ + a^-, \text{ and } u_+ = b^+ + b^- \tag{4.21}$$

With the origin chosen at the point where the external force is applied, from Eqs. (4.8) and (4.10) and Figure 4.5,

$$F_- = \frac{EA}{L}(-ika^+ + ika^-), \; F_+ = \frac{EA}{L}(-ikb^+ + ikb^-) \tag{4.22}$$

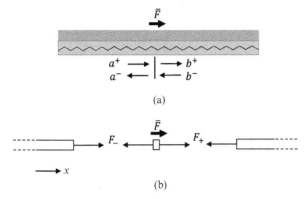

Figure 4.5 (a) Waves generated by an external axial force applied on the span and (b) free body diagram.

From Eqs. (4.19) to (4.22), the relationships between the externally applied force and the generated waves are

$$b^+ - a^+ = -q, \; b^- - a^- = q \tag{4.23}$$

where

$$q = i\frac{\bar{F}L}{2EAk} \tag{4.24}$$

is the normalized externally applied axial force.

With the availability of force generated waves, propagation relationships, and boundary reflections, forced vibration in a beam can be obtained.

Figure 4.6 illustrates a beam with an external excitation force applied at an arbitrary point G on the beam. The following wave relationships exist:

- At Boundaries C and D, from Eq. (4.11),

$$c^+ = r_C \, c^-, \; d^- = r_D \, d^+ \tag{4.25}$$

where r_C and r_D are the reflection coefficients at Boundaries C and D, respectively.

- At Point G where the external force is applied, from Eq. (4.23),

$$g_2^+ - g_1^+ = -q, \; g_2^- - g_1^- = q \tag{4.26}$$

where q is given in Eq. (4.24).

- The propagation relationships, from Eq. (4.9a), are

$$g_1^+ = f(\xi_1) \, c^+, \; c^- = f(\xi_1) \, g_1^-, \; g_2^- = f(\xi_2) \, d^-, \; d^+ = f(\xi_2) \, g_2^+ \tag{4.27}$$

where the longitudinal propagation coefficient corresponding to a distance ξ is given in Eq. (4.9b), and $\xi_1 + \xi_2 = 1$.
Putting Eqs. (4.25) to (4.27) into matrix form,

$$
\begin{bmatrix}
-1 & r_C & 0 & 0 & 0 & 0 & 0 & 0 \\
0 & 0 & r_D & -1 & 0 & 0 & 0 & 0 \\
0 & 0 & 0 & 0 & -1 & 0 & 1 & 0 \\
0 & 0 & 0 & 0 & 0 & -1 & 0 & 1 \\
f(\xi_1) & 0 & 0 & 0 & -1 & 0 & 0 & 0 \\
0 & -1 & 0 & 0 & 0 & f(\xi_1) & 0 & 0 \\
0 & 0 & 0 & f(\xi_2) & 0 & 0 & 0 & -1 \\
0 & 0 & -1 & 0 & 0 & 0 & f(\xi_2) & 0
\end{bmatrix}
\begin{bmatrix}
c^+ \\ c^- \\ d^+ \\ d^- \\ g_1^+ \\ g_1^- \\ g_2^+ \\ g_2^-
\end{bmatrix}
=
\begin{bmatrix}
0 \\ 0 \\ -q \\ q \\ 0 \\ 0 \\ 0 \\ 0
\end{bmatrix}
\tag{4.28}
$$

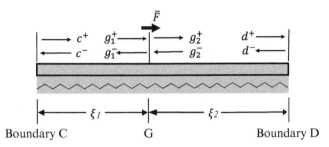

Figure 4.6 Waves in a uniform beam with an external axial force applied on the span.

Equation (4.28) can be written in the form of

$$\mathbf{A}_f \mathbf{z}_f = \mathbf{F} \tag{4.29}$$

where $\mathbf{A}_f$ is an 8 by 8 square coefficient matrix, $\mathbf{z}_f$ is a wave vector containing eight wave components, and $\mathbf{F}$ is a vector related to externally applied force of Eq. (4.28).

Solving for $\mathbf{z}_f$ from Eq. (4.29),

$$\mathbf{z}_f = \mathbf{A}_f^{-1} \mathbf{F} \tag{4.30}$$

Equation (4.30) provides the responses of all wave components, from which the deflection of any point on the beam can be obtained. For example, in Figure 4.6, the deflection of a point located between Boundary C and the excitation point G that is at a distance ξ from Point G is

$$u_1 = f(\xi)g_1^- + f(-\xi)g_1^+ \tag{4.31a}$$

or equivalently is

$$u_1 = f(\xi_1 - \xi)c^+ + f(-(\xi_1 - \xi))c^- \tag{4.31b}$$

Similarly, the deflection of a point located between Boundary D and the excitation point G in Figure 4.6 that is at a distance ξ from Point G is

$$u_2 = f(\xi)g_2^+ + f(-\xi)g_2^- \tag{4.32a}$$

or equivalently is

$$u_2 = f(\xi_2 - \xi)d^- + f(-(\xi_2 - \xi))d^+ \tag{4.32b}$$

An external force may be applied at a free end of a beam, as shown in Figure 4.7. From the free body diagram of Figure 4.7(b), the externally applied axial force is related to the internal resistant force by the following equilibrium equation

$$\bar{F} - F = 0 \tag{4.33}$$

With the origin chosen at the point where the force is applied, from Eq. (4.8) and Figure 4.7a,

$$u = a^+ + a^- \tag{4.34}$$

With the origin chosen at the point where the external force is applied, from Eqs. (4.8) and (4.10) and Figure 4.7,

$$F = \frac{EA}{L}(-ika^+ + ika^-) \tag{4.35}$$

From Eqs. (4.33) and (4.35), the relationships between an externally applied end force and its generated waves are

$$a^+ - a^- = 2q \tag{4.36}$$

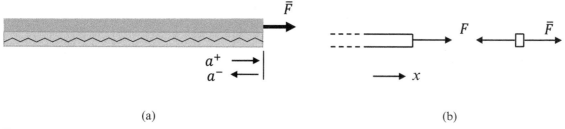

(a) (b)

Figure 4.7 (a) Waves generated by an end force on a beam and (b) free body diagram.

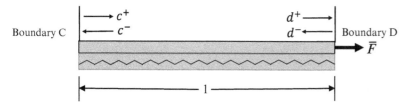

Figure 4.8 Waves in a uniform beam with an external axial force applied at a free end.

where q is given in Eq. (4.24).

Note that when there is no external force, that is, $\bar{F} = 0$, the external force applied end becomes a free end. From Eqs. (4.24) and (4.36), $a^- = a^+$. This indicates that the reflection coefficient $r = 1$, which is the reflection coefficient of a free end, as expected.

With the availability of force generated waves, propagation relationships, and boundary reflections, forced vibration response in a beam caused by an end excitation force can be obtained.

Figure 4.8 illustrates a beam with an external excitation point force applied at End D. The following wave relationships exist:

- At Boundary C, from Eq. (4.11),

$$c^+ = r_C \, c^- \tag{4.37}$$

where r_C is the reflection coefficient at Boundary C.

- At the external force applied end D, from Eq. (4.36),

$$d^+ - d^- = 2q \tag{4.38}$$

where q is given in Eq. (4.24).

- The propagation relationships, from Eq. (4.9a), are

$$d^+ = f(1) \, c^+, \; c^- = f(1) d^- \tag{4.39}$$

where the longitudinal propagation coefficient corresponding to a distance ξ is given in Eq. (4.9b), and the total length of a beam in dimensionless unit is $\xi = 1$.

Putting Eqs. (4.37) to (4.39) into matrix form,

$$\begin{bmatrix} -1 & r_C & 0 & 0 \\ 0 & 0 & 1 & -1 \\ f(1) & 0 & -1 & 0 \\ 0 & -1 & 0 & f(1) \end{bmatrix} \begin{bmatrix} c^+ \\ c^- \\ d^+ \\ d^- \end{bmatrix} = \begin{bmatrix} 0 \\ 2q \\ 0 \\ 0 \end{bmatrix} \tag{4.40}$$

Equation (4.40) can be written in the form of Eq. (4.29), and the wave components can be solved from Eq. (4.30). In which $\mathbf{A}_f$ is a 4 by 4 square coefficient matrix, $\mathbf{z}_f$ is a wave vector containing four wave components, and $\mathbf{F}$ is a vector related to externally applied force of Eq. (4.40).

With the availability of responses of all wave components, the deflection of any point on the beam can be obtained. For example, the deflection of a point that is a distance ξ away from Boundary C in Figure 4.8 is

$$u = f(\xi)c^+ + f(-\xi)c^-$$ (4.41a)

or equivalently is

$$u = f(1-\xi)d^- + f(-(1-\xi))d^+$$ (4.41b)

4.1.5 Numerical Examples

Free and forced axial vibrations of a uniform beam resting on a Winkler elastic foundation of non-dimensional stiffness K_w are analyzed using the wave vibration approach. The beam has one end fixed and the other end attached to a spring with non-dimensional flexibility coefficient X_T. Both K_w and X_T are allowed to vary, where K_w takes four different values, namely, $K_w = 0, 10, 100,$ and 1000, and X_T also takes four different values, namely, $X_T = 0, 1, 10,$ and 1000.

Free wave vibration analysis follows the procedures described in Section 4.1.3. Both the non-dimensional natural frequencies Ω_n and modeshapes are obtained. The lowest ten non-dimensional natural frequency values, corresponding to the various combination of K_w and X_T values, are read off the dB magnitude plot of the characteristic polynomial and listed in Table 4.1. A sample plot of the dB magnitude of the characteristic polynomial is presented in Figure 4.9, which corresponds to $K_w = 10$ and $X_T = 10$. The local minima in the dB magnitude of the characteristic polynomial correspond to the natural frequencies, except for the local minimum at the cut-off frequency, which is associated with a sharp change in magnitude at $\Omega = \Omega_c$. This cut-off frequency related local minimum is highlighted on the graph using a dotted vertical line.

Modeshapes of the lowest ten modes are presented in Figure 4.10, where each modeshape is normalized to unity. Each subplot contains 16 overlaid modeshape curves corresponding to $K_w = 0, 10, 100,$ and 1000, and $X_T = 0, 1, 10,$ and 1000. The stiffness of the Winkler foundation has little influence on the normalized modeshapes. However, as the flexibility coefficient varies from 0 to 1000, the modeshapes move from resembling those of a fixed-free beam to a fixed-fixed beam.

In forced vibration analysis, the receptance frequency responses (displacement/normalized force) can be obtained following the procedures described in Section 4.1.4. The external force applied point and the response observation point need to be determined first. To observe all resonant peaks for the lowest ten natural frequencies, a nodal point shall be avoided in the selection process for both the external force applied point and the response observation point. The overlaid modeshapes of the lowest ten modes corresponding to the various combinations of K_w and X_T values are presented in Figure 4.11, on which two non-nodal points are identified and marked using an asterisk and a circle, respectively. The excitation force applied point is chosen at one of the two identified non-nodal points of $\xi = 0.325$, and the response observation point is at the other non-nodal point of $\xi = 0.175$, both measured from the fixed boundary.

Figure 4.12 shows the receptance frequency response corresponding to $X_T = 1$, and $K_w = 0, 10, 100,$ and 1000, respectively. The corresponding dB magnitude of the characteristic polynomial is presented with the receptance frequency response in each subplot, with the cut-off frequency highlighted using a dotted vertical line on each plot. The non-dimensional frequency range, for both free and forced vibration analysis, is from 0 to 45. When the locations of both the force applied point and response observed point are chosen at non-nodal points of the lowest ten modes, all natural frequencies identified in the dB magnitude plots of the characteristic polynomials appear in the receptance frequency responses as resonant peaks, as expected.

The forced responses show that the local minima associated with the cut-off frequencies as marked on the graphs by dotted vertical lines do not correspond to resonant peaks. This indicates that these local minima corresponding to the cut-off frequencies do not represent the natural frequencies, they are sharp changes in magnitude caused by the cut-off frequencies and shall be distinguished from the remaining local minima.

Table 4.1 and Figure 4.12 show that the increase of the stiffness of the Winkler elastic foundation K_w results in the increase of the natural frequency Ω_n of the corresponding mode, regardless of end flexibility. Additionally, when X_T takes values 0 and 1000, the boundary becomes fixed and free accordingly.

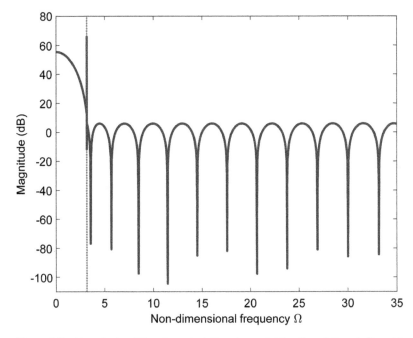

Figure 4.9 Magnitude of the characteristic polynomial for $K_w = 10$ and $X_T = 10$.

Table 4.1 Non-dimensional natural frequency Ω_n of the axially vibrating beam.

		Natural frequency Ω_n			
X_T	Mode Number	$K_w = 0$ $\Omega_c = \sqrt{K_w} = 0$	$K_w = 10$ $\Omega_c = \sqrt{K_w} = 3.1623$	$K_w = 100$ $\Omega_c = \sqrt{K_w} = 10$	$K_w = 1000$ $\Omega_c = \sqrt{K_w} = 31.6228$
0	1	3.142	4.458	10.482	31.778
	2	6.283	7.034	11.810	32.241
	3	9.425	9.941	13.741	32.997
	4	12.566	12.958	16.060	34.028
	5	15.708	16.023	18.621	35.309
	6	18.850 $n\pi$ (Fixed-Fixed Beam)	19.113	21.338	36.815
	7	21.991	22.217	24.158	38.518
	8	25.133	25.331	27.050	40.394
	9	28.274	28.451	29.991	42.420
	10	31.416	31.575	32.969	44.575
1	1	2.029	3.757	10.204	31.688
	2	4.913	5.843	11.142	32.002
	3	7.979	8.583	12.793	32.614
	4	11.086	11.528	14.930	33.510
	5	14.207	14.555	17.374	34.668
	6	17.336	17.622	20.014	36.063
	7	20.469	20.712	22.781	37.669
	8	23.604	23.815	25.635	39.461
	9	26.741	26.927	28.550	41.414
	10	29.879	30.046	31.508	43.506

(Continued)

Table 4.1 (Continued)

		Natural frequency Ω_n			
X_T	Mode Number	$K_w = 0$ $\Omega_c = \sqrt{K_w} = 0$	$K_w = 10$ $\Omega_c = \sqrt{K_w} = 3.1623$	$K_w = 100$ $\Omega_c = \sqrt{K_w} = 10$	$K_w = 1000$ $\Omega_c = \sqrt{K_w} = 31.6228$
10	1	1.632	3.559	10.132	31.665
	2	4.734	5.693	11.064	31.975
	3	7.867	8.479	12.723	32.587
	4	11.005	11.450	14.870	33.483
	5	14.144	14.493	17.322	34.642
	6	17.285	17.571	19.969	36.038
	7	20.425	20.669	22.742	37.646
	8	23.566	23.777	25.600	39.438
	9	26.707	26.894	28.518	41.392
	10	29.849	30.016	31.479	43.485
1000	1	1.571	3.531	10.123	31.662
	2	4.712	5.675	11.055	31.972
	3	7.854	8.467	12.716	32.584
	4	10.996	11.441	14.863	33.480
	5	14.137 $\dfrac{2n-1}{2}\pi$ (Fixed-Free Beam)	14.487	17.317	34.639
	6	17.279	17.566	19.964	36.036
	7	20.420	20.664	22.738	37.643
	8	23.562	23.773	25.596	39.436
	9	26.704	26.890	28.515	41.389
	10	29.845	30.012	31.476	43.483

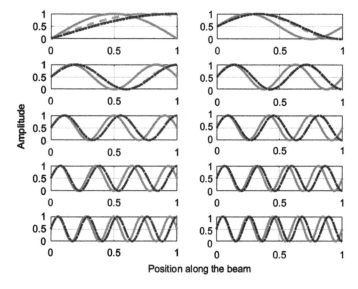

Figure 4.10 Overlaid lowest ten modeshapes: $X_T = 0$ (___); $X_T = 1$ (- - -); $X_T = 10$ (-.-.-); and $X_T = 1000$ (...).

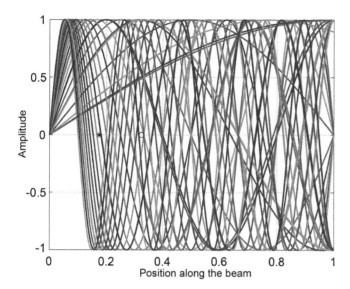

Figure 4.11 Overlaid modeshapes of the lowest ten modes.

4.2 Bending Waves in Beams

4.2.1 The Governing Equation and the Propagation Relationships

Illustrated in Figure 4.13 is a transversely vibrating uniform beam lying on a Winkler elastic foundation.

The governing equation of motion for free bending vibration is (Zhou 1993)

$$EI\frac{\partial^4 y(x,t)}{\partial x^4} + k_w y(x,t) + \rho A\frac{\partial^2 y(x,t)}{\partial t^2} = 0 \tag{4.42}$$

where x is the position along the beam axis, t is the time, $y(x,t)$ is the transverse deflection of the centerline of the beam, I is the area moment of inertia of the cross section, and A is the cross-sectional area. ρ and E are the volume mass density and Young's modulus, respectively. k_w is the transverse stiffness of the Winkler elastic foundation.

With the positive sign directions for shear force and bending moment defined in Figure 4.14, the shear force $V(x,t)$ and bending moment $M(x,t)$ at a section of the beam at location x are related to the transverse deflection $y(x,t)$ as follows

$$V(x,t) = -EI\frac{\partial^3 y(x,t)}{\partial x^3} \tag{4.43}$$

$$M(x,t) = EI\frac{\partial^2 y(x,t)}{\partial x^2} \tag{4.44}$$

In terms of non-dimensional position ξ along the beam axis where $\xi = \dfrac{x}{L}$, the governing equation of motion Eq. (4.42) becomes

$$\frac{\partial^4 y(\xi,t)}{\partial \xi^4} + \frac{k_w L^4}{EI} y(\xi,t) + \frac{\rho A L^4}{EI}\frac{\partial^2 y(\xi,t)}{\partial t^2} = 0 \tag{4.45}$$

Assuming time harmonic motion and using separation of variables, the solution to Eq. (4.45) can be written in the form of $y(\xi,t) = y_0 e^{-ik\xi} e^{i\omega t}$, where y_0 is the amplitude of the transverse deflection of the centerline of the beam, ω is the circular frequency, k is the wavenumber, and i is the imaginary unit. Substituting this expression for $y(\xi,t)$ into Eq. (4.45) gives the dispersion equation, from which the bending wavenumber is obtained,

$$k^4 = \Omega^4 - K_w \tag{4.46}$$

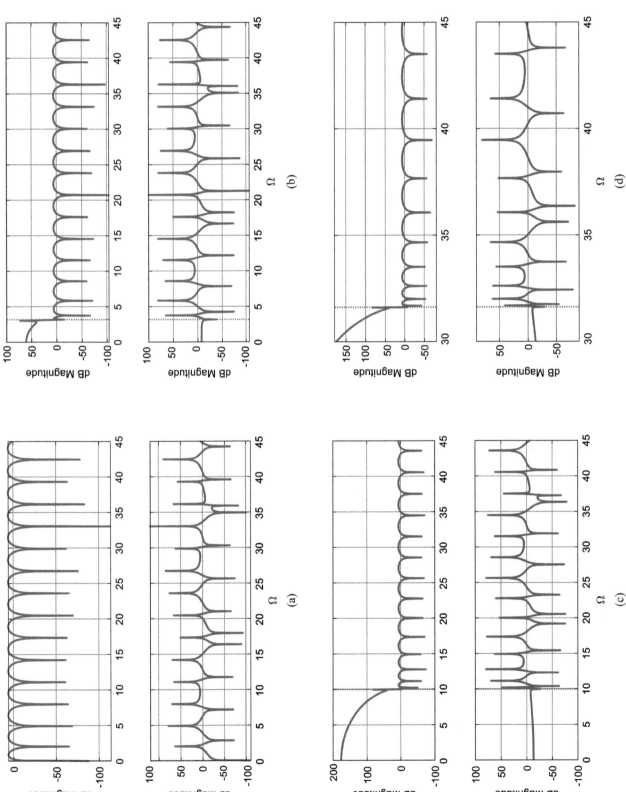

Figure 4.12 Magnitudes of the characteristic polynomials, and the receptance frequency responses corresponding to $X_T = 1$ and (a) $K_w = 0$, (b) $K_w = 10$, (c) $K_w = 100$, and (d) $K_w = 1000$.

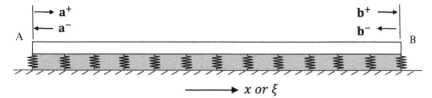

Figure 4.13 A transversely vibrating beam on a Winkler elastic foundation.

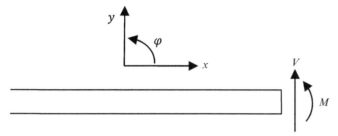

Figure 4.14 Definition of positive internal resistant shear force and bending moment.

where Ω is the non-dimensional frequency, and K_w is the non-dimensional transverse stiffness of the Winkler elastic foundation, which are related to the physical parameters of the beam and the elastic foundation by

$$\Omega^4 = \frac{\rho A}{EI} \omega^2 L^4, \ K_w = \frac{k_w L^4}{EI} \tag{4.47}$$

Equation (4.46) shows that there exists a cut-off frequency $\Omega_c = \sqrt[4]{K_w}$. The bending wavenumbers are

$$k_{1,2} = \pm k_b, \ k_{3,4} = \pm i k_b \qquad \text{for } \Omega > \Omega_c \tag{4.48a}$$

$$k_{1,2} = \pm \sqrt{i} k_b, \ k_{3,4} = \pm i \sqrt{i} k_b \qquad \text{for } \Omega < \Omega_c \tag{4.48b}$$

where $k_b = \sqrt[4]{\left|\Omega^4 - K_w\right|}$ and subscript b stands for bending waves. Equation (4.48a) corresponds to a pair of positive- and negative-going propagating and a pair of positive- and negative-going decaying waves, and Eq. (4.48b) corresponds to two pairs of positive- and negative-going decaying oscillatory waves. This indicates that a wave mode transition occurs at the cut-off frequency.

With time dependence $e^{i\omega t}$ suppressed, the solution to Eq. (4.45) can be written as

$$y(\xi) = a_1^+ e^{-ik_B\xi} + a_2^+ e^{-k_B\xi} + a_1^- e^{ik_B\xi} + a_2^- e^{k_B\xi} \tag{4.49}$$

where

$$k_B = k_b \qquad \text{for } \Omega > \Omega_c \tag{4.50a}$$

$$k_B = \sqrt{i} k_b \qquad \text{for } \Omega < \Omega_c \tag{4.50b}$$

Superscripts $+$ and $-$ and subscripts 1 and 2 in wave amplitude denote the two pairs of positive- and negative-going bending waves.

Assuming that Points A and B on the uniform beam shown in Figure 4.13 are a distance ξ apart, waves at these two points are related by

$$\mathbf{a}^- = \mathbf{f}(\xi)\mathbf{b}^-; \quad \mathbf{b}^+ = \mathbf{f}(\xi)\mathbf{a}^+ \tag{4.51}$$

where

$$\mathbf{f}(\xi) = \begin{bmatrix} e^{-ik_B\xi} & 0 \\ 0 & e^{-k_B\xi} \end{bmatrix} \tag{4.52}$$

is the bending propagation matrix for a distance ξ, and the wave vectors are

$$\mathbf{a}^+ = \begin{bmatrix} a_1^+ \\ a_2^+ \end{bmatrix}, \ \mathbf{a}^- = \begin{bmatrix} a_1^- \\ a_2^- \end{bmatrix}, \ \mathbf{b}^+ = \begin{bmatrix} b_1^+ \\ b_2^+ \end{bmatrix}, \ \mathbf{b}^- = \begin{bmatrix} b_1^- \\ b_2^- \end{bmatrix} \tag{4.53}$$

4.2.2 Wave Reflection at Classical Boundaries

According to the Euler–Bernoulli bending vibration theory, the bending slope $\psi(x,t)$ is related to the transverse deflection $y(x,t)$ by

$$\psi(x,t) = \frac{\partial y(x,t)}{\partial x} \tag{4.54}$$

Since

$$\frac{\partial y(x,t)}{\partial x} = \frac{\partial y(\xi,t)}{\partial \xi}\frac{\partial \xi}{\partial x} = \frac{1}{L}\frac{\partial y(\xi,t)}{\partial \xi}, \ \frac{\partial y^2(x,t)}{\partial x^2} = \frac{1}{L^2}\frac{\partial y^2(\xi,t)}{\partial \xi^2}, \text{ and } \frac{\partial y^3(x,t)}{\partial x^3} = \frac{1}{L^3}\frac{\partial y^3(\xi,t)}{\partial \xi^3}, \tag{4.55}$$

Equations (4.43), (4.44), and (4.54) can be written in terms of non-dimensional displacement ξ,

$$V(\xi,t) = -\frac{EI}{L^3}\frac{\partial^3 y(\xi,t)}{\partial \xi^3}, \ M(\xi,t) = \frac{EI}{L}\frac{\partial \psi(\xi,t)}{\partial \xi}, \ \psi(\xi,t) = \frac{1}{L}\frac{\partial y(\xi,t)}{\partial \xi} \tag{4.56}$$

Incident waves are reflected at a boundary, as shown in Figure 4.15. The incident waves $\mathbf{a}^+$ and the reflected waves $\mathbf{a}^-$ are related by reflection matrix $\mathbf{r}$,

$$\mathbf{a}^- = \mathbf{r}\mathbf{a}^+ \tag{4.57}$$

where $\mathbf{r}$ is determined by boundary conditions.

The reflection coefficients at classical boundaries, namely the free, clamped, and simply-supported (sometimes called pinned) boundaries, are derived.

a) Free boundary

At a free boundary, the equilibrium conditions are

$$M(\xi,t) = 0, \ V(\xi,t) = 0 \tag{4.58}$$

From Eqs. (4.49), (4.56) and (4.58),

$$\frac{EI}{L^2}\left(\left[(-ik_B)^2 a_1^+ e^{-ik_B\xi} + (-k_B)^2 a_2^+ e^{-k_B\xi} + (ik_B)^2 a_1^- e^{ik_B\xi} + (k_B)^2 a_2^- e^{k_B\xi}\right]\right) = 0 \tag{4.59a}$$

$$-\frac{EI}{L^3}\left(\left[(-ik_B)^3 a_1^+ e^{-ik_B\xi} + (-k_B)^3 a_2^+ e^{-k_B\xi} + (ik_B)^3 a_1^- e^{ik_B\xi} + (k_B)^3 a_2^- e^{k_B\xi}\right]\right) = 0 \tag{4.59b}$$

Choosing the origin at the boundary and putting Eqs. (4.59a) and (4.59b) into matrix form,

$$\begin{bmatrix} -1 & 1 \\ i & -1 \end{bmatrix}\begin{bmatrix} a_1^+ \\ a_2^+ \end{bmatrix} - \begin{bmatrix} 1 & -1 \\ i & -1 \end{bmatrix}\begin{bmatrix} a_1^- \\ a_2^- \end{bmatrix} = 0 \tag{4.60}$$

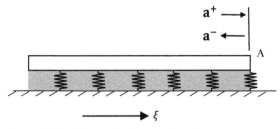

Figure 4.15 Wave reflection at a boundary.

Solving for the reflected waves $\begin{bmatrix} a_1^- \\ a_2^- \end{bmatrix}$ in terms of the incident waves $\begin{bmatrix} a_1^+ \\ a_2^+ \end{bmatrix}$,

$$\begin{bmatrix} a_1^- \\ a_2^- \end{bmatrix} = \begin{bmatrix} 1 & -1 \\ i & -1 \end{bmatrix}^{-1} \begin{bmatrix} -1 & 1 \\ i & -1 \end{bmatrix} \begin{bmatrix} a_1^+ \\ a_2^+ \end{bmatrix} = \begin{bmatrix} -i & 1+i \\ 1-i & i \end{bmatrix} \begin{bmatrix} a_1^+ \\ a_2^+ \end{bmatrix} \tag{4.61}$$

From Eqs. (4.53), (4.57), and (4.61), the reflection matrix at a free boundary is

$$\mathbf{r} = \begin{bmatrix} -i & 1+i \\ 1-i & i \end{bmatrix} \tag{4.62}$$

b) Clamped boundary

The boundary conditions at a clamped boundary are

$$y(\xi,t) = 0, \ \psi(\xi,t) = 0 \tag{4.63}$$

From Eqs. (4.49), (4.56), and (4.63),

$$a_1^+ e^{-ik_B\xi} + a_2^+ e^{-k_B\xi} + a_1^- e^{ik_B\xi} + a_2^- e^{k_B\xi} = 0 \tag{4.64a}$$

$$\frac{1}{L}\left(-ik_B a_1^+ e^{-ik_B\xi} - k_B a_2^+ e^{-k_B\xi} + ik_B a_1^- e^{ik_B\xi} + k_B a_2^- e^{k_B\xi}\right) = 0 \tag{4.64b}$$

Choosing the origin at the boundary and putting Eqs. (4.64a) and (4.64b) into matrix form,

$$\begin{bmatrix} 1 & 1 \\ -i & -1 \end{bmatrix} \begin{bmatrix} a_1^+ \\ a_2^+ \end{bmatrix} + \begin{bmatrix} 1 & 1 \\ i & 1 \end{bmatrix} \begin{bmatrix} a_1^- \\ a_2^- \end{bmatrix} = 0 \tag{4.65}$$

From Eqs. (4.53), (4.57), and (4.65), the reflection matrix at a clamped boundary is

$$\mathbf{r} = \begin{bmatrix} -i & -1-i \\ -1+i & i \end{bmatrix} \tag{4.66}$$

c) Simply-supported boundary

The boundary conditions at a simply-supported boundary are

$$y(\xi,t) = 0, \ M(\xi,t) = 0 \tag{4.67}$$

From Eqs. (4.49), (4.56) and (4.67),

$$a_1^+ e^{-ik_B\xi} + a_2^+ e^{-k_B\xi} + a_1^- e^{ik_B\xi} + a_2^- e^{k_B\xi} = 0 \tag{4.68a}$$

$$\frac{EI}{L^2}\left(\left[(-ik_B)^2 a_1^+ e^{-ik_B\xi} + (-k_B)^2 a_2^+ e^{-k_B\xi} + (ik_B)^2 a_1^- e^{ik_B\xi} + (k_B)^2 a_2^- e^{k_B\xi}\right]\right) = 0 \tag{4.68b}$$

Choosing the origin at the boundary and putting Eqs. (4.68a) and (4.68b) into matrix form,

$$\begin{bmatrix} 1 & 1 \\ -i & -1 \end{bmatrix} \begin{bmatrix} a_1^+ \\ a_2^+ \end{bmatrix} + \begin{bmatrix} 1 & 1 \\ i & 1 \end{bmatrix} \begin{bmatrix} a_1^- \\ a_2^- \end{bmatrix} = 0 \tag{4.69}$$

From Eqs. (4.53), (4.57), and (4.69), the reflection matrix at a simply-supported boundary is

$$\mathbf{r} = \begin{bmatrix} -1 & 0 \\ 0 & -1 \end{bmatrix} \tag{4.70}$$

Not surprisingly, these reflection matrices at classical boundaries are identical to those of the Euler–Bernoulli beam obtained in Chapter 3 because none of the reflection relationships are wavenumber dependent.

4.2.3 Free Wave Vibration Analysis

Free vibration analysis of a transversely vibrating uniform beam on a Winkler elastic foundation follows a similar procedure as that of a transversely vibrating Euler–Bernoulli beam described in Chapter 3.

From the wave standpoint, vibrations in a uniform beam can be viewed as waves that propagate along a uniform beam and are reflected at both boundaries. Assuming that the uniform beam shown in Figure 4.13 is of length L, in the non-dimensional coordinate system, the propagation and reflection relationships are

$$\mathbf{b}^+ = \mathbf{f}(1)\,\mathbf{a}^+;\ \mathbf{a}^- = \mathbf{f}(1)\,\mathbf{b}^-;\ \mathbf{a}^+ = \mathbf{r}_A\,\mathbf{a}^-;\ \mathbf{b}^- = \mathbf{r}_B\,\mathbf{b}^+ \tag{4.71}$$

where $\mathbf{r}_A$ and $\mathbf{r}_B$ are the reflection matrices at Boundaries A and B, respectively, and $\mathbf{f}(1)$ is the propagation matrix over the length of the uniform beam.

The relationships described in Eq. (4.71) can be written in matrix form,

$$\begin{bmatrix} -\mathbf{I} & \mathbf{r}_A & \mathbf{0} & \mathbf{0} \\ \mathbf{0} & \mathbf{0} & \mathbf{r}_B & -\mathbf{I} \\ \mathbf{f}(1) & \mathbf{0} & -\mathbf{I} & \mathbf{0} \\ \mathbf{0} & -\mathbf{I} & \mathbf{0} & \mathbf{f}(1) \end{bmatrix} \begin{bmatrix} \mathbf{a}^+ \\ \mathbf{a}^- \\ \mathbf{b}^+ \\ \mathbf{b}^- \end{bmatrix} = \mathbf{0} \tag{4.72}$$

where $\mathbf{I}$ and $\mathbf{0}$ denote an identity matrix and a zero matrix, respectively.

Setting the determinant of the coefficient matrix of Eq. (4.72) to zero gives the characteristic equation,

$$\left| \mathbf{r}_A \mathbf{f}(1)\mathbf{r}_B \mathbf{f}(1) - \mathbf{I} \right| = 0 \tag{4.73}$$

The natural frequencies of the beam are the roots of the above characteristic equation.

With the availability of natural frequencies, the corresponding modeshapes can be obtained from Eq. (4.72) by deleting one row of the coefficient matrix and solving the remaining wave components in terms of a chosen wave component, which has been described in detail in Chapter 3.

4.2.4 Force Generated Waves and Forced Wave Vibration Analysis

An externally applied force or moment has the effect of injecting vibration waves into a continuous structure. Figure 4.16 illustrates a point transverse force $\bar{Q}$, a bending moment $\bar{M}$, and waves $\mathbf{a}^\pm$ and $\mathbf{b}^\pm$ at the point where the

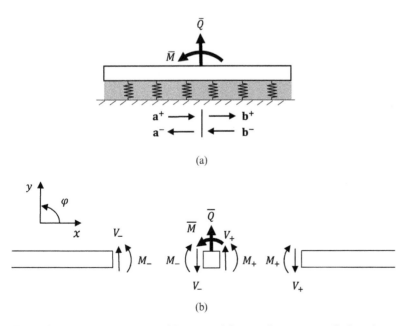

(a)

(b)

Figure 4.16 (a) Waves generated by external force and moment applied on the span and (b) free body diagram.

external force and moment are applied. These waves are injected by and related to the externally applied force and moment.

Because the bending denoting on the left and right side of the point where the external force and moment are applied are the same, denoting physical parameters on the left and right side of the point where the external force and moment are applied using subscripts − and +, respectively,

$$y_- = y_+ \, , \, \psi_- = \psi_+ \tag{4.74}$$

From the free body diagram of Figure 4.16, the externally applied transverse force and bending moment are related to the internal resistant forces and moments on the left and right side of the point where the external force and moment are applied by

$$\bar{Q} = V_- - V_+, \, \bar{M} = M_- - M_+ \tag{4.75}$$

The relationships described in Eqs. (4.74) and (4.75) are called the continuity and equilibrium equations, respectively.

From Eq. (4.49) and Figure 4.16a,

$$y_- = a_1^+ e^{-ik_B\xi} + a_2^+ e^{-k_B\xi} + a_1^- e^{ik_B\xi} + a_2^- e^{k_B\xi} \tag{4.76a}$$

$$y_+ = b_1^+ e^{-ik_B\xi} + b_2^+ e^{-k_B\xi} + b_1^- e^{ik_B\xi} + b_2^- e^{k_B\xi} \tag{4.76b}$$

From Eqs. (4.56), (4.76a), and (4.76b),

$$\psi_- = \frac{1}{L}\frac{\partial y_-}{\partial \xi} = \frac{1}{L}\left(-ik_B a_1^+ e^{-ik_B\xi} - k_B a_2^+ e^{-k_B\xi} + ik_B a_1^- e^{ik_B\xi} + k_B a_2^- e^{k_B\xi}\right) \tag{4.77a}$$

$$\psi_+ = \frac{1}{L}\frac{\partial y_+}{\partial \xi} = \frac{1}{L}\left(-ik_B b_1^+ e^{-ik_B\xi} - k_B b_2^+ e^{-k_B\xi} + ik_B b_1^- e^{ik_B\xi} + k_B b_2^- e^{k_B\xi}\right) \tag{4.77b}$$

With the origin chosen at the point where the external force and moment are applied, from Eqs. (4.74), (4.76a), (4.76b), (4.77a), and (4.77b), the continuity conditions become

$$a_1^+ + a_2^+ + a_1^- + a_2^- = b_1^+ + b_2^+ + b_1^- + b_2^- \tag{4.78a}$$

$$-ik_B a_1^+ - k_B a_2^+ + ik_B a_1^- + k_B a_2^- = -ik_B b_1^+ - k_B b_2^+ + ik_B b_1^- + k_B b_2^- \tag{4.78b}$$

Putting Eqs. (4.78) into matrix form,

$$\begin{bmatrix} 1 & 1 \\ -i & -1 \end{bmatrix}(\mathbf{a}^+ - \mathbf{b}^+) + \begin{bmatrix} 1 & 1 \\ i & 1 \end{bmatrix}(\mathbf{a}^- - \mathbf{b}^-) = \mathbf{0} \tag{4.79}$$

From Eq. (4.56) and Figure 4.16,

$$V_- = -\frac{EI}{L^3}\frac{\partial^3 y_-}{\partial \xi^3}, \, M_- = \frac{EI}{L}\frac{\partial \psi_-}{\partial \xi} = \frac{EI}{L^2}\frac{\partial^2 y_-}{\partial \xi^2} \tag{4.80a}$$

$$V_+ = -\frac{EI}{L^3}\frac{\partial^3 y_+}{\partial \xi^3}, \, M_+ = \frac{EI}{L}\frac{\partial \psi_+}{\partial \xi} = \frac{EI}{L^2}\frac{\partial^2 y_+}{\partial \xi^2} \tag{4.80b}$$

From Eqs. (4.75), (4.80a), and (4.80b),

$$\bar{Q} = -\frac{EI}{L^3}\left(\frac{\partial^3 y_-}{\partial \xi^3} - \frac{\partial^3 y_+}{\partial \xi^3}\right), \, \bar{M} = \frac{EI}{L^2}\left(\frac{\partial^2 y_-}{\partial \xi^2} - \frac{\partial^2 y_+}{\partial \xi^2}\right) \tag{4.81}$$

With the origin chosen at the point where the external force and moment are applied, from Eqs. (4.76a), (4.76b), and (4.81), the equilibrium conditions become

$$\bar{Q} = -\frac{EI}{L^3}\left(\left[(-ik_B)^3 a_1^+ + (-k_B)^3 a_2^+ + (ik_B)^3 a_1^- + (k_B)^3 a_2^-\right] - \left[(-ik_B)^3 b_1^+ + (-k_B)^3 b_2^+ + (ik_B)^3 b_1^- + (k_B)^3 b_2^-\right]\right) \tag{4.82a}$$

$$\bar{M} = \frac{EI}{L^2}\left(\left[(-ik_B)^2 a_1^+ + (-k_B)^2 a_2^+ + (ik_B)^2 a_1^- + (k_B)^2 a_2^-\right] - \left[(-ik_B)^2 b_1^+ + (-k_B)^2 b_2^+ + (ik_B)^2 b_1^- + (k_B)^2 b_2^-\right]\right) \tag{4.82b}$$

Putting Eqs. (4.82a) and (4.82b) into matrix form,

$$\begin{bmatrix} i & -1 \\ -1 & 1 \end{bmatrix}(\mathbf{a}^+ - \mathbf{b}^+) + \begin{bmatrix} -i & 1 \\ -1 & 1 \end{bmatrix}(\mathbf{a}^- - \mathbf{b}^-) = \begin{bmatrix} -q \\ m \end{bmatrix} \tag{4.83}$$

where

$$q = \frac{\bar{Q}L^3}{EIk_B^3}, \; m = \frac{\bar{M}L^2}{EIk_B^2} \tag{4.84}$$

are the normalized externally applied transverse force and bending moment.

From Eqs. (4.79) and (4.83), the relationships between externally applied forces/moments and the injected bending waves are

$$\mathbf{a}^+ - \mathbf{b}^+ = \mathbf{q} + \mathbf{m}, \; \mathbf{a}^- - \mathbf{b}^- = -\mathbf{q} + \mathbf{m} \tag{4.85a}$$

where the load vectors are

$$\mathbf{q} = \begin{bmatrix} i \\ 1 \end{bmatrix}\frac{q}{4}, \; \mathbf{m} = \begin{bmatrix} -1 \\ 1 \end{bmatrix}\frac{m}{4} \tag{4.85b}$$

With the availability of force generated waves, propagation relationships, and boundary reflections, forced vibration in a beam can be obtained.

Figure 4.17 illustrates a beam with an external transverse force and bending moment applied at an arbitrary point G that is distance ξ_1 and ξ_2 away from Boundaries C and D, respectively. The following wave relationships exist:

- At Boundaries C and D, from Eq. (4.57),

$$\mathbf{c}^+ = \mathbf{r}_C\, \mathbf{c}^-, \; \mathbf{d}^- = \mathbf{r}_D\, \mathbf{d}^+ \tag{4.86}$$

  where $\mathbf{r}_C$ and $\mathbf{r}_D$ are the reflection matrices at Boundaries C and D, respectively.

- At Point G where the external force and moment are applied, from Eq. (4.85a),

$$\mathbf{g}_1^+ - \mathbf{g}_2^+ = \mathbf{q} + \mathbf{m}, \; \mathbf{g}_1^- - \mathbf{g}_2^- = -\mathbf{q} + \mathbf{m} \tag{4.87}$$

  where $\mathbf{q}$ and $\mathbf{m}$ are given in Eq. (4.85b).

- The propagation relationships, from Eq. (4.51), are

$$\mathbf{g}_1^+ = \mathbf{f}(\xi_1)\, \mathbf{c}^+, \; \mathbf{c}^- = \mathbf{f}(\xi_1)\, \mathbf{g}_1^-; \; \mathbf{g}_2^- = \mathbf{f}(\xi_2)\, \mathbf{d}^-, \; \mathbf{d}^+ = \mathbf{f}(\xi_2)\, \mathbf{g}_2^+ \tag{4.88}$$

  where the propagation matrix $\mathbf{f}(\xi)$ for a distance ξ is given in Eq. (4.52).

Assembling Eqs. (4.86) to (4.88) into matrix form,

$$\begin{bmatrix} -\mathbf{I} & \mathbf{r}_C & \mathbf{0} & \mathbf{0} & \mathbf{0} & \mathbf{0} & \mathbf{0} & \mathbf{0} \\ \mathbf{0} & \mathbf{0} & \mathbf{r}_D & -\mathbf{I} & \mathbf{0} & \mathbf{0} & \mathbf{0} & \mathbf{0} \\ \mathbf{0} & \mathbf{0} & \mathbf{0} & \mathbf{0} & \mathbf{I} & \mathbf{0} & -\mathbf{I} & \mathbf{0} \\ \mathbf{0} & \mathbf{0} & \mathbf{0} & \mathbf{0} & \mathbf{0} & \mathbf{I} & \mathbf{0} & -\mathbf{I} \\ \mathbf{f}(\xi_1) & \mathbf{0} & \mathbf{0} & \mathbf{0} & -\mathbf{I} & \mathbf{0} & \mathbf{0} & \mathbf{0} \\ \mathbf{0} & -\mathbf{I} & \mathbf{0} & \mathbf{0} & \mathbf{0} & \mathbf{f}(\xi_1) & \mathbf{0} & \mathbf{0} \\ \mathbf{0} & \mathbf{0} & \mathbf{0} & \mathbf{f}(\xi_2) & \mathbf{0} & \mathbf{0} & \mathbf{0} & -\mathbf{I} \\ \mathbf{0} & \mathbf{0} & -\mathbf{I} & \mathbf{0} & \mathbf{0} & \mathbf{0} & \mathbf{f}(\xi_2) & \mathbf{0} \end{bmatrix} \begin{bmatrix} \mathbf{c}^+ \\ \mathbf{c}^- \\ \mathbf{d}^+ \\ \mathbf{d}^- \\ \mathbf{g}_1^+ \\ \mathbf{g}_1^- \\ \mathbf{g}_2^+ \\ \mathbf{g}_2^- \end{bmatrix} = \begin{bmatrix} \mathbf{0} \\ \mathbf{0} \\ \mathbf{q} + \mathbf{m} \\ -\mathbf{q} + \mathbf{m} \\ \mathbf{0} \\ \mathbf{0} \\ \mathbf{0} \\ \mathbf{0} \end{bmatrix} \tag{4.89}$$

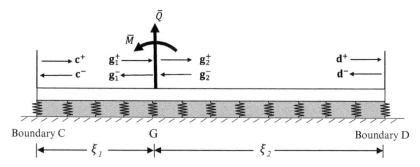

Figure 4.17 Waves in a uniform beam with external transverse force and bending moment applied on the span.

where $\mathbf{I}$ and $\mathbf{0}$ denote an identity matrix and a zero matrix, respectively.

Equation (4.89) can be expressed in the form of

$$\mathbf{A}_f \mathbf{z}_f = \mathbf{F} \tag{4.90}$$

where $\mathbf{A}_f$ is a 16 by 16 square coefficient matrix, $\mathbf{z}_f$ is a wave vector containing 16 wave components, and $\mathbf{F}$ is a vector related to externally applied force and moment of Eq. (4.89).

Solving for $\mathbf{z}_f$ from Eq. (4.90),

$$\mathbf{z}_f = \mathbf{A}_f^{-1} \mathbf{F} \tag{4.91}$$

Equation (4.91) provides the responses of all wave components, from which the deflection of any point on the beam can be obtained. For example, in Figure 4.17, the deflection of a point located between Boundary C and the excitation point G that is at a distance ξ from Point G is

$$y = \begin{bmatrix} 1 & 1 \end{bmatrix} \mathbf{f}(\xi) \mathbf{g}_1^- + \begin{bmatrix} 1 & 1 \end{bmatrix} \mathbf{f}(-\xi) \mathbf{g}_1^+ \tag{4.92a}$$

or equivalently is

$$y = \begin{bmatrix} 1 & 1 \end{bmatrix} \mathbf{f}(\xi_1 - \xi) \mathbf{c}^+ + \begin{bmatrix} 1 & 1 \end{bmatrix} \mathbf{f}(-(\xi_1 - \xi)) \mathbf{c}^- \tag{4.92b}$$

Similarly, the deflection of a point located between Boundary D and the excitation point G in Figure 4.17 that is at a distance ξ from Point G is

$$y = \begin{bmatrix} 1 & 1 \end{bmatrix} \mathbf{f}(\xi) \mathbf{g}_2^+ + \begin{bmatrix} 1 & 1 \end{bmatrix} \mathbf{f}(-\xi) \mathbf{g}_2^- \tag{4.93a}$$

or equivalently is

$$y = \begin{bmatrix} 1 & 1 \end{bmatrix} \mathbf{f}(\xi_2 - \xi) \mathbf{d}^- + \begin{bmatrix} 1 & 1 \end{bmatrix} \mathbf{f}(-(\xi_2 - \xi)) \mathbf{d}^+ \tag{4.93b}$$

An external force/moment may be applied at a free end of a beam, as shown in Figure 4.18. From the free body diagram of Figure 4.18, the externally applied force and moment are related to the internal resistant force and moment by the following equilibrium equations

$$\bar{Q} = V, \quad \bar{M} = M \tag{4.94}$$

From Figure 4.18a and Eqs. (4.49) and (4.56),

$$y(\xi) = a_1^+ e^{-ik_B \xi} + a_2^+ e^{-k_B \xi} + a_1^- e^{ik_B \xi} + a_2^- e^{k_B \xi} \tag{4.95}$$

$$\psi(\xi) = \frac{1}{L} \frac{\partial y(\xi)}{\partial \xi} = \frac{1}{L} \left(-ik_B a_1^+ e^{-ik_B \xi} - k_B a_2^+ e^{-k_B \xi} + ik_B a_1^- e^{ik_B \xi} + k_B a_2^- e^{k_B \xi} \right) \tag{4.96}$$

From Eqs. (4.56) and (4.94),

$$\bar{Q} = -\frac{EI}{L^3} \frac{\partial^3 y(\xi)}{\partial \xi^3}, \quad \bar{M} = \frac{EI}{L} \frac{\partial \psi(\xi)}{\partial \xi} = \frac{EI}{L^2} \frac{\partial^2 y(\xi)}{\partial \xi^2} \tag{4.97}$$

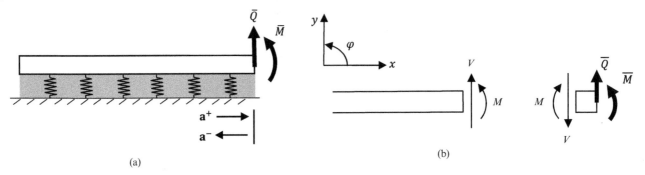

Figure 4.18 (a) Waves generated by external transverse force and bending moment applied at a free boundary and (b) free body diagram.

With the origin chosen at the point where the external force and moment are applied, from Eqs. (4.95) to (4.97), the equilibrium conditions become

$$\bar{Q} = -\frac{EI}{L^3}\left(\left[(-ik_B)^3 a_1^+ + (-k_B)^3 a_2^+ + (ik_B)^3 a_1^- + (k_B)^3 a_2^-\right]\right) \tag{4.98a}$$

$$\bar{M} = \frac{EI}{L^2}\left(\left[(-ik_B)^2 a_1^+ + (-k_B)^2 a_2^+ + (ik_B)^2 a_1^- + (k_B)^2 a_2^-\right]\right) \tag{4.98b}$$

Putting Eqs. (4.98a) and (4.98b) into matrix form,

$$\begin{bmatrix} i & -1 \\ -1 & 1 \end{bmatrix}\mathbf{a}^+ + \begin{bmatrix} -i & 1 \\ -1 & 1 \end{bmatrix}\mathbf{a}^- = \begin{bmatrix} -q \\ m \end{bmatrix} \tag{4.99}$$

where q and m are the normalized externally applied transverse force and bending moment given in Eq. (4.84).

In the special case of $\bar{Q}=0$ and $\bar{M}=0$, which represents a free boundary, from Eq. (4.99)

$$-\begin{bmatrix} i & -1 \\ -1 & 1 \end{bmatrix}\mathbf{a}^+ + \begin{bmatrix} -i & 1 \\ -1 & 1 \end{bmatrix}\mathbf{a}^- = 0 \tag{4.100}$$

From Eqs. (4.57) and (4.100), the reflection matrix is

$$\mathbf{r} = \begin{bmatrix} -i & 1+i \\ 1-i & i \end{bmatrix} \tag{4.101}$$

which, as expected, is identical to the reflection matrix of a free boundary obtained in Eq. (4.62).

With the availability of force generated waves, propagation relationships, and boundary reflections, forced vibration in a beam caused by an end force and moment can be obtained.

Figure 4.19 illustrates a beam with external excitation force and moment applied at End D. The following wave relationships exist:

- At Boundary C, from Eq. (4.57),

$$\mathbf{c}^+ = \mathbf{r}_C\, \mathbf{c}^- \tag{4.102}$$

where $\mathbf{r}_C$ is the reflection matrix at Boundary C.

- At external force and moment applied end D, from Eq. (4.99),

$$\begin{bmatrix} i & -1 \\ -1 & 1 \end{bmatrix}\mathbf{d}^+ + \begin{bmatrix} -i & 1 \\ -1 & 1 \end{bmatrix}\mathbf{d}^- = \begin{bmatrix} -q \\ m \end{bmatrix} \tag{4.103}$$

where q and m are given in Eq. (4.84).

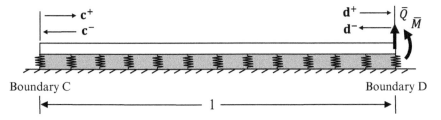

Figure 4.19 Waves in a uniform beam with external force and moment applied at a free end.

- The propagation relationships, according to Eq. (4.51), are

$$\mathbf{d}^+ = \mathbf{f}(1)\,\mathbf{c}^+, \quad \mathbf{c}^- = \mathbf{f}(1)\,\mathbf{d}^- \tag{4.104}$$

where the propagation matrix $\mathbf{f}(\xi)$ for a distance ξ is given in Eq. (4.52). The total length of a beam in dimensionless unit is $\xi = 1$.

Putting Eqs. (4.102) to (4.104) into matrix form,

$$\begin{bmatrix} -\mathbf{I} & \mathbf{r}_C & \mathbf{0} & \mathbf{0} \\ \mathbf{0} & \mathbf{0} & \mathbf{q}_{c1} & \mathbf{q}_{c2} \\ \mathbf{f}(1) & \mathbf{0} & -\mathbf{I} & \mathbf{0} \\ \mathbf{0} & -\mathbf{I} & \mathbf{0} & \mathbf{f}(1) \end{bmatrix} \begin{bmatrix} \mathbf{c}^+ \\ \mathbf{c}^- \\ \mathbf{d}^+ \\ \mathbf{d}^- \end{bmatrix} = \begin{bmatrix} \mathbf{0} \\ \mathbf{q}_0 \\ \mathbf{0} \\ \mathbf{0} \end{bmatrix} \tag{4.105}$$

where

$$\mathbf{q}_{c1} = \begin{bmatrix} i & -1 \\ -1 & 1 \end{bmatrix}, \ \mathbf{q}_{c2} = \begin{bmatrix} -i & 1 \\ -1 & 1 \end{bmatrix}, \ \mathbf{q}_0 = \begin{bmatrix} -q \\ m \end{bmatrix} \tag{4.106}$$

and $\mathbf{I}$ and $\mathbf{0}$ denote an identity matrix and a zero matrix, respectively.

Equation (4.105) can be written in the form of Eq. (4.90), and the wave components can be solved from Eq. (4.91). In which $\mathbf{A}_f$ is an 8 by 8 square coefficient matrix, $\mathbf{z}_f$ is a wave vector containing eight wave components, and $\mathbf{F}$ is a vector related to externally applied force and moment of Eq. (4.105).

With the availability of responses of all wave components, the deflection of any point on the beam can be obtained. For example, the deflection of a point located a distance ξ from Boundary C in Figure 4.19 is

$$y = \begin{bmatrix} 1 & 1 \end{bmatrix} \mathbf{f}(\xi)\mathbf{c}^+ + \begin{bmatrix} 1 & 1 \end{bmatrix} \mathbf{f}(-\xi)\mathbf{c}^- \tag{4.107a}$$

or equivalently is

$$y = \begin{bmatrix} 1 & 1 \end{bmatrix} \mathbf{f}(1-\xi)\mathbf{d}^- + \begin{bmatrix} 1 & 1 \end{bmatrix} \mathbf{f}(-(1-\xi))\mathbf{d}^+ \tag{4.107b}$$

4.2.5 Numerical Examples

Free and forced bending vibrations in a uniform beam resting on a Winkler elastic foundation are analyzed using the wave vibration approach. Both boundaries of the beam are assumed simply-supported. The non-dimensional stiffness K_w of the Winkler elastic foundation takes four different values $K_w = 0, 10, 400$, and 2000.

Figure 4.20 presents the dB magnitude of the characteristic polynomial corresponding to the various K_w values. The local minima correspond to the natural frequencies, except for the local minima highlighted by dotted vertical lines, which are sharp changes in magnitudes because of the cut-off frequency Ω_c associated with the stiffness K_w of the elastic foundation. The non-dimensional natural frequencies Ω_n are listed in Table 4.2.

The modeshapes of the lowest nine modes corresponding to the natural frequencies in Table 4.2 are plotted in Figure 4.21. Each of the nine subplots of normalized modeshapes contains four modeshape curves corresponding to $K_w = 0, 10, 400$, and 2000,

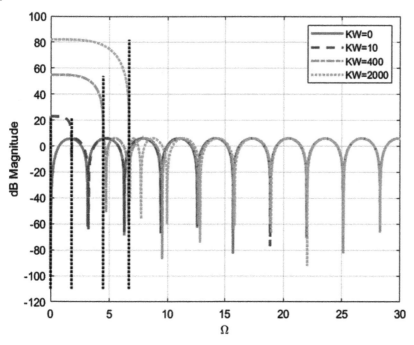

Figure 4.20 Magnitudes of the characteristic polynomials.

Table 4.2 The lowest nine non-dimensional natural frequencies of the beam.

Mode Number	Natural frequency Ω_n			
	$K_w = 0$ $\Omega_c = \sqrt[4]{K_w} = 0$	$K_w = 10$ $\Omega_c = \sqrt[4]{10} = 1.778$	$K_w = 400$ $\Omega_c = \sqrt[4]{400} = 4.472$	$K_w = 2000$ $\Omega_c = \sqrt[4]{2000} = 6.687$
1	3.142	3.219	4.723	6.767
2	6.283	6.293	6.652	7.724
3	9.425	9.428	9.542	9.972
4	12.566	12.568	12.616	12.811
5	15.708 $n\,\pi$	15.709	15.734	15.835
6	18.850	18.850	18.864	18.924
7	21.991	21.991	22.001	22.038
8	25.133	25.133	25.139	25.164
9	28.274	28.274	28.279	28.296

respectively. The four normalized modeshape curves in each subplot are identical. This indicates that the variation of stiffness of the Winkler elastic foundation has little influence on the modeshapes.

For forced vibration analysis, the external excitation and response observation locations need to be determined first. Overlaying the modeshapes of the lowest nine modes in Figure 4.22 helps with identifying the nodal points. In forced vibration analysis, the frequency range is from 0 to 30, which is kept the same as that in free vibration analysis. Because a point on the structure is either a nodal point or a non-nodal point of a certain mode, the selection of external force and moment applied points and the response observation points covers both scenarios as listed in Table 4.3 and marked on Figure 4.22, where all non-dimensional distances are measured from Boundary C of Figure 4.17.

The receptance frequency responses (displacement/normalized excitation force) are shown in Figures 4.23 and 4.24, corresponding to Test 1 and Test 2, respectively. Each plot contains three overlaid responses at the three response observation points. It can be seen that resonant responses always occur at natural frequencies, and the resonant peaks line up,

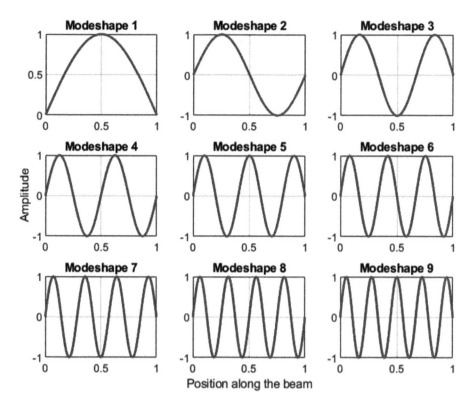

Figure 4.21 Modeshapes of the lowest nine modes.

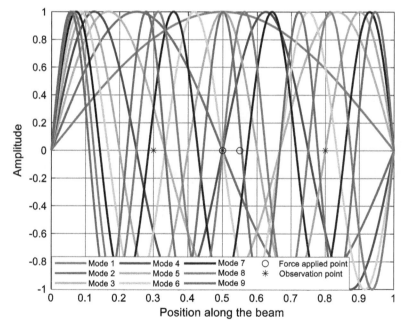

Figure 4.22 Overlaid modeshapes of the lowest nine modes.

regardless of the external excitation and response observation locations. Examining the response curves in each plot, one finds that not all resonant peaks occur at all natural frequencies. This is because the occurrence of resonant peaks in the frequency responses is closely related to the locations of both the external force applied point and the response observation point relative to a nodal point.

Recall that in Test 1, the external excitation force is applied at a non-nodal point. Among the three observation locations, Location 1 is at a non-nodal point, Location 2 is at a nodal point of all even modes, and Location 3 is a nodal point of the

Table 4.3 External force and moment applied point and responses observed points.

External force and moment applied point	Test 1	0.55 (a non-nodal point of the lowest 9 modes)
	Test 2	0.5 (a nodal point of even modes 2, 4, 6, 8...)
Responses observed points	Location 1	0.3 (a non-nodal point of the lowest 9 modes)
	Location 2	0.5 (a nodal point of even modes 2, 4, 6, 8...)
	Location 3	0.8 (a nodal point of the 5th mode)

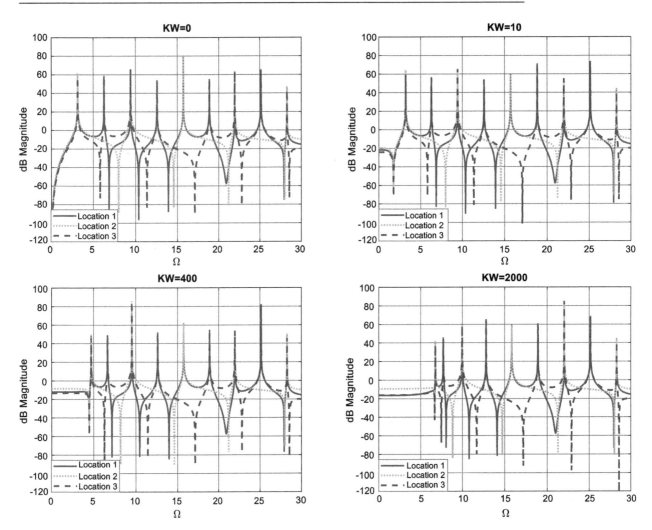

Figure 4.23 Overlaid receptance frequency responses at various observation locations of Test 1.

5th mode. This explains why at Location 1 all resonant peaks appear, at Location 2 the resonant peaks corresponding to even modes 2, 4, 6, and 8 disappear, and at Location 3 the resonant peak corresponding to the 5th mode disappears.

In Test 2, with the external excitation force applied point purposely chosen at a nodal point of all even modes, resonant peaks corresponding to Modes 2, 4, 6, and 8 all disappear, regardless of the observation locations. At Location 3, an additional resonant peak disappears, which corresponds to Mode 5. This is again because this observation point is at a nodal point of the 5th mode.

The occurrence and disappearance of resonant peaks in the frequency responses of Figures 4.23 and 4.24 are summarized in Table 4.4.

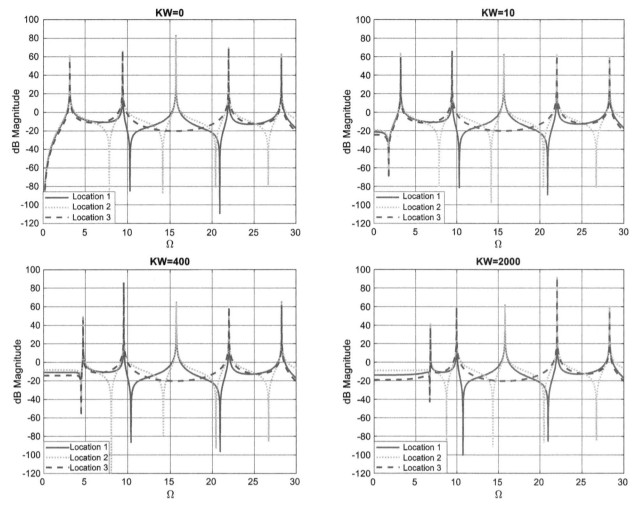

Figure 4.24 Overlaid receptance frequency responses at various observation locations of Test 2.

Table 4.4 Occurrence/disappearance of resonant peaks.

Occurrence of resonant peaks		Mode 1	Mode 2	Mode 3	Mode 4	Mode 5	Mode 6	Mode 7	Mode 8	Mode 9
Test 1	Location 1	✓	✓	✓	✓	✓	✓	✓	✓	✓
	Location 2	✓	✗	✓	✗	✓	✗	✓	✗	✓
	Location 3	✓	✓	✓	✓	✗	✓	✓	✓	✓
Test 2	Location 1	✓	✗	✓	✗	✓	✗	✓	✗	✓
	Location 2	✓	✗	✓	✗	✓	✗	✓	✗	✓
	Location 3	✓	✗	✓	✗	✗	✗	✓	✗	✓

In Figures 4.25 and 4.26, which correspond to Test 1 and Test 2 respectively, the receptance frequency responses corresponding to $K_w = 0, 10, 400$, and 2000 are overlaid at each response observation location. From these frequency responses, as well as the magnitudes of the characteristic polynomials of Figure 4.20, it can be seen that as the stiffness of the Winkler elastic foundation increases, the natural frequency of the corresponding mode increases accordingly. The influence of the stiffness of the Winkler elastic foundation on natural frequency is more significant on vibration modes whose natural frequency values are in the vicinity of the cut-off frequency. Furthermore, these receptance frequency responses show that the local minima highlighted by the dotted lines in Figure 4.20 do not correspond to resonant peaks, indicating that they are not natural frequencies. These local minima are sharp changes in magnitudes at the cut-off frequencies.

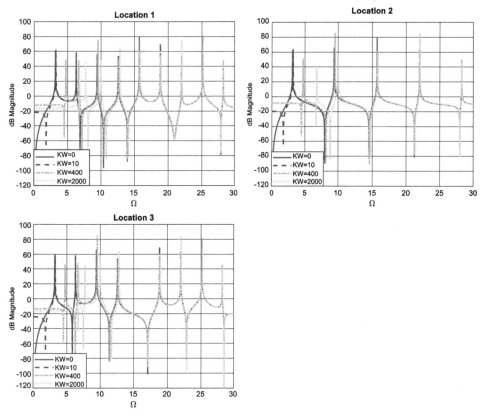

Figure 4.25 Overlaid receptance frequency responses for various foundation stiffnesses of Test 1.

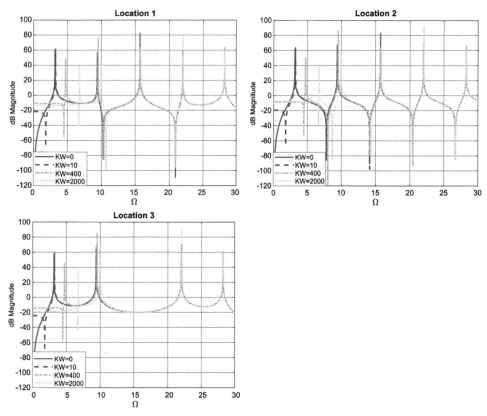

Figure 4.26 Overlaid receptance frequency responses for various foundation stiffnesses of Test 2.

References

De Rosa M. A., Maurizi M.J. Three Approaches for the Axial Vibrations of Bars on Modified Winkler Soil with Non-Classical Boundary Conditions, *Journal of Sound and Vibration*, 231(5), 1257–1269 (2000).

Zhou D. General Solution to Vibrations of Beams on Variable Winkler Elastic Foundation, *Computers and Structures*, 47(1), 83–90 (1993).

Homework Project

Wave Analysis of Free Longitudinal Vibrations of a Uniform Beam Resting on a Winkler Foundation

The figure below is a uniform steel beam resting on a Winkler Foundation. The beam has one end fixed and the other end attached to a spring. The non-dimensional stiffness of the Winkler elastic foundation is K_w, and the non-dimensional flexibility coefficient of the end spring is X_T.

Free wave vibration analysis of an axially vibrating uniform beam on a Winkler elastic foundation follows a similar procedure as that of an axially vibrating beam described in Chapter 2. Analyze free longitudinal vibration in this uniform steel beam resting on a Winkler foundation by allowing K_w to take four different values, namely, $K_w = 0, 10, 100$, and 1000; and X_T to take four different values, namely, $X_T = 0, 1, 10$, and 1000.

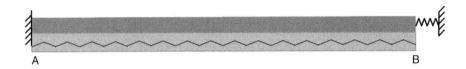

Write MATLAB script to plot the magnitude of characteristic polynomials and mark the cut-off frequency Ω_c using vertical lines on the plot. Read the longitudinal natural frequencies of the beam from the plot and compare the results with those listed in Table 4.1.

Hints on MATLAB scripts:

- Pay attention to wavenumber expressions below and above the cut-off frequency, as described in Eq. (4.7). You may use the following scripts, where "Omega" stands for the non-dimensional frequency Ω:

  ```
  if Omega<sqrt(Kw)
  k=i*k;
  else
  k=k;
  end
  ```

- To mark the cut-off frequency Ω_c using a vertical line on the magnitude plot of a characteristic polynomial, you may use the following command, where "OmegaC" stands for the value of a specific Ω_c:

  ```
  hold on, xline(OmegaC)
  ```

5

Coupled Waves in Composite Beams

Fiber-reinforced composite materials are an important development in the continuing search for lightweight materials of great strength and stiffness. Lighter structures are usually more prone to vibrations. Because of material coupling, composite beams generally vibrate with coupled bending and torsional vibration modes.

In this chapter, the concept of coupled waves is introduced. The wave-based vibration analysis approach is applicable to structures undergoing a single type of vibration and also coupled multiple types of vibrations, such as the coupled bending and torsional vibrations in composite beams.

5.1   The Governing Equations and the Propagation Relationships

Figure 5.1 shows a uniform composite beam with an unbalanced lay-up. Bending–torsion coupling is known to occur for such configurations. The governing equations of motion for free vibration of a slender composite beam, under the assumption that shear deformation, rotary inertia, and warping effects are negligibly small, are (Lottati 1985)

$$EI(x)\frac{\partial^4 w(x,t)}{\partial x^4} + K\frac{\partial^3 \theta(x,t)}{\partial x^3} + \rho(x)A(x)\frac{\partial^2 w(x,t)}{\partial t^2} = 0 \tag{5.1a}$$

$$GJ(x)\frac{\partial^2 \theta(x,t)}{\partial x^2} + K\frac{\partial^3 w(x,t)}{\partial x^3} - \rho(x)J(x)\frac{\partial^2 \theta(x,t)}{\partial t^2} = 0 \tag{5.1b}$$

where x is the position along the beam axis, t is the time, w is the transverse deflection of the centerline of the beam, θ is the torsional rotation, I is the area moment of inertia of the cross section about the z-axis, J is the polar moment of inertia (that is, the second moment of area of the cross section about the x-axis), A is the cross-sectional area, and E, G, and ρ are the Young's modulus, shear modulus, and volume mass density, respectively. Equations (5.1a) and (5.1b) are coupled through the torsional rotation and transverse deflection of the structure, and the coupling coefficient is K. Under the special situation where the coupling coefficient K is zero, Eqs. (5.1a) and (5.1b) are uncoupled, they are the equations of motion of a beam in free bending and torsional vibrations based on the Euler–Bernoulli bending vibration and the Elementary torsional vibration theories, respectively.

Assuming time harmonic motion and using separation of variables, the solutions to Eqs. (5.1a) and (5.1b) can be written in the form $w(x,t) = w_0 e^{-ikx} e^{i\omega t}$ and $\theta(x,t) = \theta_0 e^{-ikx} e^{i\omega t}$, where i is the imaginary unit, ω is the circular frequency, k is the wavenumber, and w_0 and θ_0 are the amplitudes of flexural deflection and torsional rotation, respectively. Substituting these expressions for $w(x,t)$ and $\theta(x,t)$ into Eqs. (5.1a) and (5.1b) and writing the equations in matrix form,

$$\begin{bmatrix} EIk^4 - \rho A\omega^2 & iKk^3 \\ iKk^3 & -GJk^2 + \rho J\omega^2 \end{bmatrix}\begin{bmatrix} w_0 \\ \theta_0 \end{bmatrix} = 0 \tag{5.2}$$

Mechanical Wave Vibrations: Analysis and Control, First Edition. Chunhui Mei.
© 2023 Chunhui Mei. Published 2023 by John Wiley & Sons Ltd.
Companion Website: www.wiley.com/go/Mei/MechanicalWaveVibrations

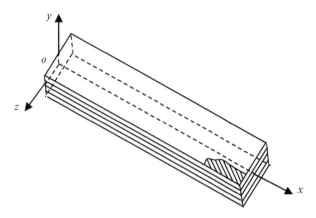

Figure 5.1 Uniform composite beam with unbalanced lay-up.

Setting the determinant of the coefficient matrix of Eq. (5.2) to zero gives the dispersion equation, which is a third degree polynomial equation in k^2,

$$\left(EIk^4 - \rho A\omega^2\right)\left(-GJk^2 + \rho J\omega^2\right) + K^2k^6 = 0 \tag{5.3}$$

Letting the coupling coefficient K be zero gives the wavenumbers of the uncoupled bending and torsional waves, denoted using subscripts b and t, respectively,

$$k_b = \sqrt[4]{\frac{\rho A\omega^2}{EI}}, \text{ and } k_t = \sqrt{\frac{\rho J}{GJ}}\omega = \sqrt{\frac{\rho}{G}}\omega \tag{5.4}$$

In terms of the uncoupled wavenumbers of Eq. (5.4), Eq. (5.3) becomes

$$-\frac{K^2}{EIGJ}k^6 + \left(k^4 - k_b^4\right)\left(k^2 - k_t^2\right) = 0 \tag{5.5}$$

Expanding Eq. (5.5) and combining like terms gives

$$\left(1 - \frac{K^2}{EIGJ}\right)k^6 - k_t^2k^4 - k_b^4k^2 + k_t^2k_b^4 = 0 \tag{5.6}$$

Equation (5.6) can be written as

$$ak^6 + bk^4 + ck^2 + d = 0 \tag{5.7}$$

where $a = 1 - \dfrac{K^2}{EIGJ}$, $b = -k_t^2$, $c = -k_b^4$, and $d = k_t^2k_b^4$.

The explicit solutions to Eq. (5.7), a cubic polynomial equation in terms of k^2, are found by introducing an intermediate parameter z,

$$z = k^2 + \frac{b}{3a} \tag{5.8}$$

Solving for k^2 in terms of z from Eq. (5.8) and substituting it into Eq. (5.7),

$$z^3 + p_1z + p_2 = 0 \tag{5.9}$$

where $p_1 = \dfrac{-b^2}{3a^2} + \dfrac{c}{a}$ and $p_2 = \dfrac{2b^3}{27a^3} + \dfrac{d}{a} - \dfrac{bc}{3a^2}$.

The roots of Eq. (5.9) are (The Mathematical Handbook Editorial Group 1997)

$$z_1 = \sqrt[3]{-\frac{p_2}{2} + \sqrt{\left(\frac{p_2}{2}\right)^2 + \left(\frac{p_1}{3}\right)^3}} + \sqrt[3]{-\frac{p_2}{2} - \sqrt{\left(\frac{p_2}{2}\right)^2 + \left(\frac{p_1}{3}\right)^3}}$$

$$z_2 = \frac{-1+i\sqrt{3}}{2}\sqrt[3]{-\frac{p_2}{2} + \sqrt{\left(\frac{p_2}{2}\right)^2 + \left(\frac{p_1}{3}\right)^3}} + \frac{-1-i\sqrt{3}}{2}\sqrt[3]{-\frac{p_2}{2} - \sqrt{\left(\frac{p_2}{2}\right)^2 + \left(\frac{p_1}{3}\right)^3}} \tag{5.10}$$

$$z_3 = \frac{-1-i\sqrt{3}}{2}\sqrt[3]{-\frac{p_2}{2} + \sqrt{\left(\frac{p_2}{2}\right)^2 + \left(\frac{p_1}{3}\right)^3}} + \frac{-1+i\sqrt{3}}{2}\sqrt[3]{-\frac{p_2}{2} - \sqrt{\left(\frac{p_2}{2}\right)^2 + \left(\frac{p_1}{3}\right)^3}}$$

From Eqs. (5.8) and (5.10),

$$k_1^2 = z_1^2 - \frac{b}{3a} \tag{5.11a}$$

$$k_2^2 = z_2^2 - \frac{b}{3a} \tag{5.11b}$$

$$k_3^2 = z_3^2 - \frac{b}{3a} \tag{5.11c}$$

In Eqs. (5.11a-c), k_1^2 and k_2^2 are positive, and k_3^2 is negative (Mei 2005a). These wavenumbers are functions of circular frequency ω, as well as the material and geometrical properties of the structure.

With time dependence $e^{i\omega t}$ suppressed, the solutions to Eqs. (5.1a) and (5.1b) are

$$\theta(x) = a_1^+ e^{-ik_1 x} + a_2^+ e^{-ik_2 x} + a_3^+ e^{-k_3 x} + a_1^- e^{ik_1 x} + a_2^- e^{ik_2 x} + a_3^- e^{k_3 x} \tag{5.12a}$$

$$w(x) = \overline{a_1^+} e^{-ik_1 x} + \overline{a_2^+} e^{-ik_2 x} + \overline{a_3^+} e^{-k_3 x} + \overline{a_1^-} e^{ik_1 x} + \overline{a_2^-} e^{ik_2 x} + \overline{a_3^-} e^{k_3 x} \tag{5.12b}$$

where k_1, k_2, and k_3 are wavenumbers of the torsional propagating, bending propagating, and bending decaying components, respectively. From Eqs. (5.10) and (5.11a-c),

$$k_1 = \sqrt{\sqrt[3]{-\frac{p_2}{2} + \sqrt{\left(\frac{p_2}{2}\right)^2 + \left(\frac{p_1}{3}\right)^3}} + \sqrt[3]{-\frac{p_2}{2} - \sqrt{\left(\frac{p_2}{2}\right)^2 + \left(\frac{p_1}{3}\right)^3}} - \frac{b}{3a}}, \tag{5.13a}$$

$$k_2 = \sqrt{\left|\frac{-1+i\sqrt{3}}{2}\sqrt[3]{-\frac{p_2}{2} + \sqrt{\left(\frac{p_2}{2}\right)^2 + \left(\frac{p_1}{3}\right)^3}} + \frac{-1-i\sqrt{3}}{2}\sqrt[3]{-\frac{p_2}{2} - \sqrt{\left(\frac{p_2}{2}\right)^2 + \left(\frac{p_1}{3}\right)^3}} - \frac{b}{3a}\right|}, \tag{5.13b}$$

$$k_3 = \sqrt{\left|\frac{-1-i\sqrt{3}}{2}\sqrt[3]{-\frac{p_2}{2} + \sqrt{\left(\frac{p_2}{2}\right)^2 + \left(\frac{p_1}{3}\right)^3}} + \frac{-1+i\sqrt{3}}{2}\sqrt[3]{-\frac{p_2}{2} - \sqrt{\left(\frac{p_2}{2}\right)^2 + \left(\frac{p_1}{3}\right)^3}} - \frac{b}{3a}\right|} \tag{5.13c}$$

The wave amplitudes $a_n^\pm$ of $\theta(x)$ and $\overline{a_n^\pm}$ of $w(x)$ in Eqs. (5.12a) and (5.12b), where $n = 1,2,3$, are related to each other. The relationships are found from Eq. (5.2),

$$\frac{\overline{a_1^+}}{a_1^+} = -iP_1, \quad \frac{\overline{a_2^+}}{a_2^+} = -iP_2, \quad \frac{\overline{a_3^+}}{a_3^+} = -P_3, \quad \frac{\overline{a_1^-}}{a_1^-} = iP_1, \quad \frac{\overline{a_2^-}}{a_2^-} = iP_2, \quad \frac{\overline{a_3^-}}{a_3^-} = P_3 \tag{5.14a}$$

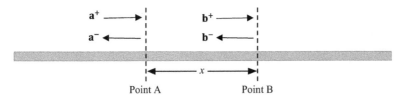

Figure 5.2 Wave propagation relationships.

where

$$P_1 = \frac{K}{EI}\left(\frac{k_1^3}{k_1^4 - k_b^4}\right), \; P_2 = \frac{K}{EI}\left(\frac{k_2^3}{k_2^4 - k_b^4}\right), \text{ and } P_3 = -\frac{K}{EI}\left(\frac{k_3^3}{k_3^4 - k_b^4}\right) \tag{5.14b}$$

Consider two points A and B on a torsionally and flexurally vibrating uniform composite beam that are a distance x apart, as shown in Figure 5.2. Waves propagate from one point to the other, with the propagation relationships determined by the appropriate wavenumber. Positive- and negative-going waves $\mathbf{a}^+$, $\mathbf{a}^-$, $\mathbf{b}^+$, and $\mathbf{b}^-$ at Points A and B are related by

$$\mathbf{b}^+ = \mathbf{f}(x)\mathbf{a}^+, \; \mathbf{a}^- = \mathbf{f}(x)\mathbf{b}^- \tag{5.15a}$$

where

$$\mathbf{a}^\pm = \begin{bmatrix} a_1^\pm \\ a_2^\pm \\ a_3^\pm \end{bmatrix}, \; \mathbf{b}^\pm = \begin{bmatrix} b_1^\pm \\ b_2^\pm \\ b_3^\pm \end{bmatrix}, \tag{5.15b}$$

are the wave components whose subscripts 1, 2, and 3 denoting the torsional propagating, bending propagating, and bending decaying wave components, respectively, and

$$\mathbf{f}(x) = \begin{bmatrix} e^{-ik_1 x} & 0 & 0 \\ 0 & e^{-ik_2 x} & 0 \\ 0 & 0 & e^{-k_3 x} \end{bmatrix} \tag{5.15c}$$

is the propagation matrix for a distance x.

5.2 Wave Reflection at Classical and Non-Classical Boundaries

Following the sign conventions defined in Figure 5.3, the internal resistant shear force $V(x,t)$, bending moment $M(x,t)$, and torque $T(x,t)$ at a section of the beam are related to the transverse deflection $w(x,t)$ and the torsional rotation $\theta(x,t)$ by

$$V(x,t) = EI\frac{\partial^3 w(x,t)}{\partial x^3} + K\frac{\partial^2 \theta(x,t)}{\partial x^2} \tag{5.16a}$$

$$M(x,t) = -EI\frac{\partial^2 w(x,t)}{\partial x^2} - K\frac{\partial \theta(x,t)}{\partial x} \tag{5.16b}$$

$$T(x,t) = K\frac{\partial^2 w(x,t)}{\partial x^2} + GJ\frac{\partial \theta(x,t)}{\partial x} \tag{5.16c}$$

Consider a non-classical boundary with spring attachments, as shown in Figure 5.4. At the boundary, incident waves $\mathbf{a}^+$ give rise to reflected waves $\mathbf{a}^-$, which are related by

$$\mathbf{a}^- = \mathbf{r}\mathbf{a}^+ \tag{5.17}$$

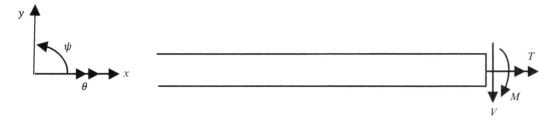

Figure 5.3 Definition of positive shear force, torque, and bending moment.

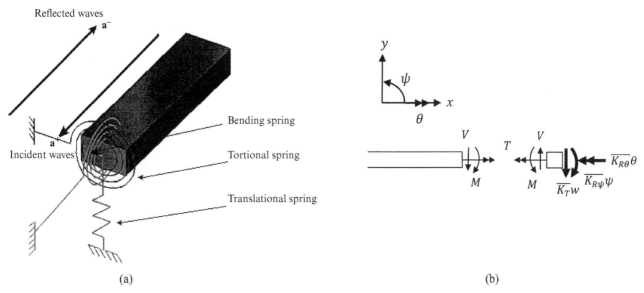

Figure 5.4 (a) Wave reflection at a non-classical boundary (Courtesy: (Mei 2005b)) and (b) free body diagram.

From the free body diagram of Figure 5.4, the equilibrium conditions at the non-classical boundary are

$$
\begin{aligned}
M\left(x,t\right)-\overline{K_{R\psi}}\psi\left(x,t\right)&=0\\
V\left(x,t\right)-\overline{K_T}w\left(x,t\right)&=0\\
-T\left(x,t\right)-\overline{K_{R\theta}}\theta\left(x,t\right)&=0
\end{aligned}
\tag{5.18}
$$

where $\psi(x,t)$ is the bending slope. $\overline{K_T}$, $\overline{K_{R\theta}}$, and $\overline{K_{R\psi}}$ are the translational, torsional, and bending stiffness of the springs, respectively.

From Eqs. (5.16a-c) and (5.18), the equilibrium conditions become

$$
\begin{aligned}
-EI\frac{\partial^2 w\left(x,t\right)}{\partial x^2}-K\frac{\partial\theta\left(x,t\right)}{\partial x}-\overline{K_{R\psi}}\psi\left(x,t\right)&=0\\
EI\frac{\partial^3 w\left(x,t\right)}{\partial x^3}+K\frac{\partial^2\theta\left(x,t\right)}{\partial x^2}-\overline{K_T}w\left(x,t\right)&=0\\
-EI\frac{\partial^2 w\left(x,t\right)}{\partial x^2}-GJ\frac{\partial\theta\left(x,t\right)}{\partial x}-\overline{K_{R\theta}}\theta\left(x,t\right)&=0
\end{aligned}
\tag{5.19}
$$

From Figure 5.4a and Eqs. (5.12b) and (5.14a),

$$
w(x)=-iP_1a_1^+e^{-ik_1x}-iP_2a_2^+e^{-ik_2x}-P_3a_3^+e^{-k_3x}+iP_1a_1^-e^{ik_1x}+iP_2a_2^-e^{ik_2x}+P_3a_3^-e^{k_3x}
\tag{5.20}
$$

In accordance with the Euler–Bernoulli bending vibration theory, bending slope $\psi=\partial w/\partial x$, from Eq. (5.20),

$$
\psi(x)=-P_1k_1a_1^+e^{-ik_1x}-P_2k_2a_2^+e^{-ik_2x}+P_3k_3a_3^+e^{-k_3x}-P_1k_1a_1^-e^{ik_1x}-P_2k_2a_2^-e^{ik_2x}+P_3k_3a_3^-e^{k_3x}
\tag{5.21}
$$

Choosing the origin at the boundary, substituting Eqs. (5.12a), (5.20), and (5.21) into Eq. (5.19), and assembling the equilibrium equations into matrix form,

$$\boldsymbol{\alpha}_{11}\mathbf{a}^- - \boldsymbol{\alpha}_{12}\mathbf{a}^+ = 0 \tag{5.22a}$$

where

$$\boldsymbol{\alpha}_{11} = \begin{bmatrix} iEIP_1k_1^2 - iKk_1 + \overline{K_{R\psi}}P_1k_1 & iEIP_2k_2^2 - iKk_2 + \overline{K_{R\psi}}P_2k_2 & -EIP_3k_3^2 - Kk_3 - \overline{K_{R\psi}}P_3k_3 \\ EIP_1k_1^3 - Kk_1^2 - i\overline{K_T}P_1 & EIP_2k_2^3 - Kk_2^2 - i\overline{K_T}P_2 & EIP_3k_3^3 + Kk_3^2 - \overline{K_T}P_3 \\ -iKP_1k_1^2 + iGJk_1 + \overline{K_{R\theta}} & -iKP_2k_2^2 + iGJk_2 + \overline{K_{R\theta}} & KP_3k_3^2 + GJk_3 + \overline{K_{R\theta}} \end{bmatrix},$$

$$\tag{5.22b}$$

$$\boldsymbol{\alpha}_{12} = \begin{bmatrix} iEIP_1k_1^2 - iKk_1 - \overline{K_{R\psi}}P_1k_1 & iEIP_2k_2^2 - iKk_2 - \overline{K_{R\psi}}P_2k_2 & -EIP_3k_3^2 - Kk_3 + \overline{K_{R\psi}}P_3k_3 \\ -EIP_1k_1^3 + Kk_1^2 - i\overline{K_T}P_1 & -EIP_2k_2^3 + Kk_2^2 - i\overline{K_T}P_2 & -EIP_3k_3^3 - Kk_3^2 - \overline{K_T}P_3 \\ -iKP_1k_1^2 + iGJk_1 - \overline{K_{R\theta}} & -iKP_2k_2^2 + iGJk_2 - \overline{K_{R\theta}} & KP_3k_3^2 + GJk_3 - \overline{K_{R\theta}} \end{bmatrix}$$

From Eqs. (5.17) and (5.22a)

$$\mathbf{r} = \boldsymbol{\alpha}_{11}^{-1}\boldsymbol{\alpha}_{12} \tag{5.23}$$

Reflection matrices at classical boundaries, namely, the simply-supported, clamped, and free boundaries, are obtained from Eqs. (5.22b) and (5.23) when spring stiffness $\overline{K_T}$, $\overline{K_{R\theta}}$, and $\overline{K_{R\psi}}$ becomes zero or infinite.

$$\mathbf{r}_s = \begin{bmatrix} 1 & 1 & 1 \\ -iP_1 & -iP_2 & -P_3 \\ -iEIP_1k_1^2 + iKk_1 & -iEIP_2k_2^2 + iKk_2 & EIP_3k_3^2 + Kk_3 \end{bmatrix}^{-1}$$

$$\begin{bmatrix} -1 & -1 & -1 \\ -iP_1 & -iP_2 & -P_3 \\ -iEIP_1k_1^2 + iKk_1 & -iEIP_2k_2^2 + iKk_2 & EIP_3k_3^2 + Kk_3 \end{bmatrix}$$

$$\mathbf{r}_c = \begin{bmatrix} 1 & 1 & 1 \\ iP_1 & iP_2 & P_3 \\ P_1k_1 & P_2k_2 & -P_3k_3 \end{bmatrix} \begin{bmatrix} -1 & -1 & -1 \\ iP_1 & iP_2 & P_3 \\ -P_1k_1 & -P_2k_2 & P_3k_3 \end{bmatrix} \tag{5.24}$$

$$\mathbf{r}_f = \begin{bmatrix} iKk_1 - iEIP_1k_1^2 & iKk_2 - iEIP_2k_2^2 & Kk_3 + EIP_3k_3^2 \\ -Kk_1^2 + EIP_1k_1^3 & -Kk_2^2 + EIP_2k_2^3 & Kk_3^2 + EIP_3k_3^3 \\ -iKP_1k_1^2 + iGJk_1 & -iKP_2k_2^2 + iGJk_2 & KP_3k_3^2 + GJk_3 \end{bmatrix}^{-1}$$

$$\begin{bmatrix} iKk_1 - iEIP_1k_1^2 & iKk_2 - iEIP_2k_2^2 & Kk_3 + EIP_3k_3^2 \\ Kk_1^2 - EIP_1k_1^3 & Kk_2^2 - EIP_2k_2^3 & -Kk_3^2 - EIP_3k_3^3 \\ -iKP_1k_1^2 + iGJk_1 & -iKP_2k_2^2 + iGJk_2 & KP_3k_3^2 + GJk_3 \end{bmatrix}$$

where $\mathbf{r}_s$, $\mathbf{r}_c$, and $\mathbf{r}_f$ denote the reflection matrices for simply-supported, clamped, and free boundaries, respectively.

5.3 Wave Reflection and Transmission at a Point Attachment

Consider a point support that exerts translational, bending, and torsional constraints to the beam, as shown in Figure 5.5. The translational, torsional, and bending stiffness of the springs are $\overline{K_T}$, $\overline{K_{R\theta}}$, and $\overline{K_{R\psi}}$, respectively.

A set of positive-going waves $\mathbf{a}^+$ is incident upon the point support and gives rise to transmitted and reflected waves $\mathbf{b}^+$ and $\mathbf{a}^-$, which are related to the incident waves by the transmission and reflection matrices $\mathbf{t}$ and $\mathbf{r}$,

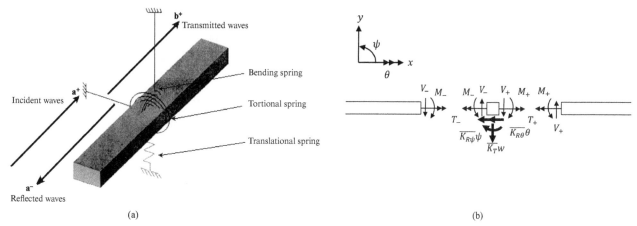

Figure 5.5 (a) Wave reflection and transmission at a discontinuity (Courtesy: (Mei 2005b)) and (b) free body diagram.

$$\mathbf{b}^+ = \mathbf{t}\mathbf{a}^+, \ \mathbf{a}^- = \mathbf{r}\mathbf{a}^+ \tag{5.25}$$

Denoting the torsional rotation, transverse deflection, and bending slope on the left and right side of the point support as $\theta_-, \theta_+, w_-, w_+, \psi_-,$ and ψ_+, from Figure 5.5a and Eqs. (5.12a), (5.20), and (5.21),

$$\begin{aligned}
\theta_- &= a_1^+ e^{-ik_1 x} + a_2^+ e^{-ik_2 x} + a_3^+ e^{-k_3 x} + a_1^- e^{ik_1 x} + a_2^- e^{ik_2 x} + a_3^- e^{k_3 x} \\
\theta_+ &= b_1^+ e^{-ik_1 x} + b_2^+ e^{-ik_2 x} + b_3^+ e^{-k_3 x} \\
w_- &= -iP_1 a_1^+ e^{-ik_1 x} - iP_2 a_2^+ e^{-ik_2 x} - P_3 a_3^+ e^{-k_3 x} + iP_1 a_1^- e^{ik_1 x} + iP_2 a_2^- e^{ik_2 x} + P_3 a_3^- e^{k_3 x}, \\
w_+ &= -iP_1 b_1^+ e^{-ik_1 x} - iP_2 b_2^+ e^{-ik_2 x} - P_3 b_3^+ e^{-k_3 x}, \\
\psi_- &= -P_1 k_1 a_1^+ e^{-ik_1 x} - P_2 k_2 a_2^+ e^{-ik_2 x} + P_3 k_3 a_3^+ e^{-k_3 x} - P_1 k_1 a_1^- e^{ik_1 x} - P_2 k_2 a_2^- e^{ik_2 x} + P_3 k_3 a_3^- e^{k_3 x} \\
\psi_+ &= -P_1 k_1 b_1^+ e^{-ik_1 x} - P_2 k_2 b_2^+ e^{-ik_2 x} + P_3 k_3 b_3^+ e^{-k_3 x}
\end{aligned} \tag{5.26}$$

The continuity in the torsional rotation, transverse deflection, and bending slope on both sides of the point support gives,

$$\theta_- = \theta_+, \ w_- = w_+, \ \psi_- = \psi_+ \tag{5.27}$$

Choosing the origin at the point support, substituting Eq. (5.26) into Eq. (5.27), and assembling the continuity equations into matrix form,

$$\boldsymbol{\beta}_{11} \mathbf{b}^+ + \boldsymbol{\beta}_{12} \mathbf{a}^- = \boldsymbol{\beta}_{13} \mathbf{a}^+ \tag{5.28a}$$

where

$$\boldsymbol{\beta}_{11} = \begin{bmatrix} -1 & -1 & -1 \\ -iP_1 & -iP_2 & -P_3 \\ -P_1 k_1 & -P_2 k_2 & P_3 k_3 \end{bmatrix}, \ \boldsymbol{\beta}_{12} = \begin{bmatrix} 1 & 1 & 1 \\ -iP_1 & -iP_2 & -P_3 \\ P_1 k_1 & P_2 k_2 & -P_3 k_3 \end{bmatrix},$$

$$\boldsymbol{\beta}_{13} = \begin{bmatrix} -1 & -1 & -1 \\ -iP_1 & -iP_2 & -P_3 \\ -P_1 k_1 & -P_2 k_2 & P_3 k_3 \end{bmatrix}. \tag{5.28b}$$

From the free body diagram of Figure 5.5, the equilibrium conditions at the point support are

$$\begin{aligned}
M_- - M_+ - \overline{K_{R\psi}} \psi(x,t) &= 0 \\
V_- - V_+ - \overline{K_T} w(x,t) &= 0 \\
T_+ - T_- - \overline{K_{R\theta}} \theta(x,t) &= 0
\end{aligned} \tag{5.29a}$$

From Eqs. (5.16a-c) and (5.29a),

$$
\left(-EI\frac{\partial^2 w_-(x,t)}{\partial x^2} - K\frac{\partial \theta_-(x,t)}{\partial x} \right) - \left(-EI\frac{\partial^2 w_+(x,t)}{\partial x^2} - K\frac{\partial \theta_+(x,t)}{\partial x} \right) - \overline{K_{R\psi}}\psi_\pm(x,t) = 0
$$

$$
\left(EI\frac{\partial^3 w_-(x,t)}{\partial x^3} + K\frac{\partial^2 \theta_-(x,t)}{\partial x^2} \right) - \left(EI\frac{\partial^3 w_+(x,t)}{\partial x^3} + K\frac{\partial^2 \theta_+(x,t)}{\partial x^2} \right) - \overline{K_T}w_\pm(x,t) = 0 \tag{5.29b}
$$

$$
\left(K\frac{\partial^2 w_+(x,t)}{\partial x^2} + GJ\frac{\partial \theta_+(x,t)}{\partial x} \right) - \left(K\frac{\partial^2 w_-(x,t)}{\partial x^2} + GJ\frac{\partial \theta_-(x,t)}{\partial x} \right) - \overline{K_{R\theta}}\theta_\pm(x,t) = 0
$$

Choosing the origin at the point support, substituting Eqs. (5.26) into (5.29b), and assembling the equilibrium equations into matrix form,

$$
\beta_{21}\mathbf{b}^+ + \beta_{22}\mathbf{a}^- = \beta_{23}\mathbf{a}^+ \tag{5.30a}
$$

where the coefficient matrices are not unique, because of the continuity in torsional rotation, transverse deflection, and bending slope on the left and right side of the point support. One example set of coefficient matrices is given below

$$
\beta_{21} = \begin{bmatrix} -iEIP_1k_1^2 + iKk_1 - \overline{K_{R\psi}}P_1k_1 & -iEIP_2k_2^2 + iKk_2 - \overline{K_{R\psi}}P_2k_2 & EIP_3k_3^2 + Kk_3 + \overline{K_{R\psi}}P_3k_3 \\ -EIP_1k_1^3 + Kk_1^2 + i\overline{K_T}P_1 & -EIP_2k_2^3 + Kk_2^2 + i\overline{K_T}P_2 & -EIP_3k_3^3 - Kk_3^2 + \overline{K_T}P_3 \\ -iKP_1k_1^2 + iGJk_1 + \overline{K_{R\theta}} & -iKP_2k_2^2 + iGJk_2 + \overline{K_{R\theta}} & KP_3k_3^2 + GJk_3 + \overline{K_{R\theta}} \end{bmatrix}
$$

$$
\beta_{22} = \begin{bmatrix} -iEIP_1k_1^2 + iKk_1 & -iEIP_2k_2^2 + iKk_2 & EIP_3k_3^2 + Kk_3 \\ EIP_1k_1^3 - Kk_1^2 & EIP_2k_2^3 - Kk_2^2 & EIP_3k_3^3 + Kk_3^2 \\ -iKP_1k_1^2 + iGJk_1 & -iKP_2k_2^2 + iGJk_2 & KP_3k_3^2 + GJk_3 \end{bmatrix}, \tag{5.30b}
$$

$$
\beta_{23} = \begin{bmatrix} -iEIP_1k_1^2 + iKk_1 & -iEIP_2k_2^2 + iKk_2 & EIP_3k_3^2 + Kk_3 \\ -EIP_1k_1^3 + Kk_1^2 & -EIP_2k_2^3 + Kk_2^2 & -EIP_3k_3^3 - Kk_3^2 \\ -iKP_1k_1^2 + iGJk_1 & -iKP_2k_2^2 + iGJk_2 & KP_3k_3^2 + GJk_3 \end{bmatrix}
$$

From Eqs. (5.25), (5.28a), and (5.30a), the reflection and transmission matrices are

$$
\mathbf{t} = \left(\beta_{21} - \beta_{22}\beta_{12}^{-1}\beta_{11} \right)^{-1} \left(\beta_{23} - \beta_{22}\beta_{12}^{-1}\beta_{13} \right),
$$

$$
\mathbf{r} = \left(\beta_{22} - \beta_{21}\beta_{11}^{-1}\beta_{12} \right)^{-1} \left(\beta_{23} - \beta_{21}\beta_{11}^{-1}\beta_{13} \right) \tag{5.31}
$$

Note that for the reflection and transmission matrices, there are other equivalent expressions to Eq. (5.31).

5.4 Free Vibration Analysis in Finite Beams – Natural Frequencies and Modeshapes

From the wave vibration standpoint, vibrations propagate along a uniform beam and are reflected and transmitted at discontinuities. With the availability of the propagation, reflection, and transmission matrices, free vibration analysis of coupled bending and torsional vibration in a composite beam involves a concise and systematic assembling process.

Figure 5.6 shows a uniform composite beam. Denoting the incident and reflected waves at Boundaries A and B using $\mathbf{a}^+$, $\mathbf{a}^-$, $\mathbf{b}^+$, and $\mathbf{b}^-$, the reflection and propagation relationships are

$$
\mathbf{a}^+ = \mathbf{r}_A\,\mathbf{a}^-; \ \mathbf{b}^- = \mathbf{r}_B\,\mathbf{b}^+; \ \mathbf{b}^+ = \mathbf{f}(L)\,\mathbf{a}^+; \ \mathbf{a}^- = \mathbf{f}(L)\,\mathbf{b}^- \tag{5.32}
$$

where $\mathbf{r}_A$ and $\mathbf{r}_B$ are the reflection matrices at Boundaries A and B, respectively. $\mathbf{f}(L)$ is the propagation matrix for a distance L.

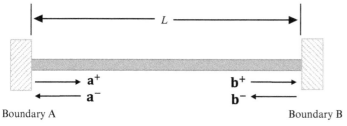

Figure 5.6 A uniform composite beam.

Assembling the equations in Eq. (5.32) into matrix form,

$$
\begin{bmatrix}
-\mathbf{I} & \mathbf{r}_A & \mathbf{0} & \mathbf{0} \\
\mathbf{0} & \mathbf{0} & \mathbf{r}_B & -\mathbf{I} \\
\mathbf{f}(L) & \mathbf{0} & -\mathbf{I} & \mathbf{0} \\
\mathbf{0} & -\mathbf{I} & \mathbf{0} & \mathbf{f}(L)
\end{bmatrix}
\begin{bmatrix}
\mathbf{a}^+ \\
\mathbf{a}^- \\
\mathbf{b}^+ \\
\mathbf{b}^-
\end{bmatrix}
= \mathbf{0}
\tag{5.33}
$$

where $\mathbf{I}$ and $\mathbf{0}$ denote an identity matrix and a zero matrix, respectively.

Equation (5.33) can be written in the form of

$$
\mathbf{A}_0 \mathbf{z}_0 = \mathbf{0}
\tag{5.34}
$$

where $\mathbf{A}_0$ is a 12 by 12 square coefficient matrix of Eq. (5.33) and $\mathbf{z}_0$ is a wave vector containing 12 wave components of Eq. (5.33).

For non-trivial solution, the determinant of the coefficient matrix is zero,

$$
|\mathbf{A}_0| = 0
\tag{5.35}
$$

From the characteristic equation obtained in Eq. (5.35), the natural frequencies of the composite beam, which are roots of the characteristic equation, can be found.

The modeshapes at a given natural frequency can be found by eliminating any one row of the coefficient matrix of Eq. (5.33), and using the remaining rows to solve the wave components in terms of a chosen wave component, as described in Chapter 3. Substituting the wave component responses into the deflection expressions of Eqs. (5.12a), (5.20), and (5.21), the modeshapes for the torsional rotation $\theta(x)$, the transverse deflection of the centerline of the beam $w(x)$, and the bending slope $\psi(x)$ are found, respectively.

5.5 Force Generated Waves and Forced Vibration Analysis of Finite Beams

Externally applied forces, moments, and torques have the effect of injecting waves into a continuous structure. Consider waves injected into a composite beam by a point transverse force $\bar{V}$, a bending moment $\bar{M}$, and a torque $\bar{T}$ shown in Figure 5.7. Denoting physical parameters on the left and right side of the point where the external excitations are applied using subscripts − and +, respectively, from the free body diagram of Figure 5.7, the following continuity and equilibrium relationships exist:

$$
\theta_- = \theta_+, \; w_- = w_+, \; \psi_- = \psi_+
\tag{5.36}
$$

$$
M_- - M_+ + \bar{M} = 0, \; V_- - V_+ + \bar{V} = 0, \; T_+ - T_- + \bar{T} = 0
\tag{5.37}
$$

From Figure 5.7a and Eqs. (5.12a), (5.20), and (5.21),

$$
\begin{aligned}
\theta_+ &= b_1^+ e^{-ik_1 x} + b_2^+ e^{-ik_2 x} + b_3^+ e^{-k_3 x} + b_1^- e^{ik_1 x} + b_2^- e^{ik_2 x} + b_3^- e^{k_3 x} \\
\theta_- &= a_1^+ e^{-ik_1 x} + a_2^+ e^{-ik_2 x} + a_3^+ e^{-k_3 x} + a_1^- e^{ik_1 x} + a_2^- e^{ik_2 x} + a_3^- e^{k_3 x} \\
w_+ &= -iP_1 b_1^+ e^{-ik_1 x} - iP_2 b_2^+ e^{-ik_2 x} - P_3 b_3^+ e^{-k_3 x} + iP_1 b_1^- e^{ik_1 x} + iP_2 b_2^- e^{ik_2 x} + P_3 b_3^- e^{k_3 x} \\
w_- &= -iP_1 a_1^+ e^{-ik_1 x} - iP_2 a_2^+ e^{-ik_2 x} - P_3 a_3^+ e^{-k_3 x} + iP_1 a_1^- e^{ik_1 x} + iP_2 a_2^- e^{ik_2 x} + P_3 a_3^- e^{k_3 x} \\
\psi_+ &= -k_1 P_1 b_1^+ e^{-ik_1 x} - k_2 P_2 b_2^+ e^{-ik_2 x} + k_3 P_3 b_3^+ e^{-k_3 x} - k_1 P_1 b_1^- e^{ik_1 x} - k_2 P_2 b_2^- e^{ik_2 x} + k_3 P_3 b_3^- e^{k_3 x} \\
\psi_- &= -k_1 P_1 a_1^+ e^{-ik_1 x} - k_2 P_2 a_2^+ e^{-ik_2 x} + k_3 P_3 a_3^+ e^{-k_3 x} - k_1 P_1 a_1^- e^{ik_1 x} - k_2 P_2 a_2^- e^{ik_2 x} + k_3 P_3 a_3^- e^{k_3 x}
\end{aligned}
\tag{5.38}
$$

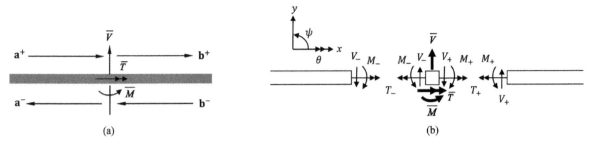

Figure 5.7 (a) Waves generated by external excitations and (b) free body diagram.

Choosing the origin at the point where the external excitations are applied, substituting Eq. (5.38) into Eq. (5.36), assembling the equations into matrix form, the continuity conditions become

$$\delta_{11}\left(\mathbf{b}^+ - \mathbf{a}^+\right) + \delta_{12}\left(\mathbf{b}^- - \mathbf{a}^-\right) = \mathbf{0}; \tag{5.39a}$$

where

$$\delta_{11} = \begin{bmatrix} 1 & 1 & 1 \\ -iP_1 & -iP_2 & -P_3 \\ -P_1 k_1 & -P_2 k_2 & P_3 k_3 \end{bmatrix}, \; \delta_{12} = \begin{bmatrix} 1 & 1 & 1 \\ iP_1 & iP_2 & P_3 \\ -P_1 k_1 & -P_2 k_2 & P_3 k_3 \end{bmatrix} \tag{5.39b}$$

From Eqs. (5.16a-c) and (5.37), the equilibrium conditions become

$$\left(-EI\frac{\partial^2 w_-(x,t)}{\partial x^2} - K\frac{\partial \theta_-(x,t)}{\partial x}\right) - \left(-EI\frac{\partial^2 w_+(x,t)}{\partial x^2} - K\frac{\partial \theta_+(x,t)}{\partial x}\right) + \bar{M} = 0$$

$$\left(EI\frac{\partial^3 w_-(x,t)}{\partial x^3} + K\frac{\partial^2 \theta_-(x,t)}{\partial x^2}\right) - \left(EI\frac{\partial^3 w_+(x,t)}{\partial x^3} + K\frac{\partial^2 \theta_+(x,t)}{\partial x}\right) + \bar{V} = 0 \tag{5.40}$$

$$\left(K\frac{\partial^2 w_+(x,t)}{\partial x^2} + GJ\frac{\partial \theta_+(x,t)}{\partial x}\right) - \left(K\frac{\partial^2 w_-(x,t)}{\partial x^2} + GJ\frac{\partial \theta_-(x,t)}{\partial x}\right) + \bar{T} = 0$$

Choosing the origin at the point where the external excitations are applied, substituting Eq. (5.38) into Eq. (5.40), assembling the equilibrium equations into matrix form,

$$\delta_{21}\left(\mathbf{b}^+ - \mathbf{a}^+\right) + \delta_{22}\left(\mathbf{b}^- - \mathbf{a}^-\right) = \mathbf{q} \tag{5.41a}$$

where $\mathbf{q} = [\bar{M} \; \bar{V} \bar{T}]^T$ with superscript T denoting the transpose of a vector and

$$\delta_{21} = \begin{bmatrix} -iEIP_1 k_1^2 + iKk_1 & -iEIP_2 k_2^2 + iKk_2 & EIP_3 k_3^2 + Kk_3 \\ EIP_1 k_1^3 - Kk_1^2 & EIP_2 k_2^3 - Kk_2^2 & EIP_3 k_3^3 + Kk_3^2 \\ -iKP_1 k_1^2 + iGJk_1 & -iKP_2 k_2^2 + iGJk_2 & KP_3 k_3^2 + GJk_3 \end{bmatrix},$$

$$\delta_{22} = \begin{bmatrix} iEIP_1 k_1^2 - iKk_1 & iEIP_2 k_2^2 - iKk_2 & -EIP_3 k_3^2 - Kk_3 \\ EIP_1 k_1^3 - Kk_1^2 & EIP_2 k_2^3 - Kk_2^2 & EIP_3 k_3^3 + Kk_3^2 \\ -iKP_1 k_1^2 + iGJk_1 & -iKP_2 k_2^2 + iGJk_2 & -KP_3 k_3^2 - GJk_3 \end{bmatrix} \tag{5.41b}$$

Solving Eqs. (5.39a) and (5.41a) gives

$$\mathbf{b}^+ - \mathbf{a}^+ = \mathbf{q}^+, \; \mathbf{b}^- - \mathbf{a}^- = \mathbf{q}^- \tag{5.42a}$$

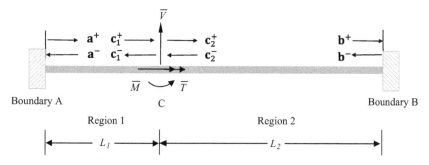

Figure 5.8   A uniform composite beam subjected to external excitations.

where

$$
\mathbf{q}^+ = -\left(\delta_{12}^{-1}\delta_{11} - \delta_{22}^{-1}\delta_{21}\right)^{-1}\delta_{22}^{-1}\mathbf{q}
$$
$$
\mathbf{q}^- = -\left(\delta_{11}^{-1}\delta_{12} - \delta_{21}^{-1}\delta_{22}\right)^{-1}\delta_{21}^{-1}\mathbf{q}
$$

(5.42b)

With the availability of force generated waves, as well as propagation, reflection, and transmission relationships, forced vibration in a composite beam can be obtained.

Figure 5.8 shows a uniform composite beam with a point transverse force $\bar{V}$, a bending moment $\bar{M}$, and a torque $\bar{T}$ applied at Point C. The externally applied force divides the structure into two regions, namely Region 1 and 2, of length L_1 and L_2, respectively. The waves at Point C, where the external excitations are applied, are denoted using $\mathbf{c}_1^+$, $\mathbf{c}_1^-$, $\mathbf{c}_2^+$, and $\mathbf{c}_2^-$.

The following wave relationships exist:

- At Boundaries A and B,

$$
\mathbf{a}^+ = \mathbf{r}_A\mathbf{a}^-, \; \mathbf{b}^- = \mathbf{r}_B\mathbf{b}^+
$$

(5.43)

where $\mathbf{r}_A$ and $\mathbf{r}_B$ are the reflection matrices at Boundaries A and B, respectively.

- At Point C, where the external excitations are applied,

$$
\mathbf{c}_2^+ - \mathbf{c}_1^+ = \mathbf{q}^+, \text{ and } \mathbf{c}_2^- - \mathbf{c}_1^- = \mathbf{q}^-
$$

(5.44)

where $\mathbf{q}^+$ and $\mathbf{q}^-$ are obtained in Eq. (5.42b).

- The propagation relationships along the uniform beam segment L_1 and L_2 are

$$
\mathbf{c}_1^+ = \mathbf{f}\left(L_1\right)\mathbf{a}^+, \; \mathbf{a}^- = \mathbf{f}\left(L_1\right)\mathbf{c}_1^-,
$$
$$
\mathbf{c}_2^- = \mathbf{f}\left(L_2\right)\mathbf{b}^-, \; \mathbf{b}^+ = \mathbf{f}\left(L_2\right)\mathbf{c}_2^+
$$

(5.45)

where $\mathbf{f}(L_1)$ and $\mathbf{f}(L_2)$ are the propagation matrices for a distance L_1 and L_2, respectively.

Assembling the wave relationships from Eqs. (5.43) to (5.45) into matrix form,

$$
\begin{bmatrix}
-\mathbf{I} & \mathbf{r}_A & \mathbf{0} & \mathbf{0} & \mathbf{0} & \mathbf{0} & \mathbf{0} & \mathbf{0} \\
\mathbf{0} & \mathbf{0} & \mathbf{r}_B & -\mathbf{I} & \mathbf{0} & \mathbf{0} & \mathbf{0} & \mathbf{0} \\
\mathbf{f}\left(L_1\right) & \mathbf{0} & \mathbf{0} & \mathbf{0} & -\mathbf{I} & \mathbf{0} & \mathbf{0} & \mathbf{0} \\
\mathbf{0} & -\mathbf{I} & \mathbf{0} & \mathbf{0} & \mathbf{0} & \mathbf{f}\left(L_1\right) & \mathbf{0} & \mathbf{0} \\
\mathbf{0} & \mathbf{0} & -\mathbf{I} & \mathbf{0} & \mathbf{0} & \mathbf{0} & \mathbf{f}\left(L_2\right) & \mathbf{0} \\
\mathbf{0} & \mathbf{0} & \mathbf{0} & \mathbf{f}\left(L_2\right) & \mathbf{0} & \mathbf{0} & \mathbf{0} & -\mathbf{I} \\
\mathbf{0} & \mathbf{0} & \mathbf{0} & \mathbf{0} & -\mathbf{I} & \mathbf{0} & \mathbf{I} & \mathbf{0} \\
\mathbf{0} & \mathbf{0} & \mathbf{0} & \mathbf{0} & \mathbf{0} & -\mathbf{I} & \mathbf{0} & \mathbf{I}
\end{bmatrix}
\begin{bmatrix}
\mathbf{a}^+ \\ \mathbf{a}^- \\ \mathbf{b}^+ \\ \mathbf{b}^- \\ \mathbf{c}_1^+ \\ \mathbf{c}_1^- \\ \mathbf{c}_2^+ \\ \mathbf{c}_2^-
\end{bmatrix}
=
\begin{bmatrix}
\mathbf{0} \\ \mathbf{0} \\ \mathbf{0} \\ \mathbf{0} \\ \mathbf{0} \\ \mathbf{0} \\ \mathbf{q}^+ \\ \mathbf{q}^-
\end{bmatrix}
$$

(5.46)

where $\mathbf{I}$ and $\mathbf{0}$ denote identity and zero matrices, respectively.

Equation (5.46) can be expressed in the form of

$$\mathbf{A}_f \mathbf{z}_f = \mathbf{F} \tag{5.47}$$

where $\mathbf{A}_f$ is a 24 by 24 square coefficient matrix, $\mathbf{z}_f$ is a wave vector containing 24 wave components, and $\mathbf{F}$ is the load vector of Eq. (5.46).

Solving for $\mathbf{z}_f$ from Eq. (5.47),

$$\mathbf{z}_f = \mathbf{A}_f^{-1} \mathbf{F} \tag{5.48}$$

Equation (5.48) provides the responses of all wave components, from which the deflections of any point on the beam can be obtained. For example, the torsional rotation at a cross section and the transverse deflection of the centerline of the beam located between Boundary A and the excitation point C that is a distance x from Point C are

$$\theta(x) = \begin{bmatrix} 1 & 1 & 1 \end{bmatrix} \mathbf{f}(x) \mathbf{c}_1^- + \begin{bmatrix} 1 & 1 & 1 \end{bmatrix} \mathbf{f}(-x) \mathbf{c}_1^+$$

$$\tag{5.49}$$

$$w(x) = \begin{bmatrix} 1 & 1 & 1 \end{bmatrix} \mathbf{f}(x) \overline{\mathbf{c}_1^-} + \begin{bmatrix} 1 & 1 & 1 \end{bmatrix} \mathbf{f}(-x) \overline{\mathbf{c}_1^+}$$

where the wave component amplitudes $\mathbf{c}_1^\pm$ of $\theta(x)$ and $\overline{\mathbf{c}_1^\pm}$ of $w(x)$ are related by Eqs. (5.14a) and (5.14b).

5.6 Numerical Examples

Free and forced vibrations in a cantilever composite beam, that is, a beam with one end clamped and the other end free, are analyzed. The material and geometrical properties of the example composite beam are: the Young's modulus, shear modulus, volume mass density, and coupling coefficient are $E = 8.418\ GN/m^2$, $G = 0.3278\ GN/m^2$, $\rho = 1347.0\ kg/m^3$, and $K = 0.1143\ Nm^2$, respectively. The cross section is a rectangular shape whose width is $12.7\ mm$ and thickness is $3.18\ mm$. The length of the cantilever beam is $L = 190.5\ mm$. In the forced responses, the external excitation is applied at $0.35\ L$, and the response is observed at $0.75\ L$, both are measured from the clamped boundary.

Figure 5.9 presents the wavenumbers for coupled torsional and bending vibration motions, which are compared with the wavenumbers of a beam, of the same material and geometrical properties, that undergoes uncoupled torsion and

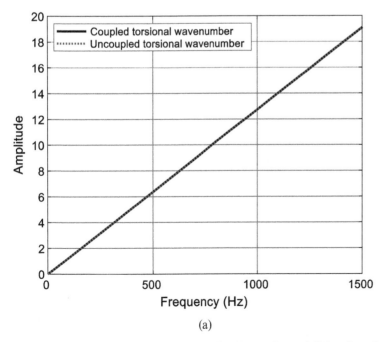

(a)

Figure 5.9 Wavenumbers of (a) torsional vibration motion and (b) bending vibration motion.

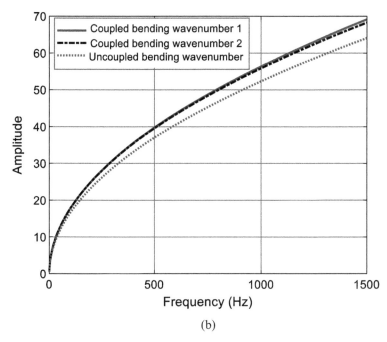

(b)

Figure 5.9 (Cont'd)

bending vibrations. The material coupling does not affect the torsional wavenumber in the observed frequency range of the example composite beam. The bending wavenumbers, however, are affected by the material coupling of the composite beam.

The natural frequencies can be obtained by plotting either the magnitudes or the real and imaginary parts of the characteristic polynomials of Eq. (5.35). The frequencies at where the magnitudes, or equivalently both the real and imaginary parts, becoming zero, are the natural frequencies, as shown in Figure 5.10. The natural frequencies can also be found by observing the local minima from the dB magnitude plots of the characteristic polynomials shown in Figure 5.11. The lowest five natural frequencies of the cantilever composite beam identified from Figures 5.10 and 5.11 are listed in Table 5.1. MATLAB scripts for free vibration analysis are available in Section 5.7.

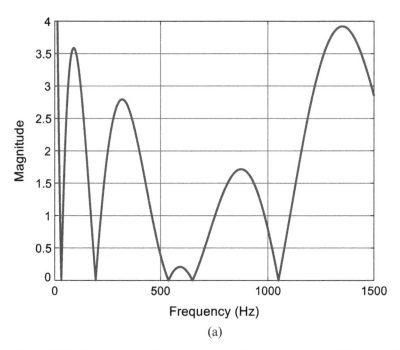

(a)

Figure 5.10 The characteristic polynomial: (a) linear magnitude and (b) real and imaginary parts.

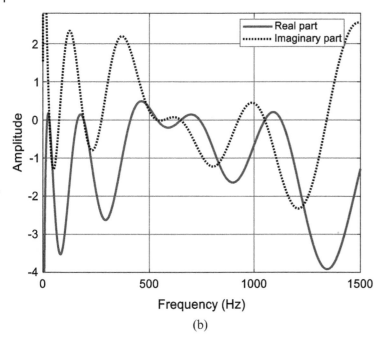

(b)

Figure 5.10 (Cont'd)

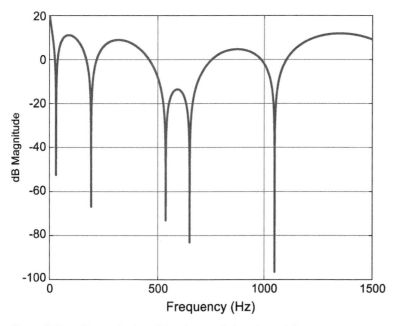

Figure 5.11 dB magnitudes of the characteristic polynomial.

Table 5.1 Natural frequencies of the cantilever composite beam.

Mode number	1	2	3	4	5
Natural frequency (*Hz*)	30.8	192.7	537.4	648.7	1049.7

The modeshapes of the lowest five modes for the uniform cantilever composite beams are presented in Figure 5.12.

The torsional rotation and flexural deflection frequency responses, of the uniform beam subjected to various external excitations, such as a point transverse force, a bending moment, and a torque excitation, are presented in Figure 5.13.

The results obtained are applicable to lower frequencies of slender composite beams. For deep composite beams, the shear deformation and rotary inertia need to be taken into account (Mei 2005b).

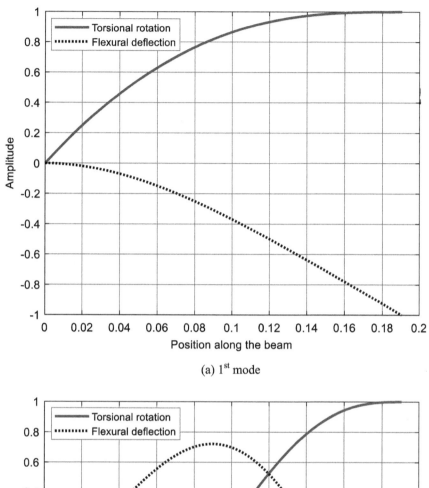

(a) 1st mode

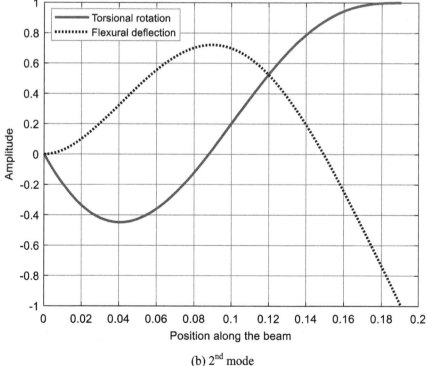

(b) 2nd mode

Figure 5.12 Torsional rotation and flexural deflection modeshapes of the lowest few modes.

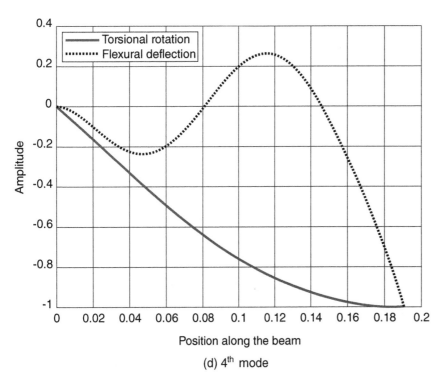

(c) 3rd mode

(d) 4th mode

Figure 5.12 (Cont'd)

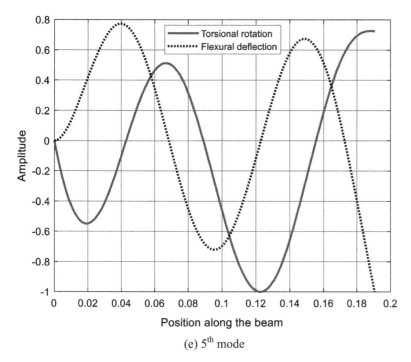

(e) 5th mode

Figure 5.12 (Cont'd)

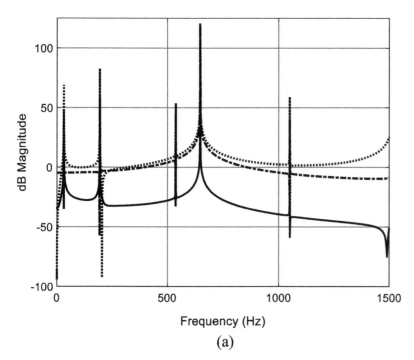

(a)

Figure 5.13 Frequency responses of the uniform beam subjected to point transverse force (__), bending moment (-.-.), and torque (...) excitations, respectively: (a) torsional rotation frequency response and (b) flexural deflection frequency response.

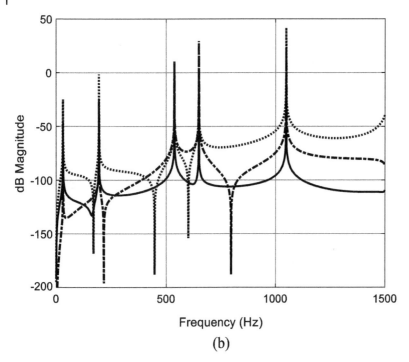

Figure 5.13 (Cont'd)

5.7 MATLAB Script

Script: Free wave vibration analysis of a uniform composite beam

```
close all
clear all

L=0.1905 % Geometrical parameters
b=0.0127
h=0.00318
A=b*h
I=b*h^3/12
J=b*h^3/12+h*b^3/12
E=8.418*10^9 % Material parameters
G=0.3278*10^9
p=1347
EI=E*I
GJ=G*J
K=0.1143
m=p*A
I2=p*J

f1=0.1 % Starting frequency in Hz
f2=1500 % Ending frequency in Hz
stepsize=0.1
freq=f1:stepsize:f2;

for n=1:length(freq)
```

```
w=2*pi*freq(n);
  kb=(m*w^2/EI)^(1/4);
kt=sqrt(I2/GJ)*w;
Kb(n)=kb;
Kt(n)=kt;

a0=1-K^2/(EI*GJ);
b0=-kt^2;
c0=-kb^4;
d0=kt^2*kb^4;

px=(-1/3/a0*b0^2+c0)/a0;
qx=(2/27/a0^2*b0^3+d0-1/3*c0*b0/a0)/a0;

%Math Handbook
tempk1=(-qx/2+sqrt((qx/2)^2+(px/3)^3))^(1/3)+(-qx/2-sqrt((qx/2)^2+(px/3)^3))^(1/3)-b0/(3*a0);
tempk2=((-1+i*sqrt(3))/2)*(-qx/2+sqrt((qx/2)^2+(px/3)^3))^(1/3)+((-1-i*sqrt(3))/2)*(-qx/2-sqrt((qx/2)^2+(px/3)^3))^
(1/3)-b0/(3*a0);
tempk3=((-1-i*sqrt(3))/2)*(-qx/2+sqrt((qx/2)^2+(px/3)^3))^(1/3)+((-1+i*sqrt(3))/2)*(-qx/2-sqrt((qx/2)^2+(px/3)^3))^
(1/3)-b0/(3*a0);

K2(n)=sqrt(abs(tempk1));    %bending propagating
K3(n)=sqrt(abs(tempk2));    %bending decaying
K1(n)=sqrt(abs(tempk3));    %Torsion

  k1=K1(n);
  k2=K2(n);
  k3=K3(n);

  P1=K/EI*k1^3/(k1^4-kb^4);
  P2=K/EI*k2^3/(k2^4-kb^4);
  P3=-K/EI*k3^3/(k3^4-kb^4);

%Reflection matrix of clamped end
xc=[1,1,1;i*P1,i*P2,P3;P1*k1,P2*k2,-P3*k3];
yc=[-1,-1,-1;i*P1,i*P2,P3;-P1*k1,-P2*k2,P3*k3];
rclamped=inv(xc)*yc;

%Reflection matrix of free end
xf=[K/EI*i*k1-i*P1*k1^2,K/EI*i*k2-i*P2*k2^2,K/EI*k3+P3*k3^2;-K/EI*k1^2+P1*k1^3,-K/EI*k2^2+P2*k2^3,
K/EI*k3^2+P3*k3^3;-K/GJ*i*P1*k1^2+i*k1,-K/GJ*i*P2*k2^2+i*k2,K/GJ*P3*k3^2+k3];
yf=[K/EI*i*k1-i*P1*k1^2,K/EI*i*k2-i*P2*k2^2,K/EI*k3+P3*k3^2;-(-K/EI*k1^2+P1*k1^3),-(-K/EI*k2^2+P2*k2^3),
-(K/EI*k3^2+P3*k3^3);-K/GJ*i*P1*k1^2+i*k1,-K/GJ*i*P2*k2^2+i*k2,K/GJ*P3*k3^2+k3];
rfree=inv(xf)*yf;

%Reflection at a simply-supported boundary
xs=[1,1,1;-i*P1,-i*P2,-P3;-i*P1*k1^2+i*K*k1/EI,-i*P2*k2^2+i*K*k2/EI,P3*k3^2+K*k3/EI];
ys=[-1,-1,-1;-i*P1,-i*P2,-P3;-i*P1*k1^2+i*K*k1/EI,-i*P2*k2^2+i*K*k2/EI,P3*k3^2+K*k3/EI];
rpinnned=inv(xs)*ys;

%Cantilever beam
ra=rclamped;
rb=rfree;
```

```
fL=[exp(-i*k1*L),0,0;0,exp(-i*k2*L),0;0,0,exp(-k3*L)];

% Forming the characteristic matrix

BigA=zeros(12,12);

%Boundary reflections
RefA=[-eye(3) ra];
RefB=[rb -eye(3)];

%Propagations along L
PropL=[fL zeros(3) -eye(3) zeros(3); zeros(3) -eye(3) zeros(3) fL];

% Wave component sequence a1+ a2+ a1− a2− b1+ b2+ b1− b2−
BigA(1:3,1:6)=RefA;
BigA(4:6,7:12)=RefB;
BigA(7:12,1:12)=PropL;

result(n)=det(BigA);

end

figure(1),plot(freq,abs(result),'r','LineWidth',2), hold on,grid on
xlabel('Frequency (Hz)'),ylabel('Magnitude')

figure(2),plot(freq,20*log10(abs(result)),'r','LineWidth',2), hold on,grid on
xlabel('Frequency (Hz)'),ylabel('dB Magnitude')

figure(3)
plot(freq/2/pi,real(result),'r','LineWidth',2), hold on
plot(freq/2/pi,imag(result),':r','LineWidth',2), hold on,grid on
xlabel('Frequency (Hz)'),ylabel('Magnitude')
legend('Real part','Imaginary part')
```

References

Lottati I.Flutter and Divergence Aeroelastic Characteristics for Composite Forward Swept Cantilevered Wing, *Journal of Aircraft*, 22, 1001–1007 (1985).

The Mathematical Handbook Editorial Group, Mathematical Handbook, Higher Education Publications, Beijing (1979).

Mei C. Effect of Material Coupling on Wave Vibration of Composite Euler–Bernoulli Beam Structures, *Journal of Sound and Vibration*, 288, 177–193 (2005a).

Mei C. Free and Forced Wave Vibration Analysis of Axially Loaded Materially Coupled Composite Timoshenko Beam Structures, *ASME Journal of Vibration and Acoustics*, 127(6), 519–529 (2005b).

Homework Project

Free Wave Vibration Analysis of a Composite Beam with an Intermediate Spring Attachment

The figures below show a uniform cantilever composite beam with a translational spring attached to its span. This discontinuity alters the propagation path of vibration waves in the beam. The goal for this project is to find out if this added discontinuity changes the natural frequencies of the uniform composite beam.

The transmission and reflection matrices **t** and **r** at the spring attached intermediate discontinuity have been derived in this chapter. Note that $K_{R\theta} = 0$ and $K_{R\psi} = 0$ because the attachment is a translational spring.

Wave relationships at the spring attached point on the uniform composite beam are

$$\mathbf{c}_1^- = \mathbf{r}\mathbf{c}_1^+ + \mathbf{t}\mathbf{c}_2^-, \mathbf{c}_2^+ = \mathbf{r}\mathbf{c}_2^- + \mathbf{t}\mathbf{c}_1^+$$

At Boundaries A and B, the incident and reflected waves are related by their corresponding reflection matrices,

$$\mathbf{a}^+ = \mathbf{r}_A\mathbf{a}^-, \mathbf{b}^- = \mathbf{r}_B\mathbf{b}^+$$

and the propagation relations along the uniform beam segments AC and CB are

$$\mathbf{c}_1^+ = \mathbf{f}(L_1)\,\mathbf{a}^+, \mathbf{a}^- = \mathbf{f}(L_1)\,\mathbf{c}_1^-,$$

$$\mathbf{c}_2^- = \mathbf{f}(L_2)\,\mathbf{b}^-, \mathbf{b}^+ = \mathbf{f}(L_2)\,\mathbf{c}_2^+$$

where $\mathbf{f}(L_1)$ and $\mathbf{f}(L_2)$ are the propagation matrices along segments AC and CB, respectively.

The material and geometrical properties of the example composite beam are as follows: the Young's modulus, shear modulus, volume mass density, and coupling coefficient are $E = 8.418\ GN/m^2$, $G = 0.3278\ GN/m^2$, $\rho = 1347.0\ kg/m^3$, and $K = 0.1143\ Nm^2$, respectively. The cross section of the beam is a rectangular shape whose width is $12.7\ mm$ and thickness is $3.18\ mm$. The length of the cantilever beam is $L = 190.5\ mm$. The point where the translational spring is attached is $0.4\ L$ measured from the clamped boundary.

Perform free vibration analysis by assembling the above wave relationships. Write MATLAB scripts to obtain the overlaid magnitudes of the characteristic polynomials of the composite beam before and after the attachment of a translational spring, corresponding to $K_T = 10^5\ N/m$ and $K_T = 10^7\ N/m$, respectively.

Does this added discontinuity change the natural frequencies of the uniform composite beam?

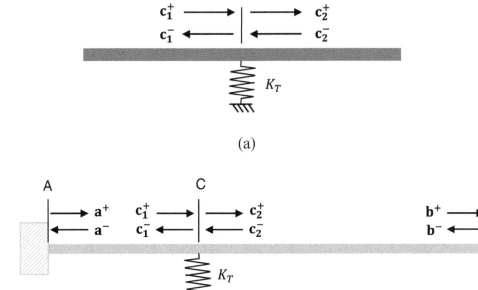

(a)

(b)

6

Coupled Waves in Curved Beams

In a curved beam, coupled vibration motions exist along the radial and tangential directions. These coupled in-plane vibrations are analyzed based on Love's vibration theory. Cut-off frequencies and dispersion relationships are studied. Wave reflections at classical and non-classical boundaries are derived. Free and forced vibrations are analyzed from the wave vibration standpoint. Natural frequencies, modeshapes, as well as steady state frequency responses, are obtained. MATLAB scripts are provided.

6.1 The Governing Equations and the Propagation Relationships

For a uniform curved beam of constant radius of curvature R, the governing equations of motion for free in-plane vibration according to Love's vibration theory are (Walsh and White 2000),

$$EI\left(\frac{1}{R}\frac{\partial^3 u(s,t)}{\partial s^3} - \frac{\partial^4 w(s,t)}{\partial s^4}\right) - \frac{EA}{R}\left(\frac{w}{R} + \frac{\partial u(s,t)}{\partial s}\right) = \rho A \frac{\partial^2 w(s,t)}{\partial t^2}$$

$$EA\left(\frac{1}{R}\frac{\partial w(s,t)}{\partial s} + \frac{\partial^2 u(s,t)}{\partial s^2}\right) + \frac{EI}{R}\left(\frac{1}{R}\frac{\partial^2 u(s,t)}{\partial s^2} - \frac{\partial^3 w(s,t)}{\partial s^3}\right) = \rho A \frac{\partial^2 u(s,t)}{\partial t^2} \quad (6.1)$$

where s is the circumferential coordinate of the centerline of the curved beam, t is the time, A is the cross-sectional area, I is the area moment of inertia of the cross section, E is the Young's modulus, and ρ is the volume mass density. $w(s,t)$ and $u(s,t)$ denote the deflections of the centerline of the beam in the radial and tangential directions, respectively. The governing equations of motion can also be written by choosing the central angle θ of the curved beam as the coordinate, where $\theta = \frac{s}{R}$ (Kang et. al. 2003).

Assuming time harmonic motion and using separation of variables, the solutions to Eq. (6.1) can be written in the form $w(s,t) = w_0 e^{-iks} e^{i\omega t}$ and $u(s,t) = u_0 e^{-iks} e^{i\omega t}$, where i is the imaginary unit, ω is the circular frequency, and k is the wavenumber. w_0 and u_0 are the amplitudes of deflections of the centerline of the beam in the radial and tangential directions, respectively. Substituting these expressions for $w(s,t)$, and $u(s,t)$ into Eq. (6.1) and putting the equations into matrix form,

$$\begin{bmatrix} -EIk^4 - \dfrac{EA}{R^2} + \rho A \omega^2 & \dfrac{ik^3 EI}{R} + \dfrac{ikEA}{R} \\ -\dfrac{ik^3 EI}{R} - \dfrac{ikEA}{R} & -EAk^2 - \dfrac{k^2 EI}{R^2} + \rho A \omega^2 \end{bmatrix} \begin{bmatrix} w_0 \\ u_0 \end{bmatrix} = 0 \quad (6.2)$$

Setting the determinant of the coefficient matrix of Eq. (6.2) to zero gives the dispersion equation, which is a cubic equation in terms of k^2,

$$ak^6 + bk^4 + ck^2 + d = 0 \quad (6.3)$$

Mechanical Wave Vibrations: Analysis and Control, First Edition. Chunhui Mei.
© 2023 Chunhui Mei. Published 2023 by John Wiley & Sons Ltd.
Companion Website: www.wiley.com/go/Mei/MechanicalWaveVibrations

where $a = 1$, $b = -\dfrac{\rho}{E}\omega^2 - \dfrac{2}{R^2}$, $c = \dfrac{1}{R^4} - \dfrac{\rho}{E}\dfrac{A}{I}\omega^2 - \dfrac{1}{R^2}\dfrac{\rho^2}{E^2}\omega^2$, and $d = -\dfrac{1}{R^2}\dfrac{\rho}{E}\dfrac{A}{I}\omega^2 + \dfrac{\rho^2}{E^2}\dfrac{A}{I}\omega^4$.

Under the special situation where the radius of curvature R approaches infinity, the curved beam becomes a straight beam, and the radial and tangential vibrations are decoupled. The dispersion equation becomes $\left(k^2 - k_b^2\right)\left(k^2 + k_b^2\right)\left(k^2 - k_l^2\right) = 0$, where $k_b = \sqrt[4]{\dfrac{\rho A}{EI}\omega^2}$ and $k_l = \sqrt{\dfrac{\rho}{E}}\omega$ are wavenumbers of bending and longitudinal waves in a straight beam by the Euler–Bernoulli bending and Elementary longitudinal vibration theories, respectively.

Under the general situation, however, values of the three wavenumbers k_1, k_2, and k_3 obtained from the dispersion equation may be real, imaginary, or complex, depending on the frequency range. In other words, wave mode transitions exist. The cut-off frequencies are associated with the ring frequency and frequencies corresponding to the roots of the discriminant Δ of the dispersion equation obtained in Eq. (6.3),

$$\Delta = 18abcd - 4b^3d + b^2c^2 - 4ac^3 - 27a^2d^2 = 0 \tag{6.4}$$

The ring frequency is defined as the nonzero frequency corresponding to wavenumber k being zero in the dispersion equation. This cut-off frequency is obtained analytically by setting k to zero in Eq. (6.3) as $\omega_c = \dfrac{1}{R}\sqrt{\dfrac{E}{\rho}}$. The cut-off frequencies associated with the roots of the discriminant equation obtained in Eq. (6.4) can be found numerically.

With time dependence $e^{i\omega t}$ suppressed, the solutions to Eq. (6.1) can be written as

$$w(s) = a_1^+ e^{-ik_1 s} + a_2^+ e^{-ik_2 s} + a_3^+ e^{-ik_3 s} + a_1^- e^{ik_1 s} + a_2^- e^{ik_2 s} + a_3^- e^{ik_3 s},$$

$$u(s) = \overline{a_1^+} e^{-ik_1 s} + \overline{a_2^+} e^{-ik_2 s} + \overline{a_3^+} e^{-ik_3 s} + \overline{a_1^-} e^{ik_1 s} + \overline{a_2^-} e^{ik_2 s} + \overline{a_3^-} e^{ik_3 s} \tag{6.5}$$

Superscripts + and − and subscripts 1, 2, and 3 in wave amplitude denote the three pairs of positive- and negative-going wave components.

The wave amplitudes a of $w(s,t)$ and $\bar{a}$ of $u(s,t)$ are related to each other, and their relationships are found from Eq. (6.2),

$$\frac{\overline{a_1^+}}{a_1^+} = P_1, \quad \frac{\overline{a_2^+}}{a_2^+} = P_2, \quad \frac{\overline{a_3^+}}{a_3^+} = P_3, \quad \frac{\overline{a_1^-}}{a_1^-} = -P_1, \quad \frac{\overline{a_2^-}}{a_2^-} = -P_2, \quad \frac{\overline{a_3^-}}{a_3^-} = -P_3 \tag{6.6a}$$

where coefficients P_n ($n = 1, 2, 3$) are

$$P_n = \frac{k_n^4 EIR^2 + EA - \rho AR^2\omega^2}{ik_n^3 EIR + ik_n EAR}, \quad \text{or equivalently} \quad P_n = \frac{ik_n^3 EIR + ik_n EAR}{-k_n^2 EAR^2 - k_n^2 EI + \rho AR^2\omega^2} \tag{6.6b}$$

Consequently, Eq. (6.5) can be rewritten as

$$w(s) = a_1^+ e^{-ik_1 s} + a_2^+ e^{-ik_2 s} + a_3^+ e^{-ik_3 s} + a_1^- e^{ik_1 s} + a_2^- e^{ik_2 s} + a_3^- e^{ik_3 s},$$

$$u(s) = P_1 a_1^+ e^{-ik_1 s} + P_2 a_2^+ e^{-ik_2 s} + P_3 a_3^+ e^{-ik_3 s} - P_1 a_1^- e^{ik_1 s} - P_2 a_2^- e^{ik_2 s} - P_3 a_3^- e^{ik_3 s} \tag{6.7}$$

Consider two points A and B on a curved beam corresponding to a span angle θ and over an arc distance s apart, as shown in Figure 6.1. Waves propagate from one point to the other, with the propagation relationships determined by the appropriate wavenumber. The positive- and negative-going waves at Points A and B are related by

$$\mathbf{b}^+ = \mathbf{f}(s)\mathbf{a}^+, \quad \mathbf{a}^- = \mathbf{f}(s)\mathbf{b}^- \tag{6.8a}$$

where

$$\mathbf{a}^\pm = \begin{bmatrix} a_1^\pm \\ a_2^\pm \\ a_3^\pm \end{bmatrix}, \quad \mathbf{b}^\pm = \begin{bmatrix} b_1^\pm \\ b_2^\pm \\ b_3^\pm \end{bmatrix} \tag{6.8b}$$

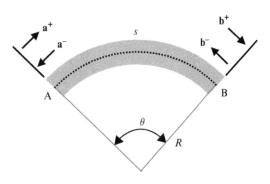

Figure 6.1 Wave propagation relationships.

are the wave vectors and

$$\mathbf{f}(s) = \begin{bmatrix} e^{-ik_1 s} & 0 & 0 \\ 0 & e^{-ik_2 s} & 0 \\ 0 & 0 & e^{-ik_3 s} \end{bmatrix}$$

(6.8c)

is the propagation matrix for an arc distance s.

6.2 Wave Reflection at Classical and Non-Classical Boundaries

By the sign convention defined in Figure 6.2, the internal resistant tangential force $N(s,t)$, radial force $V(s,t)$, and bending moment $M(s,t)$ are related to the displacements $w(s,t)$ and $u(s,t)$ of the centerline of the curved beam in the radial and tangential directions, as well as the rotation $\varphi(s,t)$ of the cross section of the curved beam due to the bending as follows

$$N(s,t) = EA\left(\frac{w(s,t)}{R} + \frac{\partial u(s,t)}{\partial s}\right)$$

$$M(s,t) = EI\frac{\partial \varphi(s,t)}{\partial s} = EI\left(-\frac{1}{R}\frac{\partial u(s,t)}{\partial s} + \frac{\partial^2 w(s,t)}{\partial s^2}\right)$$

(6.9a)

$$V(s,t) = -\frac{\partial M(s,t)}{\partial s} = EI\left(\frac{1}{R}\frac{\partial^2 u(s,t)}{\partial s^2} - \frac{\partial^3 w(s,t)}{\partial s^3}\right)$$

and

$$\varphi(s,t) = -\frac{u(s,t)}{R} + \frac{\partial w(s,t)}{\partial s}$$

(6.9b)

by Love's curved beam theory.

The reflection matrices at classical free, clamped, and simply-supported boundaries, as well as at non-classical boundaries with spring and mass attachments, are derived.

At a boundary, incident waves are reflected, as shown in Figure 6.3. The incident waves $\mathbf{a}^+$ and reflected waves $\mathbf{a}^-$ are related by reflection matrix $\mathbf{r}$

$$\mathbf{a}^- = \mathbf{r}\mathbf{a}^+$$

(6.10)

where $\mathbf{r}$ is determined by boundary conditions.

a) Free boundary

At a free boundary, the equilibrium conditions are

$$N(s,t) = 0, \ M(s,t) = 0, \ V(s,t) = 0$$

(6.11)

From Eqs. (6.7), (9.9a), and (6.11),

$$EA\left(\sum_{n=1}^{3}\left(\frac{1}{R} - ik_n P_n\right)a_n^+ e^{-ik_n s} + \sum_{n=1}^{3}\left(\frac{1}{R} - ik_n P_n\right)a_n^- e^{ik_n s}\right) = 0$$

$$-EI\left(\sum_{n=1}^{3}\left(-\frac{ik_n P_n}{R} + k_n^2\right)a_n^+ e^{-ik_n s} + \sum_{n=1}^{3}\left(-\frac{ik_n P_n}{R} + k_n^2\right)a_n^- e^{ik_n s}\right) = 0$$

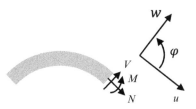

Figure 6.2 Definition of positive internal resistant tangential force, radial force, and bending moment.

$$EI\left(\sum_{n=1}^{3}\left(\frac{P_n k_n^2}{R}+ik_n^3\right)a_n^+ e^{-ik_n s}-\sum_{n=1}^{3}\left(\frac{P_n k_n^2}{R}+ik_n^3\right)a_n^- e^{ik_n s}\right)=0$$

(6.12)

Choosing the origin at the boundary and assembling Eq. (6.12) into matrix form,

$$\begin{bmatrix}\frac{1}{R}-ik_1 P_1 & \frac{1}{R}-ik_2 P_2 & \frac{1}{R}-ik_3 P_3 \\ \frac{ik_1 P_1}{R}-k_1^2 & \frac{ik_2 P_2}{R}-k_2^2 & \frac{ik_3 P_3}{R}-k_3^2 \\ \frac{P_1 k_1^2}{R}+ik_1^3 & \frac{P_2 k_2^2}{R}+ik_2^3 & \frac{P_3 k_3^2}{R}+ik_3^3\end{bmatrix}\begin{bmatrix}a_1^+ \\ a_2^+ \\ a_3^+\end{bmatrix}-\begin{bmatrix}-\left(\frac{1}{R}-ik_1 P_1\right) & -\left(\frac{1}{R}-ik_2 P_2\right) & -\left(\frac{1}{R}-ik_3 P_3\right) \\ -\left(\frac{ik_1 P_1}{R}-k_1^2\right) & -\left(\frac{ik_2 P_2}{R}-k_2^2\right) & -\left(\frac{ik_3 P_3}{R}-k_3^2\right) \\ \frac{P_1 k_1^2}{R}+ik_1^3 & \frac{P_2 k_2^2}{R}+ik_2^3 & \frac{P_3 k_3^2}{R}+ik_3^3\end{bmatrix}\begin{bmatrix}a_1^- \\ a_2^- \\ a_3^-\end{bmatrix}=0$$

(6.13)

Solving the reflected waves $\mathbf{a}^-$ in terms of the incident waves $\mathbf{a}^+$, from Eqs. (6.8b) and (6.13),

$$\mathbf{a}^-=\begin{bmatrix}-\left(\frac{1}{R}-ik_1 P_1\right) & -\left(\frac{1}{R}-ik_2 P_2\right) & -\left(\frac{1}{R}-ik_3 P_3\right) \\ -\left(\frac{ik_1 P_1}{R}-k_1^2\right) & -\left(\frac{ik_2 P_2}{R}-k_2^2\right) & -\left(\frac{ik_3 P_3}{R}-k_3^2\right) \\ \frac{P_1 k_1^2}{R}+ik_1^3 & \frac{P_2 k_2^2}{R}+ik_2^3 & \frac{P_3 k_3^2}{R}+ik_3^3\end{bmatrix}^{-1}\begin{bmatrix}\frac{1}{R}-ik_1 P_1 & \frac{1}{R}-ik_2 P_2 & \frac{1}{R}-ik_3 P_3 \\ \frac{ik_1 P_1}{R}-k_1^2 & \frac{ik_2 P_2}{R}-k_2^2 & \frac{ik_3 P_3}{R}-k_3^2 \\ \frac{P_1 k_1^2}{R}+ik_1^3 & \frac{P_2 k_2^2}{R}+ik_2^3 & \frac{P_3 k_3^2}{R}+ik_3^3\end{bmatrix}\mathbf{a}^+$$

(6.14)

From Eqs. (6.10) and (6.14), the reflection matrix at a free boundary is

$$\mathbf{r}=\begin{bmatrix}-\left(\frac{1}{R}-ik_1 P_1\right) & -\left(\frac{1}{R}-ik_2 P_2\right) & -\left(\frac{1}{R}-ik_3 P_3\right) \\ -\left(\frac{ik_1 P_1}{R}-k_1^2\right) & -\left(\frac{ik_2 P_2}{R}-k_2^2\right) & -\left(\frac{ik_3 P_3}{R}-k_3^2\right) \\ \frac{P_1 k_1^2}{R}+ik_1^3 & \frac{P_2 k_2^2}{R}+ik_2^3 & \frac{P_3 k_3^2}{R}+ik_3^3\end{bmatrix}^{-1}\begin{bmatrix}\frac{1}{R}-ik_1 P_1 & \frac{1}{R}-ik_2 P_2 & \frac{1}{R}-ik_3 P_3 \\ \frac{ik_1 P_1}{R}-k_1^2 & \frac{ik_2 P_2}{R}-k_2^2 & \frac{ik_3 P_3}{R}-k_3^2 \\ \frac{P_1 k_1^2}{R}+ik_1^3 & \frac{P_2 k_2^2}{R}+ik_2^3 & \frac{P_3 k_3^2}{R}+ik_3^3\end{bmatrix}$$

(6.15)

b) Clamped boundary

The boundary conditions at a clamped boundary are

$$w(s,t)=0,\ u(s,t)=0,\ \varphi(s,t)=0$$

(6.16)

From Eqs. (6.7) and (6.9b),

$$\sum_{n=1}^{3}a_n^+ e^{-ik_n s}+\sum_{n=1}^{3}a_n^- e^{ik_n s}=0$$

$$\sum_{n=1}^{3}P_n a_n^+ e^{-ik_n s}-\sum_{n=1}^{3}P_n a_n^- e^{ik_n s}=0$$

$$\sum_{n=1}^{3}\left(-\frac{P_n}{R}-ik_n\right)a_n^+ e^{-ik_n s}+\sum_{n=1}^{3}\left(\frac{P_n}{R}+ik_n\right)a_n^- e^{ik_n s}=0$$

(6.17)

Choosing the origin at the boundary and assembling Eq. (6.17) into matrix form,

$$\begin{bmatrix}1 & 1 & 1 \\ P_1 & P_2 & P_3 \\ -\frac{P_1}{R}-ik_1 & -\frac{P_2}{R}-ik_2 & -\frac{P_3}{R}-ik_3\end{bmatrix}\begin{bmatrix}a_1^+ \\ a_2^+ \\ a_3^+\end{bmatrix}-\begin{bmatrix}-1 & -1 & -1 \\ P_1 & P_2 & P_3 \\ -\frac{P_1}{R}-ik_1 & -\frac{P_2}{R}-ik_2 & -\frac{P_3}{R}-ik_3\end{bmatrix}\begin{bmatrix}a_1^- \\ a_2^- \\ a_3^-\end{bmatrix}=0$$

(6.18)

Solving for the reflected waves $\mathbf{a}^-$ in terms of the incident waves $\mathbf{a}^+$, from Eqs. (6.8b) and (6.18),

$$
\mathbf{a}^- = \begin{bmatrix} -1 & -1 & -1 \\ P_1 & P_2 & P_3 \\ -\dfrac{P_1}{R} - ik_1 & -\dfrac{P_2}{R} - ik_2 & -\dfrac{P_3}{R} - ik_3 \end{bmatrix}^{-1} \begin{bmatrix} 1 & 1 & 1 \\ P_1 & P_2 & P_3 \\ \dfrac{P_1}{R} - ik_1 & \dfrac{P_2}{R} - ik_2 & \dfrac{P_3}{R} - ik_3 \end{bmatrix} \mathbf{a}^+
\tag{6.19}
$$

From Eqs. (6.10) and (6.19), the reflection matrix at a clamped boundary is

$$
\mathbf{r} = \begin{bmatrix} -1 & -1 & -1 \\ P_1 & P_2 & P_3 \\ -\dfrac{P_1}{R} - ik_1 & -\dfrac{P_2}{R} - ik_2 & -\dfrac{P_3}{R} - ik_3 \end{bmatrix}^{-1} \begin{bmatrix} 1 & 1 & 1 \\ P_1 & P_2 & P_3 \\ \dfrac{P_1}{R} - ik_1 & \dfrac{P_2}{R} - ik_2 & \dfrac{P_3}{R} - ik_3 \end{bmatrix}
\tag{6.20}
$$

c) Simply-supported boundary

The boundary conditions at a simply-supported boundary are

$$
w(s,t) = 0, \; u(s,t) = 0, \; M(s,t) = 0
\tag{6.21}
$$

From Eqs. (6.7) and (6.9a),

$$
\sum_{n=1}^{3} a_n^+ e^{-ik_n s} + \sum_{n=1}^{3} a_n^- e^{ik_n s} = 0
$$

$$
\sum_{n=1}^{3} P_n a_n^+ e^{-ik_n s} - \sum_{n=1}^{3} P_n a_n^- e^{ik_n s} = 0
$$

$$
-EI\left(\sum_{n=1}^{3} \left(-\frac{ik_n P_n}{R} + k_n^2 \right) a_n^+ e^{-ik_n s} + \sum_{n=1}^{3} \left(-\frac{ik_n P_n}{R} + k_n^2 \right) a_n^- e^{ik_n s} \right) = 0
\tag{6.22}
$$

Choosing the origin at the boundary and assembling Eq. (6.22) into matrix form,

$$
\begin{bmatrix} 1 & 1 & 1 \\ P_1 & P_2 & P_3 \\ \dfrac{ik_1 P_1}{R} - k_1^2 & \dfrac{ik_2 P_2}{R} - k_2^2 & \dfrac{ik_3 P_3}{R} - k_3^2 \end{bmatrix} \begin{bmatrix} a_1^+ \\ a_2^+ \\ a_3^+ \end{bmatrix} - \begin{bmatrix} -1 & -1 & -1 \\ P_1 & P_2 & P_3 \\ -\dfrac{ik_1 P_1}{R} + k_1^2 & -\dfrac{ik_2 P_2}{R} + k_2^2 & -\dfrac{ik_3 P_3}{R} + k_3^2 \end{bmatrix} \begin{bmatrix} a_1^- \\ a_2^- \\ a_3^- \end{bmatrix} = 0
\tag{6.23}
$$

Solving the reflected waves $\mathbf{a}^-$ in terms of the incident waves $\mathbf{a}^+$, from Eqs. (6.8b), (6.10), and (6.23), the reflection matrix at a simply-supported boundary is

$$
\mathbf{r} = \begin{bmatrix} -1 & -1 & -1 \\ P_1 & P_2 & P_3 \\ -\dfrac{ik_1 P_1}{R} + k_1^2 & -\dfrac{ik_2 P_2}{R} + k_2^2 & -\dfrac{ik_3 P_3}{R} + k_3^2 \end{bmatrix}^{-1} \begin{bmatrix} 1 & 1 & 1 \\ P_1 & P_2 & P_3 \\ \dfrac{ik_1 P_1}{R} - k_1^2 & \dfrac{ik_2 P_2}{R} - k_2^2 & \dfrac{ik_3 P_3}{R} - k_3^2 \end{bmatrix}
\tag{6.24}
$$

d) Boundary with spring attachments

Figure 6.3 shows a boundary with spring attachments. The radial translational, tangential translational, and rotational stiffnesses are K_w, K_u, and K_r, respectively.

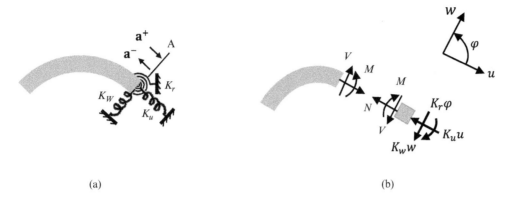

Figure 6.3 (a) Boundary with spring attachments and (b) free body diagram.

From the free body diagram of Figure 6.3,

$$
\begin{aligned}
-N(s,t) - K_u u(s,t) &= 0 \\
-V(s,t) - K_w w(s,t) &= 0 \\
-M(s,t) - K_r \varphi(s,t) &= 0
\end{aligned}
\tag{6.25}
$$

From Eqs. (6.7), (6.9a), (6.9b), and (6.25),

$$
\sum_{n=1}^{3}\left(-K_u P_n - EA\left(\frac{1}{R} - ik_n P_n\right)\right)a_n^+ e^{-ik_n s} + \sum_{n=1}^{3}\left(K_u P_n - EA\left(\frac{1}{R} - ik_n P_n\right)\right)a_n^- e^{ik_n s} = 0
$$

$$
\sum_{n=1}^{3}\left(K_r\left(\frac{P_n}{R} + ik_n\right) + EI\left(-\frac{ik_n P_n}{R} + k_n^2\right)\right)a_n^+ e^{-ik_n s} + \sum_{n=1}^{3}\left(-K_r\left(\frac{P_n}{R} + ik_n\right) + EI\left(-\frac{ik_n P_n}{R} + k_n^2\right)\right)a_n^- e^{ik_n s} = 0
$$

$$
\sum_{n=1}^{3}\left(-K_w + EI\left(\frac{P_n k_n^2}{R} + ik_n^3\right)\right)a_n^+ e^{-ik_n s} + \sum_{n=1}^{3}\left(-K_w - EI\left(\frac{P_n k_n^2}{R} + ik_n^3\right)\right)a_n^- e^{ik_n s} = 0
\tag{6.26}
$$

Choosing the origin at the boundary and assembling Eq. (6.26) into matrix form,

$$
\begin{bmatrix}
-K_u P_1 - EA\left(\frac{1}{R} - ik_1 P_1\right) & -K_u P_2 - EA\left(\frac{1}{R} - ik_2 P_2\right) & -K_u P_3 - EA\left(\frac{1}{R} - ik_3 P_3\right) \\
K_r\left(\frac{P_1}{R} + ik_1\right) + EI\left(-\frac{ik_1 P_1}{R} + k_1^2\right) & K_r\left(\frac{P_2}{R} + ik_2\right) + EI\left(-\frac{ik_2 P_2}{R} + k_2^2\right) & K_r\left(\frac{P_3}{R} + ik_3\right) + EI\left(-\frac{ik_3 P_3}{R} + k_3^2\right) \\
-K_w + EI\left(\frac{P_1 k_1^2}{R} + ik_1^3\right) & -K_w + EI\left(\frac{P_2 k_2^2}{R} + ik_2^3\right) & -K_w + EI\left(\frac{P_3 k_3^2}{R} + ik_3^3\right)
\end{bmatrix}
\begin{bmatrix} a_1^+ \\ a_2^+ \\ a_3^+ \end{bmatrix} -
$$

$$
\begin{bmatrix}
-K_u P_1 + EA\left(\frac{1}{R} - ik_1 P_1\right) & -K_u P_2 + EA\left(\frac{1}{R} - ik_2 P_2\right) & -K_u P_3 + EA\left(\frac{1}{R} - ik_3 P_3\right) \\
K_r\left(\frac{P_1}{R} + ik_1\right) - EI\left(-\frac{ik_1 P_1}{R} + k_1^2\right) & K_r\left(\frac{P_2}{R} + ik_2\right) - EI\left(-\frac{ik_2 P_2}{R} + k_2^2\right) & K_r\left(\frac{P_3}{R} + ik_3\right) - EI\left(-\frac{ik_3 P_3}{R} + k_3^2\right) \\
K_w + EI\left(\frac{P_1 k_1^2}{R} + ik_1^3\right) & K_w + EI\left(\frac{P_2 k_2^2}{R} + ik_2^3\right) & K_w + EI\left(\frac{P_3 k_3^2}{R} + ik_3^3\right)
\end{bmatrix}
\begin{bmatrix} a_1^- \\ a_2^- \\ a_3^- \end{bmatrix} = 0
\tag{6.27}
$$

From Eqs. (6.8b), (6.10), and (6.27), the reflection matrix at a spring attached boundary is

$$\mathbf{r} = \begin{bmatrix} -K_u P_1 + EA\left(\dfrac{1}{R} - ik_1 P_1\right) & -K_u P_2 + EA\left(\dfrac{1}{R} - ik_2 P_2\right) & -K_u P_3 + EA\left(\dfrac{1}{R} - ik_3 P_3\right) \\[2mm] K_r\left(\dfrac{P_1}{R} + ik_1\right) - EI\left(-\dfrac{ik_1 P_1}{R} + k_1^2\right) & K_r\left(\dfrac{P_2}{R} + ik_2\right) - EI\left(-\dfrac{ik_2 P_2}{R} + k_2^2\right) & K_r\left(\dfrac{P_3}{R} + ik_3\right) - EI\left(-\dfrac{ik_3 P_3}{R} + k_3^2\right) \\[2mm] K_w + EI\left(\dfrac{P_1 k_1^2}{R} + ik_1^3\right) & K_w + EI\left(\dfrac{P_2 k_2^2}{R} + ik_2^3\right) & K_w + EI\left(\dfrac{P_3 k_3^2}{R} + ik_3^3\right) \end{bmatrix}^{-1} \times$$

$$\begin{bmatrix} -K_u P_1 - EA\left(\dfrac{1}{R} - ik_1 P_1\right) & -K_u P_2 - EA\left(\dfrac{1}{R} - ik_2 P_2\right) & -K_u P_3 - EA\left(\dfrac{1}{R} - ik_3 P_3\right) \\[2mm] K_r\left(\dfrac{P_1}{R} + ik_1\right) + EI\left(-\dfrac{ik_1 P_1}{R} + k_1^2\right) & K_r\left(\dfrac{P_2}{R} + ik_2\right) + EI\left(-\dfrac{ik_2 P_2}{R} + k_2^2\right) & K_r\left(\dfrac{P_3}{R} + ik_3\right) + EI\left(-\dfrac{ik_3 P_3}{R} + k_3^2\right) \\[2mm] -K_w + EI\left(\dfrac{P_1 k_1^2}{R} + ik_1^3\right) & -K_w + EI\left(\dfrac{P_2 k_2^2}{R} + ik_2^3\right) & -K_w + EI\left(\dfrac{P_3 k_3^2}{R} + ik_3^3\right) \end{bmatrix} \tag{6.28}$$

MATLAB symbolic scripts for reflection matrix at an elastic boundary are available in Section 6.6.

Under the special situation where K_u, K_w, and K_r approach zero or infinity, the spring attached end is physically equivalent to one of the classical boundaries: for a free boundary, $K_u = 0$, $K_w = 0$, and $K_r = 0$; for a clamped boundary, $K_u = \infty$, $K_w = \infty$, and $K_r = \infty$; and for a simply-supported boundary, $K_u = \infty$, $K_w = \infty$, and $K_r = 0$.

e) Boundary with a mass attachment

Figure 6.4 shows a curved beam with a lumped mass and its free body diagram. Applying Newton's second law for translational and rotational motion to the lumped mass,

$$-N(s,t) = m\frac{\partial u_m^2(s,t)}{\partial t^2}, \quad -V(s,t) = m\frac{\partial w_m^2(s,t)}{\partial t^2}, \quad -M(s,t) + V(s,t)\frac{h}{2} = J_m\frac{\partial \varphi_m^2(s,t)}{\partial t^2} \tag{6.29}$$

where m is the mass, h is the width of the attached mass block, and J_m is the mass moment of inertia of the mass block about an axis passing its center of mass and normal to the plane of the curved beam. $w_m(x,t)$ and $u_m(x,t)$ are the displacements of the end mass along the radial and tangential directions, respectively. φ_m is the rotation of the mass block. $w_m(x,t)$, $u_m(x,t)$, and φ_m are related to the displacements of the centerline of the beam in the radial and tangential directions and the bending slope of the beam at the point where the mass block is attached as follows

$$w_m = w + \frac{h}{2}\varphi_m, \quad u_m = u, \quad \varphi_m = \varphi \tag{6.30}$$

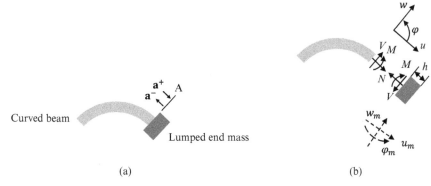

(a) (b)

Figure 6.4 (a) Boundary with a mass attachment and (b) free body diagram.

where the end mass block is modeled as a rigid body.

From Eqs. (6.7), (6.9a), (6.9b), (6.29), and (6.30),

$$\sum_{n=1}^{3}\left(mw^2 P_n + EA\left(-\frac{1}{R} + ik_n P_n\right)\right)a_n^+ e^{-ik_n s} + \sum_{n=1}^{3}\left(-mw^2 P_n + EA\left(-\frac{1}{R} + ik_n P_n\right)\right)a_n^- e^{ik_n s} = 0$$

$$\sum_{n=1}^{3}\left(-J_m w^2\left(\frac{P_n}{R} + ik_n\right) + EI\left(-\frac{ik_n P_n}{R} + k_n^2\right) - EI\frac{h}{2}\left(\frac{P_n k_n^2}{R} + ik_n^3\right)\right)a_n^+ e^{-ik_n s}$$

$$+\sum_{n=1}^{3}\left(J_m w^2\left(\frac{P_n}{R} + ik_n\right) + EI\left(-\frac{ik_n P_n}{R} + k_n^2\right) + EI\frac{h}{2}\left(\frac{P_n k_n^2}{R} + ik_n^3\right)\right)a_n^- e^{ik_n s} = 0$$

$$\sum_{n=1}^{3}\left(-mw^2\left(\frac{h}{2}\left(\frac{P_n}{R} + ik_n\right) - 1\right) + EI\left(\frac{P_n k_n^2}{R} + ik_n^3\right)\right)a_n^+ e^{-ik_n s}$$

$$+\sum_{n=1}^{3}\left(mw^2\left(\frac{h}{2}\left(\frac{P_n}{R} + ik_n\right) + 1\right) - EI\left(\frac{P_n k_n^2}{R} + ik_n^3\right)\right)a_n^- e^{ik_n s} = 0 \tag{6.31}$$

Choosing the origin at the boundary and assembling Eq. (6.31) into matrix form,

$$\begin{bmatrix} \alpha_{11} & \alpha_{12} & \alpha_{13} \\ \alpha_{21} & \alpha_{22} & \alpha_{23} \\ \alpha_{31} & \alpha_{32} & \alpha_{33} \end{bmatrix}\begin{bmatrix} a_1^+ \\ a_2^+ \\ a_3^+ \end{bmatrix} - \begin{bmatrix} \beta_{11} & \beta_{12} & \beta_{13} \\ \beta_{21} & \beta_{22} & \beta_{23} \\ \beta_{31} & \beta_{32} & \beta_{33} \end{bmatrix}\begin{bmatrix} a_1^- \\ a_2^- \\ a_3^- \end{bmatrix} = 0 \tag{6.32a}$$

where

$$\alpha_{1n} = mw^2 P_n + EA\left(-\frac{1}{R} + ik_n P_n\right), \text{ for } n = 1,2,3$$

$$\alpha_{2n} = -J_m w^2\left(\frac{P_n}{R} + ik_n\right) + EI\left(-\frac{ik_n P_n}{R} + k_n^2\right) - EI\frac{h}{2}\left(\frac{P_n k_n^2}{R} + ik_n^3\right), \text{ for } n = 1,2,3$$

$$\alpha_{3n} = -mw^2\left(\frac{h}{2}\left(\frac{P_n}{R} + ik_n\right) - 1\right) + EI\left(\frac{P_n k_n^2}{R} + ik_n^3\right), \text{ for } n = 1,2,3$$

$$\beta_{1n} = mw^2 P_n - EA\left(-\frac{1}{R} + ik_n P_n\right), \text{ for } n = 1,2,3$$

$$\beta_{2n} = -J_m w^2\left(\frac{P_n}{R} + ik_n\right) - EI\left(-\frac{ik_n P_n}{R} + k_n^2\right) - EI\frac{h}{2}\left(\frac{P_n k_n^2}{R} + ik_n^3\right), \text{ for } n = 1,2,3$$

$$\beta_{3n} = -mw^2\left(\frac{h}{2}\left(\frac{P_n}{R} + ik_n\right) + 1\right) + EI\left(\frac{P_n k_n^2}{R} + ik_n^3\right), \text{ for } n = 1,2,3 \tag{6.32b}$$

From Eqs. (6.8b), (6.10), and (6.32a), the reflection matrix at a lumped mass attached boundary is

$$\mathbf{r} = \begin{bmatrix} \beta_{11} & \beta_{12} & \beta_{13} \\ \beta_{21} & \beta_{22} & \beta_{23} \\ \beta_{31} & \beta_{32} & \beta_{33} \end{bmatrix}^{-1}\begin{bmatrix} \alpha_{11} & \alpha_{12} & \alpha_{13} \\ \alpha_{21} & \alpha_{22} & \alpha_{23} \\ \alpha_{31} & \alpha_{32} & \alpha_{33} \end{bmatrix} \tag{6.33}$$

where the elements of both matrices are given in Eq. (6.32b).

MATLAB symbolic scripts for the reflection matrix at a boundary with a rigid body mass attachment are available in Section 6.6.

Under the special situation where $h = 0$, and m and J_m both approach zero or infinity, the mass attached end is physically equivalent to a free or a clamped end, respectively.

The reflection matrices at classical boundaries are summarized in Table 6.1.

6.3 Free Vibration Analysis in a Finite Curved Beam – Natural Frequencies and Modeshapes

Free vibration analysis of a uniform curved beam from the wave standpoint involves a concise and systematic assembling process.

Figure 6.5 illustrates a curved beam of length s with a constant curvature. The following reflection and propagation relationships exist:

$$\mathbf{a}^+ = \mathbf{r}_A\,\mathbf{a}^-,\ \mathbf{b}^- = \mathbf{r}_B\,\mathbf{b}^+,\ \mathbf{b}^+ = \mathbf{f}(s)\,\mathbf{a}^+,\ \mathbf{a}^- = \mathbf{f}(s)\,\mathbf{b}^- \tag{6.34}$$

where $\mathbf{f}(s)$ is the propagation matrix for an arc distance s given in Eq. (6.8c), and $\mathbf{r}_A$ and $\mathbf{r}_B$ are the reflection matrices at Boundaries A and B, respectively.

The wave relationships in Eq. (6.34) can be written in matrix form,

$$\begin{bmatrix} -\mathbf{I} & \mathbf{r}_A & \mathbf{0} & \mathbf{0} \\ \mathbf{0} & \mathbf{0} & \mathbf{r}_B & -\mathbf{I} \\ \mathbf{f}(s) & \mathbf{0} & -\mathbf{I} & \mathbf{0} \\ \mathbf{0} & -\mathbf{I} & \mathbf{0} & \mathbf{f}(s) \end{bmatrix} \begin{bmatrix} \mathbf{a}^+ \\ \mathbf{a}^- \\ \mathbf{b}^+ \\ \mathbf{b}^- \end{bmatrix} = \mathbf{0} \tag{6.35}$$

where $\mathbf{I}$ and $\mathbf{0}$ denote an identity matrix and a zero matrix, respectively.

Table 6.1 Reflection matrices at classical boundaries.

Boundary Condition	Reflection Matrix
Free	$\begin{bmatrix} -\left(\dfrac{1}{R}-ik_1 P_1\right) & -\left(\dfrac{1}{R}-ik_2 P_2\right) & -\left(\dfrac{1}{R}-ik_3 P_3\right) \\ -\left(\dfrac{ik_1 P_1}{R}-k_1^2\right) & -\left(\dfrac{ik_2 P_2}{R}-k_2^2\right) & -\left(\dfrac{ik_3 P_3}{R}-k_3^2\right) \\ \dfrac{P_1 k_1^2}{R}+ik_1^3 & \dfrac{P_2 k_2^2}{R}+ik_2^3 & \dfrac{P_3 k_3^2}{R}+ik_3^3 \end{bmatrix}^{-1} \begin{bmatrix} \dfrac{1}{R}-ik_1 P_1 & \dfrac{1}{R}-ik_2 P_2 & \dfrac{1}{R}-ik_3 P_3 \\ \dfrac{ik_1 P_1}{R}-k_1^2 & \dfrac{ik_2 P_2}{R}-k_2^2 & \dfrac{ik_3 P_3}{R}-k_3^2 \\ \dfrac{P_1 k_1^2}{R}+ik_1^3 & \dfrac{P_2 k_2^2}{R}+ik_2^3 & \dfrac{P_3 k_3^2}{R}+ik_3^3 \end{bmatrix}$
Clamped	$\begin{bmatrix} -1 & -1 & -1 \\ P_1 & P_2 & P_3 \\ -\dfrac{P_1}{R}-ik_1 & -\dfrac{P_2}{R}-ik_2 & -\dfrac{P_3}{R}-ik_3 \end{bmatrix}^{-1} \begin{bmatrix} 1 & 1 & 1 \\ P_1 & P_2 & P_3 \\ -\dfrac{P_1}{R}-ik_1 & -\dfrac{P_2}{R}-ik_2 & -\dfrac{P_3}{R}-ik_3 \end{bmatrix}$
Simply-supported	$\begin{bmatrix} -1 & -1 & -1 \\ P_1 & P_2 & P_3 \\ -\dfrac{ik_1 P_1}{R}+k_1^2 & -\dfrac{ik_2 P_2}{R}+k_2^2 & -\dfrac{ik_3 P_3}{R}+k_3^2 \end{bmatrix}^{-1} \begin{bmatrix} 1 & 1 & 1 \\ P_1 & P_2 & P_3 \\ \dfrac{ik_1 P_1}{R}-k_1^2 & \dfrac{ik_2 P_2}{R}-k_2^2 & \dfrac{ik_3 P_3}{R}-k_3^2 \end{bmatrix}$

Figure 6.5 Wave propagation and reflection in a uniform curved beam.

Equation (6.35) can be written in the form of

$$\mathbf{A}_0 \mathbf{z}_0 = 0 \tag{6.36}$$

where $\mathbf{A}_0$ is a 12 by 12 square coefficient matrix and $\mathbf{z}_0$ is a wave vector containing 12 wave components of Eq. (6.35).

Setting the determinant of the coefficient matrix $\mathbf{A}_0$ to zero gives the characteristic equation. Natural frequencies of the curved beam are identified by finding the roots of the characteristic equation. The natural frequencies of a curved beam are normally found numerically because of the complexity of coupled wave motion.

When a distributed structure is excited at one of its natural frequencies, the motion of the structure exhibits a certain shape, namely, the modeshape. The modeshape at a given natural frequency can be found by following the same procedures as described in detail in Chapter 3, that is, by eliminating any one row of coefficient matrix $\mathbf{A}_0$ of Eq. (6.36) and using the remaining rows to solve the wave components in terms of a chosen wave component.

6.4 Force Generated Waves and Forced Vibration Analysis of Finite Curved Beams

An externally applied force or moment has the effect of injecting waves into a continuous structure. Figure 6.6 illustrates a tangential force $\bar{N}$, radial force $\bar{V}$, bending moment $\bar{M}$, and waves $\mathbf{a}^{\pm}$ and $\mathbf{b}^{\pm}$ at the point where the force and moment are applied. These waves are injected by and related to the externally applied force and moment.

Because the radial and tangential deflections, and the rotation of the cross section on the left and right side of the point where the force and moment are applied are equal, denoting physical parameters on the left and right side of the point where the force and moment are applied using subscripts $-$ and $+$, respectively,

$$w_- = w_+, \, u_- = u_+, \, \varphi_- = \varphi_+ \tag{6.37}$$

From the free body diagram of Figure 6.6, the externally applied force and moment are related to the internal resistant forces and moments on the left and right side of the point where the external force and moment are applied by

$$\bar{V} = V_- - V_+, \, \bar{N} = N_- - N_+, \, \bar{M} = M_- - M_+ \tag{6.38}$$

The relationships described in Eqs. (6.37) and (6.38) are called continuity and equilibrium equations, respectively. From Figure 6.6a and Eqs. (6.7) and (6.9b),

$$w_- = a_1^+ e^{-ik_1 s} + a_2^+ e^{-ik_2 s} + a_3^+ e^{-ik_3 s} + a_1^- e^{ik_1 s} + a_2^- e^{ik_2 s} + a_3^- e^{ik_3 s}$$

$$w_+ = b_1^+ e^{-ik_1 s} + b_2^+ e^{-ik_2 s} + b_3^+ e^{-ik_3 s} + b_1^- e^{ik_1 s} + b_2^- e^{ik_2 s} + b_3^- e^{ik_3 s}$$

$$u_- = P_1 a_1^+ e^{-ik_1 s} + P_2 a_2^+ e^{-ik_2 s} + P_3 a_3^+ e^{-ik_3 s} - P_1 a_1^- e^{ik_1 s} - P_2 a_2^- e^{ik_2 s} - P_3 a_3^- e^{ik_3 s}$$

$$u_+ = P_1 b_1^+ e^{-ik_1 s} + P_2 b_2^+ e^{-ik_2 s} + P_3 b_3^+ e^{-ik_3 s} - P_1 b_1^- e^{ik_1 s} - P_2 b_2^- e^{ik_2 s} - P_3 b_3^- e^{ik_3 s}$$

$$\varphi_- = -\frac{u_-}{R} + \frac{\partial w_-}{\partial s}$$

$$\varphi_+ = -\frac{u_+}{R} + \frac{\partial w_+}{\partial s} \tag{6.39}$$

Choosing the origin at the point where the external force and moment are applied, from Eqs. (6.37) and (6.39), the continuity conditions become

$$\sum_{n=1}^{3} a_n^+ + \sum_{n=1}^{3} a_n^- - \left(\sum_{n=1}^{3} b_n^+ + \sum_{n=1}^{3} b_n^- \right) = 0$$

$$\sum_{n=1}^{3} P_n a_n^+ - \sum_{n=1}^{3} P_n a_n^- - \left(\sum_{n=1}^{3} P_n b_n^+ - \sum_{n=1}^{3} P_n b_n^- \right) = 0$$

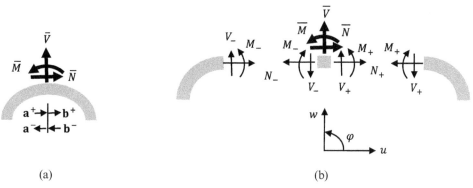

(a) (b)

Figure 6.6 (a) Waves generated by externally applied force and moment and (b) free body diagram.

$$\sum_{n=1}^{3}\left(-\frac{P_n}{R}-ik_n\right)a_n^+ +\sum_{n=1}^{3}\left(\frac{P_n}{R}+ik_n\right)a_n^- -\left(\sum_{n=1}^{3}\left(-\frac{P_n}{R}-ik_n\right)b_n^+ +\sum_{n=1}^{3}\left(\frac{P_n}{R}+ik_n\right)b_n^-\right)=0 \tag{6.40}$$

Putting Eq. (6.40) into matrix form,

$$\begin{bmatrix} 1 & 1 & 1 \\ P_1 & P_2 & P_3 \\ -\dfrac{P_1}{R}-ik_1 & -\dfrac{P_2}{R}-ik_2 & -\dfrac{P_3}{R}-ik_3 \end{bmatrix}(\mathbf{a}^+ - \mathbf{b}^+) + \begin{bmatrix} 1 & 1 & 1 \\ -P_1 & -P_2 & -P_3 \\ \dfrac{P_1}{R}+ik_1 & \dfrac{P_2}{R}+ik_2 & \dfrac{P_3}{R}+ik_3 \end{bmatrix}(\mathbf{a}^- - \mathbf{b}^-)=0 \tag{6.41}$$

From Figure 6.6 and Eq. (6.9a),

$$N_- = EA\left(\frac{w_-}{R}+\frac{\partial u_-}{\partial s}\right),\ N_+ = EA\left(\frac{w_+}{R}+\frac{\partial u_+}{\partial s}\right)$$

$$M_- = EI\frac{\partial\varphi_-}{\partial s}= EI\left(-\frac{u_-}{R}+\frac{\partial w_-}{\partial s}\right),\ M_+ = EI\frac{\partial\varphi_+}{\partial s}= EI\left(-\frac{u_+}{R}+\frac{\partial w_+}{\partial s}\right) \tag{6.42}$$

$$V_- = -\frac{\partial M_-}{\partial s}= EI\left(\frac{1}{R}\frac{\partial^2 u_-}{\partial s^2}-\frac{\partial^3 w_-}{\partial s^3}\right),\ V_+ = -\frac{\partial M_+}{\partial s}= EI\left(\frac{1}{R}\frac{\partial^2 u_+}{\partial s^2}-\frac{\partial^3 w_+}{\partial s^3}\right)$$

From Eqs. (6.38) and (6.42),

$$\overline{N} = EA\left(\frac{w_-}{R}+\frac{\partial u_-}{\partial s}\right) - EA\left(\frac{w_+}{R}+\frac{\partial u_+}{\partial s}\right)$$

$$\overline{M} = EI\left(-\frac{u_-}{R}+\frac{\partial w_-}{\partial s}\right) - EI\left(-\frac{u_+}{R}+\frac{\partial w_+}{\partial s}\right) \tag{6.43}$$

$$\overline{V} = EI\left(\frac{1}{R}\frac{\partial^2 u_-}{\partial s^2}-\frac{\partial^3 w_-}{\partial s^3}\right) - EI\left(\frac{1}{R}\frac{\partial^2 u_+}{\partial s^2}-\frac{\partial^3 w_+}{\partial s^3}\right)$$

Choosing the origin at the point where the external force and moment are applied, from Eqs. (6.39) and (6.43), the equilibrium conditions become

$$\overline{N} = \sum_{n=1}^{3}EA\left(-\frac{1}{R}+iP_nk_n\right)\left(a_n^+ - b_n^+\right)+\sum_{n=1}^{3}EA\left(-\frac{1}{R}+iP_nk_n\right)\left(a_n^- - b_n^-\right)$$

$$\overline{M} = \sum_{n=1}^{3}EI\left(-\frac{iP_nk_n}{R}+k_n^2\right)\left(a_n^+ - b_n^+\right)+\sum_{n=1}^{3}EI\left(-\frac{iP_nk_n}{R}+k_n^2\right)\left(a_n^- - b_n^-\right) \tag{6.44}$$

$$\overline{V} = \sum_{n=1}^{3}EI\left(\frac{P_nk_n^2}{R}+ik_n^3\right)\left(a_n^+ - b_n^+\right)+\sum_{n=1}^{3}EI\left(-\frac{P_nk_n^2}{R}-ik_n^3\right)\left(a_n^- - b_n^-\right)$$

Putting Eq. (6.44) into matrix form,

$$
\begin{bmatrix}
-\dfrac{1}{R}+iP_1k_1 & -\dfrac{1}{R}+iP_2k_2 & -\dfrac{1}{R}+iP_3k_3 \\[2mm]
-\dfrac{iP_1k_1}{R}+k_1^2 & -\dfrac{iP_2k_2}{R}+k_2^2 & -\dfrac{iP_3k_3}{R}+k_3^2 \\[2mm]
\dfrac{P_1k_1^2}{R}+ik_1^3 & \dfrac{P_2k_2^2}{R}+ik_2^3 & \dfrac{P_3k_3^2}{R}+ik_3^3
\end{bmatrix}\left(\mathbf{a}^+-\mathbf{b}^+\right)+
$$

$$
\begin{bmatrix}
-\dfrac{1}{R}+iP_1k_1 & -\dfrac{1}{R}+iP_2k_2 & -\dfrac{1}{R}+iP_3k_3 \\[2mm]
-\dfrac{iP_1k_1}{R}+k_1^2 & -\dfrac{iP_2k_2}{R}+k_2^2 & -\dfrac{iP_3k_3}{R}+k_3^2 \\[2mm]
-\dfrac{P_1k_1^2}{R}-ik_1^3 & -\dfrac{P_2k_2^2}{R}-ik_2^3 & -\dfrac{P_3k_3^2}{R}-ik_3^3
\end{bmatrix}\left(\mathbf{a}^--\mathbf{b}^-\right)=
\begin{bmatrix}
\dfrac{\bar{N}}{EA} \\[2mm]
\dfrac{\bar{M}}{EI} \\[2mm]
\dfrac{\bar{V}}{EI}
\end{bmatrix}
$$

(6.45)

Solving for $(\mathbf{a}^+-\mathbf{b}^+)$ and $(\mathbf{a}^--\mathbf{b}^-)$ from Eqs. (6.41) and (6.45), the relationships between injected waves and externally applied force and moment are

$$\mathbf{a}^+-\mathbf{b}^+=\mathbf{q}_V+\mathbf{q}_N+\mathbf{m}$$

$$\mathbf{a}^--\mathbf{b}^-=-\mathbf{q}_V+\mathbf{q}_N+\mathbf{m}$$

(6.46a)

where the load vectors are

$$
\mathbf{q}_V=\frac{R}{2\left(\left(-iRP_3k_1k_2+P_1P_2k_3\right)\left(k_1^2-k_2^2\right)+\left(-iRP_2k_1k_3+P_1P_3k_2\right)\left(k_3^2-k_1^2\right)+\left(-iRP_1k_2k_3+P_2P_3k_1\right)\left(k_2^2-k_3^2\right)\right)}
\begin{bmatrix}
P_2k_3-P_3k_2 \\
P_3k_1-P_1k_3 \\
P_1k_2-P_2k_1
\end{bmatrix}\frac{\bar{V}}{EI}
$$

$$
\mathbf{q}_N=\frac{1}{2R\left(P_3k_3\left(k_1^2-k_2^2\right)+P_2k_2\left(k_3^2-k_1^2\right)+P_1k_1\left(k_2^2-k_3^2\right)\right)}
\begin{bmatrix}
iR\left(k_3^2-k_2^2\right)-P_2k_2+P_3k_3 \\
iR\left(k_1^2-k_3^2\right)+P_1k_1-P_3k_3 \\
iR\left(k_2^2-k_1^2\right)-P_1k_1+P_2k_2
\end{bmatrix}\frac{\bar{N}}{EA}
$$

$$
\mathbf{m}=\frac{1}{2\left(P_3k_3\left(k_1^2-k_2^2\right)+P_2k_2\left(k_3^2-k_1^2\right)+P_1k_1\left(k_2^2-k_3^2\right)\right)}
\begin{bmatrix}
P_3k_3-P_2k_2 \\
P_1k_1-P_3k_3 \\
P_2k_2-P_1k_1
\end{bmatrix}\frac{\bar{M}}{EI}
$$

(6.46b)

With the availability of force generated waves, propagation relationships, and boundary reflections, forced vibration in a curved beam can be obtained.

Figure 6.7 illustrates a uniform curved beam with external excitations applied on the beam at an arbitrary point G that is distance s_1 and s_2 away from Boundaries C and D, respectively. The following wave relationships exist:

- At Boundaries C and D, from Eq. (6.10),

  $$\mathbf{c}^+=\mathbf{r}_C\,\mathbf{c}^-,\quad \mathbf{d}^-=\mathbf{r}_D\,\mathbf{d}^+$$

  (6.47)

  where $\mathbf{r}_C$ and $\mathbf{r}_D$ are the reflection matrices at Boundaries C and D, respectively.

- At Point G where the external force and moment are applied, from Eq. (6.46a),

  $$\mathbf{g}_1^+-\mathbf{g}_2^+=\mathbf{q}_V+\mathbf{q}_N+\mathbf{m},\quad \mathbf{g}_1^--\mathbf{g}_2^-=-\mathbf{q}_V+\mathbf{q}_N+\mathbf{m}$$

  (6.48)

  where $\mathbf{q}_V$, $\mathbf{q}_N$, and $\mathbf{m}$ are obtained in Eq. (6.46b).

- The propagation relationships, from Eq. (6.8c), are

  $$\mathbf{g}_1^+=\mathbf{f}(s_1)\,\mathbf{c}^+,\quad \mathbf{c}^-=\mathbf{f}(s_1)\,\mathbf{g}_1^-,\quad \mathbf{g}_2^-=\mathbf{f}(s_2)\,\mathbf{d}^-,\quad \mathbf{d}^+=\mathbf{f}(s_2)\,\mathbf{g}_2^+$$

  (6.49)

  where $\mathbf{f}(s)$ is the propagation matrix for an arc distance s given in Eq. (6.8c).

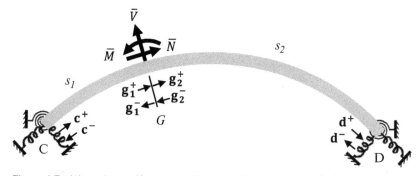

Figure 6.7 Waves in a uniform curved beam with externally applied force and moment.

Assembling the wave relationships from Eqs. (6.47) to (6.49) into matrix form,

$$
\begin{bmatrix}
-\mathbf{I} & \mathbf{r}_C & \mathbf{0} & \mathbf{0} & \mathbf{0} & \mathbf{0} & \mathbf{0} & \mathbf{0} \\
\mathbf{0} & \mathbf{0} & \mathbf{r}_D & -\mathbf{I} & \mathbf{0} & \mathbf{0} & \mathbf{0} & \mathbf{0} \\
\mathbf{0} & \mathbf{0} & \mathbf{0} & \mathbf{0} & \mathbf{I} & \mathbf{0} & -\mathbf{I} & \mathbf{0} \\
\mathbf{0} & \mathbf{0} & \mathbf{0} & \mathbf{0} & \mathbf{0} & \mathbf{I} & \mathbf{0} & -\mathbf{I} \\
\mathbf{f}(s_1) & \mathbf{0} & \mathbf{0} & \mathbf{0} & -\mathbf{I} & \mathbf{0} & \mathbf{0} & \mathbf{0} \\
\mathbf{0} & -\mathbf{I} & \mathbf{0} & \mathbf{0} & \mathbf{0} & \mathbf{f}(s_1) & \mathbf{0} & \mathbf{0} \\
\mathbf{0} & \mathbf{0} & \mathbf{0} & \mathbf{f}(s_2) & \mathbf{0} & \mathbf{0} & \mathbf{0} & -\mathbf{I} \\
\mathbf{0} & \mathbf{0} & -\mathbf{I} & \mathbf{0} & \mathbf{0} & \mathbf{0} & \mathbf{f}(s_2) & \mathbf{0}
\end{bmatrix}
\begin{bmatrix}
\mathbf{c}^+ \\
\mathbf{c}^- \\
\mathbf{d}^+ \\
\mathbf{d}^- \\
\mathbf{g}_1^+ \\
\mathbf{g}_1^- \\
\mathbf{g}_2^+ \\
\mathbf{g}_2^-
\end{bmatrix}
=
\begin{bmatrix}
\mathbf{0} \\
\mathbf{0} \\
\mathbf{q}_V + \mathbf{q}_N + \mathbf{m} \\
-\mathbf{q}_V + \mathbf{q}_N + \mathbf{m} \\
\mathbf{0} \\
\mathbf{0} \\
\mathbf{0} \\
\mathbf{0}
\end{bmatrix}
\tag{6.50}
$$

where $\mathbf{I}$ and $\mathbf{0}$ denote an identity matrix and a zero matrix, respectively.

Equation (6.50) can be expressed in the form of

$$
\mathbf{A}_f \mathbf{z}_f = \mathbf{F}
\tag{6.51}
$$

where $\mathbf{A}_f$ is a 24 by 24 square coefficient matrix, $\mathbf{z}_f$ is a wave vector containing 24 wave components, and $\mathbf{F}$ is a vector related to externally applied force and moment of Eq. (6.50).

Solving for $\mathbf{z}_f$ from Eq. (6.51),

$$
\mathbf{z}_f = \mathbf{A}_f^{-1} \mathbf{F}
\tag{6.52}
$$

Equation (6.52) provides the responses of all wave components, from which the deflections of any point on the beam are obtained. For example, in Figure 6.7, the deflections of a point located between Boundary C and the excitation point G that is an arc distance s from Point C are

$$
w = \begin{bmatrix} 1 & 1 & 1 \end{bmatrix} \mathbf{f}(s) \mathbf{c}^+ + \begin{bmatrix} 1 & 1 & 1 \end{bmatrix} \mathbf{f}(-s) \mathbf{c}^-
$$

$$
u = \begin{bmatrix} 1 & 1 & 1 \end{bmatrix} \mathbf{f}(s) \begin{bmatrix} P_1 & 0 & 0 \\ 0 & P_2 & 0 \\ 0 & 0 & P_3 \end{bmatrix} \mathbf{c}^+ + \begin{bmatrix} 1 & 1 & 1 \end{bmatrix} \mathbf{f}(-s) \begin{bmatrix} -P_1 & 0 & 0 \\ 0 & -P_2 & 0 \\ 0 & 0 & -P_3 \end{bmatrix} \mathbf{c}^-
$$

$$
\varphi = \begin{bmatrix} 1 & 1 & 1 \end{bmatrix} \mathbf{f}(s) \begin{bmatrix} -\dfrac{P_1}{R} - ik_1 & 0 & 0 \\ 0 & -\dfrac{P_2}{R} - ik_2 & 0 \\ 0 & 0 & -\dfrac{P_3}{R} - ik_3 \end{bmatrix} \mathbf{c}^+ + \begin{bmatrix} 1 & 1 & 1 \end{bmatrix} \mathbf{f}(-s) \begin{bmatrix} \dfrac{P_1}{R} + ik_1 & 0 & 0 \\ 0 & \dfrac{P_2}{R} + ik_2 & 0 \\ 0 & 0 & \dfrac{P_3}{R} + ik_3 \end{bmatrix} \mathbf{c}^-
\tag{6.53}
$$

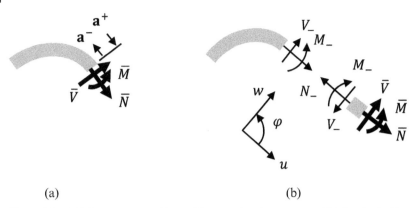

(a)   (b)

Figure 6.8 (a) Waves generated by end force and end moment and (b) free body diagram.

An external force/moment may be applied at the free end of a curved beam, as shown in Figure 6.8.

From the free body diagram of Figure 6.8, the externally applied force and moment are related to the internal resistant forces and moments by the following equilibrium equation

$$\bar{V} = V(s,t), \ \bar{N} = N(s,t), \ \bar{M} = M(s,t) \tag{6.54}$$

From Eqs. (6.9a) and (6.54),

$$
\begin{aligned}
\bar{N} &= EA\left(\frac{w(s,t)}{R} + \frac{\partial u(s,t)}{\partial s}\right) \\[4pt]
\bar{M} &= EI\left(-\frac{1}{R}\frac{\partial u(s,t)}{\partial s} + \frac{\partial^2 w(s,t)}{\partial s^2}\right) \\[4pt]
\bar{V} &= EI\left(\frac{1}{R}\frac{\partial^2 u(s,t)}{\partial s^2} - \frac{\partial^3 w(s,t)}{\partial s^3}\right)
\end{aligned}
\tag{6.55}
$$

Choosing the origin at the point where the external force and moment are applied, from Eqs. (6.7) and (6.9b), the equilibrium equations of Eqs. (6.55) become

$$
\begin{aligned}
\frac{\bar{N}}{EA} &= \sum\nolimits_{n=1}^{3}\left(\frac{1}{R} - iP_n k_n\right)\left(a_n^+ + a_n^-\right) \\[4pt]
\frac{\bar{M}}{EI} &= \sum\nolimits_{n=1}^{3}\left(-k_n^2 + i\frac{P_n k_n}{R}\right)\left(a_n^+ + a_n^-\right) \\[4pt]
\frac{\bar{V}}{EI} &= \sum\nolimits_{n=1}^{3}\left(\frac{P_n k_n^2}{R} + ik_n^3\right)\left(-a_n^+ + a_n^-\right)
\end{aligned}
\tag{6.56}
$$

Putting Eqs. (6.56) into matrix form,

$$
\begin{bmatrix}
\dfrac{1}{R} - iP_1 k_1 & \dfrac{1}{R} - iP_2 k_2 & \dfrac{1}{R} - iP_3 k_3 \\[6pt]
-k_1^2 + i\dfrac{P_1 k_1}{R} & -k_2^2 + i\dfrac{P_2 k_2}{R} & -k_3^2 + i\dfrac{P_3 k_3}{R} \\[6pt]
-\dfrac{P_1 k_1^2}{R} - ik_1^3 & -\dfrac{P_2 k_2^2}{R} - ik_2^3 & -\dfrac{P_3 k_3^2}{R} - ik_3^3
\end{bmatrix}\mathbf{a}^+ +
\begin{bmatrix}
\dfrac{1}{R} - iP_1 k_1 & \dfrac{1}{R} - iP_2 k_2 & \dfrac{1}{R} - iP_3 k_3 \\[6pt]
-k_1^2 + i\dfrac{P_1 k_1}{R} & -k_2^2 + i\dfrac{P_2 k_2}{R} & -k_3^2 + i\dfrac{P_3 k_3}{R} \\[6pt]
\dfrac{P_1 k_1^2}{R} + ik_1^3 & \dfrac{P_2 k_2^2}{R} + ik_2^3 & \dfrac{P_3 k_3^2}{R} + ik_3^3
\end{bmatrix}\mathbf{a}^- =
\begin{bmatrix}
\dfrac{\bar{N}}{EA} \\[6pt]
\dfrac{\bar{M}}{EI} \\[6pt]
\dfrac{\bar{V}}{EI}
\end{bmatrix}
\tag{6.57}
$$

In the special situation of $\bar{V} = 0$, $\bar{N} = 0$, and $\bar{M} = 0$, the boundary becomes a free boundary. From Eqs. (6.10) and (6.57), the reflection matrix at a free boundary is,

$$\mathbf{r} = - \begin{bmatrix} \dfrac{1}{R} - iP_1k_1 & \dfrac{1}{R} - iP_2k_2 & \dfrac{1}{R} - iP_3k_3 \\[2mm] -k_1^2 + i\dfrac{P_1k_1}{R} & -k_2^2 + i\dfrac{P_2k_2}{R} & -k_3^2 + i\dfrac{P_3k_3}{R} \\[2mm] \dfrac{P_1k_1^2}{R} + ik_1^3 & \dfrac{P_2k_2^2}{R} + ik_2^3 & \dfrac{P_3k_3^2}{R} + ik_3^3 \end{bmatrix}^{-1} \begin{bmatrix} \dfrac{1}{R} - iP_1k_1 & \dfrac{1}{R} - iP_2k_2 & \dfrac{1}{R} - iP_3k_3 \\[2mm] -k_1^2 + i\dfrac{P_1k_1}{R} & -k_2^2 + i\dfrac{P_2k_2}{R} & -k_3^2 + i\dfrac{P_3k_3}{R} \\[2mm] -\dfrac{P_1k_1^2}{R} - ik_1^3 & -\dfrac{P_2k_2^2}{R} - ik_2^3 & -\dfrac{P_3k_3^2}{R} - ik_3^3 \end{bmatrix} \tag{6.58}$$

This, as expected, is the same as the reflection matrix of a free boundary obtained in Eq. (6.15).

With the availability of force generated waves, propagation relationships, and boundary reflections, forced vibration in a curved beam due to external excitations applied at a free end can be obtained.

Figure 6.9 illustrates a uniform curved beam with external excitations applied at a free end D. The following wave relationships exist:

- At Boundary C, from Eq. (6.10),

$$\mathbf{c}^+ = \mathbf{r}_C \, \mathbf{c}^- \tag{6.59}$$

where $\mathbf{r}_C$ is the reflection matrix at Boundary C.

- At external excitations applied end D, from Eq. (6.57),

$$\mathbf{q}_{c1}\mathbf{d}^+ + \mathbf{q}_{c2}\mathbf{d}^- = \mathbf{q}_0 \tag{6.60a}$$

where

$$\mathbf{q}_{c1} = \begin{bmatrix} \dfrac{1}{R} - iP_1k_1 & \dfrac{1}{R} - iP_2k_2 & \dfrac{1}{R} - iP_3k_3 \\[2mm] -k_1^2 + i\dfrac{P_1k_1}{R} & -k_2^2 + i\dfrac{P_2k_2}{R} & -k_3^2 + i\dfrac{P_3k_3}{R} \\[2mm] -\dfrac{P_1k_1^2}{R} - ik_1^3 & -\dfrac{P_2k_2^2}{R} - ik_2^3 & -\dfrac{P_3k_3^2}{R} - ik_3^3 \end{bmatrix},$$

$$\mathbf{q}_{c2} = \begin{bmatrix} \dfrac{1}{R} - iP_1k_1 & \dfrac{1}{R} - iP_2k_2 & \dfrac{1}{R} - iP_3k_3 \\[2mm] -k_1^2 + i\dfrac{P_1k_1}{R} & -k_2^2 + i\dfrac{P_2k_2}{R} & -k_3^2 + i\dfrac{P_3k_3}{R} \\[2mm] \dfrac{P_1k_1^2}{R} + ik_1^3 & \dfrac{P_2k_2^2}{R} + ik_2^3 & \dfrac{P_3k_3^2}{R} + ik_3^3 \end{bmatrix}, \quad \mathbf{q}_0 = \begin{bmatrix} \dfrac{\bar{N}}{EA} \\[2mm] \dfrac{\bar{M}}{EI} \\[2mm] \dfrac{\bar{V}}{EI} \end{bmatrix} \tag{6.60b}$$

- The propagation relationships, according to Eq. (6.8a), are

$$\mathbf{d}^+ = \mathbf{f}(s)\,\mathbf{c}^+, \ \mathbf{c}^- = \mathbf{f}(s)\,\mathbf{d}^- \tag{6.61}$$

where $\mathbf{f}(s)$ is the propagation matrix for an arc distance s given in Eq. (6.8c).

Assembling the wave relationships from Eqs. (6.59) to (6.61) into matrix form,

$$\begin{bmatrix} -\mathbf{I} & \mathbf{r} & \mathbf{0} & \mathbf{0} \\ \mathbf{0} & \mathbf{0} & \mathbf{q}_{c1} & \mathbf{q}_{c2} \\ \mathbf{f}(s) & \mathbf{0} & -\mathbf{I} & \mathbf{0} \\ \mathbf{0} & -\mathbf{I} & \mathbf{0} & \mathbf{f}(s) \end{bmatrix} \begin{bmatrix} \mathbf{c}^+ \\ \mathbf{c}^- \\ \mathbf{d}^+ \\ \mathbf{d}^- \end{bmatrix} = \begin{bmatrix} \mathbf{0} \\ \mathbf{q}_0 \\ \mathbf{0} \\ \mathbf{0} \end{bmatrix} \tag{6.62}$$

Equation (6.62) can be written in the form of Eq. (6.51), and the wave components can be solved from Eq. (6.52). In which $\mathbf{A}_f$ is a 12 by 12 square coefficient matrix, $\mathbf{z}_f$ is a wave vector containing 12 wave components, and $\mathbf{F}$ is a vector related to externally applied force and moment of Eq. (6.62).

With the availability of responses of all wave components, the deflections of any point on the beam can be obtained. For example, the deflections of a point located on the curved beam that is an arc distance s from Boundary C in Figure 6.9 can be found using Eq. (6.53).

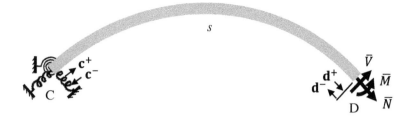

Figure 6.9 Waves in a uniform curved beam with external forces and moment applied at a free end.

6.5 Numerical Examples

Free and forced vibrations in a slender curved steel beam are analyzed. The material and geometrical properties of the curved beam are as follows: the Young's modulus and volume mass density are $E = 198.87 \, GN/m^2$ and $\rho = 7664.5 \, kg/m^3$, respectively. The curved beam is of a square cross section of side length 0.1 m. The radius of curvature R of the curved beam is 1.0 m and the span angle θ is 60°. The arc length in meters is $s = R\theta = \pi/3 \, m$. The boundaries of the curved beam are assumed clamped–clamped.

Wave mode transitions exist in coupled wave motions in a curved beam. The cut-off frequencies are associated with the ring frequency and the frequencies corresponding to the roots of the discriminant equation obtained in Eq. (6.4), as discussed in Section 6.1. The cut-off frequency corresponding to the ring frequency of the example curved beam is calculated using $f_c = \dfrac{1}{2\pi R}\sqrt{\dfrac{E}{\rho}}$ as 810.705 Hz. The cut-off frequencies associated with the roots of the discriminant of the dispersion equation are found numerically, which are 7.9 Hz and 97.5 Hz.

Figure 6.10 shows the overlaid real and imaginary dispersion relationships of the example curved beam, with the three cut-off frequencies highlighted using broken vertical lines. In Figure 6.11, the real and imaginary dispersion relationships for each wavenumber are presented separately. Wavenumber k_1, k_2, and k_3 may be real, imaginary, or complex values, depending on the frequency range. The three cut-off frequencies divide the frequency variable into four zones. Let us name them Zone I to IV in ascending order by frequency. In Zone I, all three wavenumbers are real, indicating that three pairs of propagating waves exist in Zone I. In Zone II, Wavenumber 1 is real, Wavenumbers 2 and 3 are complex. Therefore, one pair of propagating waves and two pairs of propagating decaying waves exist in Zone II. In Zone III, Wavenumber 1 is real, Wavenumbers 2 and 3 are imaginary. Hence, one pair of propagating waves and two pairs of decaying waves exist in Zone III. In Zone IV, Wavenumbers 1 and 3 are real, and Wavenumber 2 is imaginary. Consequently, two pairs of propagating waves and one pair of decaying waves exist in Zone IV. The dispersion characteristics are determined by the material property, cross-sectional dimension, and radius of curvature of the curved beam.

Natural frequencies are found by plotting the dB magnitude of the characteristic polynomial of Eq. (6.35), as shown in Figure 6.12a. To avoid missing local minima in close proximity, such as the magnified pair shown in Figure 6.12b, one needs to pay attention to frequency resolution in generating the response plot and to magnify a local minimum for closer inspection.

Local minima in the dB magnitude plot may not solely correspond to a natural frequency because of the wave mode transition. The cut-off frequencies, marked on Figure 6.12 using broken vertical lines, also correspond to local minima in the dB magnitude plot of the characteristic polynomial.

To verify a natural frequency, the real and imaginary responses at a local minimum need to be examined. This is because at a natural frequency, both the real and imaginary responses are zero. As an example, Figures 6.13a and 6.13b present the real and imaginary responses of the two local minima of Figure 6.12b. For the minimum shown in Fig. 6.13a, both the real and imaginary responses are zero, hence it corresponds to a natural frequency. The minimum shown in Fig. 6.13b, however, does not have both the real and imaginary responses being zero. As a result, it does not represent a natural frequency. This local minimum is caused by the wave mode transition at one of the cut-off frequencies identified by the broken vertical lines in Figure 6.12.

The lowest eight natural frequencies are listed in Table 6.2. Note that Modes 3 and 4 of this example curved beam are in close proximity.

The modeshapes of the lowest eight modes of vibrations for flexural motion w, longitudinal motion u, and rotation of cross section φ are presented. Figures 6.14 and 6.15 show the overlaid modeshapes by type of vibration motion and by mode number, respectively. What is of particular interest on the modeshape plots are the nodal points, which are where the amplitudes of a modeshape become zero. On the one hand, when the curved beam is excited at a non-nodal position with a frequency that is identical to a natural frequency, the nodal points of the modeshape corresponding to this natural frequency are stationary. On the other hand, if the external excitation is applied at one of the nodal points of a particular natural frequency, this specific mode will not be excited, resulting in zero contribution to the dynamic response from this mode.

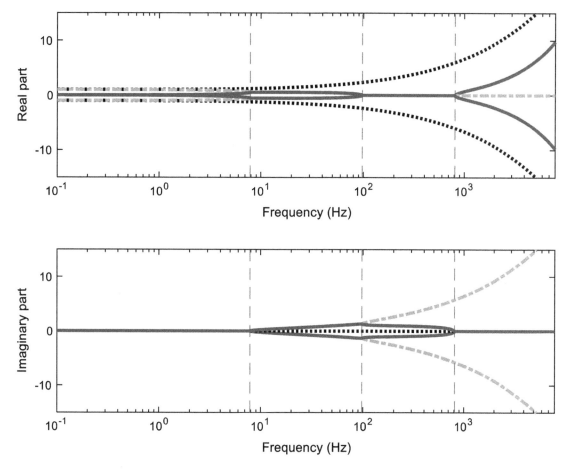

Figure 6.10 Overlaid dispersion relationships.

For forced vibration responses, the external excitation and response observation points need to be determined. Because a point on a structure is either a nodal point or a non-nodal point to a certain mode, the selections of external excitation as well as response observation location cover both scenarios and are listed in Table 6.3. These external excitation and response observation points are marked on modeshape plots of Figures 6.14 and 6.15. The locations of these points relative to the nodal points are critical in understanding forced vibration responses.

MATLAB scripts for free and forced vibration analysis are available in Section 6.6.

Figures 6.16 and 6.17 show the receptance frequency responses (displacement/force), with the external excitations applied at a non-nodal (Test A) and a nodal point (Test B). In each test, responses are observed at two different locations, as listed in Table 6.3. The frequency range for forced vibration analysis is from 0 to 8000 *Hz*, which is kept the same as that in free vibration analysis. In this frequency range, there are eight modes contributing to the dynamic responses.

The receptance frequency responses in Figures 6.16 and 6.17 show that resonant responses always occur at natural frequencies, and the resonant peaks line up regardless of the locations where the external excitations are applied or the responses are observed. Examining the response curves in both figures, however, one finds that not all resonant peaks occur at all natural frequencies. This is because the occurrence of resonant peaks in frequency responses is closely related to the locations of both the external force applied point and the response observation point relative to a nodal point.

Table 6.4 summarizes the occurrence of resonant peaks in the frequency responses of Figures 6.16 and 6.17. The locations of the external excitation applied point and/or the response observation point relative to a nodal point, which are marked on the modeshape plots of Figures 6.14 and 6.15 and are available in Table 6.3, are the reasons behind the appearance and disappearance of resonant peaks in the frequency responses.

Recall that in Test A, the external excitation is applied at a non-nodal point. Between the two observation locations, Location A1 is a non-nodal point of the lowest eight modes, while Location A2 is a nodal point of Modes 2, 3, 5, and 8 for flexural motion w; and a nodal point of Modes 1, 4, 6, and 7 for both longitudinal motion u and rotation of cross section

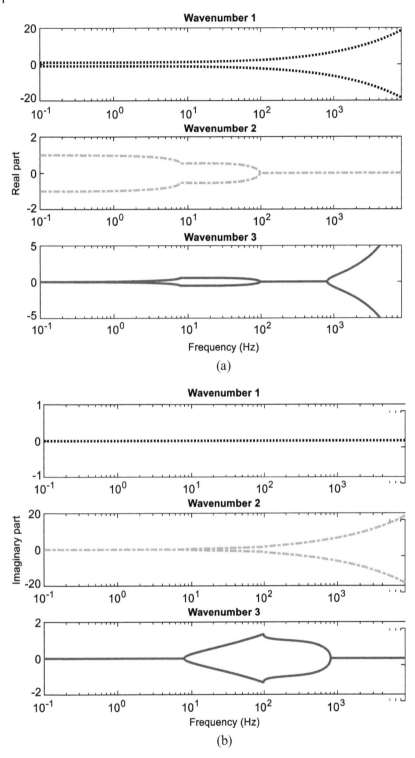

Figure 6.11 Dispersion relationships: (a) real part and (b) imaginary part.

φ. This explains why at Location A1, all resonant peaks appear; at Location A2, the resonant peaks corresponding to Modes 2, 3, 5, and 8 disappear in w; and the resonant peaks corresponding to Modes 1, 4, 6, and 7 disappear in u and φ.

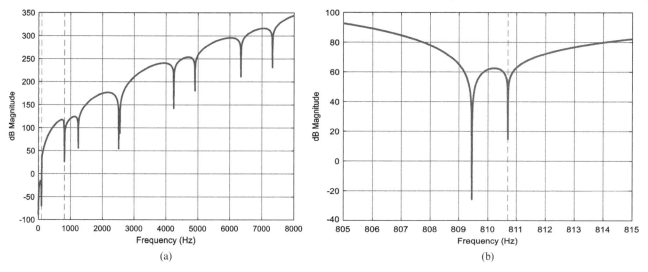

Figure 6.12 dB magnitude of the characteristic polynomial of (a) the lowest eight modes and (b) the magnified first mode.

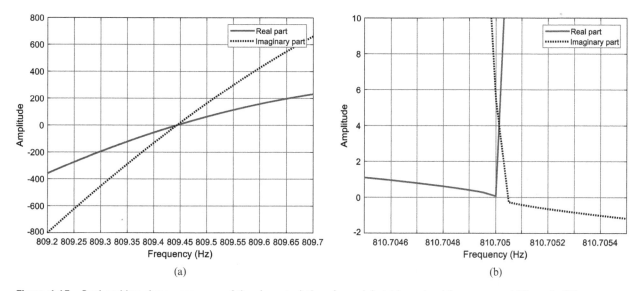

Figure 6.13 Real and imaginary responses of the characteristic polynomial at (a) a natural frequency and (b) a cut-off frequency.

In Test B, the external excitation applied point is purposely chosen at a nodal point, which is the same point as response observation point A2 in Test A. This external excitation applied point is a nodal point of Modes 2, 3, 5, and 8 for flexural motion w, and a nodal point of Modes 1, 4, 6, and 7 for both longitudinal motion u and rotation of cross section φ. None of the mentioned modes are excited; as a result, the corresponding resonant peaks disappear in the frequency responses regardless where the responses are observed. This is the case in the frequency responses of Figure 6.17.

The above studies point out the importance of nodal points. On the one hand, a nodal point needs to be avoided in identifying natural frequencies of a structure because of the disappearance of resonant peaks that results in miscounting of natural frequencies. One may avoid or reduce the related problem by exciting a structure and observing the responses at a variety of locations. On the other hand, a nodal point is an ideal location for mounting equipment on a structure from a vibration isolation standpoint because of its stationary nature.

Table 6.2 Natural frequencies (in *Hz*) of the clamped–clamped uniform curved beam.

Mode 1	Mode 2	Mode 3	Mode 4	Mode 5	Mode 6	Mode 7	Mode 8
809.44	1244.03	2518.47	2557.25	4245.09	4906.54	6346.93	7329.98

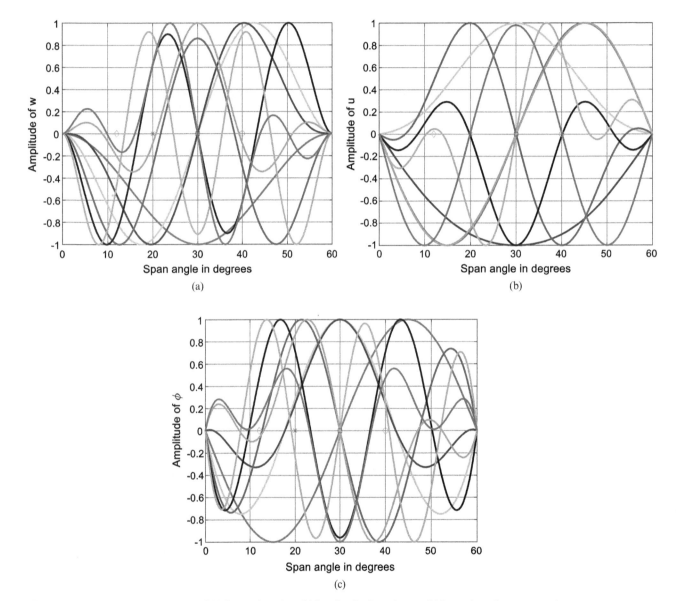

Figure 6.14 Overlaid modeshapes of (a) flexural motion, (b) longitudinal motion, and (c) rotation of a cross section.

The results obtained are applicable to slender curved beams. As a rule of thumb, the ratio between the arc length and the larger dimension of the cross section of a curved beam needs to be greater than 10.

Note that apart from Love's theory that is adopted, there are other theories for modeling in-plane vibrations in a curved beam, such as Flügge's theory (Lee et. al. 2007) and advanced theories with the effect of rotary inertia and shear deformation taken into account (Kang and Riedel 2012).

Table 6.3 External excitation and response observation points.

Test A	External excitation	Location A	s/3
			(a non-nodal point of the lowest 8 modes)
	Response observation	Location A1	s/5
			(a non-nodal point of the lowest 8 modes)
		Location A2	s/2
			(a nodal point of Modes 2, 3, 5, and 8 for w; a nodal point of Modes 1, 4, 6, and 7 for u and φ)
Test B	External excitation	Location B	s/2
			(a nodal point of modes 2, 3, 5, and 8 for w; a nodal point of Modes 1, 4, 6, and 7 for u and φ)
	Response observation	Location B1	s/5
			(a non-nodal point of the lowest eight modes)
		Location B2	2s/3
			(a non-nodal point of the lowest eight modes)

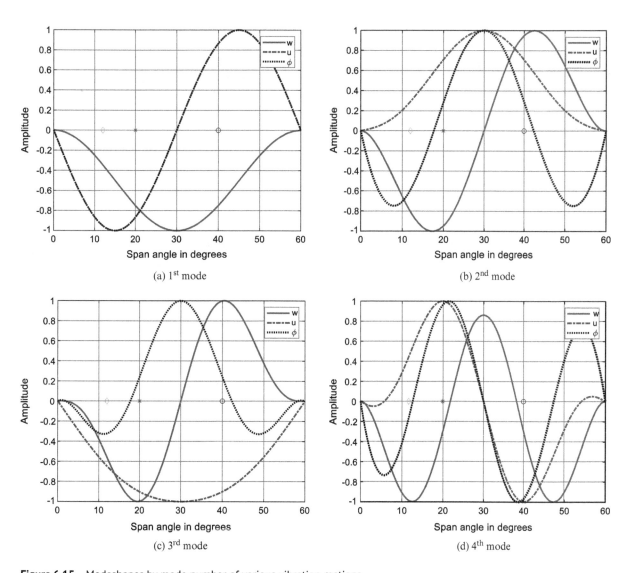

(a) 1st mode

(b) 2nd mode

(c) 3rd mode

(d) 4th mode

Figure 6.15 Modeshapes by mode number of various vibration motions.

(Continued)

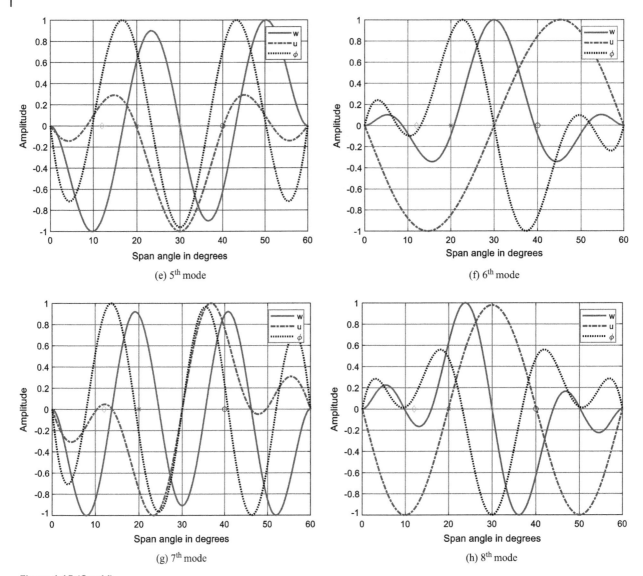

(e) 5th mode

(f) 6th mode

(g) 7th mode

(h) 8th mode

Figure 6.15 (Cont'd)

Table 6.4 Occurrence of resonant peaks.

Occurrence of resonant peaks			Mode 1	Mode 2	Mode 3	Mode 4	Mode 5	Mode 6	Mode 7	Mode 8
Test A	Location	w	✓	✓	✓	✓	✓	✓	✓	✓
(Figure 6.16)	A1	u	✓	✓	✓	✓	✓	✓	✓	✓
		φ	✓	✓	✓	✓	✓	✓	✓	✓
	Location	w	✓	✗	✗	✓	✗	✓	✓	✗
	A2	u	✗	✓	✓	✗	✓	✗	✗	✓
		φ	✗	✓	✓	✗	✓	✗	✗	✓
Test B	Locations	w	✓	✗	✗	✓	✗	✓	✓	✗
(Figure 6.17)	B1 and B2	u	✗	✓	✓	✗	✓	✗	✗	✓
		φ	✗	✓	✓	✗	✓	✗	✗	✓

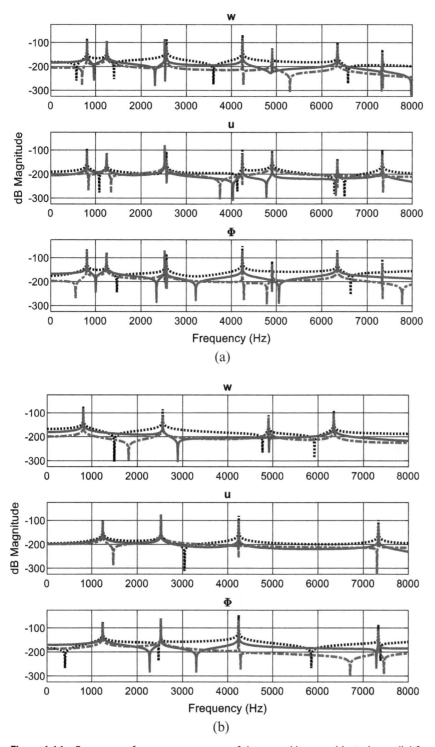

Figure 6.16 Receptance frequency responses of the curved beam subjected to radial force (___), tangential force (-·-·), and bending moment (...) excitation, respectively: responses of Test A at (a) Location A1 and (b) Location A2.

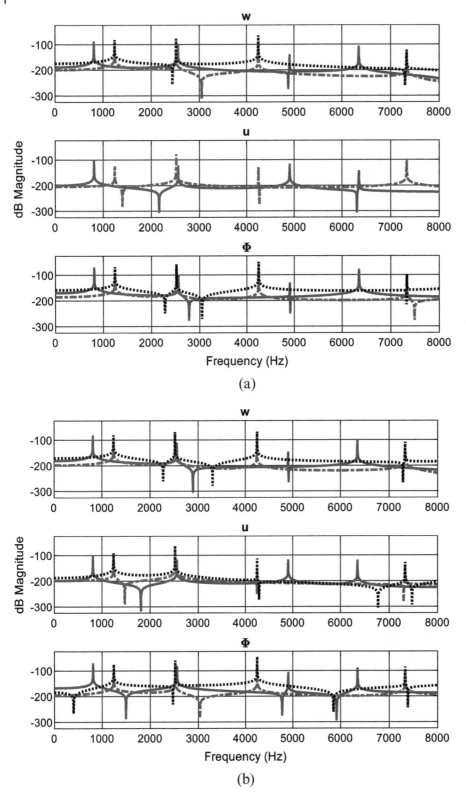

Figure 6.17 Receptance frequency responses of the curved beam subjected to radial force (___), tangential force (-.-.), and bending moment (...) excitation, respectively: responses of Test B at (a) Location B1 and (b) Location B2.

6.6 MATLAB Scripts

Script 1: Symbolic scripts for reflection matrix at an elastic boundary

```
clc, close all, clear all

syms s k1 k2 k3 ap1 ap2 ap3 an1 an2 an3 P1 P2 P3
syms E A R I Kw Ku Kr

w=ap1*exp(-i*k1*s)+ap2*exp(-i*k2*s)+ap3*exp(-i*k3*s)+an1*exp(i*k1*s)+an2*exp(i*k2*s)+an3*exp(i*k3*s)

u=P1*ap1*exp(-i*k1*s)+P2*ap2*exp(-i*k2*s)+P3*ap3*exp(-i*k3*s)-P1*an1*exp(i*k1*s)-P2*an2*exp(i*k2*s)-P3*an3*exp(i*k3*s)

w1=diff(w,s)
w2=diff(w,s,2)
w3=diff(w,s,3)

u1=diff(u,s)
u2=diff(u,s,2)

phi=-u/R+w1

N=E*A*(w/R+u1)
M=-E*I*(u1/R-w2)
V=E*I*(u2/R-w3)

Nbar=Ku*u
Vbar=Kw*w
Mbar=Kr*phi

%Spring attached BCs: -N-Nbar=0; -V-Vbar=0; -M-Mbar=0
disp('Spring attached boundary...')
N0=subs(-N-Nbar,s,0)
M0=subs(-M-Mbar,s,0)
V0=subs(-V-Vbar,s,0)

NN0=collect(N0, {'ap1', 'ap2', 'ap3', 'an1', 'an2', 'an3'})
MM0=collect(M0, {'ap1', 'ap2', 'ap3', 'an1', 'an2', 'an3'})
VV0=collect(V0, {'ap1', 'ap2', 'ap3', 'an1', 'an2', 'an3'})
```

Script 2: Symbolic scripts for reflection matrix at end mass attached boundary

```
clc, close all, clear all

syms s k1 k2 k3 ap1 ap2 ap3 an1 an2 an3 P1 P2 P3
syms E A R I Kw Ku Kr omega m Jm h

w=ap1*exp(-i*k1*s)+ap2*exp(-i*k2*s)+ap3*exp(-i*k3*s)+an1*exp(i*k1*s)+an2*exp(i*k2*s)+an3*exp(i*k3*s)
```

```
u=P1*ap1*exp(-i*k1*s)+P2*ap2*exp(-i*k2*s)+P3*ap3*exp(-i*k3*s)-P1*an1*exp(i*k1*s)-P2*an2*exp(i*k2*s)-
P3*an3*exp(i*k3*s)

w1=diff(w,s)
w2=diff(w,s,2)
w3=diff(w,s,3)

u1=diff(u,s)
u2=diff(u,s,2)

phi=-u/R+w1

phim=phi
um=u
wm=w+phim*h/2

N=E*A*(w/R+u1)
M=-E*I*(u1/R-w2)
V=E*I*(u2/R-w3)

%At mass attached boundary
%-N+m*omega^2*um=0;
%-V+m*omega^2*wm=0;
%-M+V*h/2+Jm*omega^2*phim=0
disp('End mass attached boundary...')
N0=subs(-N+m*omega^2*um,s,0)
M0=subs(-M+V*h/2+Jm*omega^2*phim,s,0)
V0=subs(-V+m*omega^2*wm,s,0)

NN0=collect(N0, {'ap1', 'ap2', 'ap3', 'an1', 'an2', 'an3'})
MM0=collect(M0, {'ap1', 'ap2', 'ap3', 'an1', 'an2', 'an3'})
VV0=collect(V0, {'ap1', 'ap2', 'ap3', 'an1', 'an2', 'an3'})
```

Script 3: Free and forced vibration analysis of a curved beam

```
clc; clear all; close all;

%Material and geometrical parameters of the steel beam
E=198.87*10^9;
p=7664.5;
R=1;            %Radius of curvature
h=0.1;          %Thickness
Theta=60        %Span angle in degrees

s=R*(Theta*pi/180)
b=h;

A = b*h;
I= 1/12*b*h^3;

%Disturbance force applied point
s1=(1/3)*s
s2=s-s1;
```

```
xforce=s1;

x1sensor=(1/5)*s;
x2sensor=s1/2;
x3sensor=s1+s2/2;

%Frequency range in Hz
f1 = 0;        %Starting frequency in Hz
f2 = 8000;     %Ending frequency in Hz
stepsize=0.1

freq = f1:stepsize:f2;

for n = 1:length(freq)

  w=2*pi*freq(n);

a0=1;
b0=-w^2*p/E-2/R^2;
c0=1/R^4-w^2*p*A/E/I-w^2*p^2/R^2/E^2;
d0=-w^2*p*A/E/I/R^2+w^4*p^2*A/E^2/I;

Delta(n)=18*a0*b0*c0*d0-4*(b0^3)*d0+(b0^2)*(c0^2)-4*a0*(c0^3)-27*(a0^2)*(d0^2);

%Sort out the sequence of wavenumbers
 ksquared=roots([a0 b0 c0 d0]);

 tempk1=sqrt(ksquared(1));
 tempk2=sqrt(ksquared(2));
 tempk3=sqrt(ksquared(3));

 if imag(tempk1)~=0
   k1=tempk2;
   k2=tempk1;
 else
   k1=tempk1;
   k2=tempk2;
 end

 k3=tempk3;

%EITHER
P1=((k1^4)*(R^2)*E*I+E*A-p*A*(R^2)*(w^2))/(i*(k1^3)*E*I*R+i*k1*E*A*R);
P2=((k2^4)*(R^2)*E*I+E*A-p*A*(R^2)*(w^2))/(i*(k2^3)*E*I*R+i*k2*E*A*R);
P3=((k3^4)*(R^2)*E*I+E*A-p*A*(R^2)*(w^2))/(i*(k3^3)*E*I*R+i*k3*E*A*R);

%OR EQUIVALENTLY
% P1=(i*(k1^3)*E*I*R+i*k1*E*A*R)/(-(k1^2)*E*A*(R^2)-(k1^2)*E*I+p*A*(R^2)*(w^2));
% P2=(i*(k2^3)*E*I*R+i*k2*E*A*R)/(-(k2^2)*E*A*(R^2)-(k2^2)*E*I+p*A*(R^2)*(w^2));
% P3=(i*(k3^3)*E*I*R+i*k3*E*A*R)/(-(k3^2)*E*A*(R^2)-(k3^2)*E*I+p*A*(R^2)*(w^2));

%Boundary reflection
```

```
AmClamped=[P1 P2 P3; 1 1 1; i*k1+P1/R i*k2+P2/R i*k3+P3/R];
ApClamped=[P1 P2 P3; -1 -1 -1; i*k1+P1/R i*k2+P2/R i*k3+P3/R];
rClamped=inv(AmClamped)*ApClamped;

AmFree=[-1/R+P1*k1*1i, -1/R+P2*k2*1i, -1/R+P3*k3*1i;k1^2-(P1*k1*1i)/R, k2^2-(P2*k2*1i)/R, k3^2-(P3*k3*1i)/R;
(P1*k1^2)/R + k1^3*1i, (P2*k2^2)/R + k2^3*1i, (P3*k3^2)/R + k3^3*1i];
ApFree=[-(-1/R+P1*k1*1i), -(-1/R+P2*k2*1i), -(-1/R+P3*k3*1i);-(k1^2-(P1*k1*1i)/R), -(k2^2-(P2*k2*1i)/R), -(k3^2-
(P3*k3*1i)/R); (P1*k1^2)/R + k1^3*1i, (P2*k2^2)/R + k2^3*1i, (P3*k3^2)/R + k3^3*1i];
rFree=inv(AmFree)*ApFree;

AmPinned=[P1, P2, P3; 1, 1, 1; k1^2-(P1*k1*1i)/R, k2^2-(P2*k2*1i)/R, k3^2-(P3*k3*1i)/R];
ApPinned=[P1, P2, P3; -1, -1, -1; -(k1^2-(P1*k1*1i)/R), -(k2^2-(P2*k2*1i)/R), -(k3^2-(P3*k3*1i)/R)];
rPinned=inv(AmPinned)*ApPinned;

%Setting boundary conditions
rA=rClamped;
rB=rClamped;

x1sensor=(1/5)*s;
x2sensor=s/2;
x3sensor=xforce+(s-xforce)/2;

fs1 = diag([exp(-i*k1*s1) exp(-i*k2*s1) exp(-i*k3*s1)]);
fs2 = diag([exp(-i*k1*s2) exp(-i*k2*s2) exp(-i*k3*s2)]);

fx1FromA=diag([exp(-i*k1*x1sensor) exp(-i*k2*x1sensor) exp(-i*k3*x1sensor)]);
fx2FromG=diag([exp(-i*k1*(x2sensor-s1)) exp(-i*k2*(x2sensor-s1)) exp(-i*k3*(x2sensor-s1))]);
fx3FromB=diag([exp(-i*k1*(s-x3sensor)) exp(-i*k2*(s-x3sensor)) exp(-i*k3*(s-x3sensor))]);

%External excitations
Vbar=1;
Nbar=0;
Mbar=0;

qVCoeff=Vbar*R/(2*E*I)/((-i*R*P3*k1*k2+P1*P2*k3)*(k1^2-k2^2)+(-i*R*P2*k1*k3+P1*P3*k2)*( k3^2-k1^2)
+(-i*R*P1*k2*k3+P2*P3*k1)*(k2^2-k3^2));
qV=qVCoeff*[P2*k3-P3*k2; P3*k1-P1*k3; P1*k2-P2*k1];

qNCoeff=Nbar/(2*R*E*A)/(P3*k3*(k1^2-k2^2)+P2*k2*(k3^2-k1^2)+P1*k1*(k2^2-k3^2));
qN=qNCoeff*[i*R*(k3^2-k2^2)-P2*k2+P3*k3; i*R*(k1^2-k3^2)+P1*k1-P3*k3;i*R*(k2^2-k1^2)-P1*k1+P2*k2];

qMCoeff=Mbar/(2*E*I)/(P3*k3*(k1^2-k2^2)+P2*k2*(k3^2-k1^2)+P1*k1*(k2^2-k3^2));
qM=qMCoeff*[-P2*k2+P3*k3; P1*k1-P3*k3; -P1*k1+P2*k2];

qpforce=qV+qN+qM;
qmforce=-qV+qN+qM;

% Forming the characteristic matrix

BigA=zeros(24,24);

%Boundary reflections
RefA=[-eye(3) rA];
```

```matlab
RefB=[rB -eye(3)];

% Force Generated Waves

BigB=[zeros(3,1);zeros(3,1);qmforce;qpforce;zeros(3,1);zeros(3,1);zeros(3,1);zeros(3,1)];

ExternalForce1=[zeros(3) eye(3) zeros(3) -eye(3)];
ExternalForce2=[eye(3) zeros(3) -eye(3) zeros(3)];

%a+ a- g1+ g1- g2+ g2- (force G) b+ b- (3 by 3 matrix form)
%Propagations along s1 and s2
PropL1=[fs1 zeros(3) -eye(3) zeros(3); zeros(3) -eye(3) zeros(3) fs1];
PropL2=[fs2 zeros(3) -eye(3) zeros(3); zeros(3) -eye(3) zeros(3) fs2];

% a1+ a2+ a3+ a1- a2- a3- g11+ g12+ g13+ g11- g12- g13- g21+ g22+ g23+
% g21- g22- g23- b1+ b2+ b3+ b1- b2- b3-

BigA(1:3,1:6)=RefA;
BigA(4:6,19:24)=RefB;
BigA(7:9,7:18)=ExternalForce1;
BigA(10:12,7:18)=ExternalForce2;

BigA(13:18,1:12)=PropL1;
BigA(19:24,13:24)=PropL2;

ForcedResponse=BigA\BigB;

% a1+ a2+ a3+ a1- a2- a3- g11+ g12+ g13+ g11- g12- g13- g21+ g22+ g23+
% g21- g22- g23- b1+ b2+ b3+ b1- b2- b3-

aplus=ForcedResponse(1:3);
aminus=ForcedResponse(4:6);

g1plus=ForcedResponse(7:9);
g1minus=ForcedResponse(10:12);
g2plus=ForcedResponse(13:15);
g2minus=ForcedResponse(16:18);

bplus=ForcedResponse(19:21);
bminus=ForcedResponse(22:24);

% w0 = an1 + an2 + an3 + ap1 + ap2 + ap3 - bn1 - bn2 - bn3 - bp1 - bp2 - bp3
% u0 = P1*ap1 - P2*an2 - P3*an3 - P1*an1 + P2*ap2 + P3*ap3 + P1*bn1 + P2*bn2 + P3*bn3 - P1*bp1 - P2*bp2 - P3*bp3
% phi0 = (P1/R + k1*1i)*an1 + (P2/R + k2*1i)*an2 + (P3/R + k3*1i)*an3 + (- P1/R - k1*1i)*ap1 + (- P2/R - k2*1i)*ap2 +
% (- P3/R - k3*1i)*ap3 + (- P1/R - k1*1i)*bn1 + (- P2/R - k2*1i)*bn2 + (- P3/R - k3*1i)*bn3 + (P1/R + k1*1i)*bp1 + (P2/R +
% k2*1i)*bp2 + (P3/R + k3*1i)*bp3

W1sensor(n)=[1 1 1]*fx1FromA*aplus+[1 1 1]*inv(fx1FromA)*aminus;
W2sensor(n)=[1 1 1]*fx2FromG*g2plus+[1 1 1]*inv(fx2FromG)*g2minus;
W3sensor(n)=[1 1 1]*fx3FromB*bminus+[1 1 1]*inv(fx3FromB)*bplus;

U1sensor(n)=[1 1 1]*fx1FromA*(diag([P1, P2, P3])*aplus)+[1 1 1]*inv(fx1FromA)*(diag([-P1, -P2, -P3])*aminus);
U2sensor(n)=[1 1 1]*fx2FromG*(diag([P1, P2, P3])*g2plus)+[1 1 1]*inv(fx2FromG)*(diag([-P1, -P2, -P3])*g2minus);
```

```
U3sensor(n)=[1 1 1]*fx3FromB*(diag([-P1, -P2, -P3])*bminus)+[1 1 1]*inv(fx3FromB)*(diag([P1, P2, P3])*bplus);

Phi1sensor(n)=[1 1 1]*fx1FromA*(diag([-P1/R-k1*1i, -P2/R-k2*1i, -P3/R-k3*1i])*aplus)+[1 1 1]*
inv(fx1FromA)*(diag([P1/R+k1*1i, P2/R+k2*1i, P3/R+k3*1i])*aminus);
Phi2sensor(n)=[1 1 1]*fx2FromG*(diag([-P1/R-k1*1i, -P2/R-k2*1i, -P3/R-k3*1i])*g2plus)+[1 1 1]*
inv(fx2FromG)*(diag([P1/R+k1*1i, P2/R+k2*1i, P3/R+k3*1i])*g2minus);
Phi3sensor(n)=[1 1 1]*fx3FromB*(diag([P1/R+k1*1i, P2/R+k2*1i, P3/R+k3*1i])*bminus)+[1 1 1]*
inv(fx3FromB)*(diag([-P1/R-k1*1i, -P2/R-k2*1i, -P3/R-k3*1i])*bplus);

result(n)=det(BigA);

end

figure(1),loglog(freq,abs(Delta)), grid on
disp('Read the roots of the discriminant of the dispersion equation for two cutoff frequencies...')
x=ginput(2)
disp('Press a key to continue with response plots...')
pause

%Cutoff frequencies
CutoffFreq1=x(1,1)
CutoffFreq2=x(2,1)
CutoffFreq0=sqrt(E/p)/(2*pi*R)

figure(2);
plot(freq,20*log10(abs(result)),':b','LineWidth',2); hold on
xline(CutoffFreq1,'--r','LineWidth',1.5),hold on, xline(CutoffFreq2,'--r','LineWidth',1.5),hold on, xline(CutoffFreq0,
'--r','LineWidth',1.5),hold on
xlabel('Frequency (Hz)')
ylabel('dB Magnitude')
grid on
hold on

figure(10);
subplot(3,1,1), plot(freq,20*log10(abs(W1sensor)),':b','LineWidth',2), title('W1'); grid on, hold on
xline(CutoffFreq1,'--r','LineWidth',1.5),hold on, xline(CutoffFreq2,'--r','LineWidth',1.5),hold on, xline(CutoffFreq0,
'--r','LineWidth',1.5),hold on

subplot(3,1,2), plot(freq,20*log10(abs(U1sensor)),':g','LineWidth',2), title('U1'); grid on, hold on
xline(CutoffFreq1,'--r','LineWidth',1.5),hold on, xline(CutoffFreq2,'--r','LineWidth',1.5),hold on, xline(CutoffFreq0,
'--r','LineWidth',1.5),hold on
ylabel('dB Magnitude')

subplot(3,1,3), plot(freq,20*log10(abs(Phi1sensor)),':k','LineWidth',2), title('\Phi1'); grid on, hold on
xline(CutoffFreq1,'--r','LineWidth',1.5),hold on, xline(CutoffFreq2,'--r','LineWidth',1.5),hold on, xline(CutoffFreq0,
'--r','LineWidth',1.5),hold on
xlabel('Frequency (Hz)')

figure(20);
subplot(3,1,1), plot(freq,20*log10(abs(W2sensor)),':b','LineWidth',2), title('W2'); grid on, hold on
xline(CutoffFreq1,'--r','LineWidth',1.5),hold on, xline(CutoffFreq2,'--r','LineWidth',1.5),hold on, xline(CutoffFreq0,
'--r','LineWidth',1.5),hold on
```

```
subplot(3,1,2), plot(freq,20*log10(abs(U2sensor))),':g','LineWidth',2), title('U2'); grid on, hold on
xline(CutoffFreq1,'--r','LineWidth',1.5),hold on, xline(CutoffFreq2,'--r','LineWidth',1.5),hold on, xline(CutoffFreq0,
'--r','LineWidth',1.5),hold on
ylabel('dB Magnitude')

subplot(3,1,3), plot(freq,20*log10(abs(Phi2sensor))),':k','LineWidth',2), title('\Phi2'); grid on, hold on
xline(CutoffFreq1,'--r','LineWidth',1.5),hold on, xline(CutoffFreq2,'--r','LineWidth',1.5),hold on, xline(CutoffFreq0,
'--r','LineWidth',1.5),hold on
xlabel('Frequency (Hz)')

figure(30);
subplot(3,1,1), plot(freq,20*log10(abs(W3sensor)),'b','LineWidth',2), title('W3'); grid on, hold on
xline(CutoffFreq1,'--r','LineWidth',1.5),hold on, xline(CutoffFreq2,'--r','LineWidth',1.5),hold on, xline(CutoffFreq0,
'--r','LineWidth',1.5),hold on

subplot(3,1,2), plot(freq,20*log10(abs(U3sensor)),'g','LineWidth',2), title('U3'); grid on, hold on
xline(CutoffFreq1,'--r','LineWidth',1.5),hold on, xline(CutoffFreq2,'--r','LineWidth',1.5),hold on, xline(CutoffFreq0,
'--r','LineWidth',1.5),hold on
ylabel('dB Magnitude')

subplot(3,1,3), plot(freq,20*log10(abs(Phi3sensor)),'k','LineWidth',2), title('\Phi3'); grid on, hold on
xline(CutoffFreq1,'--r','LineWidth',1.5),hold on, xline(CutoffFreq2,'--r','LineWidth',1.5),hold on, xline(CutoffFreq0,
'--r','LineWidth',1.5),hold on
xlabel('Frequency (Hz)')
```

References

Kang B., Riedel C.H., Tan C.A. Free Vibration Analysis of Planar Curved Beams by Wave Propagation, *Journal of Sound and Vibration*, 260, 19–44 (2003).

Kang B., Riedel C.H. Coupling of In-Plane Flexural, Tangential, and Shear Wave Modes of a Curved Beam, *ASME Journal of Vibration and Acoustics*, 134, 011001-1-13 (2012).

Lee S.K., Mace B.R., Brennan M.J. Wave Propagation, Reflection and Transmission in Curved Beams, *Journal of Sound and Vibration*, 306, 636–656 (2007).

Walsh S.J., White R.G. Vibrational Power Transmission in Curved Beams, *Journal of Sound and Vibration*, 233, 455–488 (2000).

Homework Project

Free and Forced Vibration Analysis of a Curved Beam

Free and forced vibrations in an example slender curved steel beam are analyzed in Section 6.5. The material and geometrical properties of the curved beam are as follows: the Young's modulus and volume mass density are $E = 198.87\ GN/m^2$ and $\rho = 7664.5\ kg/m^3$, respectively. The curved beam is of a square cross section whose side length is $h = 0.1\ m$. The radius of curvature R of the curved beam is $1.0\ m$ and the span angle θ is $60\,°$. The boundaries of the curved beam are clamped–clamped.

In general, the properties of a structure can be put into three categories: material, geometrical, and boundary.

In this project, vary the properties of the slender curved beam one category at a time, modify MATLAB Script 3 that is provided at the end of Chapter 6, and observe how the change in a property in each of the three categories affect the natural frequencies f_n and cut-off frequencies f_c of the curved beam.

- Task 1: Change the material property of the example slender curved beam analyzed in Section 6.5 by assuming that the curved beam is made of copper instead of steel. The Young's modulus and volume mass density for copper are $E = 103\ GN/m^2$ and $\rho = 8830\ kg/m^3$.
- Task 2: Change the boundary condition of the example slender curved beam analyzed in Section 6.5 by replacing one of the clamped boundaries with a free boundary.
- Task 3-1: Change the geometrical property of the example slender curved beam analyzed in Section 6.5 by assuming that the span angle θ is $90\,°$ instead of $60\,°$.
- Task 3-2: Change the geometrical property of the example slender curved beam analyzed in Section 6.5 by assuming that the radius of curvature R of the curved beam is $2\ m$ instead of $1\ m$.
- Task 3-3: Change the geometrical property of the example slender curved beam analyzed in Section 6.5 by assuming that the square cross section has a side length $0.08\ m$ instead of $0.1\ m$

For each task listed above,

i) Present the values of the three cut-off frequencies, that is, the two roots of the discriminant of the characteristic equation, and the cut-off frequency corresponding to the ring frequency.
ii) Obtain the overlaid magnitude plots of the characteristic polynomials of the first eight modes before and after the change of property.
iii) Obtain the overlaid receptance frequency response plots of the curved beam to include the first eight modes, with an external radial force applied at $s/3$ and with the response observed at $s/5$, both measured from the clamped end, where s is the arc length of the curved beam.
iv) Record the observations in the table below, and justify the observations.

Category	Parameters	Original value	Value for Task 1	Value for Task 2	Value for Task 3-1	Value for Task 3-2	Value for Task 3-3	Observation of change in	
								f_n (yes, no)	f_c (yes, no)
Material	$E\ (GN/m^2)$	198.87	103						
	$\rho\ (kg/m^3)$	7664.5	8830						
Boundary	Left end	Clamped							
	Right end	Clamped		Free					
Geometrical	θ (degrees)	60°			90°				
	$R\ (m)$	1.0				2			
	$H\ (m)$	0.1					0.08		

7

Flexural/Bending Vibration of Rectangular Isotropic Thin Plates with Two Opposite Edges Simply-supported

In the previous studies, the wave vibration analysis has been restricted to one-dimensional structures, that is, structures whose length is large in comparison to its cross-sectional dimensions such as diameter or width and height. In this chapter, free flexural vibrations in two-dimensional rectangular plates are analyzed from the wave vibration standpoint. The coverage is limited to situations where closed-form solutions exist for the plate, which require at least one pair of opposite edges to be simply-supported.

7.1 The Governing Equations of Motion

Consider free vibration of a uniform thin isotropic plate lying in the x–y plane shown in Figure 7.1. The equation of motion for out-of-plane vibration is (Leissa 1973)

$$D\left[\frac{\partial^4 w(x,y,t)}{\partial x^4}+2\frac{\partial^4 w(x,y,t)}{\partial x^2\partial y^2}+\frac{\partial^4 w(x,y,t)}{\partial y^4}\right]+\rho h\frac{\partial^2 w(x,y,t)}{\partial t^2}=0 \tag{7.1}$$

where x and y are the coordinates of a point on the plate and t is the time. $w(x,y,t)$ denotes the out-of-plane plate deflection along the z direction. ρ is the volume mass density and h is the thickness of the plate. The dimensions of the plate along the x and y directions are L_1 and L_2, respectively. D is the plate flexural rigidity, for isotropic and homogeneous materials, $D=\dfrac{Eh^3}{12\left(1-\gamma^2\right)}$, in which γ is the Poisson's ratio and E is the Young's modulus.

By the sign convention defined in Figure 7.1 and according to thin plate bending vibration theory, the internal resistant bending moments M_x and M_y, torques (sometimes also called twisting moments) M_{xy} and M_{yx}, and shear forces Q_x and Q_y, are related to the transverse deflection $w(x,y,t)$,

$$M_x=-D\left(\frac{\partial^2 w}{\partial x^2}+\gamma\frac{\partial^2 w}{\partial y^2}\right)$$

$$M_y=-D\left(\frac{\partial^2 w}{\partial y^2}+\gamma\frac{\partial^2 w}{\partial x^2}\right)$$

$$M_{xy}=-M_{yx}=D(1-\gamma)\frac{\partial^2 w}{\partial x\partial y}$$

$$Q_x=\frac{\partial M_x}{\partial x}+\frac{\partial M_{yx}}{\partial y}=\frac{\partial M_x}{\partial x}-\frac{\partial M_{xy}}{\partial y}$$

$$Q_y=\frac{\partial M_y}{\partial y}-\frac{\partial M_{xy}}{\partial x} \tag{7.2}$$

Mechanical Wave Vibrations: Analysis and Control, First Edition. Chunhui Mei.
© 2023 Chunhui Mei. Published 2023 by John Wiley & Sons Ltd.
Companion Website: www.wiley.com/go/Mei/MechanicalWaveVibrations

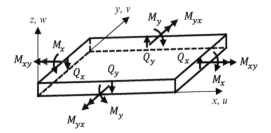

Figure 7.1 Definition of positive internal resistant bending moments, torques, and shear forces.

By analogy with beams, there are three types of classical boundaries, namely, the free, clamped, and simply-supported boundaries. The boundary conditions for these classical boundaries along the x and y directions are listed in Table 7.1, where V_x and V_y are the Kirchhoff's shear forces that are defined as

$$V_x = Q_x - \frac{\partial M_{xy}}{\partial y}$$

$$V_y = Q_y - \frac{\partial M_{xy}}{\partial x} \tag{7.3}$$

7.2 Closed-form Solutions

Assuming time harmonic motion of circular frequency ω and applying separation of variables, the solutions to Eq. (7.1), the governing equation of motion for free vibration, can be written in the form of

$$w(x,y,t) = X_x Y_y \, e^{i\omega t} \tag{7.4}$$

Substituting Eq. (7.4) into Eq. (7.1), with time dependence $e^{i\omega t}$ suppressed and using prime to denote the derivative with respect to variable x or y indicated in the subscript, the governing equation of motion becomes

$$X_x''' Y_y + 2X_x'' Y_y'' + X_x Y_y''' - \beta^4 X_x Y_y = 0 \tag{7.5}$$

where $\beta^4 = \dfrac{\rho h \omega^2}{D}$.

From Eq. (7.5), it can be seen that for the separation of variables to occur, one must have the following proportional relationships, where symbol "$\propto$" denotes "proportional to",

Table 7.1 Classical boundary conditions for out-of-plane vibration of a plate.

Classical Boundary	Edge parallel to the x axis at $y = y_0$	Edge parallel to the y axis at $x = x_0$
Clamped	$w(x, y_0, t) = 0$	$w(x_0, y, t) = 0$
	$\left. \dfrac{\partial w(x,y,t)}{\partial y} \right|_{y=y_0} = 0$	$\left. \dfrac{\partial w(x,y,t)}{\partial x} \right|_{x=x_0} = 0$
Free	$\left. V_y \right|_{y=y_0} = 0$	$\left. V_x \right|_{x=x_0} = 0$
	$\left. M_y \right|_{y=y_0} = 0$	$\left. M_x \right|_{x=x_0} = 0$
Simply-supported	$w(x, y_0, t) = 0$	$w(x_0, y, t) = 0$
	$\left. M_y \right|_{y=y_0} = 0$	$\left. M_x \right|_{x=x_0} = 0$

$$X_x''' \propto X_x'' \propto X_x \tag{7.6a}$$

or

$$Y_y''' \propto Y_y'' \propto Y_y \tag{7.6b}$$

The conditions described in Eqs. (7.6a) and (7.6b) indicate that either X_x must be in the form $e^{\pm ik_x x}$ or Y_y must be in the form $e^{\pm ik_y y}$. Without loss of generality, we assume that the case described in Eq. (7.6a) holds. The general solution to X_x is a linear combination of $e^{\pm ik_x x}$, that is

$$X_x = b^+ e^{-ik_x x} + b^- e^{ik_x x} \tag{7.7}$$

The $\pm$ sign on the power of the exponential function indicates that waves in the plates travel in both the positive and negative directions along x, and they are of the same wavenumber k_x. b^+ and b^- denote the amplitudes of the positive- and negative-going wave components.

Equation (7.7) indicates that there exists only one pair of propagating waves along the x direction. This condition is physically satisfied when the pair of opposite edges along the x direction is simply-supported, as will be shown in Section 7.3.

In summary, in order for the separation of variables to occur and hence for closed-form solutions to exist, we must have at least one pair of opposite edges along either the x or the y direction being simply-supported.

Equation (7.4) can be rewritten in the following form using separation of variables

$$w(x,y,t) = w_0 e^{-ik_x x} e^{-ik_y y} e^{i\omega t} \tag{7.8}$$

where k_x and k_y are the wavenumbers along the x and y directions, respectively, and w_0 is the amplitude of the out-of-plane deflection.

Substituting Eq. (7.8) into Eq. (7.1) gives the dispersion equation,

$$k_y^4 - 2k_x^2 k_y^2 + k_x^4 - \beta^4 = 0 \tag{7.9a}$$

which can be written in factored form as

$$\left[k_y^2 + (k_x^2 - \beta^2) \right]\left[k_y^2 + (k_x^2 + \beta^2) \right] = 0 \tag{7.9b}$$

From Eq. (7.9b), four roots are found, namely, $\pm k_{y1}$ and $\pm k_{y2}$, where

$$k_{y1} = \sqrt{-k_x^2 + \beta^2}, \text{ and } k_{y2} = \sqrt{-(k_x^2 + \beta^2)} \tag{7.9c}$$

As a result, two pairs of waves exist along the y direction,

$$Y_y = a_1^+ e^{-ik_{y1} y} + a_2^+ e^{-ik_{y2} y} + a_1^- e^{ik_{y1} y} + a_2^- e^{ik_{y2} y} \tag{7.10}$$

The $\pm$ sign on the power of the exponential function indicates that waves in the plates travel in both the positive and negative directions along y. Superscripts $+$ and $-$ and subscripts 1 and 2 in $a_{1,2}^\pm$ denote positive- and negative-going wave components along the y direction. Waves along the simply-supported x direction are described by Eq. (7.7).

For simplicity, let $k_0 = k_x$, $k_1 = k_{y1}$, and $k_2 = k_{y2}$. From Eqs. (7.4), (7.7), and (7.10), the closed-form solutions to Eq. (7.1) are

$$w(x,y,t) = (b^+ e^{-ik_0 x} + b^- e^{ik_0 x})(a_1^+ e^{-ik_1 y} + a_2^+ e^{-ik_2 y} + a_1^- e^{ik_1 y} + a_2^- e^{ik_2 y})e^{i\omega t} \tag{7.11a}$$

or in the form of Eq. (7.4), that is

$$w(x,y,t) = X_x Y_y \, e^{i\omega t} \tag{7.11b}$$

where

$$X_x = b^+ e^{-ik_0 x} + b^- e^{ik_0 x} \tag{7.11c}$$

$$Y_y = a_1^+ e^{-ik_1 y} + a_2^+ e^{-ik_2 y} + a_1^- e^{ik_1 y} + a_2^- e^{ik_2 y} \tag{7.11d}$$

Let us examine the reflection of vibration waves incident upon a pair of simply-supported boundaries along the x direction.

7.3 Wave Reflection, Propagation, and Wave Vibration Analysis along the Simply-supported x Direction

At a boundary, as shown in Figure 7.2, an incident wave b^+ gives rise to a reflected wave b^-, which are related by

$$b^- = rb^+ \tag{7.12}$$

where reflection coefficient r is determined by boundary conditions.

Choosing the origin at the boundary, the boundary conditions corresponding to a simple support boundary along the x direction, as described in Table 7.1, are

$$w(0,y,t) = 0 \text{ and } M_x\big|_{x=0} = 0 \tag{7.13}$$

From Eqs. (7.2), (7.11b), and (7.13),

$$w(0,y,t) = X_x\big|_{x=0} Y_y \, e^{i\omega t} = 0$$

$$M_x\big|_{x=0} = -D\left(\frac{\partial^2 w}{\partial x^2} + \gamma \frac{\partial^2 w}{\partial y^2}\right)\bigg|_{x=0} = -D(X_x'' Y_y + \gamma Y_y'' X_x)e^{i\omega t}\big|_{x=0} = 0 \tag{7.14}$$

From Eqs. (7.11c) and (7.14),

$$w(0,y,t) = (b^+ + b^-)Y_y \, e^{i\omega t} = 0$$

$$M_x\big|_{x=0} = -D\left[-k_0^2(b^+ + b^-)Y_y + \gamma(b^+ + b^-)Y_y''\right]e^{i\omega t} = 0 \tag{7.15}$$

From Eq. (7.15),

$$b^- = -b^+ \tag{7.16}$$

From Eqs. (7.12) and (7.16), the reflection coefficient for a simply-supported boundary along the x direction is

$$r = -1 \tag{7.17}$$

With the relationships found in Eq. (7.16), Eq. (7.11a) can be rewritten as

$$w(x,y,t) = (e^{-ik_0 x} - e^{ik_0 x})(c_1^+ e^{-ik_1 y} + c_2^+ e^{-ik_2 y} + c_1^- e^{ik_1 y} + c_2^- e^{ik_2 y})e^{i\omega t} \tag{7.18a}$$

or in compact form as

$$w(x,y,t) = (e^{-ik_0 x} - e^{ik_0 x})Y_y e^{i\omega t} \tag{7.18b}$$

where

$$Y_y = c_1^+ e^{-ik_1 y} + c_2^+ e^{-ik_2 y} + c_1^- e^{ik_1 y} + c_2^- e^{ik_2 y} \tag{7.18c}$$

Equations (7.18a) to (7.18c) describe the out-of-plane plate vibrations $w(x,y,t)$ along the x and y directions, with the pair of boundaries along the x direction simply-supported.

Vibration waves propagate along a uniform waveguide. Consider two locations A and B along the x direction on a uniform plate that are a distance x apart, as shown in Figure 7.3. Waves propagate from A to B, with the propagation relationships determined by k_0, which is the wavenumber along the x direction. The positive- and negative-going waves at Locations A and B are related by

$$b^+ = f(x)a^+, a^- = f(x)b^- \tag{7.19a}$$

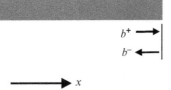

Figure 7.2 Wave reflection at a boundary along the x direction.

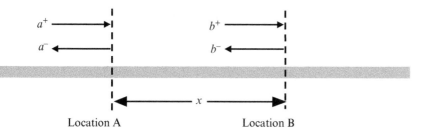

Figure 7.3 Wave propagation along the x direction.

where

$$f(x) = e^{-ik_0 x} \tag{7.19b}$$

is the propagation coefficient for a distance x.

Figure 7.4 illustrates a uniform rectangular plate of side length L_1 along the x direction.

The following reflection and propagation relationships exist:

$$a^+ = r_A\, a^-;\ b^- = r_B\, b^+;\ b^+ = f(L_1)\, a^+;\ a^- = f(L_1)\, b^- \tag{7.20}$$

where $f(L_1)$ is the propagation coefficient for a distance L_1, and r_A and r_B are the reflection coefficients at Boundaries A and B, respectively.

Assembling the relationships in Eq. (7.20) into matrix form,

$$\begin{bmatrix} -1 & r_A & 0 & 0 \\ 0 & 0 & r_B & -1 \\ f(L_1) & 0 & -1 & 0 \\ 0 & -1 & 0 & f(L_1) \end{bmatrix} \begin{bmatrix} a^+ \\ a^- \\ b^+ \\ b^- \end{bmatrix} = 0 \tag{7.21}$$

Equation (7.21) can be written in the form of,

$$\mathbf{A}_0 \mathbf{z}_0 = 0 \tag{7.22}$$

where $\mathbf{A}_0$ is a 4 by 4 square coefficient matrix and $\mathbf{z}_0$ is a wave vector containing four wave components of Eq. (7.21).

Setting the determinant of coefficient matrix $\mathbf{A}_0$ to zero gives the characteristic equation, from which the natural frequencies along the x direction can be found.

For a uniform plate of length L_1 along the x direction and with both boundaries A and B simply-supported, the propagation coefficient from Eq. (7.19b) is

$$f(L_1) = e^{-ik_0 L_1} \tag{7.23a}$$

and from Eq. (7.17) the reflection coefficients are

$$r_A = -1, \text{ and } r_B = -1 \tag{7.23b}$$

Boundary A x Boundary B

Figure 7.4 Wave propagation and reflection in a uniform plate along the x direction.

Substituting Eqs. (7.23a) and (7.23b) into Eq. (7.21) and setting the determinant of the coefficient matrix to zero gives the characteristic equation,

$$e^{-i2k_0L_1} - 1 = 0 \tag{7.24a}$$

from which

$$e^{-i2k_0L_1} = 1 \tag{7.24b}$$

or equivalently

$$e^{-i2k_0L_1} = e^{\pm 2m\pi} \tag{7.24c}$$

where $m = 0, 1, 2, 3, \ldots$.

From Eq. (7.24c), the wavenumbers k_0 corresponding to natural frequencies along the x direction are

$$(k_0)_m = \frac{m\pi}{L_1}, \text{ where } m = 1, 2, 3, \ldots \tag{7.25}$$

To allow positive-going waves to be described in the form of e^{-ik_0x}, wavenumbers k_0 take nonnegative values. In addition, the solution corresponding to $m = 0$ is excluded because it results in $k_0 = 0$. As shown in Eqs. (7.18a) or (7.18b), this corresponds to the situation of no motion in a plate, and we are clearly concerned about dynamic motions in vibration study.

The nonnegative integer m in Eq. (7.25) stands for the number of half harmonic waves along the x direction.

The wavenumbers at natural frequencies along the x direction are listed in Table 7.2.

7.4 Wave Reflection, Propagation, and Wave Vibration Analysis Along the *y* Direction

Before studying waves along the y direction, let us examine how k_1 and k_2 (the wavenumbers along the y direction) and k_0 (the wavenumber along the x direction) are related. From Eq. (7.9c),

$$k_1^2 = -k_0^2 + \beta^2, \text{ and } k_2^2 = -(k_0^2 + \beta^2) \tag{7.26}$$

It can be seen from Eq. (7.26) that k_2^2 is always negative because neither k_0 nor β are zero. As a result, the corresponding waves, $e^{\pm ik_2y}$, are decaying (sometimes called evanescent or nearfield) waves.

However, signs of k_1^2 are determined by the relative magnitudes of k_0^2 and β^2. Depending on the values of k_0 and β, the corresponding waves $e^{\pm ik_1y}$ could be propagating (when k_1 is real) or decaying waves (when k_1 is imaginary). The cut-off β value can be found by setting k_1^2 of Eq. (7.26) to zero, which gives $(\beta_c)_m = (k_0)_m = \frac{m\pi}{L_1}$, where $m = 1, 2, 3, \ldots$ are the number of half harmonic waves along the x direction. This indicates that the transition of wave mode in the y direction is determined by the mode sequence along the simply-supported x direction, namely, the number of half harmonic waves along the x direction.

Recall that $\beta^4 = \frac{\rho h \omega^2}{D}$. The cut-off circular frequency ω_c is therefore

$$(\omega_c)_m = (k_0)_m^2 \sqrt{\frac{D}{\rho h}} \tag{7.27}$$

where $m = 1, 2, 3, \ldots$.

Table 7.2 Wavenumbers at natural frequencies along the x direction.

Simply-supported boundary pair along the x direction	Wavenumbers at natural frequencies
	$(k_0)_m = \frac{m\pi}{L_1}$ where $m = 1, 2, 3, \ldots$ indicating the number of half harmonic waves along the x direction

As a result, the following situations may exist in terms of the nature of waves in the plate along the y direction:

Case 1: $\beta^2 < k_0^2$, or equivalently when $\omega < \omega_c$, two pairs of decaying waves exist in the plate, going in the positive and negative directions along the y direction;

Case 2: $\beta^2 > k_0^2$, or equivalently when $\omega > \omega_c$, a pair of propagating waves and a pair of decaying waves exist, going in the positive and negative directions along the y direction;

Case 3: $\beta^2 = k_0^2$, or equivalently when $\omega = \omega_c$, correspond to one single pair of decaying waves in the plate going in the positive and negative directions along the y direction.

Wave reflections at classical boundaries along the y direction, namely, the clamped, free, and simply-supported boundaries, are derived.

7.4.1 Wave Reflection at a Classical Boundary along the y Direction

At a boundary, incident waves give rise to reflected waves, which are related by reflection matrix $\mathbf{r}$.

a) Wave reflection at a clamped boundary along the y direction

Choosing the origin at the boundary, the boundary conditions corresponding to a clamped boundary along the y direction from Table 7.1 are

$$w(x,0,t) = 0, \quad \left.\frac{\partial w(x,y,t)}{\partial y}\right|_{y=0} = 0 \tag{7.28}$$

From Eq. (7.18a) and Eq. (7.28),

$$w(x,y,t) = \left(e^{-ik_0 x} - e^{ik_0 x}\right)\left(c_1^+ + c_2^+ + c_1^- + c_2^-\right)e^{i\omega t} = 0$$

$$\frac{\partial w(x,y,t)}{\partial y} = \left(e^{-ik_0 x} - e^{ik_0 x}\right)\left(-ik_1 c_1^+ - ik_2 c_2^+ + ik_1 c_1^- + ik_2 c_2^-\right)e^{i\omega t} = 0 \tag{7.29}$$

Choosing the origin at the clamped boundary, from Eq. (7.29),

$$c_1^+ + c_2^+ + c_1^- + c_2^- = 0$$

$$-ik_1 c_1^+ - ik_2 c_2^+ + ik_1 c_1^- + ik_2 c_2^- = 0 \tag{7.30}$$

Putting Eq. (7.30) into matrix form,

$$\begin{bmatrix} 1 & 1 \\ ik_1 & ik_2 \end{bmatrix}\begin{bmatrix} c_1^- \\ c_2^- \end{bmatrix} = \begin{bmatrix} -1 & -1 \\ ik_1 & ik_2 \end{bmatrix}\begin{bmatrix} c_1^+ \\ c_2^+ \end{bmatrix} \tag{7.31}$$

Solving for the reflected waves $\begin{bmatrix} c_1^- \\ c_2^- \end{bmatrix}$ in terms of the incident waves $\begin{bmatrix} c_1^+ \\ c_2^+ \end{bmatrix}$ from Eq. (7.31), the reflection matrix at a clamped boundary along the y direction is,

$$\mathbf{r} = \begin{bmatrix} 1 & 1 \\ ik_1 & ik_2 \end{bmatrix}^{-1}\begin{bmatrix} -1 & -1 \\ ik_1 & ik_2 \end{bmatrix} = \begin{bmatrix} \dfrac{k_1 + k_2}{k_1 - k_2} & \dfrac{2k_2}{k_1 - k_2} \\ -\dfrac{2k_1}{k_1 - k_2} & -\dfrac{k_1 + k_2}{k_1 - k_2} \end{bmatrix} \tag{7.32}$$

b) Wave reflection at a free boundary along the y direction

Choosing the origin at the boundary, the boundary conditions corresponding to a free boundary along the y direction from Table 7.1 are

$$M_y\big|_{y=0} = 0, \quad V_y\big|_{y=0} = 0 \tag{7.33}$$

From Eqs. (7.2), (7.3), and (7.33),

$$M_y\big|_{y=0} = -D\left(\frac{\partial^2 w}{\partial y^2} + \gamma\frac{\partial^2 w}{\partial x^2}\right)\Bigg|_{y=0} = 0$$

$$V_y\big|_{y=0} = \left(Q_y - \frac{\partial M_{xy}}{\partial x}\right)\Bigg|_{y=0} = \left(\frac{\partial M_y}{\partial y} - 2\frac{\partial M_{xy}}{\partial x}\right)\Bigg|_{y=0} = \left[-D\frac{\partial^3 w}{\partial y^3} - D(2-\gamma)\frac{\partial^3 w}{\partial x^2 \partial y}\right]\Bigg|_{y=0} = 0 \tag{7.34}$$

Substituting Eq. (7.18a) into Eq. (7.34),

$$-D\Big[(e^{-ik_o x} - e^{ik_o x})(-k_1^2 c_1^+ - k_2^2 c_2^+ - k_1^2 c_1^- - k_2^2 c_2^-) - \gamma k_0^2(e^{-ik_o x} - e^{ik_o x})(c_1^+ + c_2^+ + c_1^- + c_2^-)\Big]e^{i\omega t} = 0,$$

$$\Big\{-D\Big[\left(e^{-ik_o x} - e^{ik_o x}\right)(ik_1^3 c_1^+ + ik_2^3 c_2^+ - ik_1^3 c_1^- - ik_2^3 c_2^-)\Big]$$
$$+ D(2-\gamma)k_0^2\Big[\left(e^{-ik_o x} - e^{ik_o x}\right)(-ik_1 c_1^+ - ik_2 c_2^+ + ik_1 c_1^- + ik_2 c_2^-)\Big]\Big\}e^{i\omega t} = 0 \tag{7.35}$$

Choosing the origin at the free boundary, from Eq. (7.35),

$$(k_1^2 + \gamma k_0^2)c_1^+ + (k_2^2 + \gamma k_0^2)c_2^+ + (k_1^2 + \gamma k_0^2)c_1^- + (k_2^2 + \gamma k_0^2)c_2^- = 0$$

$$\Big[k_1^3 + (2-\gamma)k_0^2 k_1\Big]c_1^+ + \Big[k_2^3 + (2-\gamma)k_0^2 k_2\Big]c_2^+ - \Big[k_1^3 + (2-\gamma)k_0^2 k_1\Big]c_1^- - \Big[k_2^3 + (2-\gamma)k_0^2 k_2\Big]c_2^- = 0 \tag{7.36}$$

Putting Eq. (7.36) into matrix form,

$$\begin{bmatrix} k_1^2 + \gamma k_0^2 & k_2^2 + \gamma k_0^2 \\ k_1^3 + (2-\gamma)k_0^2 k_1 & k_2^3 + (2-\gamma)k_0^2 k_2 \end{bmatrix}\begin{bmatrix} c_1^- \\ c_2^- \end{bmatrix} = \begin{bmatrix} -(k_1^2 + \gamma k_0^2) & -(k_2^2 + \gamma k_0^2) \\ k_1^3 + (2-\gamma)k_0^2 k_1 & k_2^3 + (2-\gamma)k_0^2 k_2 \end{bmatrix}\begin{bmatrix} c_1^+ \\ c_2^+ \end{bmatrix} \tag{7.37}$$

Solving for the reflected waves $\begin{bmatrix} c_1^- \\ c_2^- \end{bmatrix}$ in terms of the incident waves $\begin{bmatrix} c_1^+ \\ c_2^+ \end{bmatrix}$ from Eq. (7.37), the reflection matrix at a free boundary along the y direction is,

$$\mathbf{r} = \begin{bmatrix} k_1^2 + \gamma k_0^2 & k_2^2 + \gamma k_0^2 \\ k_1^3 + (2-\gamma)k_0^2 k_1 & k_2^3 + (2-\gamma)k_0^2 k_2 \end{bmatrix}^{-1}\begin{bmatrix} -(k_1^2 + \gamma k_0^2) & -(k_2^2 + \gamma k_0^2) \\ k_1^3 + (2-\gamma)k_0^2 k_1 & k_2^3 + (2-\gamma)k_0^2 k_2 \end{bmatrix} \tag{7.38}$$

c) Wave reflection at a simply-supported boundary along the y direction

Choosing the origin at the boundary, the boundary conditions corresponding to a simple support along the y direction from Table 7.1 are

$$w(x,0,t) = 0, \quad M_y\big|_{y=0} = 0 \tag{7.39}$$

These expressions are obtained in Eqs. (7.30) and (7.36),

$$c_1^+ + c_2^+ + c_1^- + c_2^- = 0$$

$$(k_1^2 + \gamma k_0^2)c_1^+ + (k_2^2 + \gamma k_0^2)c_2^+ + (k_1^2 + \gamma k_0^2)c_1^- + (k_2^2 + \gamma k_0^2)c_2^- = 0 \tag{7.40}$$

Putting Eq. (7.40) into matrix form,

$$\begin{bmatrix} 1 & 1 \\ k_1^2 + \gamma k_0^2 & k_2^2 + \gamma k_0^2 \end{bmatrix} \begin{bmatrix} c_1^- \\ c_2^- \end{bmatrix} = \begin{bmatrix} -1 & -1 \\ -(k_1^2 + \gamma k_0^2) & -(k_2^2 + \gamma k_0^2) \end{bmatrix} \begin{bmatrix} c_1^+ \\ c_2^+ \end{bmatrix} \tag{7.41}$$

Solving for the reflected waves $\begin{bmatrix} c_1^- \\ c_2^- \end{bmatrix}$ in terms of the incident waves $\begin{bmatrix} c_1^+ \\ c_2^+ \end{bmatrix}$ from Eq. (7.41), the reflection matrix at a simple support along the y direction is,

$$\mathbf{r} = \begin{bmatrix} -1 & 0 \\ 0 & -1 \end{bmatrix} \tag{7.42}$$

7.4.2 Wave Propagation and Wave Vibration Analysis along the *y* Direction

Vibration waves propagate along a uniform waveguide. Consider two locations C and D along the y direction on a uniform plate that are a distance y apart. Wave propagation relationships are determined by wavenumbers k_1 and k_2 along the y direction. The positive- and negative-going waves at Locations C and D are related by

$$\mathbf{f}(y) = \begin{bmatrix} e^{-ik_1 y} & 0 \\ 0 & e^{-ik_2 y} \end{bmatrix} \tag{7.43}$$

where $\mathbf{f}(y)$ is the propagation matrix for a distance y.

For a uniform rectangular plate of side length L_2 and with Boundaries C and D along the y direction, the following wave reflection and propagation relationships exist:

$$\mathbf{c}^+ = \mathbf{r}_C \mathbf{c}^-; \ \mathbf{d}^- = \mathbf{r}_D \mathbf{d}^+; \ \mathbf{d}^+ = \mathbf{f}(L_2)\mathbf{c}^+; \ \mathbf{c}^- = \mathbf{f}(L_2)\mathbf{d}^- \tag{7.44}$$

where $\mathbf{f}(L_2)$ is the propagation matrix for a distance L_2, and $\mathbf{r}_C$ and $\mathbf{r}_D$ are the reflection matrices at boundaries C and D, respectively.

Assembling the relationships in Eq. (7.44) into matrix form,

$$\begin{bmatrix} -\mathbf{I} & \mathbf{r}_C & \mathbf{0} & \mathbf{0} \\ \mathbf{0} & \mathbf{0} & \mathbf{r}_D & -\mathbf{I} \\ \mathbf{f}(L_2) & \mathbf{0} & -\mathbf{I} & \mathbf{0} \\ \mathbf{0} & -\mathbf{I} & \mathbf{0} & \mathbf{f}(L_2) \end{bmatrix} \begin{bmatrix} \mathbf{c}^+ \\ \mathbf{c}^- \\ \mathbf{d}^+ \\ \mathbf{d}^- \end{bmatrix} = \mathbf{0} \tag{7.45}$$

where $\mathbf{I}$ and $\mathbf{0}$ denote an identity and a zero matrix, respectively.

Equation (7.45) can be written in the form of Eq. (7.22), where $\mathbf{A}_0$ is an 8 by 8 square coefficient matrix and $\mathbf{z}_0$ is a wave vector containing eight wave components of Eq. (7.45). Setting the determinant of the coefficient matrix $\mathbf{A}_0$ to zero gives the characteristic equation. Substituting the corresponding boundary reflection matrices into the characteristic equation, the natural frequencies along the y direction, corresponding to various integer multiples of half waves along the x direction, are obtained.

7.5 Numerical Examples

In the numerical examples, out-of-plane free vibrations in an isotropic and homogeneous thin uniform steel plate are analyzed. The material and geometrical properties of the plate are: the Young's modulus, Poisson's ratio, and volume mass density are $E = 198.87 \ GN/m^2$, $\gamma = 0.30$, and $\rho = 7664.5 \ kg/m^3$, respectively. The thickness of the plate is $h = 0.1 \ m$, and the length of the plate along the x direction is $L_1 = 1 \ m$. Three aspect ratios of $L_1 : L_2$ are considered, which are 2:5, 5:2, and 1:1, where L_2 is the length of the plate along the y direction.

The pair of opposite edges along the x direction of the plates is always simply-supported. However, the pair of opposite edges along the y direction is assumed of various combinations of classical boundaries:

a) S–S (that is, Simply-supported–Simply-supported);
b) C–C (that is, Clamped–Clamped);
c) F–F (that is, Free–Free);
d) C–F (that is, Clamped–Free);
e) C–S (that is, Clamped–Simply-supported);
f) S–F (that is, Simply-supported–Free).

The natural frequencies, corresponding to these six sets of example boundary pairs along the y direction, are obtained from Eq. (7.45) by substituting the propagation and boundary reflection matrices into coefficient matrix $\mathbf{A}_0$ and solving for the roots of the characteristic polynomial of matrix $\mathbf{A}_0$. The roots are read from the dB magnitude plots of the characteristic polynomials, which correspond to the x values of local minima in the dB magnitude plot of the characteristic polynomials.

The dimensional cyclic natural frequency f_n, non-dimensional natural frequency $\lambda_n = \beta_n L_1$, and $\beta_n = \dfrac{\rho h \omega_n^2}{D} = \dfrac{\rho h (2\pi f_n)^2}{D}$ are obtained and listed in Tables 7.3–7.8. Since $n = 1, 2, 3, \ldots$, there is one set of infinitely many natural frequencies for each integer number of half waves along the x direction whose opposite edges are simply-supported, that is, corresponding to each wavenumber $(k_0)_m = \dfrac{m\pi}{L_1}$, where $m = 1, 2, 3, \ldots$.

Each of the six tables (Tables 7.3 to 7.8) corresponds to one set of boundary conditions. Three aspect ratios $L_1 : L_2$, of 2:5, 5:2, and 1:1, are considered in each case. For aspect ratio $L_1 : L_2 = 2:5$, natural frequencies below $800\,Hz$ are obtained, and for aspect ratios $L_1 : L_2 = 5:2$ and 1:1, natural frequencies below $2000\,Hz$ are found. Since natural frequencies are often sorted in ascending order, the non-dimensional natural frequencies for up to the lowest 20 modes in the corresponding frequency ranges considered are separately listed as well. As a result, each table contains four sub-tables.

Figure 7.5 presents a sample set of dB magnitude plots of the characteristic polynomials of a plate with a pair of free boundaries along the y direction and with aspect ratio $L_1 : L_2 = 2:5$, corresponding to the lowest three integer multiples of half waves along the simply-supported x direction. The broken vertical lines in each graph mark the cyclic and non-dimensional cut-off frequencies or β values at each integer multiple of half waves along the x direction where wave mode transitions occur.

They are: $f_c = \dfrac{k_0^2}{2\pi}\sqrt{\dfrac{D}{\rho h}}$, $\beta_c = \dfrac{m\pi}{L_1}$, and $\lambda_c = m^2 \pi^2$, where subscript c stands for cut-off.

In Tables 7.3–7.8, natural frequencies below the corresponding cut-off frequency are marked using asterisks. Such natural frequencies are observed when at least one of the opposite edges along the y direction is a free edge, that is, when the boundary conditions are Free–Free, Clamped–Free, and Simply-supported–Free. The physical implication is that both pairs of waves along the y direction can potentially be decaying waves when there is at least one free edge along the y direction. The above observations are no coincidence.

Leissa has shown that the occurrence of natural frequencies below the cut-off frequency is, in general, dependent on the Poisson's ratio γ, the integer number of half waves along the x direction m, and the aspect ratio. The y symmetrical modes of Free–Free boundaries are an exception, where there always exists one natural frequency below the cut-off frequency for each m provided that the Poisson's ratio $\gamma \neq 0$. Leissa has also shown that, based on the properties of trigonometric and hyperbolic trigonometric functions, it is not possible for the value of a natural frequency to be below the corresponding cut-off frequency when the pair of opposite edges along the y direction does not involve a free edge, that is, when the boundary conditions are Simply-supported–Simply-supported, Clamped–Clamped, and Clamped–Simply-supported (Leissa 1973).

Recall that in the numerical examples studied in this section, the Poisson's ratio is $\gamma = 0.30$ and aspect ratios are 2:5, 5:2, and 1:1. The conditions for the occurence of natural frequencies below a cut-off frequency, corresponding to the numerical examples, are summarized in Table 7.9. For example, in the case of a pair of free boundaries along the y direction, there exist two natural frequencies, one y symmetrical mode and one y anti-symmetrical mode, below the cut-off frequency for each $m \geq 6$, 37, and, 15 at aspect ratios 2:5, 5:2, and 1:1, respectively.

Figure 7.6 presents a sample dB magnitude plot of the characteristic polynomials of a plate with a pair of free boundaries along the y direction for $m = 6$ and plate aspect ratio $L_1 : L_2 = 2:5$. Figure 7.6b, the magnified response plot around the cut-off frequency, shows that there exist two natural frequencies below the cut-off frequency, which is marked using the vertical broken line. This observation agrees well with Table 7.9, which predicts that for a pair of free boundaries along the y direction, with aspect ratio $L_1 : L_2 = 2:5$, for every $m > 5.7820$, there are two natural frequencies below the cut-off frequency, one from the y symmetrical mode and the other from the y anti-symmetrical mode.

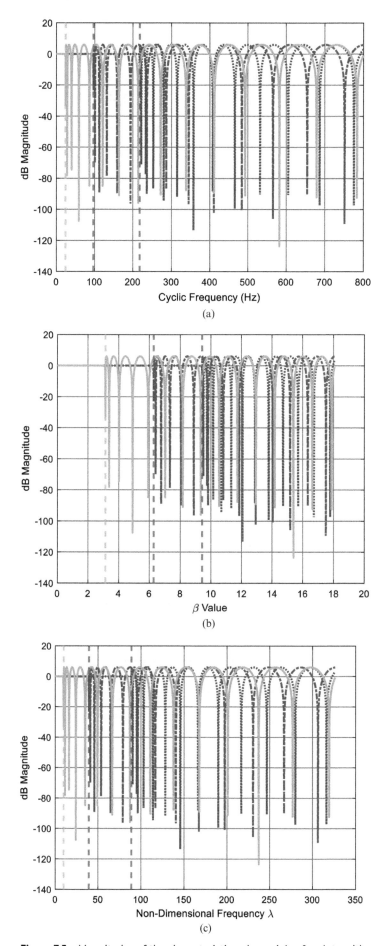

Figure 7.5 Magnitudes of the characteristic polynomials of a plate with a pair of free boundaries along the y direction and with aspect ratio $L_1 : L_2 = 2:5$, corresponding to various integer multiples of half waves along the simply – supported x direction: $m = 1$ (____), $m = 2$ (---), and $m = 3$ (...).

Table 7.3(a) Natural frequencies (below 800 Hz) of a thin plate with a pair of simple supports along the y direction.

A pair of simple supports along the x direction, $L_1 : L_2 = 2 : 5$

m	(m,n)	(1,1)	(1,2)	(1,3)	(1,4)	(1,5)	(1,6)	(1,7)	(1,8)
1	β	3.3836	4.0232	4.9073	5.9275	7.0248	8.1681	9.3406	10.5325
	λ	11.4488	16.1862	24.0818	35.1358	49.3480	66.7185	87.2473	110.9344
	f	28.0873	39.7096	59.0801	86.1998	121.0657	163.6809	214.0442	272.1558

	(m,n)	(1,9)	(1,10)	(1,11)	(1,12)	(1,13)	(1,14)		
	β	11.7380	12.9531	14.1755	15.4034	16.6356	17.8712		
	λ	137.7797	167.7833	200.9451	237.2653	276.7437	319.3804		
	f	338.0156	411.6235	492.9797	582.0841	678.9367	783.5375		

	(m,n)	(2,1)	(2,2)	(2,3)	(2,4)	(2,5)	(2,6)	(2,7)	(2,8)
2	β	6.4076	6.7672	7.3274	8.0464	8.8858	9.8147	10.8100	11.8551
	λ	41.0576	45.7950	53.6906	64.7446	78.9568	96.3273	116.8561	140.5431
	f	100.7267	112.3490	131.7195	158.8383	193.7052	236.3203	286.6837	344.7952

	(m,n)	(2,9)	(2,10)	(2,11)	(2,12)	(2,13)			
	β	12.9379	14.0496	15.1840	16.3363	17.5029			
	λ	167.3885	197.3921	230.5540	266.8741	306.3525			
	f	410.6550	484.2630	565.6192	654.7236	751.5762			

	(m,n)	(3,1)	(3,2)	(3,3)	(3,4)	(3,5)	(3,6)	(3,7)	(3,8)
3	β	9.5082	9.7541	10.1508	10.6814	11.3272	12.0696	12.8920	13.7801
	λ	90.4056	95.1430	103.0387	114.0926	128.3048	145.6754	166.2041	189.8912
	f	221.7924	233.4148	252.7853	279.9040	314.7709	357.3861	407.7494	465.8610

	(m,n)	(3,9)	(3,10)	(3,11)	(3,12)				
	β	14.7220	15.7080	16.7303	17.7826				
	λ	216.7365	246.7401	279.9020	316.2221				
	f	531.7208	605.3287	686.6849	775.7893				

	(m,n)	(4,1)	(4,2)	(4,3)	(4,4)	(4,5)	(4,6)	(4,7)	(4,8)
4	β	12.6290	12.8152	13.1197	13.5344	14.0496	14.6548	15.3392	16.0928
	λ	159.4928	164.2302	172.1259	183.1799	197.3921	214.7626	235.2914	258.9784
	f	391.2845	402.9068	422.2773	449.3961	484.2630	526.8781	577.2415	635.3530

	(m,n)	(4,9)	(4,10)						
	β	16.9063	17.7715						
	λ	285.8237	315.8273						
	f	701.2128	774.8208						

	(m,n)	(5,1)	(5,2)	(5,3)	(5,4)	(5,5)	(5,6)	(5,7)	
5	β	15.7581	15.9078	16.1540	16.4926	16.9180	17.4238	18.0033	
	λ	248.3192	253.0566	260.9524	272.0063	286.2185	303.5890	324.1178	
	f	609.2028	620.8251	640.1957	667.3144	702.1813	744.7965	795.1598	
6	None below 800 Hz								

Table 7.3(b) Natural frequencies (below 2000 *Hz*) of a thin plate with a pair of simple supports along the *y* direction.

A pair of simple supports along the *x* direction, $L_1 : L_2 = 5 : 2$

m	(m,n)	(1,1)	(1,2)	(1,3)
1	β	8.4590	16.0190	23.7705
	λ	71.5546	256.6097	565.0349
	f	175.5453	629.5419	1386.2028
	(m,n)	(2,1)	(2,2)	(2,3)
2	β	10.0580	16.9180	24.3853
	λ	101.1635	286.2185	594.6436
	f	248.1848	702.1813	1458.8422
	(m,n)	(3,1)	(3,2)	(3,3)
3	β	12.2683	18.3185	25.3770
	λ	150.5115	335.5666	643.9917
	f	369.2505	823.2471	1579.9080
	(m,n)	(4,1)	(4,2)	(4,3)
4	β	14.8189	20.1160	26.7035
	λ	219.5987	404.6538	713.0789
	f	538.7426	992.7391	1749.4000
	(m,n)	(5,1)	(5,2)	(5,3)
5	β	17.5620	22.2144	28.3179
	λ	308.4251	493.4802	801.9054
	f	756.6609	1210.6575	1967.3184
	(m,n)	(6,1)	(6,2)	
6	β	20.4204	24.5366	
	λ	416.9908	602.0459	
	f	1023.0056	1477.0021	
	(m,n)	(7,1)	(7,2)	
7	β	23.3516	27.0250	
	λ	545.2956	730.3507	
	f	1337.7765	1791.7731	
	(m,n)	(8,1)		
8	β	26.3313		
	λ	693.3397		
	f	1700.9737		
9	None below 2000 *Hz*			

Table 7.3(c) Natural frequencies (below 2000 Hz) of a thin plate with a pair of simple supports along the y direction.

A pair of simple supports along the x direction, $L_1 : L_2 = 1:1$

m	(m,n)	(1,1)	(1,2)	(1,3)	(1,4)	(1,5)	(1,6)	(1,7)	(1,8)
1	β	4.4429	7.0248	9.9346	12.9531	16.0190	19.1096	22.2144	25.3283
	λ	19.7392	49.3480	98.6960	167.7833	256.6097	365.1754	493.4802	641.5243
	f	48.4263	121.0657	242.1315	411.6235	629.5419	895.8865	1210.6575	1573.8547

(m,n)	(1,9)
β	28.4483
λ	809.3075
f	1985.4782

(m,n)	(2,1)	(2,2)	(2,3)	(2,4)	(2,5)	(2,6)	(2,7)	(2,8)
2 β	7.0248	8.8858	11.3272	14.0496	16.9180	19.8692	22.8711	25.9062
λ	49.3480	78.9568	128.3048	197.3921	286.2185	394.7842	523.0890	671.1331
f	121.0657	193.7052	314.7709	484.2630	702.1813	968.5260	1283.2969	1646.4942

(m,n)	(3,1)	(3,2)	(3,3)	(3,4)	(3,5)	(3,6)	(3,7)	(3,8)
3 β	9.9346	11.3272	13.3286	15.7080	18.3185	21.0744	23.9257	26.8418
λ	98.6960	128.3048	177.6529	246.7401	335.5666	444.1322	572.4371	720.4811
f	242.1315	314.7709	435.8367	605.3287	823.2471	1089.5917	1404.3627	1767.5599

(m,n)	(4,1)	(4,2)	(4,3)	(4,4)	(4,5)	(4,6)	(4,7)	(4,8)
4 β	12.9531	14.0496	15.7080	17.7715	20.1160	22.6543	25.3283	28.0993
λ	167.7833	197.3921	246.7401	315.8273	404.6538	513.2194	641.5243	789.5683
f	411.6235	484.2630	605.3287	774.8208	992.7391	1259.0838	1573.8547	1937.0519

(m,n)	(5,1)	(5,2)	(5,3)	(5,4)	(5,5)	(5,6)	(5,7)	
5 β	16.0190	16.9180	18.3185	20.1160	22.2144	24.5366	27.0250	
λ	256.6097	286.2185	335.5666	404.6538	493.4802	602.0459	730.3507	
f	629.5419	702.1813	823.2471	992.7391	1210.6575	1477.0021	1791.7731	
6	None below 2000 Hz							

Table 7.3(d) Up to the lowest 20 non-dimensional natural frequencies λ of a thin plate with a pair of simple supports along the y direction.

	A pair of simple supports along the x direction								
	Aspect ratio								
	$L_1 : L_2 = 2:5$			$L_1 : L_2 = 5:2$			$L_1 : L_2 = 1:1$		
Mode Sequence	(m, n)	λ Present	λ (Leissa 1973)	(m, n)	λ Present	λ (Leissa 1973)	(m, n)	λ Present	λ (Leissa 1973)
1	(1, 1)	11.4488	11.4487	(1, 1)	71.5546	71.5564	(1, 1)	19.7392	19.7392
2	(1, 2)	16.1862	16.1862	(2, 1)	101.1635	101.1634	(2, 1)	49.3480	49.3480

Table 7.3(d) (Continued)

	A pair of simple supports along the *x* direction								
	Aspect ratio								
	$L_1:L_2=2:5$				$L_1:L_2=5:2$			$L_1:L_2=1:1$	
Mode Sequence	(m, n)	λ Present	λ (Leissa 1973)	(m, n)	λ Present	λ (Leissa 1973)	(m, n)	λ Present	λ (Leissa 1973)
3	(1, 3)	24.0818	24.0818	(3, 1)	150.5115	150.5115	(1, 2)	49.3480	49.3480
4	(1, 4)	35.1358	35.1358	(4, 1)	219.5987	219.5987	(2, 2)	78.9568	78.9568
5	(2, 1)	41.0576	41.0576	(1, 2)	256.6097	256.6097	(3, 1)	98.6960	98.6960
6	(2, 2)	45.7950	45.7950	(2, 2)	286.2185	286.2185	(1, 3)	98.6960	98.6960
7	(1, 5)	49.3480	49.3480	(5, 1)	308.4251	308.4251	(3, 2)	128.3048	128.3049
8	(2, 3)	53.6906	53.6906	(3, 2)	335.5666	335.5665	(2, 3)	128.3048	128.3049
9	(2, 4)	64.7446	NA	(4, 2)	404.6538	NA	(4, 1)	167.783	167.7833
10	(1, 6)	66.7185	66.7185	(6, 1)	416.9908	416.9908	(1, 4)	167.783	
11	(2, 5)	78.9568		(5, 2)	493.4802		(3, 3)	177.6529	
12	(1, 7)	87.2473		(7,1)	545.2956		(4, 2)	197.3921	
13	(3, 1)	90.4056		(1, 3)	565.0349		(2, 4)	197.3921	
14	(3, 2)	95.1430		(2,3)	594.6436		(4, 3)	246.7401	
15	(2, 6)	96.3273		(6,2)	602.0459		(3, 4)	246.7401	
16	(3, 3)	103.0387		(3,3)	643.9917		(5,1)	256.6097	
17	(1, 8)	110.9344		(8,1)	693.3397		(1,5)	256.6097	
18	(3, 4)	114.0926		(4,3)	713.0789		(5,2)	286.2185	
19	(2, 7)	116.8561		(7,2)	730.3507		(2, 5)	286.2185	
20	(3, 5)	128.3048		(5,3)	801.9054		(4, 4)	315.8273	

Table 7.4(a) Natural frequencies (below 800 Hz) of a thin plate with a pair of clamped boundaries along the *y* direction.

A pair of simple supports along the *x* direction, $L_1:L_2=2:5$

m	(m,n)	(1,1)	(1,2)	(1,3)	(1,4)	(1,5)	(1,6)	(1,7)	(1,8)
1	β	3.4835	4.2854	5.2883	6.3836	7.5283	8.7027	9.8963	11.1031
	λ	12.1347	18.3647	27.9657	40.7500	56.6756	75.7369	97.9365	123.2784
	f	29.7700	45.0543	68.6085	99.9722	139.0425	185.8057	240.2682	302.4395
	(m,n)	(1,9)	(1,10)	(1,11)	(1,12)	(1,13)			
	β	12.3193	13.5426	14.7712	16.0039	17.2400			
	λ	151.7658	183.4016	218.1876	256.1256	297.2167			
	f	372.3279	449.9400	535.2808	628.3542	729.1632			
	(m,n)	(2,1)	(2,2)	(2,3)	(2,4)	(2,5)	(2,6)	(2,7)	(2,8)
2	β	6.4326	6.8557	7.4952	8.2915	9.1979	10.1818	11.2213	12.3015
	λ	41.3782	47.0009	56.1782	68.7486	84.6006	103.6685	125.9166	151.3268
	f	101.5133	115.3076	137.8222	168.6614	207.5511	254.3304	308.9219	371.2509

(Continued)

Table 7.4(a) (Continued)

A pair of simple supports along the *x* direction, $L_1 : L_2 = 2:5$

	(m,n)	(2,9)	(2,10)	(2,11)	(2,12)	(2,13)			
	β	13.4123	14.5466	15.6992	16.8663	18.0452			
	λ	179.8902	211.6029	246.4635	284.4721	325.6290			
	f	441.3256	519.1264	604.6502	697.8967	798.8672			
	(m,n)	(3,1)	(3,2)	(3,3)	(3,4)	(3,5)	(3,6)	(3,7)	(3,8)
3	β	9.5190	9.7952	10.2359	10.8178	11.5162	12.3091	13.1776	14.1067
	λ	90.6118	95.9465	104.7746	117.0239	132.6237	151.5143	173.6504	198.9994
	f	222.2984	235.3860	257.0440	287.0953	325.3663	371.7108	426.0175	488.2064
	(m,n)	(3,9)	(3,10)	(3,11)					
	β	15.0844	16.1014	17.1504					
	λ	227.5391	259.2544	294.1356					
	f	558.2229	636.0301	721.6044					
	(m,n)	(4,1)	(4,2)	(4,3)	(4,4)	(4,5)	(4,6)	(4,7)	(4,8)
4	β	12.6350	12.8385	13.1697	13.6179	14.1707	14.8151	15.5385	16.3296
	λ	159.6444	164.8282	173.4409	185.4480	200.8100	219.4880	241.4463	266.6547
	f	391.6565	404.3738	425.5033	454.9604	492.6482	538.4709	592.3413	654.1854
	(m,n)	(4,9)							
	β	17.1782							
	λ	295.0889							
	f	723.9431							
	(m,n)	(5,1)	(5,2)	(5,3)	(5,4)	(5,5)	(5,6)		
5	β	15.7620	15.9227	16.1866	16.5482	17.0006	17.5360		
	λ	248.4391	253.5317	262.0057	273.8427	289.0197	307.5120		
	f	609.4968	621.9906	642.7800	671.8196	709.0534	754.4208		
6	None below 800 Hz								

Table 7.4(b) Natural frequencies (below 2000 Hz) of a thin plate with a pair of clamped boundaries along the *y* direction.

A pair of simple supports along the *x* direction, $L_1 : L_2 = 5:2$

m	(m,n)	(1,1)	(1,2)	(1,3)
1	β	12.0617	19.8211	27.6358
	λ	145.4839	392.8746	763.7401
	f	356.9164	963.8411	1873.6873

Table 7.4(b) (Continued)

A pair of simple supports along the *x* direction, $L_1 : L_2 = 5 : 2$

	(m,n)	(2,1)	(2,2)	(2,3)
2	β	12.8351	20.3885	28.0758
	λ	164.7387	415.6906	788.2519
	f	404.1542	1019.8158	1933.8222

	(m,n)	(3,1)	(3,2)
3	β	14.2207	21.3379
	λ	202.2271	455.3054
	f	496.1247	1117.0031

	(m,n)	(4,1)	(4,2)
4	β	16.1588	22.6573
	λ	261.1052	513.3529
	f	640.5708	1259.4113

	(m,n)	(5,1)	(5,2)
5	β	18.4971	24.3137
	λ	342.1442	591.1580
	f	839.3840	1450.2908

	(m,n)	(6,1)	(6,2)
6	β	21.0943	26.2587
	λ	444.9682	689.5215
	f	1091.6426	1691.6066

	(m,n)	(7,1)	(7,2)
7	β	23.8540	28.4392
	λ	569.0150	808.7856
	f	1395.9674	1984.1978

	(m,n)	(8,1)
8	β	26.7177
	λ	713.8365
	f	1751.2587
9	None below 2000 Hz	

Table 7.4(c) Natural frequencies (below 2000 Hz) of a thin plate with a pair of clamped boundaries along the *y* direction.

A pair of simple supports along the *x* direction, $L_1 : L_2 = 1 : 1$

m	(m,n)	(1,1)	(1,2)	(1,3)	(1,4)	(1,5)	(1,6)	(1,7)	(1,8)
1	β	5.3806	8.3263	11.3620	14.4358	17.5304	20.6377	23.7531	26.8741
	λ	28.9509	69.3270	129.0955	208.3917	307.3161	425.9144	564.2108	722.2189
	f	71.0253	170.0803	316.7107	511.2484	753.9402	1044.8980	1384.1811	1771.8231

(Continued)

Table 7.4(c) (Continued)

A pair of simple supports along the x direction, $L_1 : L_2 = 1 : 1$

	(m,n)	(2,1)	(2,2)	(2,3)	(2,4)	(2,5)	(2,6)	(2,7)	(2,8)
2	β	7.3989	9.7255	12.4409	15.3162	18.2744	21.2817	24.3208	27.3815
	λ	54.7431	94.5853	154.7757	234.5855	333.9532	452.9124	591.5006	749.7466
	f	134.3014	232.0465	379.7120	575.5097	819.2890	1111.1322	1451.1314	1839.3571

	(m,n)	(3,1)	(3,2)	(3,3)	(3,4)	(3,5)	(3,6)	(3,7)	(3,8)
3	β	10.1102	11.8408	14.1354	16.7228	19.4748	22.3279	25.2472	28.2125
	λ	102.2162	140.2045	199.8105	279.6512	379.2685	498.5335	637.4230	795.9453
	f	250.7675	343.9644	490.1962	686.0698	930.4612	1223.0546	1563.7929	1952.6965

	(m,n)	(4,1)	(4,2)	(4,3)	(4,4)	(4,5)	(4,6)	(4,7)	
4	β	13.0517	14.3770	16.2849	18.5617	21.0722	23.7357	26.5037	
	λ	170.3465	206.6971	265.1964	344.5378	444.0366	563.3822	702.4453	
	f	417.9118	507.0910	650.6076	845.2564	1089.3572	1382.1484	1723.3125	

	(m,n)	(5,1)	(5,2)	(5,3)	(5,4)	(5,5)	(5,6)	(5,7)	
5	β	16.0815	17.1393	18.7380	20.7287	22.9946	25.4544	28.0531	
	λ	258.6136	293.7557	351.1137	429.6789	528.7539	647.9280	786.9790	
	f	634.4580	720.6723	861.3890	1054.1334	1297.1946	1589.5650	1930.6994	

	(m,n)	(5,1)	(5,2)	(5,3)	(5,4)	(5,5)	(5,6)		
6	β	19.1525	20.0270	21.3878	23.1336	25.1723	27.4309		
	λ	366.8167	401.0792	457.4395	535.1639	633.6445	752.4537		
	f	899.9132	983.9696	1122.2385	1312.9202	1554.5233	1845.9985		

	(m,n)	(5,1)	(5,2)	(5,3)	(5,4)	(5,5)			
7	β	22.2456	22.9886	24.1663	25.7087	27.5456			
	λ	494.8687	528.4767	584.0087	660.9357	758.7618			
	f	1214.0638	1296.5146	1432.7515	1621.4768	1861.4741			

	(m,n)	(5,1)	(5,2)	(5,3)	(5,4)				
8	β	25.3521	25.9968	27.0312	28.4060				
	λ	642.7268	675.8342	730.6867	806.8990				
	f	1576.8048	1658.0273	1792.5974	1979.5692				

	(m,n)	(5,1)							
9	β	28.4670							
	λ	810.3677							
	f	1988.0791							
10	None below 2000 Hz								

Table 7.4(d) Up to the lowest 20 non-dimensional natural frequency λ of a thin plate with a pair of clamped boundaries along the y direction, sorted in ascending order.

A pair of simple supports along the x direction

Aspect ratio

Mode Sequence	$L_1 : L_2 = 2:5$			$L_1 : L_2 = 5:2$			$L_1 : L_2 = 1:1$		
	(m, n)	λ Present	λ (Leissa 1973)	(m, n)	λ Present	λ (Leissa 1973)	(m, n)	λ Present	λ (Leissa 1973)
1	(1,1)	12.1347	12.1347	(1,1)	145.4839	145.4839	(1,1)	28.9509	28.9509
2	(1,2)	18.3647	18.3647	(2,1)	164.7387	164.7387	(2,1)	54.7431	54.7431
3	(1,3)	27.9657	27.9657	(3,1)	202.2271	202.2271	(1,2)	69.3270	69.3270
4	(1,4)	40.7500	40.7500	(4,1)	261.1052	261.1053	(2,2)	94.5853	94.5853
5	(2,1)	41.3782	41.3782	(5,1)	342.1442	342.1442	(3,1)	102.2162	102.2162
6	(2,2)	47.0009	47.0009	(1,2)	392.8746	392.8746	(1,3)	129.0955	129.0955
7	(2,3)	56.1782	56.1782	(2,2)	415.6906	415.6906	(3,2)	140.2045	140.2045
8	(1,5)	56.6756	56.6756	(6,1)	444.9682	444.9682	(2,3)	154.7757	154.7757
9	(2,4)	68.7486	68.7486	(3,2)	455.3054	455.3054	(4,1)	170.3465	170.3465
10	(1,6)	75.7369		(4,2)	513.3529		(3,3)	199.8105	
11	(2,5)	84.6006		(7,1)	569.0150		(4,2)	206.6971	
12	(3,1)	90.6118		(5,2)	591.1580		(1,4)	208.3917	
13	(3,2)	95.9465		(6,2)	689.5215		(2,4)	234.5855	
14	(1,7)	97.9365		(8,1)	713.8365		(5,1)	258.6136	
15	(2,6)	103.6685		(1,3)	763.7401		(4,3)	265.1964	
16	(3,3)	104.7746		(2,3)	788.2519		(3,4)	279.6512	
17	(3,4)	117.0239		(7,2)	808.7856		(5,2)	293.7557	
18	(1,8)	123.2784					(1,5)	307.3161	
19	(2,7)	125.9166					(2,5)	333.9532	
20	(3,5)	132.6237					(4,4)	344.5378	

Table 7.5(a) Natural frequencies (below 800 Hz) of a thin plate with a pair of free boundaries along the y direction.

A pair of simple supports along the x direction, $L_1 : L_2 = 2:5$

m	(m,n)	(1,1)	(1,2)	(1,3)	(1,4)	(1,5)	(1,6)	(1,7)	(1,8)
1	β	3.1241[*]	3.3222	3.8811	4.6590	5.5836	6.6083	7.6999	8.8358
	λ	9.7601[*]	11.0368	15.0626	21.7064	31.1771	43.6698	59.2880	78.0721
	f	23.9444[*]	27.0767	36.9531	53.2525	76.4869	107.1353	145.4516	191.5347

	(m,n)	(1,9)	(1,10)	(1,11)	(1,12)	(1,13)	(1,14)	(1,15)
	β	10.0017	11.1881	12.3892	13.6009	14.8204	16.0461	17.2764
	λ	100.0340	125.1747	153.4920	184.9833	219.6457	257.4769	298.4752
	f	245.4138	307.0916	376.5628	453.8204	538.8578	631.6694	732.2507

(Continued)

Table 7.5(a) (Continued)

A pair of simple supports along the x direction, $L_1 : L_2 = 2 : 5$

	(m,n)	(2,1)	(2,2)	(2,3)	(2,4)	(2,5)	(2,6)	(2,7)	(2,8)
2	β	6.2641*	6.3642	6.7039	7.2269	7.8966	8.6808	9.5547	10.4991
	λ	39.2387*	40.5035	44.9416	52.2281	62.3564	75.3561	91.2913	110.2320
	f	96.2644*	99.3674	110.2554	128.1315	152.9792	184.8716	223.9653	270.4328

	(m,n)	(2,9)	(2,10)	(2,11)	(2,12)	(2,13)	(2,14)		
	β	11.4995	12.5439	13.6235	14.7309	15.8606	17.0085		
	λ	132.2377	157.3506	185.5988	216.9987	251.5602	289.2880		
	f	324.4193	386.0291	455.3304	532.3641	617.1538	709.7218		

	(m,n)	(3,1)	(3,2)	(3,3)	(3,4)	(3,5)	(3,6)	(3,7)	(3,8)
3	β	9.4043*	9.4670	9.7065	10.0867	10.5936	11.2093	11.9171	12.7024
	λ	88.4399*	89.6234	94.2153	101.7419	112.2237	125.6478	142.0168	161.3498
	f	216.9701*	219.8735	231.1389	249.6038	275.3190	308.2523	348.4104	395.8403

	(m,n)	(3,9)	(3,10)	(3,11)	(3,12)	(3,13)			
	β	13.5527	14.4579	15.4093	16.3997	17.4231			
	λ	183.6768	209.0317	237.4467	268.9499	303.5639			
	f	450.6153	512.8184	582.5291	659.8161	744.7348			

	(m,n)	(4,1)	(4,2)	(4,3)	(4,4)	(4,5)	(4,6)	(4,7)	(4,8)
4	β	12.5444*	12.5876	12.7730	13.0685	13.4699	13.9678	14.5517	15.2119
	λ	157.3616*	158.4473	163.1483	170.7855	181.4391	195.0980	211.7527	231.4007
	f	386.0561*	388.7196	400.2525	418.9888	445.1254	478.6350	519.4940	567.6966

	(m,n)	(4,9)	(4,10)	(4,11)					
	β	15.9389	16.7244	17.5612					
	λ	254.0479	279.7066	308.3942					
	f	623.2569	686.2057	756.5851					

	(m,n)	(5,1)	(5,2)	(5,3)	(5,4)	(5,5)	(5,6)	(5,7)	(5,8)
5	β	15.6844*	15.7157	15.8680	16.1087	16.4390	16.8533	17.3454	17.9086
	λ	246.0014*	246.9839	251.7925	259.4918	270.2404	284.0336	300.8622	320.7192
	f	603.5164*	605.9269	617.7237	636.6124	662.9822	696.8210	738.1067	786.8220
6	None below 800 Hz								

Table 7.5(b) Natural frequencies (below 2000 Hz) of a thin plate with a pair of free boundaries along the y direction.

A pair of simple supports along the x direction, $L_1 : L_2 = 5 : 2$

m	(m,n)	(1,1)	(1,2)	(1,3)	(1,4)	(1,5)
1	β	3.0796*	5.7985	12.4950	19.9929	27.7253
	λ	9.4842*	33.6228	156.1247	399.7153	768.6947
	f	23.2675*	82.4870	383.0216	980.6236	1885.8425

Table 7.5(b) (Continued)

A pair of simple supports along the *x* direction, $L_1 : L_2 = 5 : 2$

	(m,n)	(2,1)	(2,2)	(2,3)	(2,4)	(2,5)
2	β	6.1938*	8.6720	14.1367	20.9977	28.4123
	λ	38.3629*	75.2037	199.8452	440.9037	807.2561
	f	94.1160*	184.4976	490.2813	1081.6712	1980.4455

	(m,n)	(3,1)	(3,2)	(3,3)	(3,4)	
3	β	9.3257*	11.4174	16.2383	22.4846	
	λ	86.9684*	130.3575	263.6809	505.5594	
	f	213.3601*	319.8068	646.8897	1240.2913	

	(m,n)	(4,1)	(4,2)	(4,3)	(4,4)	
4	β	12.4628*	14.2184	18.5622	24.3008	
	λ	155.3211*	202.1620	344.5564	590.5270	
	f	381.0501*	495.9650	845.3018	1448.7428	

	(m,n)	(5,1)	(5,2)	(5,3)	(5,4)	
5	β	15.6015*	17.0913	21.0268	26.3421	
	λ	243.4060*	292.1134	442.1257	693.9086	
	f	597.1491*	716.6433	1084.6691	1702.3694	

	(m,n)	(6,1)	(6,2)	(6,3)	(6,4)	
6	β	18.7408*	20.0232	23.5991	28.5468	
	λ	351.2170*	400.9285	556.9166	814.9214	
	f	861.6423*	983.5998	1366.2863	1999.2508	

	(m,n)	(7,1)	(7,2)	(7,3)		
7	β	21.8804*	22.9992	26.2595		
	λ	478.7525*	528.9644	689.5628		
	f	1174.5259*	1297.7110	1691.7078		

	(m,n)	(8,1)	(8,2)			
8	β	25.0202*	26.0079			
	λ	626.0122*	676.4104			
	f	1535.7988*	1659.4410			

	(m,n)	(9,1)				
9	β	28.1602*				
	λ	792.9958*				
	f	1945.4604*				
10	None below 2000 Hz					

Table 7.5(c) Natural frequencies (below 2000 *Hz*) of a thin plate with a pair of free boundaries along the *y* direction.

A pair of simple supports along the *x* direction, $L_1 : L_2 = 1 : 1$

m	(m,n)	(1,1)	(1,2)	(1,3)	(1,4)	(1,5)	(1,6)	(1,7)	(1,8)
1	β	3.1035*	4.0168	6.0602	8.6766	11.5631	14.5637	17.6183	20.7016
	λ	9.6314*	16.1348	36.7256	75.2834	133.7046	212.1019	310.4053	428.5547
	f	23.6287*	39.5835	90.0992	184.6931	328.0181	520.3506	761.5188	1051.3754

	(m,n)	(1,9)	(1,10)
	β	23.8015	26.9121
	λ	566.5134	724.2590
	f	1389.8301	1776.8283

	(m,n)	(2,1)	(2,2)	(2,3)	(2,4)	(2,5)	(2,6)	(2,7)	(2,8)
2	β	6.2406*	6.8365	8.4107	10.5369	13.0207	15.7303	18.5781	21.5115
	λ	38.9450*	46.7381	70.7401	111.0254	169.5377	247.4425	345.1472	462.7429
	f	95.5439*	114.6629	173.5471	272.3791	415.9277	607.0518	846.7514	1135.2495

	(m,n)	(2,9)	(2,10)
	β	24.4995	27.5240
	λ	600.2253	757.5694
	f	1472.5357	1858.5487

	(m,n)	(3,1)	(3,2)	(3,3)	(3,4)	(3,5)	(3,6)	(3,7)	(3,8)
3	β	9.3801*	9.8000	11.0472	12.8334	14.9919	17.4148	20.0251	22.7670
	λ	87.9867*	96.0405	122.0400	164.6959	224.7584	303.2762	401.0033	518.3364
	f	215.8582*	235.6166	299.4013	404.0493	551.4010	744.0290	983.7834	1271.6373

	(m,n)	(3,9)	(3,10)
	β	25.6017	28.5026
	λ	655.4471	812.3959
	f	1608.0115	1993.0550

	(m,n)	(4,1)	(4,2)	(4,3)	(4,4)	(4,5)	(4,6)	(4,7)	(4,8)
4	β	12.5201*	12.8369	13.8516	15.3708	17.2656	19.4405	21.8268	24.3729
	λ	156.7525*	164.7866	191.8674	236.2620	298.1026	377.9348	476.4080	594.0371
	f	384.5617*	404.2718	470.7093	579.6228	731.3366	927.1893	1168.7742	1457.3541

	(m,n)	(4,9)
	β	27.0398
	λ	731.1521
	f	1793.7391

	(m,n)	(5,1)	(5,2)	(5,3)	(5,4)	(5,5)	(5,6)	(5,7)	(5,8)
5	β	15.6602*	15.9106	16.7596	18.0673	19.7415	21.7020	23.8866	26.2479
	λ	245.2417*	253.1469	280.8849	326.4258	389.7274	470.9760	570.5703	688.9502
	f	601.6527*	621.0465	689.0963	800.8221	956.1202	1155.4477	1399.7830	1690.2049

Table 7.5(c) (Continued)

A pair of simple supports along the x direction, $L_1 : L_2 = 1 : 1$

	(m,n)	(6,1)	(6,2)	(6,3)	(6,4)	(6,5)	(6,6)	(6,7)	(6,8)
6	β	18.8004*	19.0046	19.7324	20.8728	22.3622	24.1361	26.1404	28.3315
	λ	353.4536*	361.1763	389.3691	435.6739	500.0667	582.5526	683.3217	802.6728
	f	867.1294*	886.0757	955.2412	1068.8409	1226.8161	1429.1791	1676.3965	1969.2011

	(m,n)	(7,1)	(7,2)	(7,3)	(7,4)	(7,5)	(7,6)	(7,7)	
7	β	21.9405*	22.1110	22.7474	23.7545	25.0893	26.7012	28.5445	
	λ	481.3873*	488.8975	517.4427	564.2773	629.4707	712.9549	814.7886	
	f	1180.9899*	1199.4147	1269.4448	1384.3443	1544.2836	1749.0957	1998.9248	

	(m,n)	(8,1)	(8,2)	(8,3)	(8,4)	(8,5)			
8	β	25.0807*	25.2254	25.7908	26.6905	27.8957			
	λ	629.0420*	636.3212	665.1658	712.3844	778.1714			
	f	1543.2318*	1561.0899	1631.8546	1747.6957	1909.0917			

	(m,n)	(9,1)	(9,2)						
9	β	28.2209*	28.3452						
	λ	796.4166*	803.4529						
	f	1953.8529*	1971.1149						
10	None below 2000 Hz								

Table 7.5(d) Up to the lowest 20 non-dimensional natural frequency λ of a thin plate with a pair of free boundaries along the y direction, sorted in ascending order.

	A pair of simple supports along the x direction								
	Aspect ratio								
	$L_1 : L_2 = 2 : 5$			$L_1 : L_2 = 5 : 2$			$L_1 : L_2 = 1 : 1$		
Mode Sequence	(m, n)	λ Present	λ (Leissa 1973)	(m, n)	λ Present	λ (Leissa 1973)	(m, n)	λ Present	λ (Leissa 1973)
1	(1,1)	9.7601	9.7600	(1,1)	9.4842	9.4841	(1,1)	9.6314	9.6314
2	(1,2)	11.0368	11.0368	(1,2)	33.6228	33.6228	(1,2)	16.1348	16.1348
3	(1,3)	15.0626	15.0626	(2,1)	38.3629	38.3629	(1,3)	36.7256	36.7256
4	(1,4)	21.7064	21.7064	(2,2)	75.2037	75.2037	(2,1)	38.9450	38.9450
5	(1,5)	31.1771	31.1771	(3,1)	86.9684	86.9684	(2,2)	46.7381	46.7381
6	(2,1)	39.2387	39.2387	(3,2)	130.3575	130.3576	(2,3)	70.7401	70.7401
7	(2,2)	40.5035	40.5035	(4,1)	155.3211	155.3211	(1,4)	75.2834	75.2834
8	(1,6)	43.6698	43.6698	(1,3)	156.1247	156.1248	(3,1)	87.9867	87.9867
9	(2,3)	44.9416	44.9416	(2,3)	199.8452	199.8452	(3,2)	96.0405	96.0405
10	(2,4)	52.2281		(4,2)	202.1620		(2,4)	111.0254	
11	(1,7)	59.2880		(5,1)	243.4060		(3,3)	122.0400	
12	(2,5)	62.3564		(3,3)	263.6809		(1,5)	133.7046	

(Continued)

Table 7.5(d) (Continued)

Mode Sequence	(m, n)	λ Present	λ (Leissa 1973)	(m, n)	λ Present	λ (Leissa 1973)	(m, n)	λ Present	λ (Leissa 1973)
		A pair of simple supports along the *x* direction							
		Aspect ratio							
		$L_1:L_2 = 2:5$			$L_1:L_2 = 5:2$			$L_1:L_2 = 1:1$	
13	(2,6)	75.3561		(5,2)	292.1134		(4,1)	156.7525	
14	(1,8)	78.0721		(4,3)	344.5564		(3,4)	164.6959	
15	(3,1)	88.4399		(6,1)	351.2170		(4,2)	164.7866	
16	(3,2)	89.6234		(1,4)	399.7153		(2,5)	169.5377	
17	(2,7)	91.2913		(6,2)	400.9285		(4,3)	191.8674	
18	(3,3)	94.2153		(2,4)	440.9037		(1,6)	212.1019	
19	(1,9)	100.0340		(5,3)	442.1257		(3,5)	224.7584	
20	(3,4)	101.7419		(7,1)	478.7525		(4,4)	236.2620	

Table 7.6(a) Natural frequencies (below 800 Hz) of a thin plate with clamped and free boundaries along the *y* direction.

A pair of simple supports along the *x* direction, $L_1:L_2 = 2:5$

m	(m,n)	(1,1)	(1,2)	(1,3)	(1,4)	(1,5)	(1,6)	(1,7)	(1,8)
1	β	3.1920	3.6883	4.4830	5.4426	6.4994	7.6159	8.7702	9.9495
	λ	10.1888	13.6036	20.0971	29.6219	42.2425	58.0019	76.9170	98.9932
	f	24.9962	33.3738	49.3044	72.6715	103.6338	142.2963	188.7008	242.8605
	(m,n)	**(1,9)**	**(1,10)**	**(1,11)**	**(1,12)**	**(1,13)**	**(1,14)**		
	β	11.1459	12.3545	13.5718	14.7959	16.0251	17.2582		
	λ	124.2318	152.6325	184.1949	218.9184	256.8026	297.8470		
	f	304.7784	374.4541	451.8863	537.0737	630.0150	730.7094		
	(m,n)	**(2,1)**	**(2,2)**	**(2,3)**	**(2,4)**	**(2,5)**	**(2,6)**	**(2,7)**	**(2,8)**
2	β	6.2959	6.5574	7.0409	7.7003	8.4916	9.3812	10.3444	11.3633
	λ	39.6382	42.9993	49.5740	59.2944	72.1080	88.0069	107.0057	129.1235
	f	97.2446	105.4904	121.6202	145.4672	176.9029	215.9078	262.5177	316.7794
	(m,n)	**(2,9)**	**(2,10)**	**(2,11)**	**(2,12)**	**(2,13)**			
	β	12.4248	13.5194	14.6398	15.7807	16.9380			
	λ	254.3760	182.7738	214.3238	249.0298	286.8942			
	f	378.7314	448.3999	525.8016	610.9461	703.8390			
	(m,n)	**(3,1)**	**(3,2)**	**(3,3)**	**(3,4)**	**(3,5)**	**(3,6)**	**(3,7)**	**(3,8)**
3	β	9.4246*	9.6009	9.9352	10.4135	11.0151	11.7198	12.5094	13.3688
	λ	88.8232*	92.1781	98.7079	108.4406	121.3327	137.3526	156.4839	178.7235
	f	217.9104*	226.1409	242.1606	266.0379	297.6662	336.9677	383.9027	438.4633

Table 7.6(a) (Continued)

A pair of simple supports along the x direction, $L_1 : L_2 = 2 : 5$

	(m,n)	(3,9)	(3,10)	(3,11)	(3,12)			
	β	14.2855	15.2496	16.2528	17.2886			
	λ	204.0761	232.5499	264.1535	298.8948			
	f	500.6609	570.5157	648.0491	733.2801			
	(m,n)	(4,1)	(4,2)	(4,3)	(4,4)	(4,5)	(4,6)	(4,7)	(4,8)
4	β	12.5590*	12.6935	12.9465	13.3159	13.7915	14.3617	15.0152	15.7411
	λ	157.7289*	161.1244	167.6124	177.3145	190.2050	206.2587	225.4555	247.7818
	f	386.9570*	395.2872	411.2042	435.0065	466.6309	506.0156	553.1112	607.8842
	(m,n)	(4,9)	(4,10)					
	β	16.5297	17.3724					
	λ	273.2302	301.7992					
	f	670.3171	740.4054					
	(m,n)	(5,1)	(5,2)	(5,3)	(5,4)	(5,5)	(5,6)	(5,7)
5	β	15.6955*	15.8059	16.0088	16.3080	16.6979	17.1717	17.7223
	λ	246.3501*	249.8255	256.2808	265.9509	278.8187	294.8671	314.0799
	f	604.3720*	612.8982	628.7350	652.4587	684.0274	723.3990	770.5337
6	None below 800 Hz						

Table 7.6(b) Natural frequencies (below 2000 Hz) of a thin plate with clamped and free boundaries along the y direction.

A pair of simple supports along the x direction, $L_1 : L_2 = 5 : 2$

m	(m,n)	(1,1)	(1,2)	(1,3)	(1,4)
1	β	5.5342	12.2253	19.9108	27.6806
	λ	30.6277	149.4569	396.4404	766.2154
	f	75.1390	366.6634	972.5892	1879.7600
	(m,n)	(2,1)	(2,2)	(2,3)	(2,4)
2	β	7.6211	13.5208	20.7034	28.2459
	λ	58.0804	182.8110	428.6304	797.8318
	f	142.4890	448.4912	1051.5610	1957.3248
	(m,n)	(3,1)	(3,2)	(3,3)	
3	β	10.2736	15.3302	21.9395	
	λ	105.5470	235.0155	481.3410	
	f	258.9389	576.5648	1180.8763	
	(m,n)	(4,1)	(4,2)	(4,3)	
4	β	13.1570	17.4706	23.5287	
	λ	173.1060	305.2218	553.5986	
	f	424.6818	748.8022	1358.1461	

(Continued)

Table 7.6(b) (Continued)

A pair of simple supports along the x direction, $L_1 : L_2 = 5 : 2$

	(m,n)	(5,1)	(5,2)	(5,3)
5	β	16.1443	19.8455	25.3921
	λ	260.6371	393.8427	644.7603
	f	639.4222	966.2162	1581.7937
	(m,n)	(6,1)	(6,2)	(6,3)
6	β	19.1846	22.3934	27.4707
	λ	368.0478	501.4649	754.6402
	f	902.9336	1230.2464	1851.3626
	(m,n)	(7,1)	(7,2)	
7	β	22.2549	25.0697	
	λ	495.2818	628.4901	
	f	1215.0772	1541.8778	
	(m,n)	(8,1)	(8,2)	
8	β	25.3437	27.8413	
	λ	642.3045	775.1376	
	f	1575.7688	1901.6488	
	(m,n)	(9,1)		
9	β	28.4446		
	λ	809.0948		
	f	1984.9563		
10	None below 2000 Hz			

Table 7.6(c) Natural frequencies (below 2000 Hz) of a thin plate with clamped and free boundaries along the y direction.

A pair of simple supports along the x direction, $L_1 : L_2 = 1 : 1$

m	(m,n)	(1,1)	(1,2)	(1,3)	(1,4)	(1,5)	(1,6)	(1,7)	(1,8)
1	β	3.5619	5.7502	8.5087	11.4642	14.5003	17.5746	20.6697	23.7774
	λ	12.6874	33.0651	72.3976	131.4287	210.2589	308.8664	427.2376	565.3638
	f	31.1260	81.1187	177.6133	322.4347	515.8291	757.7435	1048.1441	1387.0099
	(m,n)	(1,9)							
	β	26.8931							
	λ	723.2400							
	f	1774.3283							
	(m,n)	(2,1)	(2,2)	(2,3)	(2,4)	(2,5)	(2,6)	(2,7)	(2,8)
2	β	6.4577	7.9382	10.1569	12.7425	15.5284	18.4287	21.3978	24.4108
	λ	41.7019	63.0148	103.1617	162.3715	241.1313	339.6155	457.8657	595.8862
	f	102.3075	154.5946	253.0870	398.3467	591.5687	833.1804	1123.2842	1461.8906

Table 7.6(c) (Continued)

A pair of simple supports along the *x* direction, $L_1 : L_2 = 1:1$

	(m,n)	(2,9)						
	β	27.4531						
	λ	753.6729						
	f	1848.9894						

	(m,n)	(3,1)	(3,2)	(3,3)	(3,4)	(3,5)	(3,6)	(3,7)	(3,8)
3	β	9.5190	10.5781	12.3602	14.5828	17.0807	19.7568	22.5514	25.4268
	λ	90.6114	111.8964	152.7733	212.6587	291.7509	390.3327	508.5661	646.5228
	f	222.2973	274.5160	374.7996	521.7166	715.7540	957.6051	1247.6677	1586.1176

	(m,n)	(3,9)
	β	28.3590
	λ	804.2304
	f	1973.0225

	(m,n)	(4,1)	(4,2)	(4,3)	(4,4)	(4,5)	(4,6)	(4,7)	(4,8)
4	β	12.6215	13.4323	14.8840	16.7925	19.0162	21.4604	24.0616	26.7766
	λ	159.3022	180.4269	221.5345	281.9890	361.6159	460.5478	578.9622	716.9871
	f	390.8169	442.6422	543.4917	691.8051	887.1540	1129.8642	1420.3709	1758.9879

	(m,n)	(5,1)	(5,2)	(5,3)	(5,4)	(5,5)	(5,6)	(5,7)	(5,8)
5	β	15.7397	16.3935	17.6022	19.2507	21.2291	23.4530	25.8609	28.4081
	λ	247.7389	268.7457	309.8377	370.5899	450.6750	550.0432	668.7858	807.0221
	f	607.7790	659.3152	760.1263	909.1700	1105.6431	1349.4237	1640.7356	1979.8713

	(m,n)	(6,1)	(6,2)	(6,3)	(6,4)	(6,5)	(6,6)	(6,7)
6	β	18.8655	19.4128	20.4414	21.8793	23.6450	25.6678	27.8917
	λ	355.9077	376.8580	417.8495	478.7050	559.0857	658.8353	777.9445
	f	873.1501	924.5477	1025.1123	1174.4094	1371.6078	1616.3238	1908.5351

	(m,n)	(7,1)	(7,2)	(7,3)	(7,4)	(7,5)	(7,6)
7	β	21.9955	22.4667	23.3586	24.6267	26.2109	28.0542
	λ	483.8020	504.7517	545.6228	606.4752	687.0116	787.0368
	f	1186.9139	1238.3098	1338.5792	1487.8687	1685.4490	1930.8413

	(m,n)	(8,1)	(8,2)	(8,3)	(8,4)
8	β	25.1280[*]	25.5424	26.3281	27.4585
	λ	631.4180[*]	652.4158	693.1696	753.9677
	f	1549.0610[*]	1600.5748	1700.5565	1849.7126

	(m,n)	(9,1)
9	β	28.2622[*]
	λ	798.7532[*]
	f	1959.5851[*]
10	None below 2000 Hz	

Table 7.6(d) Up to the lowest 20 non-dimensional natural frequencies λ of a thin plate with clamped and free boundaries along the y direction, sorted in ascending order.

Mode Sequence	\multicolumn{9}{c}{A pair of simple supports along the x direction}

	\multicolumn{9}{c}{Aspect ratio}

Mode Sequence	\multicolumn{3}{c}{$L_1:L_2=2:5$}	\multicolumn{3}{c}{$L_1:L_2=5:2$}	\multicolumn{3}{c}{$L_1:L_2=1:1$}
	(m, n)	λ Present	λ (Leissa 1973)	(m, n)	λ Present	λ (Leissa 1973)	(m, n)	λ Present	λ (Leissa 1973)
1	(1,1)	10.1888	10.1888	(1,1)	30.6277	30.6277	(1,1)	12.6874	12.6874
2	(1,2)	13.6036	13.6036	(2,1)	58.0804	58.0804	(1,2)	33.0651	33.0651
3	(1,3)	20.0971	20.0971	(3,1)	105.5470	105.5470	(2,1)	41.7019	41.7019
4	(1,4)	29.6219	29.6219	(1,2)	149.4569	149.4569	(2,2)	63.0148	63.0148
5	(2,1)	39.6382	39.6382	(4,1)	173.1060	173.1060	(1,3)	72.3976	72.3976
6	(1,5)	42.2425	42.2425	(2,2)	182.8110	182.8110	(3,1)	90.6114	90.6114
7	(2,2)	42.9993	42.9993	(3,2)	235.0155	235.0155	(2,3)	103.1617	103.1617
8	(2,3)	49.5740	49.5740	(5,1)	260.6371	260.6371	(3,2)	111.8964	111.8964
9	(1,6)	58.0019	58.0019	(4,2)	305.2218	305.2218	(1,4)	131.4287	131.4287
10	(2,4)	59.2944		(6,1)	368.0478		(3,3)	152.7733	
11	(2,5)	72.1080		(5,2)	393.8427		(4,1)	159.3022	
12	(1,7)	76.9170		(1,3)	396.4404		(2,4)	162.3715	
13	(2,6)	88.0069		(2,3)	428.6304		(4,2)	180.4269	
14	(3,1)	88.8232		(3,3)	481.3410		(1,5)	210.2589	
15	(3,2)	92.1781		(7,1)	495.2818		(3,4)	212.6587	
16	(3,3)	98.7079		(6,2)	501.4649		(4,3)	221.5345	
17	(1,8)	98.9932		(4,3)	553.5986		(2,5)	241.1313	
18	(2,7)	107.0057		(7,2)	628.4901		(5,1)	247.7389	
19	(3,4)	108.4406		(8,1)	642.3045		(5,2)	268.7457	
20	(3,5)	121.3327		(5,3)	644.7603		(4,4)	281.9890	

Table 7.7(a) Natural frequencies (below 800 Hz) of a thin plate with simple support and clamped boundaries along the y direction.

A pair of simple supports along the x direction, $L_1:L_2=2:5$

m	(m,n)	(1,1)	(1,2)	(1,3)	(1,4)	(1,5)	(1,6)	(1,7)	(1,8)
1	β	3.4279	4.1457	5.0909	6.1508	7.2733	8.4332	9.6169	10.8166
	λ	11.7502	17.1871	25.9171	37.8317	52.9007	71.1180	92.4838	116.9996
	f	28.8269	42.1653	63.5826	92.8127	129.7816	174.4742	226.8909	287.0358
	(m,n)	(1,9)	(1,10)	(1,11)	(1,12)	(1,13)			
	β	12.0278	13.2472	14.4728	15.7033	16.9375			
	λ	144.6672	175.4878	209.4626	246.5923	286.8777			
	f	354.9128	430.5252	513.8757	604.9662	703.7984			
	(m,n)	(2,1)	(2,2)	(2,3)	(2,4)	(2,5)	(2,6)	(2,7)	(2,8)
2	β	6.4193	6.8090	7.4076	8.1648	9.0378	9.9947	11.0126	12.0758
	λ	41.2070	46.3620	54.8720	66.6637	81.6824	99.8937	121.2779	145.8246
	f	101.0934	113.7401	134.6177	163.5464	200.3918	245.0696	297.5317	357.7521

Table 7.7(a) (Continued)

A pair of simple supports along the *x* direction, $L_1 : L_2 = 2 : 5$

	(m,n)	(2,9)	(2,10)	(2,11)	(2,12)	(2,13)		
	β	13.1730	14.2964	15.4402	16.6001	17.7731		
	λ	173.5283	204.3865	238.3982	275.5633	315.8817		
	f	425.7178	501.4224	584.8636	676.0407	774.9541		
	(m,n)	(3,1)	(3,2)	(3,3)	(3,4)	(3,5)	(3,6)	(3,7)	(3,8)
3	β	9.5134	9.7738	10.1918	10.7474	11.4191	12.1866	13.0321	13.9408
	λ	90.5040	95.5275	103.8723	115.5062	130.3962	148.5137	169.8362	194.3470
	f	222.0339	234.3580	254.8305	283.3720	319.9017	364.3495	416.6600	476.7925
	(m,n)	(3,9)	(3,10)	(3,11)	(3,12)			
	β	14.9008	15.9025	16.9384	18.0025			
	λ	222.0344	252.8905	286.9099	324.0892			
	f	544.7182	620.4175	703.8775	795.0896			
	(m,n)	(4,1)	(4,2)	(4,3)	(4,4)	(4,5)	(4,6)	(4,7)	(4,8)
4	β	12.6319	12.8265	13.1439	13.5750	14.1087	14.7332	15.4369	16.2091
	λ	159.5661	164.5193	172.7627	184.2804	199.0543	217.0659	238.2982	262.7364
	f	391.4642	403.6160	423.8395	452.0960	488.3409	532.5289	584.6182	644.5725
	(m,n)	(4,9)	(4,10)					
	β	17.0402	17.9216					
	λ	290.3683	321.1841					
	f	712.3619	787.9625					
	(m,n)	(5,1)	(5,2)	(5,3)	(5,4)	(5,5)	(5,6)	
5	β	15.7600	15.9150	16.1699	16.5197	16.9584	17.4788	
	λ	248.3775	253.2878	261.4654	272.9017	287.5861	305.5069	
	f	609.3458	621.3923	641.4543	669.5110	705.5363	749.5017	
6	None below 800 Hz							

Table 7.7(b) Natural frequencies (below 2000 Hz) of a thin plate with simple support and clamped boundaries along the *y* direction.

A pair of simple supports along the *x* direction, $L_1 : L_2 = 5 : 2$

m	(m,n)	(1,1)	(1,2)	(1,3)
1	β	10.1942	17.9107	25.6995
	λ	103.9227	320.7921	660.4644
	f	254.9540	787.0008	1620.3206

(Continued)

Table 7.7(b) (Continued)

A pair of simple supports along the *x* direction, $L_1 : L_2 = 5 : 2$

		(2,1)	(2,2)	(2,3)
	(m,n)	**(2,1)**	**(2,2)**	**(2,3)**
2	β	11.3286	18.6209	26.2176
	λ	128.3382	346.7382	687.3646
	f	314.8528	850.6546	1686.3149
	(m,n)	**(3,1)**	**(3,2)**	**(3,3)**
3	β	13.1294	19.7754	27.0674
	λ	172.3804	391.0659	732.6417
	f	422.9016	959.4038	1797.3935
	(m,n)	**(4,1)**	**(4,2)**	**(4,3)**
4	β	15.4029	21.3242	28.2275
	λ	237.2502	454.7221	796.7927
	f	582.0471	1115.5720	1954.7754
	(m,n)	**(5,1)**	**(5,2)**	
5	β	17.9712	23.2022	
	λ	322.9642	538.3404	
	f	792.3297	1320.7132	
	(m,n)	**(6,1)**	**(6,2)**	
6	β	20.7182	25.3421	
	λ	429.2419	642.2234	
	f	1053.0614	1575.5698	
	(m,n)	**(7,1)**	**(7,2)**	
7	β	23.5761	27.6849	
	λ	555.8318	766.4542	
	f	1363.6249	1880.3458	
8	None below 2000 Hz			

Table 7.7(c) Natural frequencies (below 2000 Hz) of a thin plate with simple support and clamped boundaries along the *y* direction.

A pair of simple supports along the *x* direction, $L_1 : L_2 = 1 : 1$

m	(m,n)	(1,1)	(1,2)	(1,3)	(1,4)	(1,5)	(1,6)	(1,7)	(1,8)
1	β	4.8628	7.6581	10.6409	13.6908	16.7727	19.8724	22.9829	26.1007
	λ	23.6463	58.6464	113.2281	187.4367	281.3221	394.9104	528.2152	681.2441
	f	58.0116	143.8774	277.7831	459.8393	690.1690	968.8357	1295.8729	1671.2995

Table 7.7(c) (Continued)

A pair of simple supports along the x direction, $L_1 : L_2 = 1:1$

	(m,n)	(2,1)	(2,2)	(2,3)	(2,4)	(2,5)	(2,6)	(2,7)	(2,8)
2	β	7.1885	9.2809	11.8678	14.6729	17.5899	20.5713	23.5931	26.6418
	λ	51.6743	86.1345	140.8456	215.2944	309.4032	423.1776	556.6340	709.7868
	f	126.7728	211.3141	345.5371	528.1828	759.0605	1038.1837	1365.5930	1741.3235

	(m,n)	(3,1)	(3,2)	(3,3)	(3,4)	(3,5)	(3,6)	(3,7)	(3,8)
3	β	10.0135	11.5668	13.7154	16.2024	18.8871	21.6942	24.5813	27.5233
	λ	100.2698	133.7910	188.1134	262.5172	356.7226	470.6387	604.2425	757.5338
	f	245.9924	328.2300	461.4996	644.0348	875.1493	1154.6202	1482.3912	1858.4613

	(m,n)	(4,1)	(4,2)	(4,3)	(4,4)	(4,5)	(4,6)	(4,7)	
4	β	12.9984	14.2030	15.9834	18.1542	20.5835	23.1865	25.9093	
	λ	168.9585	201.7247	255.4686	329.5742	423.6817	537.6136	671.2902	
	f	414.5068	494.8923	626.7425	808.5460	1039.4204	1318.9301	1646.8796	

	(m,n)	(5,1)	(5,2)	(5,3)	(5,4)	(5,5)	(5,6)	(5,7)	
5	β	16.0482	17.0224	18.5189	20.4120	22.5946	24.9867	27.5316	
	λ	257.5440	289.7624	342.9499	416.6480	510.5148	624.3353	757.9869	
	f	631.8340	710.8756	841.3608	1022.1647	1252.4485	1531.6849	1859.5729	

	(m,n)	(5,1)	(5,2)	(5,3)	(5,4)	(5,5)	(5,6)		
6	β	19.1298	19.9441	21.2246	22.8858	24.8458	27.0358		
	λ	365.9495	397.7682	450.4816	523.7600	617.3154	730.9345		
	f	897.7858	975.8468	1105.1687	1284.9431	1514.4630	1793.2053		

	(m,n)	(5,1)	(5,2)	(5,3)	(5,4)	(5,5)			
7	β	22.2293	22.9273	24.0413	25.5122	27.2782			
	λ	494.1406	525.6598	577.9829	650.8727	744.0978			
	f	1212.2777	1289.6038	1417.9684	1596.7892	1825.4989			

	(m,n)	(5,1)	(5,2)	(5,3)	(5,4)				
8	β	25.3397	25.9497	26.9331	28.2478				
	λ	642.0999	673.3881	725.3916	797.9370				
	f	1575.2668	1652.0263	1779.6069	1957.5827				

	(m,n)	(5,1)							
9	β	28.4573							
	λ	809.8175							
	f	1986.7293							
10	None below 2000 Hz								

Table 7.7(d) Up to the lowest 20 non-dimensional natural frequency λ of a thin plate with simple support and clamped boundaries along the y direction, sorted in ascending order.

	A pair of simple supports along the x direction								
	Aspect ratio								
	$L_1:L_2=2:5$			$L_1:L_2=5:2$			$L_1:L_2=1:1$		
Mode Sequence	(m, n)	λ Present	λ (Leissa 1973)	(m, n)	λ Present	λ (Leissa 1973)	(m, n)	λ Present	λ (Leissa 1973)
1	(1,1)	11.7502	11.7502	(1,1)	103.9227	103.9227	(1,1)	23.6463	23.6463
2	(1,2)	17.1871	17.1872	(2,1)	128.3382	128.3382	(2,1)	51.6743	51.6743
3	(1,3)	25.9171	25.9171	(3,1)	172.3804	172.3804	(1,2)	58.6464	58.6464
4	(1,4)	37.8317	37.8317	(4,1)	237.2502	237.2502	(2,2)	86.1345	86.1345
5	(2,1)	41.2070	41.2070	(1,2)	320.7921	320.7921	(3,1)	100.2698	100.2698
6	(2,2)	46.3620	46.3620	(5,1)	322.9642	322.9642	(1,3)	113.2281	113.2281
7	(1,5)	52.9007	52.9007	(2,2)	346.7382	346.7382	(3,2)	133.7910	133.7910
8	(2,3)	54.8720	54.8720	(3,2)	391.0659	391.0659	(2,3)	140.8456	140.8456
9	(2,4)	66.6637	66.6637	(6,1)	429.2419	429.2420	(4,1)	168.9585	168.9585
10	(1,6)	71.1180		(4,2)	454.7221		(1,4)	187.4367	
11	(2,5)	81.6824		(5,2)	538.3404		(3,3)	188.1134	
12	(3,1)	90.5040		(7,1)	555.8318		(4,2)	201.7247	
13	(1,7)	92.4838		(6,2)	642.2234		(2,4)	215.2944	
14	(3,2)	95.5275		(1,3)	660.4644		(4,3)	255.4686	
15	(2,6)	99.8937		(2,3)	687.3646		(5,1)	257.5440	
16	(3,3)	103.8723		(3,3)	732.6417		(3,4)	262.5172	
17	(3,4)	115.5062		(7,2)	766.4542		(1,5)	281.3221	
18	(1,8)	116.9996		(4,3)	796.7927		(5,2)	289.7624	
19	(2,7)	121.2779					(2,5)	309.4032	
20	(3,5)	130.3962					(4,4)	329.5742	

Table 7.8(a) Natural frequencies (below $800\,Hz$) of a thin plate with simple support and free boundaries along the y direction.

A pair of simple supports along the x direction, $L_1:L_2=2:5$

m	(m,n)	(1,1)	(1,2)	(1,3)	(1,4)	(1,5)	(1,6)	(1,7)	(1,8)
1	β	3.1821	3.6135	4.3404	5.2496	6.2720	7.3654	8.5042	9.6728
	λ	10.1259	13.0570	18.8390	27.5580	39.3377	54.2497	72.3220	93.5628
	f	24.8419	32.0329	46.2179	67.6082	96.5073	133.0910	177.4280	229.5380
	(m,n)	(1,9)	(1,10)	(1,11)	(1,12)	(1,13)	(1,14)		
	β	10.8615	12.0645	13.2777	14.4985	15.7252	16.9564		
	λ	117.9729	145.5514	176.2965	210.2068	247.2810	287.5182		
	f	289.4235	357.0819	432.5090	515.7014	606.6558	705.3699		
	(m,n)	(2,1)	(2,2)	(2,3)	(2,4)	(2,5)	(2,6)	(2,7)	(2,8)
2	β	6.2938	6.5343	6.9839	7.6059	8.3622	9.2213	10.1590	11.1571
	λ	39.6118	42.6964	48.7745	57.8502	69.9267	85.0325	103.2059	124.4810
	f	97.1799	104.7472	119.6588	141.9242	171.5514	208.6107	253.1955	305.3899

Table 7.8(a) (Continued)

A pair of simple supports along the *x* direction, $L_1 : L_2 = 2:5$

(m,n)	(2,9)	(2,10)	(2,11)	(2,12)	(2,13)	(2,14)		
β	12.2018	13.2827	14.3920	15.5238	16.6736	17.8380		
λ	148.8833	176.4294	207.1290	240.9880	278.0093	318.1943		
f	365.2562	432.8350	508.1507	591.2171	682.0416	780.6276		

	(m,n)	(3,1)	(3,2)	(3,3)	(3,4)	(3,5)	(3,6)	(3,7)	(3,8)
3	β	9.4238[*]	9.5904	9.9069	10.3624	10.9391	11.6188	12.3847	13.2225
	λ	88.8088[*]	91.9756	98.1467	107.3800	119.6642	134.9959	153.3817	174.8355
	f	217.8750[*]	225.6443	240.7837	263.4358	293.5727	331.1862	376.2922	428.9248

(m,n)	(3,9)	(3,10)	(3,11)	(3,12)
β	14.1200	15.0671	16.0554	17.0783
λ	199.3746	227.0165	257.7770	291.6686
f	489.1266	556.9408	632.4055	715.5519

	(m,n)	(4,1)	(4,2)	(4,3)	(4,4)	(4,5)	(4,6)	(4,7)	(4,8)
4	β	12.5587[*]	12.6876	12.9300	13.2850	13.7434	14.2950	14.9295	15.6368
	λ	157.7202[*]	160.9744	167.1856	176.4905	188.8805	204.3484	222.8909	244.5090
	f	386.9358[*]	394.9194	410.1572	432.9851	463.3815	501.3290	546.8193	599.8551

(m,n)	(4,9)	(4,10)
β	16.4076	17.2336
λ	269.2078	296.9954
f	660.4488	728.6203

	(m,n)	(5,1)	(5,2)	(5,3)	(5,4)	(5,5)	(5,6)	(5,7)
5	β	15.6954[*]	15.8021	15.9981	16.2875	16.6652	17.1253	17.6612
	λ	246.3448[*]	249.7076	255.9386	265.2819	277.7303	293.2776	311.9190
	f	604.3588[*]	612.6089	627.8954	650.8175	681.3571	719.4993	765.2324
6	None below 800 Hz							

Table 7.8(b) Natural frequencies (below 2000 Hz) of a thin plate with simple support and free boundaries along the *y* direction.

A pair of simple supports along the *x* direction, $L_1 : L_2 = 5:2$

m	(m,n)	(1,1)	(1,2)	(1,3)	(1,4)
1	β	4.3360	10.4989	18.0159	25.7513
	λ	18.8009	110.2259	324.5718	663.1282
	f	46.1244	270.4178	796.2735	1626.8558

(Continued)

Table 7.8(b) (Continued)

A pair of simple supports along the x direction, $L_1 : L_2 = 5:2$

	(m,n)	(2,1)	(2,2)	(2,3)	(2,4)
2	β	7.1092	12.1504	18.9863	26.4109
	λ	50.5405	147.6317	360.4790	697.5334
	f	123.9912	362.1857	884.3649	1711.2634

	(m,n)	(3,1)	(3,2)	(3,3)	(3,4)
3	β	10.0116	14.2734	20.4421	27.4571
	λ	100.2321	203.7304	417.8795	753.8900
	f	245.8999	499.8127	1025.1859	1849.5220

	(m,n)	(4,1)	(4,2)	(4,3)
4	β	13.0039	16.6562	22.2461
	λ	169.1026	277.4280	494. 8908
	f	414.8603	680.6156	1214.1181

	(m,n)	(5,1)	(5,2)	(5,3)
5	β	16.0462	19.2135	24.3007
	λ	257.4791	369.1594	590.5231
	f	631.6747	905.6607	1448.7333

	(m,n)	(6,1)	(6,2)	(6,3)
6	β	19.1173	21.8974	26.5428
	λ	365.4724	479.4971	704.5213
	f	896.6153	1176.3527	1728.4056

	(m,n)	(7,1)	(7,2)
7	β	22.2065	24.6749
	λ	493.1288	608.8494
	f	1209.7953	1493.6932

	(m,n)	(8,1)	(8,2)
8	β	25.3075	27.5222
	λ	640.4704	757.4724
	f	1571.2692	1858.3109

	(m,n)	(9,1)
9	β	28.4167
	λ	807.5091
	f	1981.0661
10	None below 2000 Hz	

Table 7.8(c) Natural frequencies (below 2000 Hz) of a thin plate with simple support and free boundaries along the *y* direction.

A pair of simple supports along the *x* direction, $L_1 : L_2 = 1 : 1$

m	(m,n)	(1,1)	(1,2)	(1,3)	(1,4)	(1,5)	(1,6)	(1,7)	(1,8)
1	β	3.4183	5.2684	7.8652	10.7557	13.7620	16.8207	19.9068	23.0088
	λ	11.6845	27.7564	61.8606	115.6857	189.3924	282.9366	396.2815	529.4052
	f	28.6657	68.0948	151.7629	283.8124	464.6372	694.1297	972.1994	1298.7923

	(m,n)	(1,9)							
	β	26.1208							
	λ	682.2945							
	f	1673.8765							

	(m,n)	(2,1)	(2,2)	(2,3)	(2,4)	(2,5)	(2,6)	(2,7)	(2,8)
2	β	6.4185	7.6854	9.7203	12.1864	14.8992	17.7542	20.6945	23.6882
	λ	41.1966	59.0655	94.4837	148.5093	221.9847	315.2129	428.2617	561.1312
	f	101.0679	144.9057	231.7973	364.3385	544.5961	773.3134	1050.6566	1376.6259

	(m,n)	(2,9)							
	β	26.7172							
	λ	713.8073							
	f	1751.1871							

	(m,n)	(3,1)	(3,2)	(3,3)	(3,4)	(3,5)	(3,6)	(3,7)	(3,8)
3	β	9.5023	10.4364	12.0681	14.1657	16.5707	19.1802	21.9274	24.7688
	λ	90.2941	108.9185	145.6382	200.6682	274.5869	367.8803	480.8112	613.4923
	f	221.5189	267.2102	357.2948	492.3003	673.6455	902.5225	1179.5764	1505.0837

	(m,n)	(3,9)							
	β	27.6760							
	λ	765.9633							
	f	1879.1416							

	(m,n)	(4,1)	(4,2)	(4,3)	(4,4)	(4,5)	(4,6)	(4,7)	(4,8)
4	β	12.6127	13.3453	14.6837	16.4808	18.6098	20.9783	23.5207	26.1904
	λ	159.0803	178.0960	215.6100	271.6152	346.3233	440.0897	553.2255	685.9356
	f	390.2725	436.9239	528.9570	666.3550	849.6367	1079.6743	1357.2308	1682.8093

	(m,n)	(5,1)	(5,2)	(5,3)	(5,4)	(5,5)	(5,6)	(5,7)	(5,8)
5	β	15.7345	16.3356	17.4597	19.0148	20.9055	23.0533	25.3976	27.8928
	λ	247.5740	266.8523	304.8407	361.5637	437.0416	531.4532	645.0367	778.0065
	f	607.3744	654.6700	747.8672	887.0259	1072.1963	1303.8168	1582.4717	1908.6871

	(m,n)	(6,1)	(6,2)	(6,3)	(6,4)	(6,5)	(6,6)	(6,7)	
6	β	18.8621	19.3720	20.3362	21.6976	23.3856	25.3364	27.4965	
	λ	355.7804	375.2733	413.5622	470.7842	546.8876	641.9320	756.0596	
	f	872.8379	920.6599	1014.5941	1154.9773	1341.6821	1574.8550	1854.8447	

(Continued)

Table 7.8(c) (Continued)

A pair of simple supports along the x direction, $L_1 : L_2 = 1 : 1$

	(m,n)	(7,1)	(7,2)	(7,3)	(7,4)	(7,5)	(7,6)
7	β	21.9932	22.4365	23.2784	24.4838	26.0008	27.7782
	λ	483.7014	503.3943	541.8855	599.4578	676.0401	771.6293
	f	1186.6670	1234.9797	1329.4105	1470.6528	1658.5326	1893.0420

	(m,n)	(8,1)	(8,2)	(8,3)	(8,4)		
8	β	25.1264*	25.5192	26.2653	27.3439		
	λ	631.3371*	651.2320	689.8667	747.6914		
	f	1548.8624*	1597.6706	1692.4534	1834.3151		

	(m,n)	(9,1)					
9	β	28.2611*					
	λ	798.6873*					
	f	1959.4235*					
10	None below 2000 Hz						

Table 7.8(d) Up to the lowest 20 non-dimensional natural frequency λ of a thin plate with simple support and free boundaries along the y direction, sorted in ascending order.

	A pair of simple supports along the x direction								
	Aspect ratio								
	$L_1 : L_2 = 2:5$			$L_1 : L_2 = 5:2$			$L_1 : L_2 = 1:1$		
Mode Sequence	(m, n)	λ Present	λ (Leissa 1973)	(m, n)	λ Present	λ (Leissa 1973)	(m, n)	λ Present	λ (Leissa 1973)
1	(1,1)	10.1259	10.1259	(1,1)	18.8009	18.8009	(1,1)	11.6845	11.6845
2	(1,2)	13.0570	13.0570	(2,1)	50.5405	50.5405	(1,2)	27.7564	27.7563
3	(1,3)	18.8390	18.8390	(3,1)	100.2321	100.2321	(2,1)	41.1966	41.1967
4	(1,4)	27.5580	27.5580	(1,2)	110.2259	110.2259	(2,2)	59.0655	59.0655
5	(1,5)	39.3377	39.3377	(2,2)	147.6317	147.6317	(1,3)	61.8606	61.8606
6	(2,1)	39.6118	39.6118	(4,1)	169.1026	169.1026	(3,1)	90.2941	90.2941
7	(2,2)	42.6964	42.6964	(3,2)	203.7304	203.7304	(2,3)	94.4837	94.4837
8	(2,3)	48.7745	48.7745	(5,1)	257.4791	257.4791	(3,2)	108.9185	108.9185
9	(1,6)	54.2497	54.2497	(4,2)	277.4280	277.4280	(1,4)	115.6857	115.6857
10	(2,4)	57.8502		(1,3)	324.5718		(3,3)	145.6382	
11	(2,5)	69.9267		(2,3)	360.4790		(2,4)	148.5093	
12	(1,7)	72.3220		(6,1)	365.4724		(4,1)	159.0803	
13	(2,6)	85.0325		(5,2)	369.1594		(4,2)	178.0960	
14	(3,1)	88.8088		(3,3)	417.8795		(1,5)	189.3924	
15	(3,2)	91.9756		(6,2)	479.4971		(3,4)	200.6682	
16	(1,8)	93.5628		(7,1)	493.1288		(4,3)	215.6100	
17	(3,3)	98.1467		(4,3)	494.8908		(2,5)	221.9847	
18	(2,7)	103.2059		(5,3)	590.5231		(5,1)	247.5740	
19	(3,4)	107.3800		(7,2)	608.8494		(5,2)	266.8523	
20	(1,9)	117.9729		(8,1)	640.4704		(4,4)	271.6152	

Table 7.9 Occurrence of natural frequencies below the cut-off frequency.

| Boundaries along y | General condition | Specific m values by aspect ratio $L_1 : L_2$ | | |
		2:5	5:2	1:1
	Poisson's ratio $\gamma = 0.30$			
	Conditions for natural frequencies below the cut-off frequency			
F-F	y symmetrical mode	One natural frequency below the cut-off frequency for each m exists		
	y anti-symmetrical mode	$m\dfrac{L_2}{L_1} > 14.455$ (Leissa 1973)	$m > 5.7820$	$m > 36.1375$	$m > 14.455$
C-F		$m\dfrac{L_2}{L_1} > 7.353$ (Leissa 1973)	$m > 2.9412$	$m > 18.3825$	$m > 7.3530$
S-F		$m\dfrac{L_2}{L_1} > 7.228$ (Leissa, 1973)	$m > 2.8912$	$m > 18.0700$	$m > 7.2280$
C-C	Not possible			
S-S	Not possible			
C-S	Not possible			

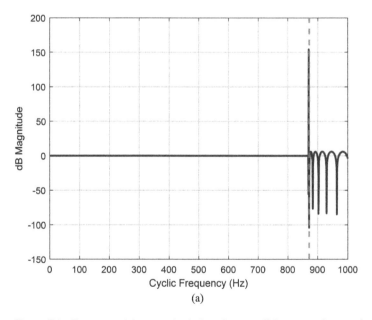

Figure 7.6 Two natural frequencies below the cut-off frequency for $m = 6$ and plate aspect ratio $L_1: L_2 = 2:5$: (a) the original plot and (b) the plot magnified around the cut-off frequency identified by the vertical broken line.

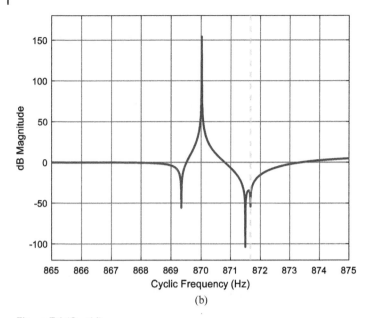

(b)

Figure 7.6 (Cont'd)

Reference

Leissa A.W. The Free Vibration of Rectangular Plates, *Journal of Sound and Vibration*, 31(3), 257–293 (1973).

8

In-Plane Vibration of Rectangular Isotropic Thin Plates with Two Opposite Edges Simply-supported

In this chapter, free in-plane vibrations in two-dimensional rectangular plates are analyzed from the wave vibration standpoint. The coverage is limited to situations where closed-form solutions exist for the plate, which require at least one pair of opposite edges to be simply-supported. This pair of simple supports can be both Type I supports, or both Type II supports, or a mixture of Type I and Type II supports. The general situations of nonzero wavenumbers as well as the special situations of a wavenumber taking zero value are thoroughly studied.

8.1 The Governing Equations of Motion

Consider free vibration of a uniform thin isotropic plate lying in the x–y plane, as shown in Figure 8.1. The equations of motion of the plate are (Liu and Xing 2011)

$$\frac{\partial^2 u(x,y,t)}{\partial x^2} + \gamma_1 \frac{\partial^2 u(x,y,t)}{\partial y^2} + \gamma_2 \frac{\partial^2 v(x,y,t)}{\partial x \partial y} + \gamma_1 \beta^2 u(x,y,t) = 0$$

$$\frac{\partial^2 v(x,y,t)}{\partial y^2} + \gamma_1 \frac{\partial^2 v(x,y,t)}{\partial x^2} + \gamma_2 \frac{\partial^2 u(x,y,t)}{\partial x \partial y} + \gamma_1 \beta^2 v(x,y,t) = 0 \tag{8.1}$$

where x and y are the coordinates and t is the time. $u(x,y,t)$ and $v(x,y,t)$ denote the in-plane plate displacements along the x and y directions, respectively. $\gamma_1 = \frac{1-\gamma}{2}$, $\gamma_2 = \frac{1+\gamma}{2}$, and $\beta^2 = \omega^2 \frac{\rho}{G}$, in which γ is the Poisson's ratio, ρ is the volume mass density, G is the shear modulus of the plate, and ω is the circular frequency.

The normal and shear stresses are related to the in-plane plate displacements, following the sign conventions defined in Figure 8.1, as follows:

$$\sigma_x = \frac{Eh}{1-\gamma^2}\left(\frac{\partial u}{\partial x} + \gamma \frac{\partial v}{\partial y}\right)$$

$$\sigma_y = \frac{Eh}{1-\gamma^2}\left(\frac{\partial v}{\partial y} + \gamma \frac{\partial u}{\partial x}\right) \tag{8.2}$$

$$\tau_{xy} = \frac{Eh}{2(1+\gamma)}\left(\frac{\partial u}{\partial y} + \frac{\partial v}{\partial x}\right) = Gh\left(\frac{\partial u}{\partial y} + \frac{\partial v}{\partial x}\right)$$

where E is the Young's modulus and h is the thickness of the plate. For isotropic and homogeneous materials, the shear modulus G is given by $G = \frac{E}{2(1+\gamma)}$.

Mechanical Wave Vibrations: Analysis and Control, First Edition. Chunhui Mei.
© 2023 Chunhui Mei. Published 2023 by John Wiley & Sons Ltd.
Companion Website: www.wiley.com/go/Mei/MechanicalWaveVibrations

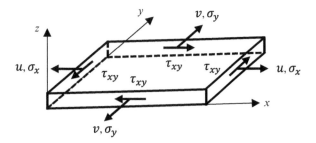

Figure 8.1 Definition of positive internal resistant stresses along the x and y directions.

Table 8.1 Four classical boundary conditions for in-plane vibration of a plate.

Classical Boundary	Edge parallel to the x axis at $y = y_0$	Edge parallel to the y axis at $x = x_0$
Clamped	$u(x,y_0,t) = 0$	$u(x_0,y,t) = 0$
	$v(x,y_0,t) = 0$	$v(x_0,y,t) = 0$
Free	$\sigma_y \big|_{y=y_0} = 0$	$\sigma_x \big|_{x=x_0} = 0$
	$\tau_{xy} \big|_{y=y_0} = 0$	$\tau_{xy} \big|_{x=x_0} = 0$
Type I Simple Support	$v(x,y_0,t) = 0$	$u(x_0,y,t) = 0$
	$\tau_{xy} \big|_{y=y_0} = 0$	$\tau_{xy} \big|_{x=x_0} = 0$
Type II Simple Support	$u(x,y_0,t) = 0$	$v(x_0,y,t) = 0$
	$\sigma_y \big|_{y=y_0} = 0$	$\sigma_x \big|_{x=x_0} = 0$

In general, there are three types of classical boundaries: free, clamped, and simply-supported. For in-plane vibration of a plate, a simply-supported boundary with the normal displacement and shear stress forbidden, is called a Type I simple support in this book. The simply-supported boundary proposed by Gorman (2006), which has the shear deformation and normal stress forbidden, is named a Type II simple support. Boundary conditions of the four types of classical boundaries along the x and y directions are listed in Table 8.1.

8.2 Closed-form Solutions

From Eq. (8.1), it can be seen that the equations of motion are coupled through in-plane plate displacements along the x and y directions, and are symmetrical in terms of the two variables $u(x,y,t)$ and $v(x,y,t)$.

Assuming time harmonic motion and applying separation of variables, the solutions to Eq. (8.1) can be written in the form of

$$u(x,y,t) = u_0 e^{-ik_x x} e^{-ik_y y} e^{i\omega t}$$

$$v(x,y,t) = v_0 e^{-ik_x x} e^{-ik_y y} e^{i\omega t} \tag{8.3a}$$

or in the form of

$$u(x,y,t) = U_x(x)U_y(y)\, e^{i\omega t}$$

$$v(x,y,t) = V_x(x)V_y(y)\, e^{i\omega t} \tag{8.3b}$$

where i is the imaginary unit, u_0 and v_0 are the amplitudes, and k_x and k_y are the wavenumbers along the x and y directions, respectively. Substituting Eq. (8.3a) into Eq. (8.1) and putting the equations into matrix form,

$$\begin{bmatrix} -k_x^2 - \gamma_1 k_y^2 + \gamma_1 \beta^2 & -k_x k_y \gamma_2 \\ -k_x k_y \gamma_2 & -k_y^2 - \gamma_1 k_x^2 + \gamma_1 \beta^2 \end{bmatrix} \begin{bmatrix} u_0 \\ v_0 \end{bmatrix} = \begin{bmatrix} 0 \\ 0 \end{bmatrix} \tag{8.4}$$

Setting the determinant of the coefficient matrix of Eq. (8.4) to zero gives the dispersion equation,

$$\left(-k_x^2 - \gamma_1 k_y^2 + \gamma_1 \beta^2\right)\left(-k_y^2 - \gamma_1 k_x^2 + \gamma_1 \beta^2\right) - \left(-k_x k_y \gamma_2\right)^2 = 0 \tag{8.5}$$

which is a symmetrical quadratic equation in terms of either k_x^2 or k_y^2. The roots, in terms of k_x^2 and k_y^2, respectively, are

$$k_{y1}^2 = -k_x^2 + \beta^2, \text{ and } k_{y2}^2 = -k_x^2 + \gamma_1 \beta^2 \tag{8.6a}$$

$$k_{x1}^2 = -k_y^2 + \beta^2, \text{ and } k_{x2}^2 = -k_y^2 + \gamma_1 \beta^2 \tag{8.6b}$$

The two pairs of wave components along the y and x directions, respectively, can be expressed as

$$U_y(y) = a_1^+ e^{-ik_{y1}y} + a_2^+ e^{-ik_{y2}y} + a_1^- e^{ik_{y1}y} + a_2^- e^{ik_{y2}y}$$

$$V_y(y) = \overline{a_1^+} e^{-ik_{y1}y} + \overline{a_2^+} e^{-ik_{y2}y} + \overline{a_1^-} e^{ik_{y1}y} + \overline{a_2^-} e^{ik_{y2}y} \tag{8.7a}$$

$$U_x(x) = b_1^+ e^{-ik_{x1}x} + b_2^+ e^{-ik_{x2}x} + b_1^- e^{ik_{x1}x} + b_2^- e^{ik_{x2}x}$$

$$V_x(x) = \overline{b_1^+} e^{-ik_{x1}x} + \overline{b_2^+} e^{-ik_{x2}x} + \overline{b_1^-} e^{ik_{x1}x} + \overline{b_2^-} e^{ik_{x2}x} \tag{8.7b}$$

Superscripts $+$ and $-$ and subscripts 1 and 2 in $a_{1,2}^\pm$, $\overline{a_{1,2}^\pm}$, $b_{1,2}^\pm$, and $\overline{b_{1,2}^\pm}$ represent the positive- and negative-going wave components along the y and x directions.

Substituting Eq. (8.3b) into Eq. (8.1), with the time dependent term suppressed and using prime to denote the derivative with respect to variable x or y indicated in the subscript, the governing equations become

$$U_x'' U_y + \gamma_1 U_x U_y'' + \gamma_2 V_x' V_y' + \gamma_1 \beta^2 U_x U_y = 0$$

$$V_y'' V_x + \gamma_1 V_y V_x'' + \gamma_2 U_x' U_y' + \gamma_1 \beta^2 V_x V_y = 0 \tag{8.8}$$

Equation (8.8) shows that for the separation of variables to occur, the following proportional relationships must hold, where symbol "$\propto$" denotes "proportional to",

$$U_x'' \propto V_x' \propto U_x, \text{ or/and } U_y'' \propto V_y' \propto U_y \tag{8.9a}$$

and

$$V_x'' \propto U_x' \propto V_x, \text{ or/and } V_y'' \propto U_y' \propto V_y \tag{8.9b}$$

The conditions described by Eqs. (8.9a) and (8.9b) are physically satisfied only when at least one pair of opposite edges, either along the x or the y direction, is simply-supported. This pair of opposite edges can be formed by Type I, or Type II, or a mixture of Type I and II, which is discussed in Section 8.3.

Without loss of generality, we assume that the opposite edges along the x direction are simply-supported. The solution $U_x(x)$ and $V_x(x)$ must be in the form of

$$U_x(x) = b^+ e^{-ik_0 x} + b^- e^{ik_0 x}$$

$$V_x(x) = \overline{b^+} e^{-ik_0 x} + \overline{b^-} e^{ik_0 x} \tag{8.10}$$

Under such a situation, the closed-form solutions to Eq. (8.1) are

$$u(x,y,t) = \left(b^+ e^{-ik_0 x} + b^- e^{ik_0 x}\right)\left(a_1^+ e^{-ik_1 y} + a_2^+ e^{-ik_2 y} + a_1^- e^{ik_1 y} + a_2^- e^{ik_2 y}\right) e^{i\omega t}$$

$$v(x,y,t) = \left(\overline{b^+} e^{-ik_0 x} + \overline{b^-} e^{ik_0 x}\right)\left(\overline{a_1^+} e^{-ik_1 y} + \overline{a_2^+} e^{-ik_2 y} + \overline{a_1^-} e^{ik_1 y} + \overline{a_2^-} e^{ik_2 y}\right) e^{i\omega t} \tag{8.11a}$$

or in the form of Eq. (8.3b),

$$u(x,y,t) = U_x(x)U_y(y)\, e^{i\omega t}$$

$$v(x,y,t) = V_x(x)V_y(y)\, e^{i\omega t} \tag{8.11b}$$

with

$$U_x(x) = b^+ e^{-ik_0 x} + b^- e^{ik_0 x}$$

$$U_y(y) = a_1^+ e^{-ik_1 y} + a_2^+ e^{-ik_2 y} + a_1^- e^{ik_1 y} + a_2^- e^{ik_2 y}$$

$$V_x(x) = \overline{b^+} e^{-ik_0 x} + \overline{b^-} e^{ik_0 x}$$

$$V_y(y) = \overline{a_1^+} e^{-ik_1 y} + \overline{a_2^+} e^{-ik_2 y} + \overline{a_1^-} e^{ik_1 y} + \overline{a_2^-} e^{ik_2 y} \tag{8.11c}$$

where k_0 and $k_{1,2}$ are the wavenumbers along the x and y directions, respectively. From Eq. (8.6a),

$$k_1^2 = -k_0^2 + \beta^2, \text{ and } k_2^2 = -k_0^2 + \gamma_1 \beta^2 \tag{8.11d}$$

The reflection of vibration waves incident upon both types of simply-supported boundaries along the x direction is studied first.

8.3 Wave Reflection, Propagation, and Wave Vibration Analysis along the Simply-supported x Direction

8.3.1 Wave Reflection at a Simply-supported Boundary along the x Direction

At a boundary, as shown in Figure 8.2, incident waves $\mathbf{b}^+$ give rise to reflected waves $\mathbf{b}^-$, which are related by

$$\mathbf{b}^- = \mathbf{r}\mathbf{b}^+ \tag{8.12}$$

where the reflection matrix $\mathbf{r}$ is determined by boundary conditions.

a) Simple Support – Type I

Choosing the origin at the boundary, the boundary conditions corresponding to Type I simple support along the x direction in Table 8.1 are

$$u(0,y,t) = 0, \text{ and } \tau_{xy}\big|_{x=0} = 0 \tag{8.13}$$

From Eqs. (8.2), (8.11b), and (8.13),

$$u(0,y,t) = U_x(0)U_y(y)e^{i\omega t} = 0$$

$$\tau_{xy}\bigg|_{x=0} = Gh\left(\frac{\partial u}{\partial y} + \frac{\partial v}{\partial x}\right)\bigg|_{x=0} = Gh\left[U_x(0)\frac{\partial U_y(y)}{\partial y} + \frac{\partial V_x(x)}{\partial x}\bigg|_{x=0} V_y(y)\right]e^{i\omega t} = 0 \tag{8.14}$$

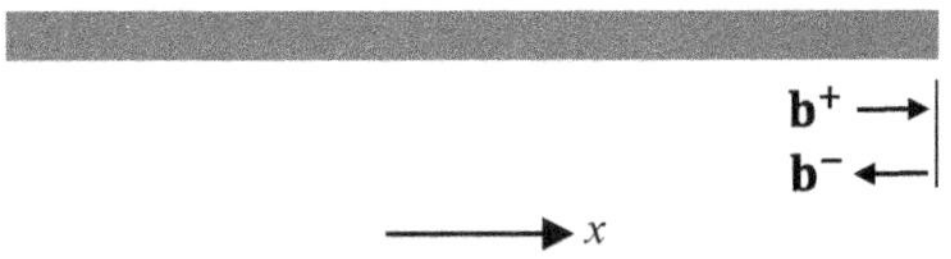

$$\begin{array}{c} \mathbf{b}^+ \longrightarrow | \\ \mathbf{b}^- \longleftarrow | \end{array}$$

$\longrightarrow x$

Figure 8.2 Wave reflection at a boundary along the x direction.

From Eqs. (8.11c) and (8.14),

$$u(0,y,t) = (b^+ + b^-)U_y(y)e^{i\omega t} = 0 \tag{8.15a}$$

$$\tau_{xy}\Big|_{x=0} = Gh\left[(b^+ + b^-)\frac{\partial U_y(y)}{\partial y} + \left(-ik_0\overline{b^+} + ik_0\overline{b^-}\right)V_y(y)\right]e^{i\omega t} = 0 \tag{8.15b}$$

From Eqs. (8.15a) and (8.15b),

$$b^- = -b^+, \text{ and } \overline{b^-} = \overline{b^+} \tag{8.16}$$

Putting Eq. (8.16) into matrix form,

$$\begin{bmatrix} b^- \\ \overline{b^-} \end{bmatrix} = \begin{bmatrix} -1 & 0 \\ 0 & 1 \end{bmatrix}\begin{bmatrix} b^+ \\ \overline{b^+} \end{bmatrix} \tag{8.17}$$

From Eqs. (8.12) and (8.17), the reflection matrix at a Type I simple support is

$$\mathbf{r} = \begin{bmatrix} -1 & 0 \\ 0 & 1 \end{bmatrix} \tag{8.18}$$

where the incident and reflected waves are $\mathbf{b}^+ = \begin{bmatrix} b^+ \\ \overline{b^+} \end{bmatrix}$ and $\mathbf{b}^- = \begin{bmatrix} b^- \\ \overline{b^-} \end{bmatrix}$, respectively.

With the relationships in Eq. (8.16), Eq. (8.11a) can be rewritten as

$$u(x,y,t) = \left(e^{-ik_0x} - e^{ik_0x}\right)\left(c_1^+ e^{-ik_1y} + c_2^+ e^{-ik_2y} + c_1^- e^{ik_1y} + c_2^- e^{ik_2y}\right)e^{i\omega t}$$

$$v(x,y,t) = \left(e^{-ik_0x} + e^{ik_0x}\right)\left(\overline{c_1^+} e^{-ik_1y} + \overline{c_2^+} e^{-ik_2y} + \overline{c_1^-} e^{ik_1y} + \overline{c_2^-} e^{ik_2y}\right)e^{i\omega t} \tag{8.19a}$$

or in compact form as

$$u(x,y,t) = \left(e^{-ik_0x} - e^{ik_0x}\right)U_y(y)e^{i\omega t}$$

$$v(x,y,t) = \left(e^{-ik_0x} + e^{ik_0x}\right)V_y(y)e^{i\omega t} \tag{8.19b}$$

where

$$U_y(y) = c_1^+ e^{-ik_1y} + c_2^+ e^{-ik_2y} + c_1^- e^{ik_1y} + c_2^- e^{ik_2y}$$

$$V_y(y) = \overline{c_1^+} e^{-ik_1y} + \overline{c_2^+} e^{-ik_2y} + \overline{c_1^-} e^{ik_1y} + \overline{c_2^-} e^{ik_2y} \tag{8.19c}$$

The in-plane plate vibrations $u(x,y,t)$ and $v(x,y,t)$ along the x and y directions, with a Type I simply-supported boundary along the x direction, are described by Eq. (8.19a), or equivalently by Eqs. (8.19b) and (8.19c).

b) Simple Support – Type II

Choosing the origin at the boundary, the boundary conditions corresponding to Type II simple support along the x direction in Table 8.1 are

$$v(0,y,t) = 0, \text{ and } \sigma_x\Big|_{x=0} = 0 \tag{8.20}$$

From Eqs. (8.2), (8.11b), and (8.20),

$$v(0,y,t) = V_x(0)V_y(y)e^{i\omega t} = 0$$

$$\sigma_x\Big|_{x=0} = \frac{Eh}{1-\gamma^2}\left(\frac{\partial u}{\partial x} + \gamma\frac{\partial v}{\partial y}\right)\Big|_{x=0} = \frac{Eh}{1-\gamma^2}\left[\frac{\partial U_x(x)}{\partial x}\Big|_{x=0} U_y(y) + \gamma V_x(0)\frac{\partial V_y(y)}{\partial y}\right]e^{i\omega t} = 0 \tag{8.21}$$

From Eqs. (8.11c) and (8.21),

$$v(0,y,t) = \left(\overline{b^+} + \overline{b^-}\right)V_y(y)e^{i\omega t} = 0 \tag{8.22a}$$

$$\sigma_x\Big|_{x=0} = \frac{Eh}{1-\gamma^2}\left[\left(-ik_0 b^+ + ik_0 b^-\right)U_y(y) + \gamma\left(\overline{b^+} + \overline{b^-}\right)\frac{\partial V_y(y)}{\partial y}\right]e^{i\omega t} = 0 \tag{8.22b}$$

From Eqs. (8.22a) and (8.22b),

$$\overline{b^-} = -\overline{b^+}, \; b^- = b^+ \tag{8.23}$$

Putting Eq. (8.23) into matrix form,

$$\begin{bmatrix} \overline{b^-} \\ b^- \end{bmatrix} = \begin{bmatrix} 1 & 0 \\ 0 & -1 \end{bmatrix}\begin{bmatrix} \overline{b^+} \\ b^+ \end{bmatrix} \tag{8.24}$$

From Eqs. (8.12) and (8.24), the reflection matrix at a Type II simple support is

$$\mathbf{r} = \begin{bmatrix} 1 & 0 \\ 0 & -1 \end{bmatrix} \tag{8.25}$$

where the incident and reflected waves are $\mathbf{b}^+ = \begin{bmatrix} \overline{b^+} \\ b^+ \end{bmatrix}$ and $\mathbf{b}^- = \begin{bmatrix} \overline{b^-} \\ b^- \end{bmatrix}$, respectively.

With the relationships in Eq. (8.23), Eq. (8.11a) can be rewritten as

$$u(x,y,t) = \left(e^{-ik_0 x} + e^{ik_0 x}\right)\left(c_1^+ e^{-ik_1 y} + c_2^+ e^{-ik_2 y} + c_1^- e^{ik_1 y} + c_2^- e^{ik_2 y}\right)e^{i\omega t}$$

$$v(x,y,t) = \left(e^{-ik_0 x} - e^{ik_0 x}\right)\left(\overline{c_1^+} e^{-ik_1 y} + \overline{c_2^+} e^{-ik_2 y} + \overline{c_1^-} e^{ik_1 y} + \overline{c_2^-} e^{ik_2 y}\right)e^{i\omega t} \tag{8.26a}$$

or in compact form as

$$u(x,y,t) = \left(e^{-ik_0 x} + e^{ik_0 x}\right)U_y(y)e^{i\omega t}$$

$$v(x,y,t) = \left(e^{-ik_0 x} - e^{ik_0 x}\right)V_y(y)e^{i\omega t} \tag{8.26b}$$

where

$$U_y(y) = c_1^+ e^{-ik_1 y} + c_2^+ e^{-ik_2 y} + c_1^- e^{ik_1 y} + c_2^- e^{ik_2 y}$$

$$V_y(y) = \overline{c_1^+} e^{-ik_1 y} + \overline{c_2^+} e^{-ik_2 y} + \overline{c_1^-} e^{ik_1 y} + \overline{c_2^-} e^{ik_2 y} \tag{8.26c}$$

The in-plane plate vibrations $u(x,y,t)$ and $v(x,y,t)$ along the x and y directions, with a Type II simply-supported boundary along the x direction, are described by Eq. (8.26a), or equivalently by Eqs. (8.26b) and (8.26c).

8.3.2   Wave Propagation and Wave Vibration Analysis along the *x* Direction

Consider two locations A and B along the *x* direction on a uniform plate that are a distance *x* apart, as shown in Figure 8.3. Waves propagate from A to B, with the propagation relationships determined by k_0, the wavenumber along the *x* direction. The positive- and negative-going waves at Locations A and B are related by

$$\mathbf{b}^+ = \mathbf{f}(x)\mathbf{a}^+, \ \mathbf{a}^- = \mathbf{f}(x)\mathbf{b}^- \tag{8.27a}$$

where

$$\mathbf{a}^+ = \begin{bmatrix} a_1^+ \\ a_1^+ \end{bmatrix}, \ \mathbf{a}^- = \begin{bmatrix} a_1^- \\ a_1^- \end{bmatrix}, \ \mathbf{b}^+ = \begin{bmatrix} b_1^+ \\ b_1^+ \end{bmatrix}, \ \mathbf{b}^- = \begin{bmatrix} b_1^- \\ b_1^- \end{bmatrix} \tag{8.27b}$$

are the wave vectors and

$$\mathbf{f}(x) = \begin{bmatrix} e^{-ik_0 x} & 0 \\ 0 & e^{-ik_0 x} \end{bmatrix} \tag{8.27c}$$

is the propagation matrix for a distance *x*.

Figure 8.4 illustrates a uniform rectangular plate of side length L_1 along the *x* direction.

The following reflection and propagation relationships exist:

$$\mathbf{a}^+ = \mathbf{r}_A\, \mathbf{a}^-; \ \mathbf{b}^- = \mathbf{r}_B\, \mathbf{b}^+; \ \mathbf{b}^+ = \mathbf{f}(L_1)\, \mathbf{a}^+; \ \mathbf{a}^- = \mathbf{f}(L_1)\, \mathbf{b}^- \tag{8.28}$$

where $\mathbf{r}_A$ and $\mathbf{r}_B$ are the reflection matrices at Boundaries A and B, respectively, and $\mathbf{f}(L_1)$ is the propagation matrix for a distance L_1.

Assembling the wave relationships in Eq. (8.28) into matrix form,

$$\begin{bmatrix} -\mathbf{I} & \mathbf{r}_A & \mathbf{0} & \mathbf{0} \\ \mathbf{0} & \mathbf{0} & \mathbf{r}_B & -\mathbf{I} \\ \mathbf{f}(L_1) & \mathbf{0} & -\mathbf{I} & \mathbf{0} \\ \mathbf{0} & -\mathbf{I} & \mathbf{0} & \mathbf{f}(L_1) \end{bmatrix} \begin{bmatrix} \mathbf{a}^+ \\ \mathbf{a}^- \\ \mathbf{b}^+ \\ \mathbf{b}^- \end{bmatrix} = \mathbf{0} \tag{8.29}$$

where $\mathbf{I}$ and $\mathbf{0}$ denote identity and zero matrices, respectively.

Equation (8.29) can be written in the form of

$$\mathbf{A}_0 \mathbf{z}_0 = 0 \tag{8.30}$$

where $\mathbf{A}_0$ is an 8 by 8 square coefficient matrix and $\mathbf{z}_0$ is a wave vector containing eight wave components of Eq. (8.29).

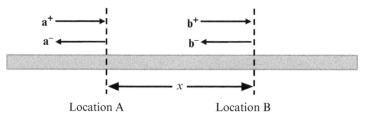

Location A   Location B

Figure 8.3   Wave propagation along the *x* direction.

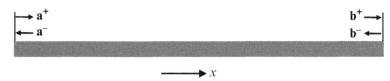

Boundary A   Boundary B

Figure 8.4   Wave propagation and reflection in a uniform plate along the *x* direction.

Setting the determinant of the coefficient matrix $\mathbf{A}_0$ to zero, the characteristic equation is obtained. From which the natural frequencies along the x direction can be found.

There are three possible combinations of simple support along the x direction of a uniform plate: (a) both opposite edges are Type I simple supports, (b) both opposite edges are Type II simple supports, and (c) between the two opposite edges, one is a Type I simple support and the other is a Type II simple support.

Wave vibrations along the x direction for each of the above possible combinations are analyzed.

a) Both opposite edges are of Type I simple supports

When both opposite edges A and B along the x direction are of Type I simple supports, from Eq. (8.18),

$$\mathbf{r}_A = \mathbf{r}_B = \begin{bmatrix} -1 & 0 \\ 0 & 1 \end{bmatrix} \tag{8.31}$$

Substituting Eq. (8.31) into Eq. (8.29) and setting the determinant of the coefficient matrix $\mathbf{A}_0$ to zero, the characteristic equation is

$$\left(e^{-i2k_0 L_1} - 1 \right)^2 = 0 \tag{8.32}$$

From Eq. (8.32),

$$e^{-i2k_0 L_1} = 1 \tag{8.33a}$$

or equivalently

$$e^{-i2k_0 L_1} = e^{\pm 2m\pi} \tag{8.33b}$$

where $m = 0, 1, 2, 3, \dots$.

From Eq. (8.33a) or (8.33b), wavenumbers k_0 corresponding to natural frequencies along the x direction are

$$(k_0)_m = \frac{m\pi}{L_1}, \text{ where } m = 0, 1, 2, 3, \dots \tag{8.34}$$

k_0 takes non-negative values for consistency of allowing positive-going waves to be described in the form of $e^{-ik_0 x}$. The non-negative integer m stands for the number of half harmonic waves along the x direction. The special situation of $m = 0$, and therefore, $k_0 = 0$, is sometimes referred to as the one-dimensional mode of in-plane plate vibration that is discussed in Section 8.5.

b) Both opposite edges are of Type II simple supports

When both opposite edges A and B along the x direction are of Type II simple supports, from Eq. (8.25),

$$\mathbf{r}_A = \mathbf{r}_B = \begin{bmatrix} 1 & 0 \\ 0 & -1 \end{bmatrix} \tag{8.35}$$

Substituting Eq. (8.35) into Eq. (8.29) and setting the determinant of the coefficient matrix $\mathbf{A}_0$ to zero gives the same characteristic equation as Eq. (8.32). Hence, wavenumbers k_0 corresponding to natural frequencies along the x direction are also given by Eq. (8.34), the same as when both opposite edges are of Type I simple supports.

c) One edge is a Type I simple support and the other edge is a Type II simple support

Between the two opposite edges A and B, there is a possibility that one is of Type I simple support and the other is of Type II simple support, that is,

$$\mathbf{r}_A = \begin{bmatrix} -1 & 0 \\ 0 & 1 \end{bmatrix}, \text{ and } \mathbf{r}_B = \begin{bmatrix} 1 & 0 \\ 0 & -1 \end{bmatrix} \tag{8.36a}$$

or

$$\mathbf{r}_A = \begin{bmatrix} 1 & 0 \\ 0 & -1 \end{bmatrix}, \text{ and } \mathbf{r}_B = \begin{bmatrix} -1 & 0 \\ 0 & 1 \end{bmatrix} \tag{8.36b}$$

Table 8.2 Wavenumbers at natural frequencies along the x direction.

Simply-supported boundary pair along the x direction	Wavenumbers at natural frequencies
Same type	$(k_0)_m = \dfrac{m\pi}{L_1}$, where $m = 0, 1, 2, 3, \ldots$
Mixed type	$(k_0)_m = \dfrac{(2m-1)\pi}{2L_1}$, where $m = 1, 2, 3, \ldots$

Substituting either Eq. (8.36a) or Eq. (8.36b) into Eq. (8.29) and setting the determinant of the coefficient matrix $\mathbf{A}_0$ to zero gives the same characteristic equation,

$$\left(e^{-i2k_0L_1} + 1\right)^2 = 0 \tag{8.37}$$

From Eq. (8.37),

$$e^{-i2k_0L_1} = -1 \tag{8.38a}$$

or equivalently

$$e^{-i2k_0L_1} = e^{\pm\pi\pm2m\pi} \tag{8.38b}$$

where $m = 0, 1, 2, 3, \ldots$.

From Eq. (8.38a) or (8.38b), wavenumbers k_0 corresponding to natural frequencies along the x direction are

$$(k_0)_m = \frac{(2m-1)\pi}{2L_1}, \text{ where } m = 1, 2, 3, \ldots \tag{8.39}$$

k_0 takes non-negative values to allow positive-going waves to be described in the form of e^{-ik_0x}. Integer $(2m-1)$ represents the number of quarter harmonic waves along the x direction.

The wavenumbers at natural frequencies, corresponding to the various combinations of Type I and Type II simple supports along the x direction, are listed in Table 8.2.

8.4   Wave Reflection, Propagation, and Wave Vibration Analysis along the *y* Direction

Before studying waves along the y direction, let us recall an important finding from the analysis in Section 8.3 on wave reflection at a simply-supported boundary along the x direction. Regardless of whether it is a Type I or II simple support along the x direction, the reflection of an incident shear or extensional wave is either "1" or "−1", and there always exists a sign difference in the reflection coefficients between the shear and extensional waves incident upon the same type of simply-supported boundary. This allows Eq. (8.11a) to be rewritten, corresponding to Type I or II simple supports along the x direction, respectively, as follows

$$u(x,y,t) = \left(e^{-ik_0x} \mp e^{ik_0x}\right)\left(c_1^+ e^{-ik_1y} + c_2^+ e^{-ik_2y} + c_1^- e^{ik_1y} + c_2^- e^{ik_2y}\right)e^{i\omega t}$$

$$v(x,y,t) = \left(e^{-ik_0x} \pm e^{ik_0x}\right)\left(\overline{c_1^+} e^{-ik_1y} + \overline{c_2^+} e^{-ik_2y} + \overline{c_1^-} e^{ik_1y} + \overline{c_2^-} e^{ik_2y}\right)e^{i\omega t} \tag{8.40}$$

The wave amplitudes $\overline{c_{1,2}^\pm}$ of $v(x,y,t)$ and $c_{1,2}^\pm$ of $u(x,y,t)$ are related to each other, and their relationships can be found from Eq. (8.4) as

$$\overline{\frac{c_1^+}{c_1^+}} = P_1, \ \overline{\frac{c_2^+}{c_2^+}} = P_2, \ \overline{\frac{c_1^-}{c_1^-}} = -P_1, \ \overline{\frac{c_2^-}{c_2^-}} = -P_2 \tag{8.41a}$$

where coefficients P_1 and P_2 are

$$P_1 = \frac{k_0 k_1 \gamma_2}{-k_1^2 - \gamma_1 k_0^2 + \gamma_1 \beta^2} = -\frac{k_0}{k_1}, \text{ and } P_2 = \frac{-k_0^2 - \gamma_1 k_2^2 + \gamma_1 \beta^2}{k_0 k_2 \gamma_2} = \frac{k_2}{k_0} \tag{8.41b}$$

Recall that $k_0 = k_x$, $k_1 = k_{y1}$, and $k_2 = k_{y2}$.

When $k_0 \neq 0$, $k_1 \neq 0$, and $k_2 \neq 0$, P_1 and P_2 do not go to zero or infinity. Under such a situation, Eq. (8.40) can be rewritten as

$$u(x,y,t) = \left(e^{-ik_0 x} \mp e^{ik_0 x}\right)\left(c_1^+ e^{-ik_1 y} + c_2^+ e^{-ik_2 y} + c_1^- e^{ik_1 y} + c_2^- e^{ik_2 y}\right)e^{i\omega t}$$

$$v(x,y,t) = \left(e^{-ik_0 x} \pm e^{ik_0 x}\right)\left(P_1 c_1^+ e^{-ik_1 y} + P_2 c_2^+ e^{-ik_2 y} - P_1 c_1^- e^{ik_1 y} - P_2 c_2^- e^{ik_2 y}\right)e^{i\omega t} \tag{8.42}$$

It can be seen from Eq. (8.11d) that the signs of k_1^2 and k_2^2 are dependent upon the relative magnitudes of k_0^2, β^2, and $\gamma_1 \beta^2$. In other words, for a plate of given material properties, wavenumbers k_1 and k_2 along the y direction may be real or imaginary, depending on the value of wavenumber k_0 along the x direction and the frequency range.

When a wavenumber is real, the corresponding pair of positive- and negative-going waves are propagating waves; an imaginary wavenumber corresponds to a pair of positive- and negative-going decaying waves.

For the majority of materials, the absolute value of Poisson's ratio γ is smaller than one. It follows that γ_1 is a positive number smaller than one. From Eq. (8.11d), it can be seen that k_1^2 is, in general, greater than k_2^2. As a result, the following situations may exist in terms of the nature of waves in the plate along the y direction.

Case 1: $k_1^2 > 0$ and $k_2^2 > 0$ correspond to two pairs of propagating waves in the plate, going in the positive and negative directions along the y direction;

Case 2: $k_1^2 > 0$ and $k_2^2 < 0$ correspond to a pair of propagating waves and a pair of decaying waves, going in the positive and negative directions along the y direction;

Case 3: $k_1^2 < 0$ and $k_2^2 < 0$ correspond to two pairs of decaying waves in the plate, going in the positive and negative directions along the y direction.

The cut-off frequencies associated with wave mode transitions are related to the phase speeds $C_S = \sqrt{\dfrac{G}{\rho}}$ and $C_L = \sqrt{\dfrac{E}{\rho\left(1-\gamma^2\right)}}$ of the shear and extensional waves in an isotropic and homogeneous plate, respectively, because

$$\beta^2 = \omega^2 \frac{\rho}{G} = \frac{\omega^2}{C_s^2}, \text{ and } \gamma_1 \beta^2 = \gamma_1 \omega^2 \frac{\rho}{G} = \frac{\omega^2}{C_L^2} \tag{8.43}$$

Setting Eq. (8.11d) to zero, the cut-off frequencies ω_{c1} and ω_{c2} in terms of the phase speeds are: $\omega_{c1} = k_0 C_S$ and $\omega_{c2} = k_0 C_L$. For isotropic and homogeneous materials, $\omega_{c1} / \omega_{c2} = C_s / C_L = \sqrt{\gamma_1}$.

Wave reflections at classical boundaries along the y direction, namely, the clamped, free, simply-supported Type I, and simply-supported Type II boundaries, are derived. The general situations of $k_0 \neq 0$ are considered below, where coefficient P_2 in Eq. (8.42) does not become infinite. The special situation of $k_0 = 0$ is discussed in Section 8.5.

8.4.1 Wave Reflection at a Classical Boundary along the y Direction

a) Wave reflection at a clamped boundary along the y direction

Choosing the origin at the boundary, the boundary conditions corresponding to a clamped boundary along the y direction from Table 8.1 are

$$u(x,0,t) = 0, \text{ and } v(x,0,t) = 0 \tag{8.44}$$

From Eqs. (8.42) and (8.44),

$$u(x,0,t) = \left(e^{-ik_0x} \mp e^{ik_0x}\right)\left(c_1^+ + c_2^+ + c_1^- + c_2^-\right)e^{i\omega t} = 0$$

$$v(x,0,t) = \left(e^{-ik_0x} \pm e^{ik_0x}\right)\left(P_1c_1^+ + P_2c_2^+ - P_1c_1^- - P_2c_2^-\right)e^{i\omega t} = 0 \tag{8.45}$$

From Eq. (8.45),

$$c_1^+ + c_2^+ + c_1^- + c_2^- = 0$$

$$P_1c_1^+ + P_2c_2^+ - P_1c_1^- - P_2c_2^- = 0 \tag{8.46}$$

Putting Eq. (8.46) into matrix form,

$$\begin{bmatrix} 1 & 1 \\ P_1 & P_2 \end{bmatrix}\begin{bmatrix} c_1^- \\ c_2^- \end{bmatrix} = \begin{bmatrix} -1 & -1 \\ P_1 & P_2 \end{bmatrix}\begin{bmatrix} c_1^+ \\ c_2^+ \end{bmatrix} \tag{8.47}$$

Solving for the reflected waves $\begin{bmatrix} c_1^- \\ c_2^- \end{bmatrix}$ in terms of the incident waves $\begin{bmatrix} c_1^+ \\ c_2^+ \end{bmatrix}$ from Eq. (8.47), the reflection matrix at a clamped boundary along the *y* direction is

$$\mathbf{r} = \begin{bmatrix} 1 & 1 \\ P_1 & P_2 \end{bmatrix}^{-1}\begin{bmatrix} -1 & -1 \\ P_1 & P_2 \end{bmatrix} = \begin{bmatrix} \dfrac{P_1 + P_2}{P_1 - P_2} & \dfrac{2P_2}{P_1 - P_2} \\ -\dfrac{2P_1}{P_1 - P_2} & -\dfrac{P_1 + P_2}{P_1 - P_2} \end{bmatrix} \tag{8.48}$$

b) Wave reflection at a free boundary along the *y* direction

Choosing the origin at the boundary, the boundary conditions corresponding to a free boundary along the *y* direction from Table 8.1 are

$$\sigma_y\big|_{y=0} = 0, \text{ and } \tau_{xy}\big|_{y=0} = 0 \tag{8.49}$$

From Eqs. (8.2), (8.42), and (8.49),

$$\sigma_y\big|_{y=0} = \frac{Ehe^{i\omega t}}{1-\gamma^2}\left\{\left[\left(e^{-ik_0x} \pm e^{ik_0x}\right)\left(-ik_1P_1c_1^+ - ik_2P_2c_2^+ - ik_1P_1c_1^- - ik_2P_2c_2^-\right)\right.\right.$$
$$\left.\left. - ik_0\gamma\left(e^{-ik_0x} \pm e^{ik_0x}\right)\left(c_1^+ + c_2^+ + c_1^- + c_2^-\right)\right]\right\} = 0$$

$$\tau_{xy}\big|_{y=0} = Ghe^{i\omega t}\left\{\left[\left(e^{-ik_0x} \mp e^{ik_0x}\right)\left(-ik_1c_1^+ - ik_2c_2^+ + ik_1c_1^- + ik_2c_2^-\right)\right.\right.$$
$$\left.\left. - ik_0\left(e^{-ik_0x} \mp e^{ik_0x}\right)\left(P_1c_1^+ + P_2c_2^+ - P_1c_1^- - P_2c_2^-\right)\right]\right\} = 0 \tag{8.50}$$

From Eq. (8.50),

$$\left(-ik_1P_1c_1^+ - ik_2P_2c_2^+ - ik_1P_1c_1^- - ik_2P_2c_2^-\right) - ik_0\gamma\left(c_1^+ + c_2^+ + c_1^- + c_2^-\right) = 0$$

$$\left(-ik_1c_1^+ - ik_2c_2^+ + ik_1c_1^- + ik_2c_2^-\right) - ik_0\left(P_1c_1^+ + P_2c_2^+ - P_1c_1^- - P_2c_2^-\right) = 0 \tag{8.51}$$

Putting Eq. (8.51) into matrix form,

$$\begin{bmatrix} k_1P_1 + k_0\gamma & k_2P_2 + k_0\gamma \\ k_1 + k_0P_1 & k_2 + k_0P_2 \end{bmatrix}\begin{bmatrix} c_1^- \\ c_2^- \end{bmatrix} = \begin{bmatrix} -k_1P_1 - k_0\gamma & -k_2P_2 - k_0\gamma \\ k_1 + k_0P_1 & k_2 + k_0P_2 \end{bmatrix}\begin{bmatrix} c_1^+ \\ c_2^+ \end{bmatrix} \tag{8.52}$$

Solving for the reflected waves $\begin{bmatrix} c_1^- \\ c_2^- \end{bmatrix}$ in terms of the incident waves $\begin{bmatrix} c_1^+ \\ c_2^+ \end{bmatrix}$ from Eq. (8.52), the reflection matrix at a free boundary along the y direction is

$$\mathbf{r} = \begin{bmatrix} k_1 P_1 + k_0\gamma & k_2 P_2 + k_0\gamma \\ k_1 + k_0 P_1 & k_2 + k_0 P_2 \end{bmatrix}^{-1} \begin{bmatrix} -k_1 P_1 - k_0\gamma & -k_2 P_2 - k_0\gamma \\ k_1 + k_0 P_1 & k_2 + k_0 P_2 \end{bmatrix} \tag{8.53}$$

c) Wave reflection at a Type I simply-supported boundary along the y direction

Choosing the origin at the boundary, the boundary conditions corresponding to a Type I simple support along the y direction from Table 8.1 are

$$v(x,0,t) = 0, \text{ and } \tau_{xy}\big|_{y=0} = 0 \tag{8.54}$$

These expressions are obtained in Eqs. (8.46) and (8.51),

$$P_1 c_1^+ + P_2 c_2^+ - P_1 c_1^- - P_2 c_2^- = 0$$

$$\left(-ik_1 c_1^+ - ik_2 c_2^+ + ik_1 c_1^- + ik_2 c_2^-\right) - ik_0\left(P_1 c_1^+ + P_2 c_2^+ - P_1 c_1^- - P_2 c_2^-\right) = 0 \tag{8.55}$$

Putting Eq. (8.55) into matrix form,

$$\begin{bmatrix} P_1 & P_2 \\ k_1 + k_0 P_1 & k_2 + k_0 P_2 \end{bmatrix} \begin{bmatrix} c_1^- \\ c_2^- \end{bmatrix} = \begin{bmatrix} P_1 & P_2 \\ k_1 + k_0 P_1 & k_2 + k_0 P_2 \end{bmatrix} \begin{bmatrix} c_1^+ \\ c_2^+ \end{bmatrix} \tag{8.56}$$

Solving for the reflected waves $\begin{bmatrix} c_1^- \\ c_2^- \end{bmatrix}$ in terms of the incident waves $\begin{bmatrix} c_1^+ \\ c_2^+ \end{bmatrix}$ from Eq. (8.56), the reflection matrix at a Type I simple support along the y direction is

$$\mathbf{r} = \begin{bmatrix} 1 & 0 \\ 0 & 1 \end{bmatrix} \tag{8.57}$$

d) Wave reflection at a Type II simply-supported boundary along the y direction

Choosing the origin at the boundary, the boundary conditions corresponding to a Type II simple support along the y direction from Table 8.1 are

$$u(x,0,t) = 0, \text{ and } \sigma_y\big|_{y=0} = 0 \tag{8.58}$$

These expressions are obtained in Eqs. (8.46) and (8.51),

$$c_1^+ + c_2^+ + c_1^- + c_2^- = 0$$

$$\left(-ik_1 P_1 c_1^+ - ik_2 P_2 c_2^+ - ik_1 P_1 c_1^- - ik_2 P_2 c_2^-\right) - ik_0\gamma\left(c_1^+ + c_2^+ + c_1^- + c_2^-\right) = 0 \tag{8.59}$$

Putting Eq. (8.59) into matrix form,

$$\begin{bmatrix} 1 & 1 \\ k_1 P_1 + k_0\gamma & k_2 P_2 + k_0\gamma \end{bmatrix} \begin{bmatrix} c_1^- \\ c_2^- \end{bmatrix} = \begin{bmatrix} -1 & -1 \\ -k_1 P_1 - k_0\gamma & -k_2 P_2 - k_0\gamma \end{bmatrix} \begin{bmatrix} c_1^+ \\ c_2^+ \end{bmatrix} \tag{8.60}$$

Solving for the reflected waves $\begin{bmatrix} c_1^- \\ c_2^- \end{bmatrix}$ in terms of the incident waves $\begin{bmatrix} c_1^+ \\ c_2^+ \end{bmatrix}$ from Eq. (8.60), the reflection matrix at a Type II simple support along the y direction is

$$\mathbf{r} = \begin{bmatrix} -1 & 0 \\ 0 & -1 \end{bmatrix} \tag{8.61}$$

The above analysis shows that, under the general situations of $k_0 \neq 0$, the reflection matrices at classical boundaries along the y direction, in general, are not dependent upon the type of simple supports along the x direction.

8.4.2 Wave Propagation and Wave Vibration Analysis along the *y* Direction

Wave propagation and wave vibration analysis in a uniform plate along the y direction follows a similar procedure as that along the x direction described in Section 8.3.2. Wave propagation relationships are determined by wavenumbers k_1 and k_2 along the y direction. The positive- and negative-going waves that are a distance y apart along the y direction are related by the propagation matrix,

$$\mathbf{f}(y) = \begin{bmatrix} e^{-ik_1 y} & 0 \\ 0 & e^{-ik_2 y} \end{bmatrix} \tag{8.62}$$

For a uniform rectangular plate of side length L_2 and with Boundaries C and D along the y direction, the following reflection and propagation relationships exist:

$$\mathbf{c}^+ = \mathbf{r}_C\, \mathbf{c}^-; \quad \mathbf{d}^- = \mathbf{r}_D\, \mathbf{d}^+; \quad \mathbf{d}^+ = \mathbf{f}(L_2)\, \mathbf{c}^+; \quad \mathbf{c}^- = \mathbf{f}(L_2)\, \mathbf{d}^- \tag{8.63}$$

where $\mathbf{r}_C$ and $\mathbf{r}_D$ are the reflection matrices at Boundaries C and D, respectively, and $\mathbf{f}(L_2)$ is the propagation matrix for a distance L_2.

Assembling the wave relationships in Eq. (8.63) into matrix form,

$$\begin{bmatrix} -\mathbf{I} & \mathbf{r}_C & \mathbf{0} & \mathbf{0} \\ \mathbf{0} & \mathbf{0} & \mathbf{r}_D & -\mathbf{I} \\ \mathbf{f}(L_2) & \mathbf{0} & -\mathbf{I} & \mathbf{0} \\ \mathbf{0} & -\mathbf{I} & \mathbf{0} & \mathbf{f}(L_2) \end{bmatrix} \begin{bmatrix} \mathbf{c}^+ \\ \mathbf{c}^- \\ \mathbf{d}^+ \\ \mathbf{d}^- \end{bmatrix} = \mathbf{0} \tag{8.64}$$

where $\mathbf{I}$ and $\mathbf{0}$ denote identity and zero matrices, respectively.

Equation (8.64) can be written in the form of Eq. (8.30), where $\mathbf{A}_0$ is an 8 by 8 square coefficient matrix and $\mathbf{z}_0$ is a wave vector containing eight wave components of Eq. (8.64). Setting the determinant of the coefficient matrix $\mathbf{A}_0$ to zero, the characteristic equation is obtained. From which the natural frequencies along the y direction, of various combinations of classical boundary conditions corresponding to various integer multiples of half or quarter waves along the x direction, are obtained.

From Eqs. (8.57) and (8.61), the reflection matrices at Type I and Type II simple supports along the y direction are associated with a 2 by 2 identity matrix. The situation of a pair of simply-supported boundaries along the y direction is therefore unique and deserves additional attention, which is addressed in Section 8.6.

8.5 Special Situation of $k_0 = 0$: Wave Reflection, Propagation, and Wave Vibration Analysis along the *y* Direction

The analysis in Section 8.3.2 shows that when both the opposite edges along the x direction are of the same type of simple supports, either both Type I or both Type II, it is possible for $k_0 = 0$. This happens when $m = 0$, as shown in Eq. (8.34). From Eq. (8.41b), $k_0 = 0$ leads coefficient P_2 in Eq. (8.42) to infinity, this indicates that Eq. (8.42) is not an appropriate form of solutions to Eq. (8.1). Under such a situation, the analysis of in-plane plate vibration needs to be based on Eq. (8.40).

For Type I simple support along the x direction, substituting $k_0 = 0$ into Eq. (8.40),

$$u(x,y,t) = 0 \tag{8.65a}$$

$$v(x,y,t)=2\left(\overline{c_1^+}e^{-ik_1y}+\overline{c_2^+}e^{-ik_2y}+\overline{c_1^-}e^{ik_1y}+\overline{c_2^-}e^{ik_2y}\right)e^{i\omega t} \tag{8.65b}$$

From Eqs. (8.41b), $\overline{c_1^{\pm}}=0$ when $k_0=0$. Consequently, Eq. (8.65b) becomes

$$v(x,y,t)=2\left(\overline{c_2^+}e^{-ik_2y}+\overline{c_2^-}e^{ik_2y}\right)e^{i\omega t} \tag{8.65c}$$

Similarly, for Type II simple support along the x direction, substituting $k_0=0$ into (8.40),

$$v(x,y,t)=0 \tag{8.66a}$$

$$u(x,y,t)=2(c_1^+e^{-ik_1y}+c_2^+e^{-ik_2y}+c_1^-e^{ik_1y}+c_2^-e^{ik_2y})e^{i\omega t} \tag{8.66b}$$

It is known from Eqs. (8.41a) and (8.41b) that $\dfrac{c_2^+}{c_2^+}=-\dfrac{c_2^-}{c_2^-}=\dfrac{1}{P_2}=\dfrac{k_0}{k_2}=0$, when $k_0=0$. This gives $c_2^{\pm}=0$. Consequently, Eq. (8.66b) becomes

$$u(x,y,t)=2\left(c_1^+e^{-ik_1y}+c_1^-e^{ik_1y}\right)e^{i\omega t} \tag{8.66c}$$

It is seen from Eqs. (8.65a-c) and (8.66a-c) that, under the special situations of $k_0=0$, wave reflections at classical boundaries along the y direction need to be derived separately for the two types of simple supports along the x direction. This is different from the general situation where $k_0\neq0$.

Equations (8.65a) and (8.66a) also show that under the special situation of $m=0$, and therefore $k_0=0$, either $u(x,y,t)$ or $v(x,y,t)$ vanishes. This explains why such a situation is sometimes referred to as the one-dimensional mode of in-plane plate vibration.

8.5.1 Wave Reflection at a Classical Boundary along the y Direction Corresponding to a Pair of Type I Simple Supports along the x Direction When $k_0=0$

a) Wave reflection at a clamped boundary

Choosing the origin at the boundary, the boundary conditions corresponding to a clamped boundary along the y direction are given in Eq. (8.44). From Eqs. (8.65a) and (8.65c), the only condition that needs to be satisfied is

$$v(x,0,t)=2\left(\overline{c_2^+}+\overline{c_2^-}\right)e^{i\omega t}=0 \tag{8.67}$$

From Eq. (8.67),

$$\overline{c_2^+}+\overline{c_2^-}=0 \tag{8.68}$$

Solving for the reflected wave $\overline{c_2^-}$ in terms of the incident wave $\overline{c_2^+}$ from Eq. (8.68), the reflection coefficient at a clamped boundary along the y direction for $k_0=0$ corresponding to a pair of Type I simple supports along the x direction is

$$r=-1 \tag{8.69}$$

b) Wave reflection at a free boundary

Choosing the origin at the boundary, the boundary conditions corresponding to a free boundary along the y direction are given in Eq. (8.49). From Eqs. (8.2), (8.49), (8.65a), and (8.65c),

$$\tau_{xy}=0$$

$$\sigma_y\big|_{y=0}=\frac{Eh}{1-\gamma^2}\left[2(ik_2)\left(-\overline{c_2^+}+\overline{c_2^-}\right)\right]e^{i\omega t}=0 \tag{8.70}$$

From Eq. (8.70),

$$-\overline{c_2^+} + \overline{c_2^-} = 0 \tag{8.71}$$

Solving for the reflected wave $\overline{c_2^-}$ in terms of the incident wave $\overline{c_2^+}$ from Eq. (8.71), the reflection coefficient at a free boundary along the y direction for $k_0 = 0$ corresponding to a pair of Type I simple supports along the x direction is

$$r = 1 \tag{8.72}$$

c) Wave reflection at a Type I simple support

Choosing the origin at the boundary, the boundary conditions corresponding to a Type I simple support along the y direction are given in Eq. (8.54). From Eqs. (8.2), (8.54), (8.65a), and (8.65c),

$$\tau_{xy} = 0$$

$$v(x,0,t) = 2\left(\overline{c_2^+} + \overline{c_2^-}\right)e^{i\omega t} \tag{8.73}$$

From Eq. (8.73),

$$\overline{c_2^+} + \overline{c_2^-} = 0 \tag{8.74}$$

Solving for the reflected wave $\overline{c_2^-}$ in terms of the incident wave $\overline{c_2^+}$ from Eq. (8.74), the reflection coefficient at a Type I simply-supported boundary along the y direction for $k_0 = 0$ corresponding to a pair of Type I simple supports along the x direction is

$$r = -1 \tag{8.75}$$

d) Wave reflection at a Type II simple support

Choosing the origin at the boundary, the boundary conditions corresponding to a Type II simple support along the y direction are given in Eqs. (8.58). From Eqs. (8.2), (8.58), (8.65a), and (8.65c),

$$u(x,y,t) = 0$$

$$\sigma_y\big|_{y=0} = \frac{Eh}{1-\gamma^2}\left[2(ik_2)\left(-\overline{c_2^+} + \overline{c_2^-}\right)\right]e^{i\omega t} = 0 \tag{8.76}$$

From Eq. (8.76),

$$-\overline{c_2^+} + \overline{c_2^-} = 0 \tag{8.77}$$

Solving for the reflected wave $\overline{c_2^-}$ in terms of the incident wave $\overline{c_2^+}$ from Eq. (8.77), the reflection coefficient at a Type II simply-supported boundary along the y direction for $k_0 = 0$ corresponding to a pair of Type I simple supports along the x direction is

$$r = 1 \tag{8.78}$$

8.5.2 Wave Reflection at a Classical Boundary along the *y* Direction Corresponding to a Pair of Type II Simple Supports along the *x* Direction When $k_0 = 0$

a) Wave reflection at a clamped boundary

Choosing the origin at the boundary, the boundary conditions corresponding to a clamped boundary along the y direction are given in Eq. (8.44). From Eqs. (8.66a) and (8.66c), the only condition that needs to be satisfied is

$$u(x,0,t) = 2(c_1^+ + c_1^-)e^{i\omega t} = 0 \tag{8.79}$$

From Eq. (8.79),

$$c_1^+ + c_1^- = 0 \tag{8.80}$$

Solving for the reflected wave c_1^- in terms of the incident wave c_1^+ from Eq. (8.80), the reflection coefficient at a clamped boundary along the y direction for $k_0 = 0$ corresponding to a pair of Type II simple supports along the x direction is

$$r = -1 \tag{8.81}$$

b) Wave reflection at a free boundary along the y direction

Choosing the origin at the boundary, the boundary conditions corresponding to a free boundary along the y direction are given in Eq. (8.49). From Eqs. (8.2), (8.49), (8.66a), and (8.66c),

$$\sigma_y = 0$$

$$\tau_{xy}\big|_{y=0} = Gh\left[2(ik_1)\left(-c_1^+ + c_1^-\right)\right]e^{i\omega t} = 0 \tag{8.82}$$

From Eq. (8.82),

$$-c_1^+ + c_1^- = 0 \tag{8.83}$$

Solving for the reflected wave c_1^- in terms of the incident wave c_1^+ from Eq. (8.83), the reflection coefficient at a free boundary along the y direction for $k_0 = 0$ corresponding to a pair of Type II simple supports along the x direction is

$$r = 1 \tag{8.84}$$

c) Wave reflection at a Type I simple support

Choosing the origin at the boundary, the boundary conditions corresponding to a Type I simple support along the y direction are given in Eq. (8.54). From Eqs. (8.2), (8.54), (8.66a), and (8.66c),

$$v(x,y,t) = 0$$

$$\tau_{xy}\big|_{y=0} = Gh\left[2(ik_1)\left(-c_1^+ + c_1^-\right)\right]e^{i\omega t} = 0 \tag{8.85}$$

From Eq. (8.85),

$$-c_1^+ + c_1^- = 0 \tag{8.86}$$

Solving for the reflected wave c_1^- in terms of the incident wave c_1^+ from Eq. (8.86), the reflection coefficient at a Type I simply-supported boundary along the y direction for $k_0 = 0$ corresponding to a pair of Type II simple supports along the x direction is

$$r = 1 \tag{8.87}$$

d) Wave reflection at a Type II simple support

Choosing the origin at the boundary, the boundary conditions corresponding to a Type II simple support along the y direction are given in Eq. (8.58). From Eqs. (8.2), (8.58), (8.66a), and (8.66c),

$$\sigma_y = 0$$

$$u(x,0,t) = 2(c_1^+ + c_1^-)e^{i\omega t} = 0 \tag{8.88}$$

From Eq. (8.88),

$$c_1^+ + c_1^- = 0 \tag{8.89}$$

Solving for the reflected wave c_1^- in terms of the incident wave c_1^+ from Eq. (8.89), the reflection coefficient at a Type II simply-supported boundary along the y direction for $k_0 = 0$ corresponding to a pair of Type II simple supports along the x direction is

$$r = -1 \tag{8.90}$$

8.5.3 Wave Propagation and Wave Vibration Analysis along the *y* Direction When *k₀* = 0

Wave propagation and wave vibration analysis in a uniform plate along the y direction follows a similar procedure as described in Section 8.4.2. Wave propagation relationships along the y direction are determined by either wavenumber k_1 or k_2, as shown in Eqs. (8.65c) and (8.66c). When it is a pair of Type I simple supports along the x direction, the propagation coefficient along the y direction for $k_0 = 0$ is

$$f(y) = e^{-ik_2 y} \tag{8.91a}$$

For a pair of Type II simple supports along the x direction, the propagation coefficient along the y direction for $k_0 = 0$ is

$$f(y) = e^{-ik_1 y} \tag{8.91b}$$

For a uniform rectangular plate of side length L_2 and with Boundaries C and D along the y direction, the following wave reflection and propagation relationships exist:

$$c^+ = r_C \, c^-; \; d^- = r_D \, d^+; \; d^+ = f(L_2) \, c^+; \; c^- = f(L_2) \, d^- \tag{8.92}$$

where the propagation coefficients corresponding to a pair of Type I or Type II simple supports along the x direction, respectively, are

$$f(L_2) = e^{-ik_2 L_2} \tag{8.93a}$$

$$f(L_2) = e^{-ik_1 L_2} \tag{8.93b}$$

Assembling the wave relationships in Eq. (8.92) into matrix form,

$$\begin{bmatrix} -1 & r_C & 0 & 0 \\ 0 & 0 & r_D & -1 \\ f(L_2) & 0 & -1 & 0 \\ 0 & -1 & 0 & f(L_2) \end{bmatrix} \begin{bmatrix} c^+ \\ c^- \\ d^+ \\ d^- \end{bmatrix} = 0 \tag{8.94}$$

Setting the determinant of the coefficient matrix to zero, the characteristic equation is

$$r_C r_D \, (f(L_2))^2 - 1 = 0 \tag{8.95}$$

Substituting Eq. (8.93a) or (8.93b) into Eq. (8.95), the characteristic equations corresponding to a pair of Type I or Type II simple supports along the x direction, respectively, are

$$r_C r_D \, e^{-i2k_2 L_2} - 1 = 0 \tag{8.96a}$$

$$r_C r_D \, e^{-i2k_1 L_2} - 1 = 0 \tag{8.96b}$$

From the analysis in Sections 8.5.1 and 8.5.2, it is seen that the reflection coefficients at classical boundaries along the y direction for $k_0 = 0$ corresponding to either a pair of Type I or a pair of Type II simple supports along the x direction are either 1 or -1. Depending on the boundary conditions of the opposite edges along the y direction, the product of r_C and r_D takes either 1 or -1.

Equations (8.96a) and (8.96b) can be written in a combined form as follows

$$e^{-i2k_l L_2} = 1, \quad \text{for } r_C r_D = 1 \tag{8.97a}$$

$$e^{-i2k_l L_2} = 1, \quad \text{for } r_C r_D = -1 \tag{8.97b}$$

where subscript $l = 1$ or 2 for a pair of Type II or a pair of Type I simple supports along the x direction, respectively.

Equations (8.97a) and (8.97b) can be rearranged as

$$e^{-i2(k_l)_n L_2} = 1 = e^{\pm 2n\pi}, \quad \text{for } r_C r_D = 1 \tag{8.98a}$$

$$e^{-i2(k_l)_n L_2} = -1 = e^{\pm\pi\pm 2n\pi}, \quad \text{for } r_C r_D = -1 \tag{8.98b}$$

where $n = 0, 1, 2, 3, \ldots$

From Eq. (8.98a) and (8.98b) the wavenumbers corresponding to natural frequencies along the y direction are

$$(k_l)_n = \frac{n\pi}{L_2}, \quad \text{for } r_C r_D = 1 \tag{8.99a}$$

$$(k_l)_n = \frac{(2n-1)\pi}{2L_2}, \quad \text{for } r_C r_D = -1 \tag{8.99b}$$

Table 8.3 Boundary conditions and natural frequencies along the y direction when $k_0 = 0$.

Simple supports along the x direction	Reflection coefficient "r"	Type	Boundary pair	Wavenumber at natural frequency	Natural frequency (rad/s)
Type I at Both Ends	$r_C = -1$	Same type	C, C		
	$r_F = 1$		F, F		
	$r_{SI} = -1$		SI, SI	$k_2 = \dfrac{n\pi}{L_2}$,	$\omega_n = k_2 C_L$
	$r_{SII} = 1$		SII, SII	where $n = 1, 2, 3, \ldots$	or equivalently
		Different type, same sign in "r"	C, SI		$\omega_n = \beta C_s$,
			F, SII		
		Different type, opposite signs in "r"	C, F	$k_2 = \dfrac{(2n-1)\pi}{2L_2}$,	$\beta = k_2/\sqrt{\gamma_1}$
			C, SII		
			F, SI	where $n = 1, 2, 3, \ldots$	
			SI, SII		
Type II at Both Ends	$r_C = -1$	Same type	C, C		
	$r_F = 1$		F, F	$k_1 = \dfrac{n\pi}{L_2}$,	
	$r_{SI} = 1$		SI, SI		$\omega_n = k_1 C_S$
	$r_{SII} = -1$		SII, SII	where $n = 1, 2, 3, \ldots$	or equivalently
		Different type, same sign in "r"	C, SII		$\omega_n = \beta C_s$,
			F, SI		
		Different type, opposite signs in "r"	C, F	$k_1 = \dfrac{(2n-1)\pi}{2L_2}$,	$\beta = k_1$
			C, SI		
			F, SII	where $n = 1, 2, 3, \ldots$	
			SI, SII		

Header spanning "Type", "Boundary pair", "Wavenumber at natural frequency", "Natural frequency": **Classical boundary* pairs along the y direction**

*Clamped, free, Type I simple support, and Type II simple support boundaries along the y direction are denoted using letters C, F, SI, and SII, respectively.

where subscript $l = 1$ or 2 for a pair of Type II or a pair of Type I simple supports along the x direction, respectively, and $n = 1, 2, 3, \ldots$ Note that the solution to Eq. (8.98a) at $n = 0$ is excluded because it results in all wavenumbers being zero. From Eq. (8.40), this corresponds to a rigid body motion. This is not physically possible because of the simple support constraints along the x direction of the plate.

Summarized in Table 8.3 are wavenumbers k_1 and k_2 corresponding to natural frequencies along the y direction for one-dimensional in-plane plate vibration when $k_0 = 0$. Since the reflection coefficients associated with one-dimensional in-plane plate vibrations are either 1 or -1, $r_C r_D = 1$ holds when the reflection coefficients of the pair of opposite edges along the y direction are of the same sign. When the reflection coefficients of the pair of opposite edges along the y direction are of opposite signs, $r_C r_D = -1$. Possible combinations of classical boundaries and their corresponding wavenumbers at natural frequencies are listed in Table 8.3.

The natural frequencies ω_n along the y direction can be found from Eqs. (8.11d) and (8.43) by setting $k_0 = 0$ for Type I or Type II simple supports along the x direction, respectively,

$$\omega_n = \left(k_2\right)_n C_L = \begin{cases} \dfrac{n\pi}{L_2} C_L, & \text{for } r_C r_D = 1 \\[2mm] \dfrac{(2n-1)\pi}{L_2} C_L, & \text{for } r_C r_D = -1 \end{cases} \tag{8.100a}$$

$$\omega_n = \left(k_1\right)_n C_S = \begin{cases} \dfrac{n\pi}{L_2} C_S, & \text{for } r_C r_D = 1 \\[2mm] \dfrac{(2n-1)\pi}{L_2} C_S, & \text{for } r_C r_D = -1 \end{cases} \tag{8.100b}$$

where $n = 1, 2, 3, \ldots$ Recall that the phase speeds for the shear and extensional in-plane waves in an isotropic and homogeneous plate are $C_S = \sqrt{\dfrac{G}{\rho}}$ and $C_L = \sqrt{\dfrac{E}{\rho\left(1-\gamma^2\right)}}$, respectively.

8.6 Wave Reflection, Propagation, and Wave Vibration Analysis with a Pair of Simply-supported Boundaries along the y Direction When $k_0 \neq 0$

The reflection matrices at Type I and Type II simply-supported boundaries along the y direction are unique because they are associated with a 2 by 2 identity matrix, as obtained in Eqs. (8.57) and (8.61).

This section is devoted to plates with all four edges simply-supported by either Type I or II simple supports. Only the situation of $k_0 \neq 0$ needs to be studied in this section because the situation of $k_0 = 0$ is covered in Section 8.5.

8.6.1 Wave Reflection, Propagation, and Wave Vibration Analysis with a Pair of Simply-supported Boundaries along the y Direction When $k_0 \neq 0$, $k_1 \neq 0$, and $k_2 \neq 0$

There are three possible combinations of a simple support along the y direction of a plate: (a) both opposite edges are of Type I simple supports, (b) both opposite edges are Type II simple supports, and (c) between the two opposite edges, one is a Type I simple support and the other is a Type II simple support.

a) Both opposite edges are Type I simple supports

When both opposite edges along the y direction are Type I simple supports, from Eq. (8.57),

$$\mathbf{r}_A = \mathbf{r}_B = \begin{bmatrix} 1 & 0 \\ 0 & 1 \end{bmatrix} \tag{8.101}$$

Substituting Eq. (8.101) into Eq. (8.64) and setting the determinant of the coefficient matrix to zero, the characteristic equation is

$$\left(e^{i2k_1L_2} - 1\right)\left(e^{i2k_2L_2} - 1\right) = 0 \tag{8.102}$$

From Eq. (8.102),

$$e^{i2k_1L_2} = 1, \text{ or } e^{i2k_2L_2} = 1 \tag{8.103a}$$

or equivalently

$$e^{i2k_1L_2} = e^{\pm 2n\pi}, \text{ or } e^{i2k_2L_2} = e^{\pm 2n\pi} \tag{8.103b}$$

where $n = 0, 1, 2, 3, \ldots$.

From Eq. (8.103b), the wavenumbers corresponding to natural frequencies along the y direction are

$$(k_1)_n = \frac{n\pi}{L_2}, \text{ or } (k_2)_n = \frac{n\pi}{L_2}, \text{ where } n = 0, 1, 2, 3, \ldots \tag{8.104}$$

k_1 and k_2 take non-negative values for consistency of allowing positive-going waves being described in the form of e^{-ik_1y} or e^{-ik_2y}. In this section, $n = 0$ is excluded because the current discussions are on the situations where none of the wavenumbers take a zero value.

Equation (8.104) does indicate, however, that when both opposite edges along the y direction are of Type I simple supports, there is a possibility for $k_1 = 0$ and/or $k_2 = 0$, which happens at $n = 0$. This situation is discussed in Section 8.6.2.

b) Both opposite edges are Type II simple supports

When both opposite edges along the y direction are of Type II simple supports, from Eq. (8.61),

$$\mathbf{r}_A = \mathbf{r}_B = \begin{bmatrix} -1 & 0 \\ 0 & -1 \end{bmatrix} \tag{8.105}$$

Substituting Eq. (8.105) into Eq. (8.64) and setting the determinant of the coefficient matrix to zero, the same characteristic equation as Eq. (8.102) is obtained. Consequently wavenumbers k_1 and k_2 corresponding to natural frequencies along the y direction are also given by Eq. (8.104), the same as when both opposite edges are of Type I simple supports. As a result, there is also a possibility for $k_1 = 0$ and/or $k_2 = 0$, when both opposite edges along the y direction are of Type II simple supports.

c) One edge is a Type I simple support and the other is a Type II simple support

Between the two opposite edges, there is a possibility that one is a Type I simple support, and the other is a Type II simple support, that is

$$\mathbf{r}_A = \begin{bmatrix} 1 & 0 \\ 0 & 1 \end{bmatrix}, \text{ and } \mathbf{r}_B = \begin{bmatrix} -1 & 0 \\ 0 & -1 \end{bmatrix} \tag{8.106a}$$

or

$$\mathbf{r}_A = \begin{bmatrix} -1 & 0 \\ 0 & -1 \end{bmatrix}, \text{ and } \mathbf{r}_B = \begin{bmatrix} 1 & 0 \\ 0 & 1 \end{bmatrix} \tag{8.106b}$$

Substituting either Eq. (8.106a) or (8.106b) into Eq. (8.64) and setting the determinant of the coefficient matrix to zero gives the same characteristic equation,

$$\left(e^{i2k_1L_2} + 1\right)\left(e^{i2k_2L_2} + 1\right) = 0 \tag{8.107}$$

From Eq. (8.107),

$$e^{i2k_1L_2} = -1, \text{ or } e^{i2k_2L_2} = -1 \tag{8.108a}$$

or equivalently

$$e^{i2k_1L_2} = e^{\pm\pi\pm2n\pi}, \text{ or } e^{i2k_2L_2} = e^{\pm\pi\pm2n\pi} \tag{8.108b}$$

where $n = 0, 1, 2, 3, \dots$.

From Eq. (8.108a) or (8.108b), the wavenumbers corresponding to natural frequencies along the y direction are

$$(k_1)_n = \frac{(2n-1)\pi}{2L_2}, \text{ or } (k_2)_n = \frac{(2n-1)\pi}{2L_2}, \text{ where } n = 1, 2, 3, \dots \tag{8.109}$$

k_1 and k_2 again take non-negative values to allow positive-going waves to be described in the form of e^{-ik_1y} or e^{-ik_2y}. Neither k_1 nor k_2 takes a zero value when there is a mixed type of simple supports on opposite edges along the y direction, as shown in Eq. (8.109). In conclusion, there is a possibility for $k_1 = 0$ and/or $k_2 = 0$, only under the situation where both opposite edges along the y direction are of the same type of simple supports, either both of Type I or both of Type II.

8.6.2 Wave Reflection, Propagation, and Wave Vibration Analysis with a Pair of Simply-supported Boundaries along the y Direction When $k_0 = 0$, and either $k_1 = 0$ or $k_2 = 0$

The analysis in Section 8.6.1 shows that when both the opposite edges along the y direction are of the same type simple supports, either both Type I or both Type II, it is possible for $k_1 = 0$ and/or $k_2 = 0$. This happens at $n = 0$, as shown in Eq. (8.104).

From Eq. (8.11d), it is clear that k_1 and k_2 cannot be zero simultaneously. Consequently, two special situations exist, one is $k_1 = 0$, $k_2 \neq 0$, and $k_0 \neq 0$; and the other is $k_2 = 0$, $k_1 \neq 0$, and $k_0 \neq 0$.

a) Special situation A: $k_1 = 0$, $k_2 \neq 0$, and $k_0 \neq 0$

Under the situation of $k_1 = 0$, $k_2 \neq 0$, and $k_0 \neq 0$, it is known from Eqs. (8.41a) and (8.41b) that $\dfrac{c_1^{\pm}}{c_1^{\pm}} = 0$, which gives $c_1^{\pm} = 0$. Substituting $c_1^{\pm} = 0$ and $k_1 = 0$ into Eq. (8.40),

$$u(x,y,t) = \left(e^{-ik_0x} \mp e^{ik_0x}\right)\left(c_2^+ e^{-ik_2y} + c_2^- e^{ik_2y}\right)e^{i\omega t}$$

$$v(x,y,t) = \left(e^{-ik_0x} \pm e^{ik_0x}\right)\left(\overline{c_1^+} + \overline{c_1^-} + \overline{c_2^+} e^{-ik_2y} + \overline{c_2^-} e^{ik_2y}\right)e^{i\omega t} \tag{8.110}$$

To satisfy the boundary conditions of a pair of Type I simple supports along the y direction, from Table 8.1 and Eq. (8.2), the following equations must hold simultaneously,

$$v(x,y_0,t) = 0, \text{ and } \left.\frac{\partial u}{\partial y}\right|_{y=y_0} = 0 \tag{8.111}$$

at both $y_0 = 0$ and $y_0 = L_2$.

Substituting Eq. (8.110) into Eq. (8.111),

$$-c_2^+ e^{-ik_2y_0} + c_2^- e^{ik_2y_0} = 0$$

$$\overline{c_1^+} + \overline{c_1^-} + \overline{c_2^+} e^{-ik_2y_0} + \overline{c_2^-} e^{ik_2y_0} = 0 \tag{8.112}$$

It is known from Eq. (8.41a) that $\overline{c_2^{\pm}} = \pm P_2 c_2^{\pm}$, from which Eq. (8.112) can be rewritten as

$$c_2^+ e^{-ik_2y_0} - c_2^- e^{ik_2y_0} = 0 \tag{8.113a}$$

$$\overline{c_1^+} + \overline{c_1^-} + P_2\left(c_2^+ e^{-ik_2y_0} - c_2^- e^{ik_2y_0}\right) = 0 \tag{8.113b}$$

Equation (8.113a) can be satisfied at $y_0 = 0$ and $y_0 = L_2$ when $c_2^- = c_2^+$ (that is, the reflection coefficient $r = 1$) and $k_2 L_2 = n\pi$, where $n = 1, 2, 3...$. However, Eq. (8.113b) cannot be satisfied at $y_0 = 0$ and $y_0 = L_2$. As a result, the special situation A, of $k_1 = 0$, $k_2 \neq 0$, and $k_0 \neq 0$, does not exist when it is a pair of Type I simple supports along the y direction.

To satisfy the boundary conditions of a pair of Type II simple supports along the y direction, from Table 8.1 and Eq. (8.2), the following equations must hold simultaneously,

$$u(x, y_0, t) = 0, \text{ and } \left. \frac{\partial v}{\partial y} \right|_{y = y_0} = 0 \tag{8.114}$$

at both $y_0 = 0$ and $y_0 = L_2$.

Substituting Eq. (8.110) into Eq. (8.114),

$$c_2^+ e^{-ik_2 y_0} + c_2^- e^{ik_2 y_0} = 0$$

$$\overline{c_2^+} e^{-ik_2 y_0} - \overline{c_2^-} e^{ik_2 y_0} = 0 \tag{8.115}$$

It is known from Eq. (8.41a) that $\overline{c_2^\pm} = \pm P_2 c_2^\pm$, from which Eq. (8.115) can be rewritten as

$$c_2^+ e^{-ik_2 y_0} + c_2^- e^{ik_2 y_0} = 0 \tag{8.116a}$$

$$P_2 (c_2^+ e^{-ik_2 y_0} + c_2^- e^{ik_2 y_0}) = 0 \tag{8.116b}$$

Both Eqs. (8.116a) and (8.116b) are satisfied at $y_0 = 0$ and $y_0 = L_2$ when $c_2^- = -c_2^+$ (that is, the reflection coefficient $r = -1$) and $k_2 L_2 = n\pi$, where $n = 1, 2, 3...$.

From the above analysis, it is concluded that the special situation A, of $k_1 = 0$, $k_2 \neq 0$, and $k_0 \neq 0$ exists, and can only exist, when it is a pair of Type II simple supports along the y direction.

b) Special situation B: $k_2 = 0$, $k_1 \neq 0$, and $k_0 \neq 0$

Under the situation of $k_2 = 0$, $k_1 \neq 0$, and $k_0 \neq 0$, it is known from Eqs. (8.41a) and (8.41b) that $\dfrac{\overline{c_2^\pm}}{c_2^\pm} = 0$. This gives $\overline{c_2^\pm} = 0$. Substituting $\overline{c_2^\pm} = 0$ and $k_2 = 0$ into Eq. (8.40),

$$u(x, y, t) = \left(e^{-ik_0 x} \mp e^{ik_0 x} \right) \left(c_1^+ e^{-ik_1 y} + c_1^- e^{ik_1 y} + c_2^+ + c_2^- \right) e^{i\omega t}$$

$$v(x, y, t) = \left(e^{-ik_0 x} \pm e^{ik_0 x} \right) \left(\overline{c_1^+} e^{-ik_1 y} + \overline{c_1^-} e^{ik_1 y} \right) e^{i\omega t} \tag{8.117}$$

To satisfy the boundary conditions of a pair of Type I simple supports along the y direction, from Table 8.1 and Eq. (8.2), the following equations must hold simultaneously,

$$v(x, y_0, t) = 0, \text{ and } \left. \frac{\partial u}{\partial y} \right|_{y = y_0} = 0 \tag{8.118}$$

at both $y_0 = 0$ and $y_0 = L_2$.

Substituting Eq. (8.117) into Eq. (8.118),

$$\overline{c_1^+} e^{-ik_1 y_0} + \overline{c_1^-} e^{ik_1 y_0} = 0$$

$$c_1^+ e^{-ik_1 y_0} - c_1^- e^{ik_1 y_0} = 0 \tag{8.119}$$

It is known from Eqs. (8.41a) that $\overline{c_1^\pm} = \pm P_1 c_1^\pm$, from which Eq. (8.119) can be rewritten as

$$P_1 (c_1^+ e^{-ik_1 y_0} - c_1^- e^{ik_1 y_0}) = 0 \tag{8.120a}$$

$$c_1^+ e^{-ik_1 y_0} - c_1^- e^{ik_1 y_0} = 0 \tag{8.120b}$$

Table 8.4 Natural frequencies with a pair of simple supports along the y direction.

Simple supports along the y direction	Type of boundary	Wavenumber at natural frequency $k_0 \neq 0$			Natural frequency (rad/s)
		$k_1 \neq 0$; $k_2 \neq 0$	$k_1 = 0$; $k_2 \neq 0$	$k_1 \neq 0$; $k_2 = 0$	
Same type at both ends	A pair of Type I	$\beta = \sqrt{k_1^2 + k_0^2}$, and	Not applicable	$\beta = k_0 / \sqrt{\gamma_1}$	$\omega_n = \beta C_s$
	A pair of Type II	$\beta = \sqrt{(k_2^2 + k_0^2)/\gamma_1}$ where $(k_1)_n = \dfrac{n\pi}{L_2}$, and $(k_2)_q = \dfrac{q\pi}{L_2}$	$\beta = k_0$	Not applicable	
Mixed type at both ends		$\beta = \sqrt{k_1^2 + k_0^2}$, and $\beta = \sqrt{(k_2^2 + k_0^2)/\gamma_1}$ where $(k_1)_n = \dfrac{(2n-1)\pi}{2L_2}$, and $(k_2)_q = \dfrac{(2q-1)\pi}{2L_2}$	Not applicable	Not applicable	

Both Eqs. (8.120a) and (8.120b) are satisfied at $y_0 = 0$ and $y_0 = L_2$ when $c_1^- = c_1^+$ (that is, the reflection coefficient $r = 1$) and $k_1 L_2 = n\pi$, where $n = 1, 2, 3....$ As a result, the special situation B of $k_2 = 0$, $k_1 \neq 0$, and $k_0 \neq 0$ can exist when it is a pair of Type I simple supports along the y direction.

To satisfy the boundary conditions of a pair of Type II simple supports along the y direction, from Table 8.1 and Eq. (8.2), the following equations must hold simultaneously,

$$u(x, y_0, t) = 0, \text{ and } \left.\frac{\partial v}{\partial y}\right|_{y=y_0} = 0 \tag{8.121}$$

at both $y_0 = 0$ and $y_0 = L_2$.

Substituting Eq. (8.117) into Eq. (8.121),

$$\overline{c_1^+} e^{-ik_1 y} - \overline{c_1^-} e^{ik_1 y} = 0$$

$$c_1^+ e^{-ik_1 y} + c_1^- e^{ik_1 y} + c_2^+ + c_2^- = 0 \tag{8.122}$$

It is known from Eq. (8.41a) that $\overline{c_1^\pm} = \pm P_1 c_1^\pm$, from which Eq. (8.122) can be rewritten as

$$c_1^+ e^{-ik_1 y} + c_1^- e^{ik_1 y} = 0 \tag{8.123a}$$

$$c_1^+ e^{-ik_1 y} + c_1^- e^{ik_1 y} + c_2^+ + c_2^- = 0 \tag{8.123b}$$

Equation (8.123a) can be satisfied at $y_0 = 0$ and $y_0 = L_2$ when $c_1^- = -c_1^+$ (that is, the reflection coefficient $r = -1$) and $k_1 L_2 = n\pi$, where $n = 1, 2, 3....$ However, Eq. (8.123b) cannot be satisfied at $y_0 = 0$ and $y_0 = L_2$. As a result, the special situation B of $k_2 = 0$, $k_1 \neq 0$, and $k_0 \neq 0$ does not exist when it is a pair of Type II simple supports along the y direction.

From the above analysis, it is concluded that the special situation B of $k_2 = 0$, $k_1 \neq 0$, and $k_0 \neq 0$ exists, and can only exist, when it is a pair of Type I simple supports along the y direction.

The natural frequencies of the general and special situations of having simple supports along the y direction are obtained from Eqs. (8.11d), (8.43), (8.104), and (8.109), and are summarized in Table 8.4, in which n and q are positive integers.

8.7 Numerical Examples

In the numerical examples, an isotropic and homogeneous thin uniform steel plate with simply-supported edges along the x direction is considered. The material properties of the plate are: the Young's modulus, Poisson's ratio, and mass density are $E = 198.87 \ GN/m^2$, $\gamma = 0.30$, and $\rho = 7664.5 \ kg/m^3$, respectively. The shear modulus is calculated by $G = \dfrac{E}{2(1+\gamma)}$. Three examples are considered.

8.7.1 Example 1: Two Pairs of the Same Type of Simple Supports along the *x* and *y* Directions

In the first example considered, the two pairs of edges of the thin plate along the x and y directions are of the same type of simple supports, respectively. This includes the following four cases:

- Case 1: Type I – Type I along the x direction, and Type I – Type I along the y direction;
- Case 2: Type II – Type II along the x direction, and Type II – Type II along the y direction;
- Case 3: Type I – Type I along the x direction, and Type II – Type II along the y direction;
- Case 4: Type II – Type II along the x direction, and Type I – Type I along the y direction.

In addition to the general situation, the appropriate special situations listed in Tables 8.2 to 8.4 need to be considered. This is because when there is a pair of the same type simple supports along the x direction, it is possible for $k_0 = 0$; when there is a pair of Type I simple supports along the y direction, it is possible for $k_2 = 0$; and when there is a pair of Type II simple supports along the y direction, it is possible for $k_1 = 0$.

a) The general situation of $k_0 \neq 0$, $k_1 \neq 0$, and $k_2 \neq 0$

$$(k_0)_m = \frac{m\pi}{L_1}, \ (k_1)_n = \frac{n\pi}{L_2}, \ (k_2)_q = \frac{q\pi}{L_2},$$

$$\beta = \sqrt{k_1^2 + k_0^2}, \text{ and } \beta = \sqrt{(k_2^2 + k_0^2)/\gamma_1};$$

b) The special situation of $k_0 = 0$, $k_1 \neq 0$, and $k_2 \neq 0$
   - For a pair of Type I simple supports along the x direction:

$$(k_2)_n = \frac{n\pi}{L_2}, \text{ and } \beta = \sqrt{(k_2^2 + k_0^2)/\gamma_1} = \frac{k_2}{\sqrt{\gamma_1}};$$

   - For a pair of Type II simple supports along the x direction:

$$(k_1)_n = \frac{n\pi}{L_2}, \text{ and } \beta = \sqrt{k_1^2 + k_0^2} = k_1;$$

c) The special situation of $k_0 \neq 0$, and either $k_1 = 0$ or $k_2 = 0$
   - For a pair of Type I simple supports along the y direction:

$$k_1 \neq 0, \text{ and } k_2 = 0 : (k_0)_m = \frac{m\pi}{L_1}, \text{ and } \beta = \frac{k_0}{\sqrt{\gamma_1}};$$

   - For a pair of Type II simple supports along the y direction:

Table 8.5(a)   Natural frequencies of the square plate in Example 1 under the general situation where $k_0 \neq 0$, $k_1 \neq 0$, and $k_2 \neq 0$.

A pair of Type I or Type II simple supports along the *x* or *y* directions

		(m,n)	(1,1)	(1,2)	(1,3)	(1,4)	(1,5)	(1,6)	(1,7)	(1,8)
		β	4.4429	7.0248	9.9346	12.9531	16.0190	19.1096	22.2144	25.3283
	n varies	Ω_1	2.6284	4.1559	5.8774	7.6632	9.4770	11.3054	13.1422	14.9844
		Ω_2	1.4142	2.2361	3.1623	4.1231	5.0990	6.0828	7.0711	8.0623
m = 1		ω_n	14035	22192	31384	40920	50605	60368	70176	80013
		(m,q)	(1,1)	(1,2)	(1,3)	(1,4)	(1,5)	(1,6)	(1,7)	(1,8)
		β	7.5098	11.8741	16.7925	21.8948	27.0771	32.3011	37.5492	42.8127
	q varies	Ω_1	4.4429	7.0248	9.9346	12.9531	16.0190	19.1096	22.2144	25.3283
		Ω_2	2.3905	3.7796	5.3452	6.9693	8.6189	10.2817	11.9523	13.6277
		ω_n	23724	37511	53048	69167	85538	102041	118620	135247
		(m,n)	(2,1)	(2,2)	(2,3)	(2,4)	(2,5)	(2,6)	(2,7)	(2,8)
		β	7.0248	8.8858	11.3272	14.0496	16.9180	19.8692	22.8711	25.9062
	n varies	Ω_1	4.1559	5.2569	6.7012	8.3119	10.0088	11.7548	13.5307	15.3263
		Ω_2	2.2361	2.8284	3.6056	4.4721	5.3852	6.3246	7.2801	8.2462
m = 2		ω_n	22192	28071	35783	44383	53445	62768	72251	81839
		(m,q)	(2,1)	(2,2)	(2,3)	(2,4)	(2,5)	(2,6)	(2,7)	(2,8)
		β	11.8741	15.0197	19.1464	23.7482	28.5966	33.5850	38.6593	43.7895
	q varies	Ω_1	7.0248	8.8858	11.3272	14.0496	16.9180	19.8692	22.8711	25.9062
		Ω_2	3.7796	4.7809	6.0945	7.5593	9.1026	10.6904	12.3056	13.9386
		ω_n	37511	47448	60484	75022	90338	106097	122127	138333
		(m,n)	(3,1)	(3,2)	(3,3)	(3,4)	(3,5)	(3,6)	(3,7)	(3,8)
		β	9.9346	11.3272	13.3286	15.7080	18.3185	21.0744	23.9257	26.8418
	n varies	Ω_1	5.8774	6.7012	7.8853	9.2930	10.8374	12.4678	14.1546	15.8798
		Ω_2	3.1623	3.6056	4.2426	5.0000	5.8310	6.7082	7.6158	8.5440
m = 3		ω_n	31384	35783	42106	49622	57869	66575	75582	84794
		(m,q)	(3,1)	(3,2)	(3,3)	(3,4)	(3,5)	(3,6)	(3,7)	(3,8)
		β	16.7925	19.1464	22.5295	26.5513	30.9639	35.6223	40.4417	45.3709
	q varies	Ω_1	9.9346	11.3272	13.3286	15.7080	18.3185	21.0744	23.9257	26.8418
		Ω_2	5.3452	6.0945	7.1714	8.4515	9.8561	11.3389	12.8730	14.4420
		ω_n	53048	60484	71172	83877	97816	112533	127757	143329
		(m,n)	(4,1)	(4,2)	(4,3)	(4,4)	(4,5)	(4,6)	(4,7)	(4,8)
		β	12.9531	14.0496	15.7080	17.7715	20.1160	22.6543	25.3283	28.0993
	n varies	Ω_1	7.6632	8.3119	9.2930	10.5138	11.9008	13.4025	14.9844	16.6237
		Ω_2	4.1231	4.4721	5.0000	5.6569	6.4031	7.2111	8.0623	8.9443
m = 4		ω_n	40920	44383	49622	56141	63547	71566	80013	88767
		(m,q)	(4,1)	(4,2)	(4,3)	(4,4)	(4,5)	(4,6)	(4,7)	(4,8)
		β	21.8948	23.7482	26.5513	30.0394	34.0023	38.2928	42.8127	47.4964
	q varies	Ω_1	12.9531	14.0496	15.7080	17.7715	20.1160	22.6543	25.3283	28.0993
		Ω_2	6.9693	7.5593	8.4515	9.5618	10.8233	12.1890	13.6277	15.1186
		ω_n	69167	75022	83877	94896	107415	120969	135247	150043
		(m,n)	(5,1)	(5,2)	(5,3)	(5,4)	(5,5)	(5,6)	(5,7)	(5,8)

(Continued)

Table 8.5(a) (Continued)

A pair of Type I or Type II simple supports along the *x* or *y* directions

		(m,n)	(1,1)	(1,2)	(1,3)	(1,4)	(1,5)	(1,6)	(1,7)	(1,8)
		β	16.0190	16.9180	18.3185	20.1160	22.2144	24.5366	27.0250	29.6377
	n varies	Ω_1	9.4770	10.0088	10.8374	11.9008	13.1422	14.5161	15.9882	17.5339
		Ω_2	5.0990	5.3852	5.8310	6.4031	7.0711	7.8102	8.6023	9.4340
m = 5		ω_n	50605	53445	57869	63547	70176	77512	85373	93627
		(m,q)	(5,1)	(5,2)	(5,3)	(5,4)	(5,5)	(5,6)	(5,7)	(5,8)
		β	27.0771	28.5966	30.9639	34.0023	37.5492	41.4745	45.6806	50.0969
	q varies	Ω_1	16.0190	16.9180	18.3185	20.1160	22.2144	24.5366	27.0250	29.6377
		Ω_2	8.6189	9.1026	9.8561	10.8233	11.9523	13.2017	14.5406	15.9463
		ω_n	85538	90338	97816	107415	118620	131020	144307	158258

Table 8.5(b) Natural frequencies of the square plate in Example 1 under the special situation where $k_0 = 0$, $k_1 \neq 0$, and $k_2 \neq 0$.

A pair of Type I simple supports along the *x* direction

		(m,n)	(0,1)	(0,2)	(0,3)	(0,4)	(0,5)	(0,6)	(0,7)	(0,8)
m = 0		β	5.3103	10.6205	15.9308	21.2410	26.5513	31.8616	37.1718	42.4821
n varies		Ω_1	π	2π	3π	4π	5π	6π	7π	8π
		Ω_2	1.6903	3.3806	5.0709	6.7612	8.4515	10.1419	11.8322	13.5225
		ω_n	16775	33551	50326	67101	83877	100652	117428	134203

A pair of Type II simple supports along the *x* direction

		(m,n)	(0,1)	(0,2)	(0,3)	(0,4)	(0,5)	(0,6)	(0,7)	(0,8)
m = 0		β	π	2π	3π	4π	5π	6π	7π	8π
n varies		Ω_1	1.8586	3.7172	5.5758	7.4344	9.2930	11.1515	13.0101	14.8687
		Ω_2	1	2	3	4	5	6	7	8
		ω_n	9924	19849	29773	39698	49622	59547	69471	79396

Table 8.5(c) Natural frequencies of the square plate in Example 1 under the special situation where $k_0 \neq 0$, either $k_1 = 0$ or $k_2 = 0$.

A pair of Type I simple supports along the *y* direction ($k_1 \neq 0$ and $k_2 = 0$)

		(m,n)	(1,0)	(2,0)	(3,0)	(4,0)	(5,0)	(6,0)	(7,0)	(8,0)
n = 0		β	5.3103	10.6205	15.9308	21.2410	26.5513	31.8616	37.1718	42.4821
m varies		Ω_1	π	2π	3π	4π	5π	6π	7π	8π
		Ω_2	1.6903	3.3806	5.0709	6.7612	8.4515	10.1419	11.8322	13.5225
		ω_n	16775	33551	50326	67101	83877	100652	117428	134203

Table 8.5(c) (Continued)

A pair of Type II simple supports along the y direction ($k_1 = 0$ and $k_2 \neq 0$)

	(m,n)	(1,0)	(2,0)	(3,0)	(4,0)	(5,0)	(6,0)	(7,0)	(8,0)
n = 0	β	π	2π	3π	4π	5π	6π	7π	8π
m varies	Ω_1	1.8586	3.7172	5.5758	7.4344	9.2930	11.1515	13.0101	14.8687
	Ω_2	1	2	3	4	5	6	7	8
	ω_n	9924	19849	29773	39698	49622	59547	69471	79396

Table 8.6(a) Natural frequencies of the rectangular plate in Example 1 under the general situation where $k_0 \neq 0$, $k_1 \neq 0$, and $k_2 \neq 0$.

		(m,n)	(1,1)	(1,2)	(1,3)	(1,4)	(1,5)	(1,6)	(1,7)	(1,8)
		β	3.5124	6.4766	9.5548	12.6642	15.7863	18.9149	22.0472	25.1818
	n varies	Ω_1	4.1559	7.6632	11.3054	14.9844	18.6786	22.3804	26.0866	29.7955
		Ω_2	2.2361	4.1231	6.0828	8.0623	10.0499	12.0416	14.0357	16.0312
$m=1$		ω_n	11096	20460	30184	40007	49870	59753	69648	79550
		(m,q)	(1,1)	(1,2)	(1,3)	(1,4)	(1,5)	(1,6)	(1,7)	(1,8)
		β	5.9371	10.9474	16.1505	21.4063	26.6837	31.9720	37.2665	42.5650
	q varies	Ω_1	7.0248	12.9531	19.1096	25.3283	31.5726	37.8298	44.0944	50.3636
		Ω_2	3.7796	6.9693	10.2817	13.6277	16.9874	20.3540	23.7246	27.0977
		ω_n	18755	34583	51020	67624	84295	101001	117727	134465
		(m,n)	(2,1)	(2,2)	(2,3)	(2,4)	(2,5)	(2,6)	(2,7)	(2,8)
		β	4.4429	7.0248	9.9346	12.9531	16.0190	19.1096	22.2144	25.3283
	n varies	Ω_1	5.2569	8.3119	11.7548	15.3263	18.9540	22.6107	26.2844	29.9689
		Ω_2	2.8284	4.4721	6.3246	8.2462	10.1980	12.1655	14.1421	16.1245
$m=2$		ω_n	14035	22192	31384	40920	50605	60368	70176	80013
		(m,q)	(2,1)	(2,2)	(2,3)	(2,4)	(2,5)	(2,6)	(2,7)	(2,8)
		β	7.5098	11.8741	16.7925	21.8948	27.0771	32.3011	37.5492	42.8127
	q varies	Ω_1	8.8858	14.0496	19.8692	25.9062	32.0381	38.2191	44.4288	50.6567
		Ω_2	4.7809	7.5593	10.6904	13.9386	17.2378	20.5635	23.9046	27.2554
		ω_n	23724	37511	53048	69167	85538	102041	118620	135247
		(m,n)	(3,1)	(3,2)	(3,3)	(3,4)	(3,5)	(3,6)	(3,7)	(3,8)
		β	5.6636	7.8540	10.5372	13.4209	16.3996	19.4297	22.4904	25.5707
	n varies	Ω_1	6.7012	9.2930	12.4678	15.8798	19.4043	22.9895	26.6110	30.2557
		Ω_2	3.6056	5.0000	6.7082	8.5440	10.4403	12.3693	14.3178	16.2788
$m=3$		ω_n	17892	24811	33288	42397	51807	61379	71048	80779
		(m,q)	(3,1)	(3,2)	(3,3)	(3,4)	(3,5)	(3,6)	(3,7)	(3,8)
		β	9.5732	13.2757	17.8112	22.6854	27.7204	32.8421	38.0157	43.2224

(Continued)

Table 8.6(a) (Continued)

		(m,n)	(1,1)	(1,2)	(1,3)	(1,4)	(1,5)	(1,6)	(1,7)	(1,8)
	q varies	Ω_1	11.3272	15.7080	21.0744	26.8418	32.7992	38.8594	44.9808	51.1414
		Ω_2	6.0945	8.4515	11.3389	14.4420	17.6473	20.9080	24.2015	27.5162
		ω_n	30242	41938	56266	71664	87570	103750	120093	136542
		(m,n)	(4,1)	(4,2)	(4,3)	(4,4)	(4,5)	(4,6)	(4,7)	(4,8)
		β	7.0248	8.8858	11.3272	14.0496	16.9180	19.8692	22.8711	25.9062
	n varies	Ω_1	8.3119	10.5138	13.4025	16.6237	20.0176	23.5095	27.0615	30.6527
		Ω_2	4.4721	5.6569	7.2111	8.9443	10.7703	12.6491	14.5602	16.4924
m = 4		ω_n	22192	28071	35783	44383	53445	62768	72251	81839
		(m,q)	(4,1)	(4,2)	(4,3)	(4,4)	(4,5)	(4,6)	(4,7)	(4,8)
		β	11.8741	15.0197	19.1464	23.7482	28.5966	33.5850	38.6593	43.7895
	q varies	Ω_1	14.0496	17.7715	22.6543	28.0993	33.8360	39.7384	45.7423	51.8125
		Ω_2	7.5593	9.5618	12.1890	15.1186	18.2052	21.3809	24.6113	27.8773
		ω_n	37511	47448	60484	75022	90338	106097	122127	138333
		(m,n)	(5,1)	(5,2)	(5,3)	(5,4)	(5,5)	(5,6)	(5,7)	(5,8)
		β	8.4590	10.0580	12.2683	14.8189	17.5620	20.4204	23.3516	26.3313
	n varies	Ω_1	10.0088	11.9008	14.5161	17.5339	20.7797	24.1617	27.6299	31.1557
		Ω_2	5.3852	6.4031	7.8102	9.4340	11.1803	13.0000	14.8661	16.7631
m = 5		ω_n	26722	31774	38756	46813	55479	64509	73769	83182
		(m,q)	(5,1)	(5,2)	(5,3)	(5,4)	(5,5)	(5,6)	(5,7)	(5,8)
		β	14.2983	17.0011	20.7372	25.0485	29.6853	34.5167	39.4714	44.5081
	q varies	Ω_1	16.9180	20.1160	24.5366	29.6377	35.1241	40.8407	46.7031	52.6627
		Ω_2	9.1026	10.8233	13.2017	15.9463	18.8982	21.9740	25.1282	28.3347
		ω_n	45169	53707	65510	79129	93777	109040	124692	140603

Table 8.6(b)   Natural frequencies of the rectangular plate in Example 1 under the special situation where $k_0 = 0$, $k_1 \neq 0$, and $k_2 \neq 0$.

A pair of Type I simple supports along the *x* direction

	(m,n)	(0,1)	(0,2)	(0,3)	(0,4)	(0,5)	(0,6)	(0,7)	(0,8)
m = 0	β	5.3103	10.6205	15.9308	21.2410	26.5513	31.8616	37.1718	42.4821
n varies	Ω_1	2π	4π	6π	8π	10π	12π	14π	16π
	Ω_2	3.3806	6.7612	10.1419	13.5225	16.9031	20.2837	23.6643	27.0449
	ω_n	16775	33551	50326	67101	83877	100652	117428	134203

A pair of Type II simple supports along the *x* direction

	(m,n)	(0,1)	(0,2)	(0,3)	(0,4)	(0,5)	(0,6)	(0,7)	(0,8)
m = 0	β	π	2π	3π	4π	5π	6π	7π	8π
n varies	Ω_1	3.7172	7.4344	11.1515	14.8687	18.5859	22.3031	26.0203	29.7375
	Ω_2	2	4	6	8	10	12	14	16
	ω_n	9924	19849	29773	39698	49622	59547	69471	79396

Table 8.6(c)   Natural frequencies of the rectangular plate in Example 1 under the special situation where $k_0 \neq 0$, $k_1 = 0$, and $k_2 \neq 0$.

A pair of Type I simple supports along the y direction ($k_1 \neq 0$ and $k_2 = 0$)

	(m,n)	(1,0)	(2,0)	(3,0)	(4,0)	(5,0)	(6,0)	(7,0)	(8,0)
n = 0	β	2.6551	5.3103	7.9654	10.6205	13.2757	15.9308	18.5859	21.2410
m varies	Ω_1	π	2π	3π	4π	5π	6π	7π	8π
	Ω_2	1.6903	3.3806	5.0709	6.7612	8.4515	10.1419	11.8322	13.5225
	ω_n	8388	16775	25163	33551	41938	50326	58714	67101

A pair of Type II simple supports along the y direction ($k_1 = 0$ and $k_2 \neq 0$)

	(m,n)	(1,0)	(2,0)	(3,0)	(4,0)	(5,0)	(6,0)	(7,0)	(8,0)
n = 0	β	$\pi/2$	π	$3\pi/2$	2π	$5\pi/2$	3π	$7\pi/2$	4π
m varies	Ω_1	1.8586	3.7172	5.5758	7.4344	9.2930	11.1515	13.0101	14.8687
	Ω_2	1	2	3	4	5	6	7	8
	ω_n	4962	9924	14887	19849	24811	29773	34736	39698

$$k_1 = 0, \text{ and } k_2 \neq 0: (k_0)_m = \frac{m\pi}{L_1}, \text{ and } \beta = \sqrt{k_1^2 + k_0^2} = k_0.$$

where m, n, and q are positive integers.

Tables 8.5 and 8.6 list the natural frequencies of the general and special situations of a square and a rectangular thin plate, corresponding to aspect ratios $L_1 : L_2 = 1.0\,m\,/\,1.0\,m$ and $L_1 : L_2 = 2.0\,m : 1.0\,m$, respectively. For comparison purposes, in addition to dimensional circular frequency ω_n in rad/s, two non-dimensional frequencies Ω_1 and Ω_2 are introduced, namely, $\Omega_1 = \beta L_1 \sqrt{\gamma_1}$ and $\Omega_2 = \beta L_1 / \pi$.

Tables 8.5 and 8.6 each consist of three subsets of tables labeled (a), (b), and (c). Tables 8.5(a) and 8.6(a) list the natural frequencies from the general situation of $k_0 \neq 0$, $k_1 \neq 0$, and $k_2 \neq 0$. Tables 8.5(b) and 8.6(b) list the natural frequencies for the special situation of $k_0 = 0$, $k_1 \neq 0$, and $k_2 \neq 0$. Tables 8.5(c) and 8.6(c) list the natural frequencies for the special situation of $k_0 \neq 0$, either $k_1 = 0$ or $k_2 = 0$.

The natural frequencies in Tables 8.5(b) and 8.5(c) are identical, and they represent repeated natural frequencies. This is as expected because they are the natural frequencies of a uniform square plate with the same type of simple supports along all four edges.

Natural frequencies are often sorted in ascending order. Table 8.7 shows an example list of non-dimensional natural frequency Ω_1 from Tables 8.5 and 8.6 of the square and rectangular uniform thin plates with all four edges of Type II simple supports.

8.7.2   Example 2: One Pair of the Same Type Simple Supports along the x Direction, Various Combinations of Classical Boundaries on Opposite Edges along the y Direction

In the second example, a square thin plate, that is, $L_1 : L_2 = 1.0\,m : 1.0\,m$, is considered. The pair of opposite edges of the square plate along the x direction are still of the same type of simple supports, that is, either Type I – Type I or Type II – Type II. However, the pair of opposite edges along the y direction is of various combinations of free, clamped, and simply-supported boundaries. The following example combinations of boundary conditions along the y direction are considered:

a) SII – C (that is, Type II simple support – Clamped);
b) SI – C (that is, Type I simple support – Clamped);
c) C – C (that is, Clamped – Clamped);

Table 8.7 Non-dimensional natural frequency Ω_1 of the square and rectangular thin plates in Example 1 with all four edges of Type II simple supports, sorted in ascending order.

(m, n) or (m,q*)	Ω_1 Present	Ω_1 (Bardell et. al. 1996)	(m, n) or (m,q)	Ω_1 Present	Ω_1 (Bardell et. al. 1996)
	$L_1:L_2=1:1$			$L_1:L_2=2:1$	
(0,1)	1.8586	1.859	(1,0)	1.8586	1.859
(1,0)	1.8586	1.859	(0,1)	3.7172	3.717
(1,1)	2.6284	2.628	(2,0)	3.7172	3.717
(0,2)	3.7172	3.717	(1,1)	4.1559	4.156
(2,0)	3.7172	3.717	(2,1)	5.2569	5.257
(1,2)	4.1559	4.156	(3,0)	5.5758	5.576
(2,1)	4.1559	4.156	(3,1)	6.7012	6.701
(1,1*)	4.4429	4.443	(1,1*)	7.0248	7.025
(2,2)	5.2569	5.257	(0,2)	7.4344	7.434
(0,3)	5.5758	5.576	(4,0)	7.4344	7.434
(3,0)	5.5758	5.576	(1,2)	7.6632	7.663
(1,3)	5.8774	5.877	(2,2)	8.3119	8.312
(3,1)	5.8774	5.877	(4,1)	8.3119	8.312
(2,3)	6.7012	6.701	(2,1*)	8.8858	8.886
(3,2)	6.7012	6.701	(3,2)	9.2930	9.293
(1,2*)	7.0248	7.025	(5,0)	9.2930	9.293
(2,1*)	7.0248	7.025	(5,1)	10.0088	NA
(0,4)	7.4344	NA	(4,2)	10.5138	
(4,0)	7.4344		(0,3)	11.1515	
(1,4)	7.6632		(6,0)	11.1515	
(4,1)	7.6632		(1,3)	11.3054	
(3,3)	7.8853		(3,1*)	11.3272	
(2,4)	8.3119		(2,3)	11.7548	
(4,2)	8.3119		(5,2)	11.9008	
(2,2*)	8.8858		(3,3)	12.4678	
(3,4)	9.2930		(1,2*)	12.9531	
(4,3)	9.2930		(7,0)	13.0101	
(0,5)	9.2930		(4,3)	13.4025	
(5,0)	9.2930		(2,2*)	14.0496	
(1,5)	9.4770		(4,1*)	14.0496	
(5,1)	9.4770		(5,3)	14.5161	
(1,3*)	9.9346		(0,4)	14.8687	
(3,1*)	9.9346		(8,0)	14.8687	
(2,5)	10.0088		(1,4)	14.9844	
(5,2)	10.0088		(2,4)	15.3263	

Table 8.8 Natural frequencies of the square plate in Example 2 under the special situation where $k_0 = 0$, $k_1 \neq 0$, and $k_2 \neq 0$.

Grouped boundary in the y direction	(m,n)	(0,1)	(0,2)	(0,3)	(0,4)	(0,5)	(0,6)	(0,7)	(0,8)
A pair of Type I simple supports along the x direction									
Group I-1 (SI-C; SII-F; C-C; F-F)	β	5.3103	10.6205	15.9308	21.2410	26.5513	31.8616	37.1718	42.4821
	Ω_1	π	2π	3π	4π	5π	6π	7π	8π
	Ω_2	1.6903	3.3806	5.0709	6.7612	8.4515	10.1419	11.8322	13.5225
	ω_n	16775	33551	50326	67101	83877	100652	117428	134203
Group I-2 (SI-F; SII-C; C-F)	β	2.6551	7.9654	13.2757	18.5859	23.8962	29.2064	34.5167	39.8270
	Ω_1	0.5π	1.5π	2.5π	3.5π	4.5π	5.5π	6.5π	7.5π
	Ω_2	0.8452	2.5355	4.2258	5.9161	7.6064	9.2967	10.9870	12.6773
	ω_n	8388	25163	41938	58714	75489	92264	109040	125815
A pair of Type II simple supports along the x direction									
Group II-1 (SI-F; SII-C; C-C; F-F)	β	π	2π	3π	4π	5π	6π	7π	8π
	Ω_1	1.8586	3.7172	5.5758	7.4344	9.2930	11.1515	13.0101	14.8687
	Ω_2	1	2	3	4	5	6	7	8
	ω_n	9924	19849	29773	39698	49622	59547	69471	79396
Group II-2 (SI-C; SII-F; C-F)	β	0.5π	1.5π	2.5π	3.5π	4.5π	5.5π	6.5π	7.5π
	Ω_1	0.9293	2.7879	4.6465	6.5051	8.3637	10.2223	12.0808	13.9394
	Ω_2	0.5	1.5	2.5	3.5	4.5	5.5	6.5	7.5
	ω_n	4962	14887	24811	34736	44660	54584	64509	74433

d) SII – F (that is, Type II simple support – Free);
e) SI – F (that is, Type I simple support – Free);
f) C – F (that is, Clamped – Free);
g) F – F (that is, Free – Free).

In addition to the general situation of $k_0 \neq 0$, $k_1 \neq 0$, and $k_2 \neq 0$, the special situation of $k_0 = 0$ needs to be considered in this example because when there is a pair of the same type of simple supports along the x direction, there exists a possibility for $k_0 = 0$. Note that the other special situation of either $k_1 = 0$ or $k_2 = 0$ is not applicable in this example because it only happens when there is a pair of simple supports of the same type along the y direction.

The natural frequencies corresponding to the special situation of $k_0 = 0$, as shown in Table 8.3, are dependent on the type of the pair of simple supports along the x direction. Table 8.3 also shows that, from the wave vibration stand point, the natural frequencies under the special situation of $k_0 = 0$ are determined by the signs of reflection coefficients of the classical boundaries on the opposite edges along the y direction. As a result, for the same type of x direction pair of simple supports, natural frequencies corresponding to the various combinations of classical boundaries along the y direction under the special situation of $k_0 = 0$ fall into only two simple groups, namely, with the reflection coefficients at the opposite edges along the y direction being of the same or of opposite signs, as summarized in Table 8.8.

The natural frequencies corresponding to the general situation are obtained from the dB magnitude plot of the characteristic polynomial by substituting the corresponding boundary conditions along the y direction into Eq. (8.64) and setting the determinant of the coefficient matrix to zero. The natural frequencies of the plate corresponding to various integer multiples of half waves along the x direction are the roots of the characteristic polynomial, which correspond to the x values of local minima in the dB magnitude plot of the characteristic polynomial.

Figure 8.5 presents a sample dB magnitude plot of the characteristic polynomials corresponding to the first three integer multiples of half waves along the x direction, for the general situation of a combination of SII – C (that is, Type II simple

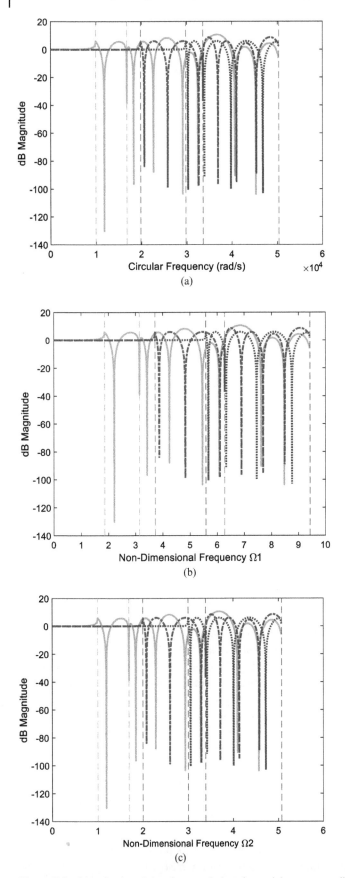

Figure 8.5 Magnitudes of the characteristic polynomials corresponding to various integer multiples of half waves *m* along the *x* direction: *m* = 1 (___), *m* = 2 (-·-·), and *m* = 3 (...).

Table 8.9 Natural frequencies of the square plate in Example 2 corresponding to the general situation, with same type simple supports along the *x* and various combinations of classical boundaries along the *y* direction.

Boundary condition	Vibration mode								
SI – C	(m, n)	(1,1)	(1,2)	(2,1)	(1,3)	(1,4)	(2,2)	(3,1)	(2,3)	(1,5)
	Ω_1	2.8106	3.4504	4.2646	4.3271	5.1210	5.4645	5.9308	6.3987	6.5976
	Ω_2	1.1522	1.8565	2.2945	2.3282	2.7601	2.9401	3.1910	3.4428	3.5498
	ω_n	15008	18424	22772	23106	27393	29179	31669	34168	35229
	(m, n)	(2,4)	(3,2)	(1,6)	(2,5)	(4,1)				
	Ω_1	6.8460	6.8547	7.1636	7.6435	7.6939				
	Ω_2	3.6835	3.6881	3.8543	4.1125	4.1396				
	ω_n	36556	36602	38252	40814	41084				
SII – C	(m, n)	(1,1)	(1,2)	(2,1)	(1,3)	(2,2)	(1,4)	(3,1)	(1,5)	(2,3)
	Ω_1	2.2053	3.4220	3.8654	4.2411	4.8251	5.4492	5.6674	6.0750	6.1038
	Ω_2	1.1865	1.8412	2.0798	2.2819	2.5961	2.9319	3.0493	3.2686	3.2841
	ω_n	11776	18272	20640	22646	25765	29097	30262	32439	32593
	(m, n)	(3,2)	(2,4)	(3,3)	(4,1)					
	Ω_1	6.3378	6.8994	7.4498	7.5003					
	Ω_2	3.4100	3.7122	4.0083	4.0355					
	ω_n	33842	36841	39780	40050					
C – C	(m, n)	(1,1)	(1,2)	(2,1)	(1,3)	(2,2)	(1,4)	(3,1)	(2,3)	(2,4)
	Ω_1	3.2752	3.4946	4.4105	4.9572	5.6212	5.6289	5.9978	6.8439	6.9008
	Ω_2	1.7622	1.8803	2.3731	2.6672	3.0244	3.0286	3.2271	3.6823	3.7129
	ω_n	17489	18661	23551	26471	30016	30056	32027	36545	36849
	(m, n)	(3,2)	(1,5)	(1,6)	(4,1)					
	Ω_1	7.0256	7.1408	7.6186	7.7308					
	Ω_2	3.7801	3.8421	4.0991	4.1595					
	ω_n	37515	38130	40682	41281					
SI – F	(m, n)	(1,1)	(1,2)	(2,1)	(1,3)	(1,4)	(2,2)	(3,1)	(1,5)	(2,3)
	Ω_1	1.8663	3.0594	3.4232	3.4251	4.2583	4.6087	5.1106	5.4050	5.8816
	Ω_2	1.0041	1.6461	1.8419	1.8429	2.2912	2.4797	2.7497	2.9081	3.1646
	ω_n	9966	16337	18279	18289	22739	24610	27289	28861	31406
	(m, n)	(1,6)	(3,2)	(2,4)	(4,1)					
	Ω_1	6.1165	6.1251	6.3361	6.8117					
	Ω_2	3.2909	3.2955	3.4091	3.6650					
	ω_n	32661	32706	33834	36373					
SII – F	(m, n)	(1,1)	(1,2)	(2,1)	(1,3)	(2,2)	(1,4)	(3,1)	(1,5)	(2,3)
	Ω_1	1.6243	2.6284	3.3908	3.4491	4.0375	4.3056	5.1067	5.1617	5.2569
	Ω_2	0.8739	1.4142	1.8244	1.8558	2.1723	2.3166	2.7476	2.7772	2.8284
	ω_n	8673	14035	18106	18417	21559	22991	27269	27562	28071
	(m, n)	(3,2)	(2,4)	(1,6)	(3,3)	(4,1)				
	Ω_1	5.7375	6.3132	6.5551	6.6422	6.8113				
	Ω_2	3.0870	3.3968	3.5269	3.5738	3.6647				
	ω_n	30637	33711	35003	35468	36371				

(Continued)

Table 8.9 (Continued)

Boundary condition	Vibration mode									
C – F	(m, n)	(1,1)	(1,2)	(2,1)	(1,3)	(1,4)	(2,2)	(3,1)	(1,5)	(2,3)
	Ω_1	1.9775	3.0615	3.4391	4.1624	4.4610	4.8093	5.1125	5.9880	6.0608
	Ω_2	1.0640	1.6472	1.8504	2.2396	2.4002	2.5876	2.7507	3.2218	3.2610
	ω_n	10560	16348	18364	22226	23821	25680	27300	31974	32363
	(m, n)	(3,2)	(2,4)	(1,6)	(4,1)					
	Ω_1	6.2343	6.3995	6.6333	6.8119					
	Ω_2	3.3543	3.4432	3.5690	3.6651					
	ω_n	33290	34172	35420	36374					
F – F	(m, n)	(1,1)	(1,2)	(2,1)	(1,3)	(1,4)	(2,2)	(1,5)	(3,1)	(3,2)
	Ω_1	1.4075	2.6284	3.2486	3.3639	3.4987	3.7326	5.0349	5.0359	5.2134
	Ω_2	0.7573	1.4142	1.7479	1.8099	1.8825	2.0083	2.7090	2.7095	2.8051
	ω_n	7516	14035	17347	17962	18682	19931	26885	26890	27839
	(m, n)	(2,3)	(1,6)	(2,4)	(3,3)	(4,1)				
	Ω_1	5.2569	5.5523	6.1188	6.5691	6.7816				
	Ω_2	2.8284	2.9873	3.2922	3.5344	3.6488				
	ω_n	28071	29648	32673	35077	36212				

Table 8.10 An example list of non-dimensional natural frequency Ω_2 of the square plate in Example 2 corresponding to a pair of Type II simple supports along the *x* direction.

Boundary condition	Vibration mode									
SI – C	(m, n)	(0,1)	(1,1)	(0,2)	(1,2)	(2,1)	(1,3)	(0,3)	(1,4)	(2,2)	(3,1)
	Ω_2	0.5	1.1522	1.5	1.8565	2.2945	2.3282	2.5	2.7601	2.9401	3.1910
	(m, n)	(2,3)	(0,4)	(1,5)	(2,4)	(3,2)	(1,6)	(2,5)	(4,1)		
	Ω_2	3.4428	3.5	3.5498	3.6835	3.6881	3.8543	4.1125	4.1396		
SII – C	(m, n)	(0,1)	(1,1)	(1,2)	(0,2)	(2,1)	(1,3)	(2,2)	(1,4)	(0,3)	(3,1)
	Ω_2	1	1.1865	1.8412	2	2.0798	2.2819	2.5961	2.9319	3	3.0493
	(m, n)	(1,5)	(2,3)	(3,2)	(2,4)	(0,4)	(3,3)	(4,1)			
	Ω_2	3.2686	3.2841	3.4100	3.7122	4	4.0083	4.0355			
C – C	(m, n)	(0,1)	(1,1)	(1,2)	(0,2)	(2,1)	(1,3)	(0,3)	(2,2)	(1,4)	(3,1)
	Ω_2	1	1.7622	1.8803	2	2.3731	2.6672	3	3.0244	3.0286	3.2271
	(m, n)	(2,3)	(2,4)	(3,2)	(1,5)	(0,4)	(1,6)	(4,1)			
	Ω_2	3.6823	3.7129	3.7801	3.8421	4	4.0991	4.1595			
SI – F	(m, n)	(0,1)	(1,1)	(1,2)	(2,1)	(1,3)	(0,2)	(1,4)	(2,2)	(3,1)	(1,5)
	Ω_2	1	1.0041	1.6461	1.8419	1.8429	2	2.2912	2.4797	2.7497	2.9081
	(m, n)	(0,3)	(2,3)	(1,6)	(3,2)	(2,4)	(4,1)				
	Ω_2	3	3.1646	3.2909	3.2955	3.4091	3.6650				
SII – F	(m, n)	(0,1)	(1,1)	(1,2)	(0,2)	(2,1)	(1,3)	(2,2)	(1,4)	(0,3)	(3,1)
	Ω_2	0.5	0.8739	1.4142	1.5	1.8244	1.8558	2.1723	2.3166	2.5	2.7476
	(m, n)	(1,5)	(2,3)	(3,2)	(2,4)	(0,4)	(1,6)	(3,3)	(4,1)		
	Ω_2	2.7772	2.8284	3.0870	3.3968	3.5	3.5269	3.5738	3.6647		

Table 8.10 (Continued)

Boundary condition	Vibration mode									
C – F	(m, n)	(0,1)	(1,1)	(0,2)	(1,2)	(2,1)	(1,3)	(1,4)	(0,3)	(2,2)	(3,1)
	Ω_2	0.5	1.0640	1.5	1.6472	1.8504	2.2396	2.4002	2.5	2.5876	2.7507
	(m, n)	(1,5)	(2,3)	(3,2)	(2,4)	(0,4)	(1,6)	(4,1)			
	Ω_2	3.2218	3.2610	3.3543	3.4432	3.5	3.5690	3.6651			
F – F	(m, n)	(1,1)	(0,1)	(1,2)	(2,1)	(1,3)	(1,4)	(0,2)	(2,2)	(1,5)	(3,1)
	Ω_2	0.7573	1	1.4142	1.7479	1.8099	1.8825	2	2.0083	2.7090	2.7095
	(m, n)	(3,2)	(2,3)	(1,6)	(0,3)	(2,4)	(3,3)	(4,1)			
	Ω_2	2.8051	2.8284	2.9873	3	3.2922	3.5344	3.6488			

support – Clamped) boundaries along the y direction. The broken vertical lines in the graphs mark the cut-off frequencies at which wave mode transitions occur. The natural frequencies corresponding to the general situations of the example combinations of boundary conditions along the y direction, which are obtained from identifying roots of the characteristic polynomials in the corresponding dB magnitude plots, are listed in Table 8.9.

To obtain the natural frequencies of the plate, the natural frequencies corresponding to both the general situation from Table 8.9 and the special situation from Table 8.8 need to be put together. Table 8.10 shows an example list of non-dimensional natural frequency Ω_2 corresponding to a pair of Type II simple supports along the x direction, with natural frequencies sorted in ascending order up to the fourth integer multiples of half waves along the x direction, that is, $m = 4$.

8.7.3 Example 3: One Pair of Mixed Type Simple Supports along the *x* Direction, Various Combinations of Classical Boundaries on Opposite Edges along the *y* Direction

In this example, a square thin plate of the same geometrical and material properties and the same example combinations of boundary conditions along the y direction as that of Example 2 in Section 8.7.2 is studied. The difference is that the opposite edges along the x direction are now of mixed type simple supports.

Because neither the x nor the y direction involves the same type of simple supports, only the general situation applies. As a result, the natural frequencies are obtained from the dB magnitude plot of the characteristic polynomial by substituting the corresponding boundary conditions along the y direction into Eq. (8.64) and setting the determinant of the coefficient matrix to zero. The natural frequencies of the plate corresponding to various odd integer multiples of quarter waves along the x direction are the roots of the characteristic polynomial, which correspond to the x values of local minima in the dB magnitude plot of the characteristic polynomial.

Figure 8.6 presents a sample plot of the characteristic polynomials corresponding to the first three odd integer multiples of quarter waves along the x direction, for the combination of SII – C (that is, Type II Simple Support – Clamped) boundaries along the y direction. The broken vertical lines in the graphs mark the cut-off frequencies at which wave mode transitions occur. The natural frequencies are obtained from identifying roots of the characteristic polynomials in the corresponding dB magnitude plots, and are listed in Table 8.11. The natural frequencies are sorted in ascending order up to the fourth odd integer multiples of quarter waves along the x direction, that is, $m = 7$.

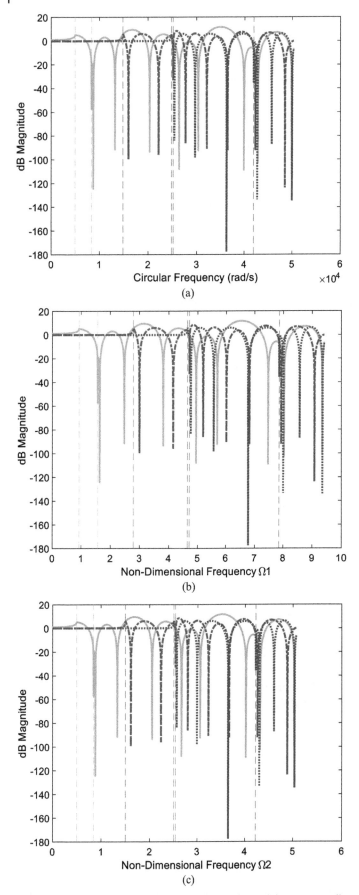

Figure 8.6 Magnitudes of the characteristic polynomials corresponding to various odd integer multiples of quarter waves *m* along the *x* direction: *m* = 1 (___), *m* = 3 (-··-), and *m* = 5 (...).

Table 8.11 Natural frequencies of the square plate in Example 3 with mixed type simple supports along the *x* direction and various combinations of classical boundaries along the *y* direction.

Boundary condition	Vibration mode									
SI – C	(m, n)	(1,1)	(1,2)	(3,1)	(1,3)	(3,2)	(1,4)	(5,1)	(3,3)	(3,4)	(5,2)
	Ω_1	1.7473	2.8144	3.5128	3.5704	4.6819	4.7481	5.0789	5.1161	5.9211	6.1372
	Ω_2	0.9401	1.5143	1.8900	1.9210	2.5191	2.5547	2.7327	2.7527	3.1858	3.3021
	ω_n	9330	15028	18758	19065	25000	25354	27120	27319	31617	32771
	(m, n)	(1,5)	(1,6)	(7,1)							
	Ω_1	6.3734	6.6694	6.8052							
	Ω_2	3.4292	3.5884	3.6615							
	ω_n	34033	35613	36338							
SII – C	(m, n)	(1,1)	(1,2)	(3,1)	(1,3)	(3,2)	(5,1)	(1,4)	(3,3)	(5,2)	(1,5)
	Ω_1	1.6376	2.4786	2.9989	3.8093	4.1530	4.7599	4.9447	5.1880	5.5551	5.6826
	Ω_2	0.8811	1.3336	1.6135	2.0496	2.2345	2.5610	2.6605	2.7914	2.9889	3.0575
	ω_n	8744	13235	16014	20341	22176	25417	26404	27703	29663	30344
	(m, n)	(3,4)	(7,1)								
	Ω_1	6.0132	6.5818								
	Ω_2	3.2353	3.5413								
	ω_n	32109	35145								
C – C	(m, n)	(1,1)	(1,2)	(3,1)	(1,3)	(3,2)	(5,1)	(1,4)	(3,3)	(3,4)	(5,2)
	Ω_1	2.3861	3.0908	3.7499	4.1326	4.6831	5.1752	5.5416	5.9152	6.0206	6.3312
	Ω_2	1.2838	1.6630	2.0176	2.2235	2.5197	2.7845	2.9816	3.1826	3.2393	3.4065
	ω_n	12741	16504	20024	22067	25006	27635	29591	31586	32149	33807
	(m, n)	(1,5)	(7,1)								
	Ω_1	6.5481	6.8541								
	Ω_2	3.5231	3.6878								
	ω_n	34965	36599								
SI – F	(m, n)	(1,1)	(1,2)	(1,3)	(3,1)	(1,4)	(3,2)	(5,1)	(3,3)	(1,5)	(3,4)
	Ω_1	1.3142	1.6819	2.5174	2.6067	3.8010	3.9427	4.2631	4.7041	4.9368	5.1857
	Ω_2	0.7071	0.9050	1.3545	1.4025	2.0451	2.1213	2.2937	2.5310	2.6562	2.7901
	ω_n	7018	8981	13443	13919	20297	21053	22764	25119	26361	27691
	(m, n)	(5,2)	(1,6)	(7,1)							
	Ω_1	5.3376	5.6936	5.9607							
	Ω_2	2.8719	3.0634	3.2071							
	ω_n	28502	30402	31829							
SII – F	(m, n)	(1,1)	(1,2)	(3,1)	(1,3)	(3,2)	(1,4)	(5,1)	(3,3)	(1,5)	(5,2)
	Ω_1	0.7038	1.7494	2.5179	2.7761	3.2846	3.5872	4.2516	4.5309	4.7517	4.8667
	Ω_2	0.3787	0.9412	1.3548	1.4937	1.7672	1.9301	2.2876	2.4378	2.5566	2.6185
	ω_n	3758	9341	13445	14824	17539	19155	22703	24194	25373	25987
	(m, n)	(3,4)	(5,3)	(3,5)	(7,1)						
	Ω_1	5.1254	5.9257	5.9498	5.9594						
	Ω_2	2.7577	3.1883	3.2012	3.2064						
	ω_n	27369	31642	31770	31822						

(Continued)

Table 8.11 (Continued)

Boundary condition	Vibration mode										
C – F	(m, n)	(1,1)	(1,2)	(3,1)	(1,3)	(3,2)	(5,1)	(1,4)	(3,3)	(3,4)	(1,5)
	Ω_1	1.5985	1.7596	2.6500	3.2377	4.1247	4.2687	4.3881	5.1143	5.1914	5.2425
	Ω_2	0.8600	0.9468	1.4258	1.7420	2.2193	2.2967	2.3610	2.7517	2.7932	2.8207
	ω_n	8535	9396	14151	17288	22025	22794	23431	27309	27721	27994
	(m, n)	(5,2)	(7,1)								
	Ω_1	5.4887	5.9613								
	Ω_2	2.9532	3.2074								
	ω_n	29308	31832								
F – F	(m, n)	(1,1)	(1,2)	(3,1)	(1,3)	(1,4)	(3,2)	(5,1)	(1,5)	(5,2)	(3,3)
	Ω_1	0.5140	1.4804	2.3337	2.4440	2.9865	3.1368	4.1486	4.1963	4.4411	4.5061
	Ω_2	0.2766	0.7965	1.2556	1.3150	1.6069	1.6878	2.2321	2.2578	2.3895	2.4245
	ω_n	2745	7905	12462	13050	15947	16750	22153	22407	23714	24061
	(m, n)	(3,4)	(1,6)	(5,3)	(7,1)						
	Ω_1	4.7043	5.5022	5.8973	5.9128						
	Ω_2	2.5311	2.9604	3.1730	3.1813						
	ω_n	25120	29380	31490	31573						

References

Bardell N.S., Langley R.S., Dunsdon J.M. On the Free In-Plane Vibration of Isotropic Plates, *Journal of Sound and Vibration*, 191(3), 459–467 (1996).

Gorman D. Exact Solutions for the Free In-Plane Vibration of Rectangular Plates with Two Opposite Edges Simply Supported, *Journal of Sound and Vibration*, 294 (1–2), 131–161 (2006).

Liu B., Xing Y. Exact Solutions for Free In-Plane Vibrations of Rectangular Plates, *Acta Mechica Solida Sinica*, 24, 556–567 (2011).

9

Bending Waves in Beams Based on Advanced Vibration Theories

The Euler–Bernoulli bending vibration theory neglects the effects of rotary inertia and shear distortion. As a result, the Euler–Bernoulli bending vibration model is not valid for higher frequencies, typically when the transverse dimensions are not negligible with respect to the wavelength. Rayleigh introduced the effect of rotary inertia into the bending vibration theory and the Shear model took into account the effect of transverse shear deformation. Timoshenko extended the Euler–Bernoulli, Rayleigh, and Shear bending vibration models to include both the effect of rotary inertia and transverse shear deformation (Meirovitch 1997). These various advanced bending vibration theories are introduced, and free and forced vibrations are analyzed from the wave standpoint. MATLAB scripts for numerical simulations are provided.

9.1 The Governing Equations and the Propagation Relationships

9.1.1 Rayleigh Bending Vibration Theory

According to the Rayleigh bending vibration theory, the governing equation of motion for free bending vibration of a uniform beam, as shown in Figure 9.1, is

$$EI\frac{\partial^4 y(x,t)}{\partial x^4} + \rho A\frac{\partial^2 y(x,t)}{\partial t^2} - \rho I\frac{\partial^4 y(x,t)}{\partial x^2 \partial t^2} = 0 \tag{9.1}$$

where $y(x,t)$ is the bending deflection of the centerline of the beam, x is the position along the beam axis, t is the time, A is the cross-sectional area, I is the area moment of inertia of the cross section, E is the Young's modulus, and ρ is the volume mass density of the beam.

Assuming time harmonic motion and using separation of variables, the solution to Eq. (9.1) can be written in the form $y(x,t) = y_0 e^{-ikx} e^{i\omega t}$, where y_0 is the amplitude of the bending deflection of the centerline of the beam, ω is the circular frequency, k is the wavenumber, and i is the imaginary unit. Substituting this expression for $y(x,t)$ into Eq. (9.1) gives the dispersion equation, from which the bending wavenumber is obtained,

$$k = \pm\left[\frac{1}{2}\left(\frac{C_r}{C_b}\right)^2 \omega^2 \pm \sqrt{\frac{\omega^2}{C_b^2} + \frac{1}{4}\left(\frac{C_r}{C_b}\right)^4 \omega^4}\right]^{\frac{1}{2}} \tag{9.2}$$

where $C_b = \sqrt{\frac{EI}{\rho A}}$ and $C_r = \sqrt{\frac{\rho I}{\rho A}} = \sqrt{\frac{I}{A}}$. Subscripts b and r indicate that the coefficients C_b and C_r are related to bending and rotary inertia, respectively. $C_r = 0$ corresponds to the situation of rotary inertia being neglected. Under such a special situation, the governing equation given in Eq. (9.1) and the bending wavenumber expressions described in Eq. (9.2) become identical to those by the Euler–Bernoulli bending vibration theory covered in Chapter 3.

The wavenumbers obtained in Eq. (9.2) are functions of circular frequency ω as well as the material and geometrical properties of the structure. The $\pm$ signs outside the brackets of Eq. (9.2) indicate that waves in the beam travel in both the positive and negative directions. The waves are dispersive because the phase velocity ω/k is frequency dependent.

Mechanical Wave Vibrations: Analysis and Control, First Edition. Chunhui Mei.
© 2023 Chunhui Mei. Published 2023 by John Wiley & Sons Ltd.
Companion Website: www.wiley.com/go/Mei/MechanicalWaveVibrations

Figure 9.1 A uniform beam.

Equation (9.2) shows that, in the absence of damping, one pair of wavenumbers is always real, which corresponds to positive- and negative-going propagating waves. The other pair of wavenumbers is always imaginary, which corresponds to positive- and negative-going decaying (which are sometimes called evanescent or nearfield) waves.

With time dependence $e^{i\omega t}$ suppressed, the solution to Eq. (9.1) can be written as

$$y = a_1^+ e^{-ik_1 x} + a_2^+ e^{-k_2 x} + a_1^- e^{ik_1 x} + a_2^- e^{k_2 x} \tag{9.3}$$

where superscripts $+$ and $-$ and subscripts 1 and 2 in wave amplitude denote the positive- and negative-going propagating and decaying bending waves, and

$$k_1 = \left[\frac{1}{2} \left(\frac{C_r}{C_b} \right)^2 \omega^2 + \sqrt{\frac{\omega^2}{C_b^2} + \frac{1}{4} \left(\frac{C_r}{C_b} \right)^4 \omega^4} \right]^{\frac{1}{2}}, \quad k_2 = \left\| \frac{1}{2} \left(\frac{C_r}{C_b} \right)^2 \omega^2 - \sqrt{\frac{\omega^2}{C_b^2} + \frac{1}{4} \left(\frac{C_r}{C_b} \right)^4 \omega^4} \right\|^{\frac{1}{2}} \tag{9.4}$$

Consider two points A and B on a uniform beam that are a distance x apart, as shown in Figure 9.2. Waves propagate from one point to the other, with the propagation relationships determined by the appropriate wavenumber. The positive- and negative-going waves at Points A and B are related by

$$\mathbf{b}^+ = \mathbf{f}(x)\mathbf{a}^+, \quad \mathbf{a}^- = \mathbf{f}(x)\mathbf{b}^- \tag{9.5}$$

where

$$\mathbf{a}^+ = \begin{bmatrix} a_1^+ \\ a_2^+ \end{bmatrix}, \quad \mathbf{a}^- = \begin{bmatrix} a_1^- \\ a_2^- \end{bmatrix}, \quad \mathbf{b}^+ = \begin{bmatrix} b_1^+ \\ b_2^+ \end{bmatrix}, \quad \mathbf{b}^- = \begin{bmatrix} b_1^- \\ b_2^- \end{bmatrix} \tag{9.6}$$

are the wave vectors and

$$\mathbf{f}(x) = \begin{bmatrix} e^{-ik_1 x} & 0 \\ 0 & e^{-k_2 x} \end{bmatrix} \tag{9.7}$$

is the bending propagation matrix for a distance x.

9.1.2 Shear Bending Vibration Theory

For a uniform beam subjected to no external force, as shown in Figure 9.1, the governing equations of motion for free bending vibration by the Shear bending vibration theory are

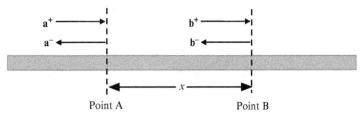

Figure 9.2 Wave propagation relationships.

$$GA\kappa\left[\frac{\partial\psi(x,t)}{\partial x}-\frac{\partial^2 y(x,t)}{\partial x^2}\right]+\rho A\frac{\partial^2 y(x,t)}{\partial t^2}=0 \tag{9.8a}$$

$$EI\frac{\partial^2 \psi(x,t)}{\partial x^2}+GA\kappa\left[\frac{\partial y(x,t)}{\partial x}-\psi(x,t)\right]=0 \tag{9.8b}$$

where $\psi(x,t)$ is the slope due to bending, $\partial y(x,t)/\partial x$ is the slope of the centerline of the beam, and $\partial y(x,t)/\partial x-\psi(x,t)$ is the shear angle. G and κ are the shear modulus and shear coefficient, respectively.

Equations (9.8a) and (9.8b) are coupled through the bending slope and the bending deflection of the structure.

Eliminating $\psi(x,t)$ from Eqs. (9.8a) and (9.8b), the equation of motion of the bending deflection of the centerline of the beam $y(x,t)$ is

$$EI\frac{\partial^4 y(x,t)}{\partial x^4}+\rho A\frac{\partial^2 y(x,t)}{\partial t^2}-\frac{\rho EI}{G\kappa}\frac{\partial^4 y(x,t)}{\partial x^2 \partial t^2}=0 \tag{9.9a}$$

Similarly, eliminating $y(x,t)$ from Eqs. (9.8a) and (9.8b), the equation of motion of the bending slope $\psi(x,t)$ is

$$EI\frac{\partial^4 \psi(x,t)}{\partial x^4}+\rho A\frac{\partial^2 \psi(x,t)}{\partial t^2}-\frac{\rho EI}{G\kappa}\frac{\partial^4 \psi(x,t)}{\partial x^2 \partial t^2}=0 \tag{9.9b}$$

which is governed by the same differential equation as Eq. (9.9a).

Assuming time harmonic motion and using separation of variables, the solutions can be written in the form $y(x,t)=y_0 e^{-ikx}e^{i\omega t}$ and $\psi(x,t)=\psi_0 e^{-ikx}e^{i\omega t}$, where i is the imaginary unit, ω is the circular frequency, and k is the wavenumber. y_0 and ψ_0 are the amplitudes of the bending deflection of the centerline of the beam and the bending slope, respectively. Substituting these expressions for $y(x,t)$ and $\psi(x,t)$ into Eqs. (9.8a) and (9.8b) and putting the equations into matrix form,

$$\begin{bmatrix} k_1^2 GA\kappa-\rho A\omega^2 & -ik_1 GA\kappa \\ -ik_1 GA\kappa & -k_1^2 EI-GA\kappa \end{bmatrix}\begin{bmatrix} y_0 \\ \psi_0 \end{bmatrix}=0 \tag{9.10}$$

Setting the determinant of the coefficient matrix of Eq. (9.10) to zero gives the dispersion equation, from which the bending wavenumbers are solved,

$$k=\pm\left[\frac{1}{2}\left(\frac{1}{C_s}\right)^2\omega^2\pm\sqrt{\frac{\omega^2}{C_b^2}+\frac{1}{4}\left(\frac{1}{C_s}\right)^4\omega^4}\right]^{\frac{1}{2}} \tag{9.11}$$

where $C_s=\sqrt{\dfrac{GA\kappa}{\rho A}}$ and subscript s indicates that coefficient C_s is related to the shear effect. An infinite shear rigidity, that is, $C_s=\infty$, corresponds to zero shear deformation.

The bending wavenumbers obtained in Eq. (9.11) are functions of the circular frequency ω as well as the material and geometrical properties of the structure. The $\pm$ signs outside the brackets of Eq. (9.11) indicate that waves in the beam travel in both the positive and negative directions. The waves are dispersive because the phase velocity ω/k is frequency dependent. Equation (9.11) shows that, in the absence of damping, one pair of wavenumbers is always real, which corresponds to positive- and negative-going propagating waves. The other pair of wavenumbers is always imaginary, which corresponds to positive- and negative-going decaying waves.

With time dependence $e^{i\omega t}$ suppressed, the solutions to Eqs. (9.8a) and (9.8b) can be written as

$$y(x)=a_1^+ e^{-ik_1 x}+a_2^+ e^{-k_2 x}+a_1^- e^{ik_1 x}+a_2^- e^{k_2 x} \tag{9.12a}$$

$$\psi(x)=\overline{a_1^+}e^{-ik_1 x}+\overline{a_2^+}e^{-k_2 x}+\overline{a_1^-}e^{ik_1 x}+\overline{a_2^-}e^{k_2 x} \tag{9.12b}$$

where the wavenumbers are

$$k_1 = \left[\frac{1}{2}\left(\frac{1}{C_s}\right)^2 \omega^2 + \sqrt{\frac{\omega^2}{C_b^2} + \frac{1}{4}\left(\frac{1}{C_s}\right)^4 \omega^4} \right]^{\frac{1}{2}}, \quad k_2 = \left[\left|\frac{1}{2}\left(\frac{1}{C_s}\right)^2 \omega^2 - \sqrt{\frac{\omega^2}{C_b^2} + \frac{1}{4}\left(\frac{1}{C_s}\right)^4 \omega^4}\right| \right]^{\frac{1}{2}} \tag{9.13}$$

The wave amplitudes a of $y(x)$ and $\bar{a}$ of $\psi(x)$ are related and their relationships are found from Eq. (9.10),

$$\frac{\overline{a_1^+}}{a_1^+} = -iP, \quad \frac{\overline{a_1^-}}{a_1^-} = iP, \quad \frac{\overline{a_2^+}}{a_2^+} = -N, \quad \frac{\overline{a_2^-}}{a_2^-} = N \tag{9.14}$$

where

$$P = k_1\left(1 - \frac{\omega^2}{k_1^2 C_s^2}\right), \quad N = k_2\left(1 + \frac{\omega^2}{k_2^2 C_s^2}\right) \tag{9.15}$$

9.1.3 Timoshenko Bending Vibration Theory

For a uniform beam subjected to no external force, as shown in Figure 9.1, the governing equations of motion for free bending vibration by the Timoshenko's bending vibration theory, which takes into account both the rotary inertia and shear deformation, are

$$GA\kappa\left[\frac{\partial\psi(x,t)}{\partial x} - \frac{\partial^2 y(x,t)}{\partial x^2}\right] + \rho A\frac{\partial^2 y(x,t)}{\partial t^2} = 0 \tag{9.16a}$$

$$EI\frac{\partial^2\psi(x,t)}{\partial x^2} + GA\kappa\left[\frac{\partial y(x,t)}{\partial x} - \psi(x,t)\right] - \rho I\frac{\partial^2\psi(x,t)}{\partial t^2} = 0 \tag{9.16b}$$

Eliminating $\psi(x,t)$ from Eqs. (9.16a) and (9.16b), the governing equation of motion of the bending deflection of the centerline of the beam $y(x,t)$ is

$$EI\frac{\partial^4 y(x,t)}{\partial x^4} + \rho A\frac{\partial^2 y(x,t)}{\partial t^2} - \rho I\left(1 + \frac{E}{G\kappa}\right)\frac{\partial^4 y(x,t)}{\partial x^2 \partial t^2} + \frac{\rho^2 I}{G\kappa}\frac{\partial^4 y(x,t)}{\partial t^4} = 0 \tag{9.17a}$$

Similarly, eliminating $y(x,t)$ from Eqs. (9.16a) and (9.16b), the equation of motion of the bending slope $\psi(x,t)$ is

$$EI\frac{\partial^4\psi(x,t)}{\partial x^4} + \rho A\frac{\partial^2\psi(x,t)}{\partial t^2} - \rho I\left(1 + \frac{E}{G\kappa}\right)\frac{\partial^4\psi(x,t)}{\partial x^2 \partial t^2} + \frac{\rho^2 I}{G\kappa}\frac{\partial^4\psi(x,t)}{\partial t^4} = 0 \tag{9.17b}$$

which is governed by the same differential equation as Eq. (9.17a).

Assuming time harmonic motion and using separation of variables, the solutions can be written in the form $y(x,t) = y_0 e^{-ikx} e^{i\omega t}$ and $\psi(x,t) = \psi_0 e^{-ikx} e^{i\omega t}$, where i is the imaginary unit, ω is the circular frequency, and k is the wavenumber. y_0 and ψ_0 are the amplitudes of the bending deflection of the centerline of the beam and the bending slope, respectively. Substituting these expressions for $y(x,t)$ and $\psi(x,t)$ into Eqs. (9.16a) and (9.16b) and putting the equations into matrix form,

$$\begin{bmatrix} k_1^2 GA\kappa - \rho A\omega^2 & -ik_1 GA\kappa \\ -ik_1 GA\kappa & -k_1^2 EI - GA\kappa + \rho I\omega^2 \end{bmatrix}\begin{bmatrix} y_0 \\ \psi_0 \end{bmatrix} = 0 \tag{9.18}$$

Setting the determinant of the coefficient matrix of Eq. (9.18) to zero gives the dispersion equation, from which the bending wavenumbers are solved,

$$k = \pm\left\{ \frac{1}{2}\left[\left(\frac{1}{C_s}\right)^2 + \left(\frac{C_r}{C_b}\right)^2\right]\omega^2 + \sqrt{\frac{\omega^2}{C_b^2} + \frac{1}{4}\left[\left(\frac{1}{C_s}\right)^2 - \left(\frac{C_r}{C_b}\right)^2\right]^2 \omega^4} \right\}^{\frac{1}{2}} \tag{9.19a}$$

$$k = \pm \left\{ \frac{1}{2}\left[\left(\frac{1}{C_s}\right)^2 + \left(\frac{C_r}{C_b}\right)^2\right]\omega^2 - \sqrt{\frac{\omega^2}{C_b^2} + \frac{1}{4}\left[\left(\frac{1}{C_s}\right)^2 - \left(\frac{C_r}{C_b}\right)^2\right]^2 \omega^4} \right\}^{\frac{1}{2}} \tag{9.19b}$$

The wavenumbers obtained in Eqs. (9.19a) and (9.19b) are functions of circular frequency ω as well as the material and geometrical properties of the structure. The $\pm$ signs outside the bracket of Eqs. (9.19a) and (9.19b) indicate that waves in the beam travel in both the positive and negative directions. The waves are dispersive because the phase velocity ω/k is frequency dependent. Equations (9.19a) and (9.19b) show that, in the absence of damping, the pair of wavenumbers obtained in Eq. (9.19a) is always real, which corresponds to positive- and negative-going propagating waves. However, the pair of wavenumbers obtained in Eq. (9.19b) can be either imaginary or real, depending on the frequency range under consideration. This means that the pair of waves is either decaying waves (at low frequencies) or propagating waves (at high frequencies). There exists a wave mode transition at a cut-off frequency ω_c, which is found by setting Eq. (9.19b) to zero,

$$\omega_c = \frac{C_s}{C_r} \tag{9.20}$$

Below the cut-off frequency ω_c, a pair of propagating waves and a pair of decaying waves exist in the beam; while above the cut-off frequency ω_c, both pairs of waves are propagating waves. In audio frequency applications, the former case is overwhelmingly the most common.

With time dependence $e^{i\omega t}$ suppressed, for the most common situation of $\omega < \omega_c$, the solutions $y(x)$ and $\psi(x)$ to Eq. (9.16a) and (9.16b) are given by the same expressions of Eqs. (9.12a) and (9.12b), with wavenumbers

$$k_1 = \left\{ \frac{1}{2}\left[\left(\frac{1}{C_s}\right)^2 + \left(\frac{C_r}{C_b}\right)^2\right]\omega^2 + \sqrt{\frac{\omega^2}{C_b^2} + \frac{1}{4}\left[\left(\frac{1}{C_s}\right)^2 - \left(\frac{C_r}{C_b}\right)^2\right]^2 \omega^4} \right\}^{\frac{1}{2}} \tag{9.21a}$$

$$k_2 = \left\{ \left|\frac{1}{2}\left[\left(\frac{1}{C_s}\right)^2 + \left(\frac{C_r}{C_b}\right)^2\right]\omega^2 - \sqrt{\frac{\omega^2}{C_b^2} + \frac{1}{4}\left[\left(\frac{1}{C_s}\right)^2 - \left(\frac{C_r}{C_b}\right)^2\right]^2 \omega^4}\right| \right\}^{\frac{1}{2}} \tag{9.21b}$$

For $\omega > \omega_c$, the solutions $y(x)$ and $\psi(x)$ to Eqs. (9.16a) and (9.16b), in terms of the wavenumbers obtained in Eqs. (9.21a) and (9.21b), are

$$y(x) = a_1^+ e^{-ik_1 x} + a_2^+ e^{-ik_2 x} + a_1^- e^{ik_1 x} + a_2^- e^{ik_2 x}, \tag{9.22a}$$

$$\psi(x) = \overline{a_1^+} e^{-ik_1 x} + \overline{a_2^+} e^{-ik_2 x} + \overline{a_1^-} e^{ik_1 x} + \overline{a_2^-} e^{ik_2 x} \tag{9.22b}$$

Equations (9.22a) and (9.22b) can be obtained by replacing k_2 in Eqs. (9.12a) and (9.12b) with ik_2, which reflects the wave mode transition at the cut-off frequency.

The wave amplitudes a of $y(x)$ and $\overline{a}$ of $\psi(x)$ are related to each other and these relationships can be found from Eq. (9.18). For $\omega < \omega_c$, the coefficients of wave components of $y(x)$ and those of $\psi(x)$ are given by the same expressions obtained in Eqs. (9.14) and (9.15), with wavenumbers given in Eqs. (9.21a) and (9.21b). For $\omega > \omega_c$, the relationships between coefficients of wave components of $y(x)$ and $\psi(x)$ are still described by Eq. (9.14), however, k_2 in Eq. (9.15) needs to be replaced with ik_2 because of the wave mode transition at the cut-off frequency (Mei and Mace 2005).

Among the four engineering bending vibration theories, namely, the Euler–Bernoulli, Rayleigh, Shear, and Timoshenko bending vibration theories, the first three theories are special situations of the Timoshenko bending vibration theory. This is because the Timoshenko bending vibration theory takes into account the effects of both rotary inertia and shear deformation, the Shear bending vibration theory considers shear deformation but with rotary inertia neglected, the Rayleigh bending vibration theory considers rotary inertia but assumes zero shear deformation or equivalently infinite shear rigidity, and the Euler–Bernoulli bending vibration theory considers neither shear deformation nor rotary inertia.

Writing the governing equations of motion of Eqs. (9.1), (9.9a), and (9.17a) in terms of the bending, shear, and rotary inertia related coefficients C_b, C_s, and C_r, it is seen that the equations of motion of a beam based on the Shear bending vibration

Table 9.1 Comparison of decoupled governing equations of motion of the four bending vibration theories.

Model	Equation of motion for free vibration	C_r	C_s
Timoshenko	$C_b^2 \dfrac{\partial^4 y}{\partial x^4} + \dfrac{\partial^2 y}{\partial t^2} - \left(C_r^2 + \dfrac{C_b^2}{C_s^2}\right)\dfrac{\partial^4 y}{\partial x^2 \partial t^2} + \dfrac{C_r^2}{C_s^2}\dfrac{\partial^4 y}{\partial t^4} = 0$	$(0,\infty)$	$(0,\infty)$
Shear	$C_b^2 \dfrac{\partial^4 y}{\partial x^4} + \dfrac{\partial^2 y}{\partial t^2} - \dfrac{C_b^2}{C_s^2}\dfrac{\partial^4 y}{\partial x^2 \partial t^2} = 0$	0	$(0,\infty)$
Rayleigh	$C_b^2 \dfrac{\partial^4 y}{\partial x^4} + \dfrac{\partial^2 y}{\partial t^2} - C_r^2 \dfrac{\partial^4 y}{\partial x^2 \partial t^2} = 0$	$(0,\infty)$	∞
Euler–Bernoulli	$C_b^2 \dfrac{\partial^4 y}{\partial x^4} + \dfrac{\partial^2 y}{\partial t^2} = 0$	0	∞

Table 9.2 Comparison of wavenumbers from the four bending vibration theories.

Model	Wavenumbers	C_r	C_s
Timoshenko	$\left\| \dfrac{1}{2}\left[\left(\dfrac{1}{C_s}\right)^2 + \left(\dfrac{C_r}{C_b}\right)^2\right]\omega^2 \pm \sqrt{\dfrac{\omega^2}{C_b^2} + \dfrac{1}{4}\left[\left(\dfrac{1}{C_s}\right)^2 - \left(\dfrac{C_r}{C_b}\right)^2\right]^2 \omega^4} \right\|^{\frac{1}{2}}$	$(0,\infty)$	$(0,\infty)$
Shear	$\left\| \dfrac{1}{2}\left(\dfrac{1}{C_s}\right)^2 \omega^2 \pm \sqrt{\dfrac{\omega^2}{C_b^2} + \dfrac{1}{4}\left(\dfrac{1}{C_s}\right)^4 \omega^4} \right\|^{\frac{1}{2}}$	0	$(0,\infty)$
Rayleigh	$\left\| \dfrac{1}{2}\left(\dfrac{C_r}{C_b}\right)^2 \omega^2 \pm \sqrt{\dfrac{\omega^2}{C_b^2} + \dfrac{1}{4}\left(\dfrac{C_r}{C_b}\right)^4 \omega^4} \right\|^{\frac{1}{2}}$	$(0,\infty)$	∞
Euler–Bernoulli	$\sqrt[4]{\dfrac{\omega^2}{C_b^2}}$	0	∞

model, Rayleigh bending vibration model, and Euler–Bernoulli bending vibration model can be obtained from the Timoshenko bending vibration model by setting C_r to zero (that is, neglecting the rotary inertia), C_s to infinite (that is, neglecting the shear effect by assuming infinite shear rigidity), and setting both C_r to zero and C_s to infinite, respectively, as shown in Table 9.1.

Similarly, as shown in Table 9.2, the wavenumbers of the Shear, Rayleigh, and Euler–Bernoulli bending vibration theories can be obtained from the wavenumber expression of the Timoshenko bending vibration theory by letting $C_r = 0$, $C_s = \infty$, and both $C_r = 0$ and $C_s = \infty$, respectively.

Note that among the four engineering bending vibration theories, the Timoshenko bending vibration theory is the only theory that predicts a cut-off frequency and a wave mode transition.

9.2 Wave Reflection at Classical and Non-Classical Boundaries

Figure 9.3 shows the sign convention adopted in the analysis that follows.

9.2.1 Rayleigh Bending Vibration Theory

By the sign convention defined in Figure 9.3 and according to the Rayleigh bending vibration theory, the internal resistant shear force $V(x,t)$ and bending moment $M(x,t)$ are related to the bending deflection $y(x,t)$ and bending slope $\psi(x,t)$ as follows

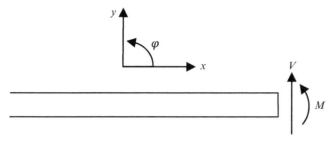

Figure 9.3 Definition of positive internal resistant shear force and bending moment.

$$V(x,t) = -EI\frac{\partial^3 y(x,t)}{\partial x^3} + \rho I\frac{\partial^3 y(x,t)}{\partial x \partial t^2} \tag{9.23a}$$

$$M(x,t) = EI\frac{\partial^2 y(x,t)}{\partial x^2} \tag{9.23b}$$

Comparing to the Euler–Bernoulli bending vibration theory, the resistant shear force contains an additional term to take into account the rotary inertia. The internal resistant bending moment remains the same because both the Rayleigh and Euler–Bernoulli bending vibration theories neglect shear deformation by assuming infinite shear rigidity, which results in $\psi(x,t) = dy(x,t)/dx$.

The reflection matrices at classical free, clamped, and pinned boundaries, as well as at non-classical boundaries with spring and mass attachments, are derived.

Figure 9.4 illustrates a boundary with translational and rotational constraints. Incident waves are reflected at the boundary, the incident waves $\mathbf{a}^+$ and reflected waves $\mathbf{a}^-$ are related by the reflection matrix $\mathbf{r}$,

$$\mathbf{a}^- = \mathbf{r}\mathbf{a}^+ \tag{9.24}$$

where $\mathbf{r}$ is determined by boundary conditions.

a) Free boundary

At a free boundary, the equilibrium conditions are

$$M(x,t) = 0, V(x,t) = 0 \tag{9.25}$$

From Eqs. (9.3), (9.4), (9.23a), (9.23b), and (9.25),

$$EI\left[(-ik_1)^2 a_1^+ e^{-ik_1 x} + (-k_2)^2 a_2^+ e^{-k_2 x} + (ik_1)^2 a_1^- e^{ik_1 x} + (k_2)^2 a_2^- e^{k_2 x}\right] = 0$$

$$-EI\left[(-ik_1)^3 a_1^+ e^{-ik_1 x} + (-k_2)^3 a_2^+ e^{-k_2 x} + (ik_1)^3 a_1^- e^{ik_1 x} + (k_2)^3 a_2^- e^{k_2 x}\right] \tag{9.26}$$

$$-\rho I\omega^2\left(-ik_1 a_1^+ e^{-ik_1 x} - k_2 a_2^+ e^{-k_2 x} + ik_1 a_1^- e^{ik_1 x} + k_2 a_2^- e^{k_2 x}\right) = 0$$

Choosing the origin at the boundary and putting Eq. (9.26) into matrix form,

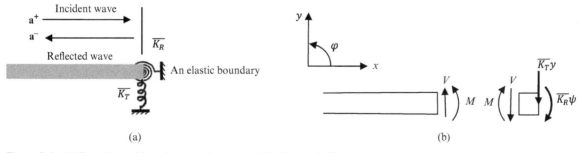

(a)   (b)

Figure 9.4 (a) Boundary with spring attachments and (b) free body diagram.

$$\begin{bmatrix} -k_1^2 & k_2^2 \\ ik_1^3 - ik_1\left(\dfrac{C_r}{C_b}\omega\right)^2 & -k_2^3 - k_2\left(\dfrac{C_r}{C_b}\omega\right)^2 \end{bmatrix}\begin{bmatrix} a_1^+ \\ a_2^+ \end{bmatrix} - \begin{bmatrix} k_1^2 & -k_2^2 \\ ik_1^3 - ik_1\left(\dfrac{C_r}{C_b}\omega\right)^2 & -k_2^3 - k_2\left(\dfrac{C_r}{C_b}\omega\right)^2 \end{bmatrix}\begin{bmatrix} a_1^- \\ a_2^- \end{bmatrix} = 0 \tag{9.27}$$

Solving for the reflected waves $\mathbf{a}^- = \begin{bmatrix} a_1^- \\ a_2^- \end{bmatrix}$ in terms of the incident waves $\mathbf{a}^+ = \begin{bmatrix} a_1^+ \\ a_2^+ \end{bmatrix}$,

$$\begin{bmatrix} a_1^- \\ a_2^- \end{bmatrix} = \begin{bmatrix} k_1^2 & -k_2^2 \\ ik_1^3 - ik_1\left(\dfrac{C_r}{C_b}\omega\right)^2 & -k_2^3 - k_2\left(\dfrac{C_r}{C_b}\omega\right)^2 \end{bmatrix}^{-1}\begin{bmatrix} -k_1^2 & k_2^2 \\ ik_1^3 - ik_1\left(\dfrac{C_r}{C_b}\omega\right)^2 & -k_2^3 - k_2\left(\dfrac{C_r}{C_b}\omega\right)^2 \end{bmatrix}\begin{bmatrix} a_1^+ \\ a_2^+ \end{bmatrix} \tag{9.28}$$

From Eqs. (9.24) and (9.28), the reflection matrix at a free boundary by the Rayleigh bending vibration theory is

$$\mathbf{r} = \begin{bmatrix} k_1^2 & -k_2^2 \\ ik_1^3 - ik_1\left(\dfrac{C_r}{C_b}\omega\right)^2 & -k_2^3 - k_2\left(\dfrac{C_r}{C_b}\omega\right)^2 \end{bmatrix}^{-1}\begin{bmatrix} -k_1^2 & k_2^2 \\ ik_1^3 - ik_1\left(\dfrac{C_r}{C_b}\omega\right)^2 & -k_2^3 - k_2\left(\dfrac{C_r}{C_b}\omega\right)^2 \end{bmatrix} \tag{9.29}$$

In the situation where $k_1 = k_2$ and $C_r = 0$, Eq. (9.29) gives the reflection matrix at a free boundary by the Euler–Bernoulli bending vibration theory, which is $\mathbf{r} = \begin{bmatrix} -i & 1+i \\ 1-i & i \end{bmatrix}$.

b) Clamped boundary

The boundary conditions at a clamped boundary are

$$y(x,t) = 0, \quad \psi(x,t) = \frac{\partial y(x,t)}{\partial x} = 0 \tag{9.30}$$

From Eq. (9.3), (9.4), and (9.30),

$$\begin{aligned} a_1^+ e^{-ik_1 x} + a_2^+ e^{-k_2 x} + a_1^- e^{ik_1 x} + a_2^- e^{k_2 x} &= 0 \\ -ik_1 a_1^+ e^{-ik_1 x} - k_2 a_2^+ e^{-k_2 x} + ik_1 a_1^- e^{ik_1 x} + k_2 a_2^- e^{k_2 x} &= 0 \end{aligned} \tag{9.31}$$

Choosing the origin at the boundary and putting Eq. (9.31) into matrix form,

$$\begin{bmatrix} 1 & 1 \\ ik_1 & k_2 \end{bmatrix}\begin{bmatrix} a_1^+ \\ a_2^+ \end{bmatrix} - \begin{bmatrix} -1 & -1 \\ ik_1 & k_2 \end{bmatrix}\begin{bmatrix} a_1^- \\ a_2^- \end{bmatrix} = 0 \tag{9.32}$$

From Eqs. (9.6), (9.24), and (9.32), the reflection matrix at a clamped boundary by the Rayleigh bending vibration theory is

$$\mathbf{r} = \begin{bmatrix} -1 & -1 \\ ik_1 & k_2 \end{bmatrix}^{-1}\begin{bmatrix} 1 & 1 \\ ik_1 & k_2 \end{bmatrix} = \frac{1}{k_2 - ik_1}\begin{bmatrix} -k_2 - ik_1 & -2k_2 \\ i2k_1 & k_2 + ik_1 \end{bmatrix} \tag{9.33}$$

In the situation of $k_1 = k_2$, Eq. (9.33) gives the reflection matrix at a clamped boundary by the Euler–Bernoulli bending vibration theory, which is $\mathbf{r} = \begin{bmatrix} -i & -1-i \\ -1+i & i \end{bmatrix}$.

c) Pinned boundary

The boundary conditions at a pinned boundary are

$$y(x,t) = 0, \quad M(x,t) = 0 \tag{9.34}$$

From Eqs. (9.3), (9.4), (9.23a), (9.23b), and (9.34),

$$a_1^+ e^{-ik_1 x} + a_2^+ e^{-k_2 x} + a_1^- e^{ik_1 x} + a_2^- e^{k_2 x} = 0$$

$$EI\left(\left[(-ik_1)^2 a_1^+ e^{-ik_1 x} + (-k_2)^2 a_2^+ e^{-k_2 x} + (ik_1)^2 a_1^- e^{ik_1 x} + (k_2)^2 a_2^- e^{k_2 x}\right]\right) = 0 \tag{9.35}$$

Choosing the origin at the boundary and putting Eq. (9.35) into matrix form,

$$\begin{bmatrix} 1 & 1 \\ -k_1^2 & k_2^2 \end{bmatrix}\begin{bmatrix} a_1^+ \\ a_2^+ \end{bmatrix} - \begin{bmatrix} -1 & -1 \\ k_1^2 & -k_2^2 \end{bmatrix}\begin{bmatrix} a_1^- \\ a_2^- \end{bmatrix} = 0 \tag{9.36}$$

From Eqs. (9.6), (9.24), and (9.36), the reflection matrix at a pinned boundary by the Rayleigh bending vibration theory is

$$\mathbf{r} = \begin{bmatrix} -1 & 0 \\ 0 & -1 \end{bmatrix} \tag{9.37}$$

The reflection matrix at a pinned boundary by the Rayleigh bending vibration theory is the same as that by the Euler–Bernoulli bending vibration theory. This is as expected because the rotary inertia considered by Rayleigh only affects the resistant shear force expression, which does not play a role in the boundary conditions of a pinned boundary.

d) Boundary with spring attachments

Figure 9.4 shows a boundary with spring attachments. The translational and rotational stiffness are $\overline{K_T}$ and $\overline{K_R}$, respectively.

From the free body diagram of Figure 9.4,

$$-V(x,t) - \overline{K_T} y(x,t) = 0, \ - M(x,t) - \overline{K_R}\psi(x,t) = 0 \tag{9.38}$$

Recall that $\psi(x,t) = \dfrac{\partial y(x,t)}{\partial x}$ by the Rayleigh bending vibration theory. From Eqs. (9.3), (9.4), (9.23a), (9.23b), and (9.38), with the origin chosen at the spring attached boundary,

$$EI\left[(-ik_1)^3 a_1^+ + (-k_2)^3 a_2^+ + (ik_1)^3 a_1^- + (k_2)^3 a_2^-\right]$$
$$+ \rho I \omega^2 \left(-ik_1 a_1^+ - k_2 a_2^+ + ik_1 a_1^- + k_2 a_2^-\right) - \overline{K_T}\left(a_1^+ + a_2^+ + a_1^- + a_2^-\right) = 0 \tag{9.39a}$$

$$-EI\left[(-ik_1)^2 a_1^+ + (-k_2)^2 a_2^+ + (ik_1)^2 a_1^- + (k_2)^2 a_2^-\right] - \overline{K_R}(-ik_1 a_1^+ - k_2 a_2^+ + ik_1 a_1^- + k_2 a_2^-) = 0 \tag{9.39b}$$

Putting Eqs. (9.39a) and (9.39b) into matrix form,

$$\begin{bmatrix} -ik_1\rho I\omega^2 + iEIk_1^3 - \overline{K_T} & -k_2\rho I\omega^2 - EIk_2^3 - \overline{K_T} \\ EIk_1^2 + i\overline{K_R}k_1 & -EIk_2^2 + \overline{K_R}k_2 \end{bmatrix}\begin{bmatrix} a_1^+ \\ a_2^+ \end{bmatrix} +$$
$$\begin{bmatrix} ik_1\rho I\omega^2 - iEIk_1^3 - \overline{K_T} & k_2\rho I\omega^2 + EIk_2^3 - \overline{K_T} \\ EIk_1^2 - i\overline{K_R}k_1 & -EIk_2^2 - \overline{K_R}k_2 \end{bmatrix}\begin{bmatrix} a_1^- \\ a_2^- \end{bmatrix} = 0 \tag{9.40a}$$

or equivalently

$$\begin{bmatrix} ik_1^3 - ik_1\left(\dfrac{C_r}{C_b}\omega\right)^2 - \dfrac{\overline{K_T}}{EI} & -k_2^3 - k_2\left(\dfrac{C_r}{C_b}\omega\right)^2 - \dfrac{\overline{K_T}}{EI} \\ k_1^2 + ik_1\dfrac{\overline{K_R}}{EI} & -k_2^2 + k_2\dfrac{\overline{K_R}}{EI} \end{bmatrix}\begin{bmatrix} a_1^+ \\ a_2^+ \end{bmatrix} +$$
$$\begin{bmatrix} -ik_1^3 + ik_1\left(\dfrac{C_r}{C_b}\omega\right)^2 - \dfrac{\overline{K_T}}{EI} & k_2^3 + k_2\left(\dfrac{C_r}{C_b}\omega\right)^2 - \dfrac{\overline{K_T}}{EI} \\ k_1^2 - ik_1\dfrac{\overline{K_R}}{EI} & -k_2^2 - k_2\dfrac{\overline{K_R}}{EI} \end{bmatrix}\begin{bmatrix} a_1^- \\ a_2^- \end{bmatrix} = 0 \tag{9.40b}$$

From Eqs. (9.6), (9.24), and (9.40a), the reflection matrix at a spring attached boundary by the Rayleigh bending vibration theory is

$$\mathbf{r} = -\begin{bmatrix} -ik_1\rho I\omega^2 + iEIk_1^3 + \overline{K_T} & -k_2\rho I\omega^2 - EIk_2^3 + \overline{K_T} \\ -EIk_1^2 + i\overline{K_R}k_1 & EIk_2^2 + \overline{K_R}k_2 \end{bmatrix}^{-1} \begin{bmatrix} -ik_1\rho I\omega^2 + iEIk_1^3 - \overline{K_T} & -k_2\rho I\omega^2 - EIk_2^3 - \overline{K_T} \\ EIk_1^2 + i\overline{K_R}k_1 & -EIk_2^2 + \overline{K_R}k_2 \end{bmatrix} \tag{9.41a}$$

or equivalently from Eqs. (9.6), (9.24), and (9.40b), the reflection matrix at a spring attached boundary by the Rayleigh bending vibration theory is

$$\mathbf{r} = \begin{bmatrix} ik_1^3 - ik_1\left(\dfrac{C_r}{C_b}\omega\right)^2 + \dfrac{\overline{K_T}}{EI} & -k_2^3 - k_2\left(\dfrac{C_r}{C_b}\omega\right)^2 + \dfrac{\overline{K_T}}{EI} \\ -k_1^2 + ik_1\dfrac{\overline{K_R}}{EI} & k_2^2 + k_2\dfrac{\overline{K_R}}{EI} \end{bmatrix}^{-1} \begin{bmatrix} ik_1^3 - ik_1\left(\dfrac{C_r}{C_b}\omega\right)^2 - \dfrac{\overline{K_T}}{EI} & -k_2^3 - k_2\left(\dfrac{C_r}{C_b}\omega\right)^2 - \dfrac{\overline{K_T}}{EI} \\ k_1^2 + ik_1\dfrac{\overline{K_R}}{EI} & -k_2^2 + k_2\dfrac{\overline{K_R}}{EI} \end{bmatrix} \tag{9.41b}$$

When either $\rho I = 0$ in Eq. (9.41a) or equivalently $C_r = 0$ in Eq. (9.41b), the reflection matrix at a boundary with spring attachments by the Euler–Bernoulli bending vibration theory is obtained, recall that $k_1 = k_2$ by the Euler–Bernoulli bending vibration theory.

Under the special situation where $\overline{K_T}$ and $\overline{K_R}$ approach zero or infinity, the spring attached boundary is physically equivalent to one of the classical boundaries.

- Free boundary when $\overline{K_T} = 0$ and $\overline{K_R} = 0$

Substituting $\overline{K_T} = 0$ and $\overline{K_R} = 0$ into Eq. (9.41b), the same reflection matrix for a free boundary as Eq. (9.29) is obtained.

- Clamped boundary when $\overline{K_T} = \infty$ and $\overline{K_R} = \infty$

Dividing both sides of Eqs. (9.39a) and (9.39b) by $\overline{K_T}$ and $\overline{K_R}$, respectively, substituting $\overline{K_T} = \infty$ and $\overline{K_R} = \infty$ in these two equations, assembling them into matrix form, and solving the matrix equation for the reflection matrix, the same reflection matrix for a clamped boundary as Eq. (9.33) is obtained.

- Pinned boundary for when $\overline{K_T} = \infty$ and $\overline{K_R} = 0$

Dividing both sides of Eq. (9.39a) by $\overline{K_T}$, assembling this equation and Eq. (9.39b) into matrix form, substituting $\overline{K_T} = \infty$ and $\overline{K_R} = 0$ into the matrix equation, and solving the matrix equation for the reflection matrix, the same reflection matrix for a pinned boundary as Eq. (9.37) is obtained.

Viscous damping effect, if it exists, can be taken into account by adding a frequency dependent imaginary term to the spring stiffness, that is, by introducing a dynamic spring stiffness. Denoting viscous damping constants for translational and rotational motion by $\overline{C_T}$ and $\overline{C_R}$, one may add $i\omega\overline{C_T}$ and $i\omega\overline{C_R}$ to the translational and rotational stiffness $\overline{K_T}$ and $\overline{K_R}$, respectively.

e) Boundary with a mass block attachment

Figure 9.5 shows a boundary with a mass block attachment and its free body diagram. Applying Newton's second law for translational and rotational motions to the mass block,

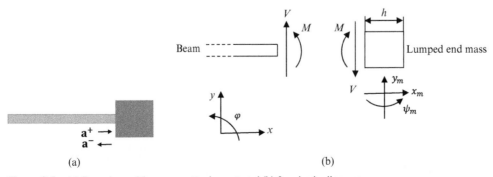

Figure 9.5 (a) Boundary with a mass attachment and (b) free body diagram.

$$-V = m\frac{\partial^2 y_m(x,t)}{\partial t^2}, \quad -M + V\frac{h}{2} = J_m\frac{\partial^2 \psi_m(x,t)}{\partial t^2} \tag{9.42}$$

where m is the mass and h is the width of the attached mass block, and J_m is the mass moment of inertia of the mass block about an axis normal to the x–y plane and passing its center of mass. y_m and ψ_m are the transverse displacement and rotation of the mass block, which are related to the bending deflection and bending slope of the beam at the point where the mass block is attached,

$$y_m = y + \frac{h}{2}\psi_m, \quad \text{and} \quad \psi_m = \psi = \frac{\partial y}{\partial x} \tag{9.43}$$

where the mass block is modeled as a rigid body.

From Eqs. (9.3), (9.4), (9.23a), (9.23b), (9.42), and (9.43), choosing the point where the mass block is attached as the origin,

$$\left[iEIk_1^3 + m\omega^2\left(1 - ik_1\frac{h}{2}\right) - ik_1\rho I\omega^2\right]a_1^+ + \left[-EIk_2^3 + m\omega^2\left(1 - k_2\frac{h}{2}\right) - k_2\rho I\omega^2\right]a_2^+$$
$$+ \left[-iEIk_1^3 + m\omega^2\left(1 + ik_1\frac{h}{2}\right) + ik_1\rho I\omega^2\right]a_1^- + \left[EIk_2^3 + m\omega^2\left(1 + k_2\frac{h}{2}\right) + k_2\rho I\omega^2\right]a_2^- = 0 \tag{9.44a}$$

$$\left[-\left(iEIk_1^3 - ik_1\rho I\omega^2\right)\frac{h}{2} + EIk_1^2 - ik_1 J_m\omega^2\right]a_1^+ + \left[\left(EIk_2^3 + k_2\rho I\omega^2\right)\frac{h}{2} - EIk_2^2 - k_2 J_m\omega^2\right]a_2^+$$
$$+ \left[\left(iEIk_1^3 - ik_1\rho I\omega^2\right)\frac{h}{2} + EIk_1^2 + ik_1 J_m\omega^2\right]a_1^- + \left[-\left(EIk_2^3 + k_2\rho I\omega^2\right)\frac{h}{2} - EIk_2^2 + k_2 J_m\omega^2\right]a_2^- = 0 \tag{9.44b}$$

Putting Eqs. (9.44a) and (9.44b) into matrix form,

$$\begin{bmatrix} iEIk_1^3 + m\omega^2\left(1 - ik_1\frac{h}{2}\right) - ik_1\rho I\omega^2 & -EIk_2^3 + m\omega^2\left(1 - k_2\frac{h}{2}\right) - k_2\rho I\omega^2 \\ -\left(iEIk_1^3 - ik_1\rho I\omega^2\right)\frac{h}{2} + EIk_1^2 - ik_1 J_m\omega^2 & \left(EIk_2^3 + k_2\rho I\omega^2\right)\frac{h}{2} - EIk_2^2 - k_2 J_m\omega^2 \end{bmatrix}\begin{bmatrix} a_1^+ \\ a_2^+ \end{bmatrix}$$
$$+ \begin{bmatrix} -iEIk_1^3 + m\omega^2\left(1 + ik_1\frac{h}{2}\right) + ik_1\rho I\omega^2 & EIk_2^3 + m\omega^2\left(1 + k_2\frac{h}{2}\right) + k_2\rho I\omega^2 \\ \left(iEIk_1^3 - ik_1\rho I\omega^2\right)\frac{h}{2} + EIk_1^2 + ik_1 J_m\omega^2 & -\left(EIk_2^3 + k_2\rho I\omega^2\right)\frac{h}{2} - EIk_2^2 + k_2 J_m\omega^2 \end{bmatrix}\begin{bmatrix} a_1^- \\ a_2^- \end{bmatrix} = 0 \tag{9.45a}$$

or equivalently

$$\begin{bmatrix} ik_1^3 + \frac{m}{EI}\omega^2\left(1 - ik_1\frac{h}{2}\right) - ik_1\left(\frac{C_r}{C_b}\omega\right)^2 & -k_2^3 + \frac{m}{EI}\omega^2\left(1 - k_2\frac{h}{2}\right) - k_2\left(\frac{C_r}{C_b}\omega\right)^2 \\ -\left[ik_1^3 - ik_1\left(\frac{C_r}{C_b}\omega\right)^2\right]\frac{h}{2} + k_1^2 - ik_1\frac{J_m}{EI}\omega^2 & \left[k_2^3 + k_2\left(\frac{C_r}{C_b}\omega\right)^2\right]\frac{h}{2} - k_2^2 - k_2\frac{J_m}{EI}\omega^2 \end{bmatrix}\begin{bmatrix} a_1^+ \\ a_2^+ \end{bmatrix}$$
$$+ \begin{bmatrix} -ik_1^3 + \frac{m}{EI}\omega^2\left(1 + ik_1\frac{h}{2}\right) + ik_1\left(\frac{C_r}{C_b}\omega\right)^2 & k_2^3 + \frac{m}{EI}\omega^2\left(1 + k_2\frac{h}{2}\right) + k_2\left(\frac{C_r}{C_b}\omega\right)^2 \\ \left[ik_1^3 - ik_1\left(\frac{C_r}{C_b}\omega\right)^2\right]\frac{h}{2} + k_1^2 + ik_1\frac{J_m}{EI}\omega^2 & -\left[k_2^3 + k_2\left(\frac{C_r}{C_b}\omega\right)^2\right]\frac{h}{2} - k_2^2 + k_2\frac{J_m}{EI}\omega^2 \end{bmatrix}\begin{bmatrix} a_1^- \\ a_2^- \end{bmatrix} = 0 \tag{9.45b}$$

From Eqs. (9.6), (9.24), and (9.45a), the reflection matrix at a mass block attached boundary by the Rayleigh bending vibration theory is

$$\mathbf{r} = \begin{bmatrix} iEIk_1^3 - m\omega^2\left(1 + ik_1\dfrac{h}{2}\right) - ik_1\rho I\omega^2 & -EIk_2^3 - m\omega^2\left(1 + k_2\dfrac{h}{2}\right) - k_2\rho I\omega^2 \\[2mm] -\left(iEIk_1^3 - ik_1\rho I\omega^2\right)\dfrac{h}{2} - EIk_1^2 - ik_1 J_m\omega^2 & \left(EIk_2^3 + k_2\rho I\omega^2\right)\dfrac{h}{2} + EIk_2^2 - k_2 J_m\omega^2 \end{bmatrix}^{-1}$$

$$\begin{bmatrix} iEIk_1^3 + m\omega^2\left(1 - ik_1\dfrac{h}{2}\right) - ik_1\rho I\omega^2 & -EIk_2^3 + m\omega^2\left(1 - k_2\dfrac{h}{2}\right) - k_2\rho I\omega^2 \\[2mm] -\left(iEIk_1^3 - ik_1\rho I\omega^2\right)\dfrac{h}{2} + EIk_1^2 - ik_1 J_m\omega^2 & \left(EIk_2^3 + k_2\rho I\omega^2\right)\dfrac{h}{2} - EIk_2^2 - k_2 J_m\omega^2 \end{bmatrix} \tag{9.46a}$$

or equivalently from Eqs. (9.6), (9.24), and (9.45b), the reflection matrix at a mass block attached boundary by the Rayleigh bending vibration theory is

$$\mathbf{r} = \begin{bmatrix} ik_1^3 - \dfrac{m}{EI}\omega^2\left(1 + ik_1\dfrac{h}{2}\right) - ik_1\left(\dfrac{C_r}{C_b}\omega\right)^2 & -k_2^3 - \dfrac{m}{EI}\omega^2\left(1 + k_2\dfrac{h}{2}\right) - k_2\left(\dfrac{C_r}{C_b}\omega\right)^2 \\[2mm] -\left[ik_1^3 - ik_1\left(\dfrac{C_r}{C_b}\omega\right)^2\right]\dfrac{h}{2} - k_1^2 - ik_1\dfrac{J_m}{EI}\omega^2 & \left[k_2^3 + k_2\left(\dfrac{C_r}{C_b}\omega\right)^2\right]\dfrac{h}{2} + k_2^2 - k_2\dfrac{J_m}{EI}\omega^2 \end{bmatrix}^{-1}$$

$$\begin{bmatrix} ik_1^3 + \dfrac{m}{EI}\omega^2\left(1 - ik_1\dfrac{h}{2}\right) - ik_1\left(\dfrac{C_r}{C_b}\omega\right)^2 & -k_2^3 + \dfrac{m}{EI}\omega^2\left(1 - k_2\dfrac{h}{2}\right) - k_2\left(\dfrac{C_r}{C_b}\omega\right)^2 \\[2mm] -\left[ik_1^3 - ik_1\left(\dfrac{C_r}{C_b}\omega\right)^2\right]\dfrac{h}{2} + k_1^2 - ik_1\dfrac{J_m}{EI}\omega^2 & \left[k_2^3 + k_2\left(\dfrac{C_r}{C_b}\omega\right)^2\right]\dfrac{h}{2} - k_2^2 - k_2\dfrac{J_m}{EI}\omega^2 \end{bmatrix} \tag{9.46b}$$

When either $\rho I = 0$ in Eq. (9.46a) or equivalently $C_r = 0$ in Eq. (9.46b), the reflection matrix at a boundary with a mass block attachment by the Euler–Bernoulli bending vibration theory is obtained, recall that $k_1 = k_2$ by the Euler–Bernoulli bending vibration theory.

Under the special situation where $h = 0$, and m and J_m approach either zero or infinity, the mass attached boundary is physically equivalent to a free, clamped, or pinned boundary.

- Free boundary when $m = 0$ and $J_m = 0$

Substituting $m = 0$, $J_m = 0$, and $h = 0$ into Eq. (9.46b), the same reflection matrix for a free boundary as Eq. (9.29) is obtained.

- Clamped boundary when $m = \infty$ and $J_m = \infty$

Dividing both sides of Eq. (9.44a) and Eq. (9.44b) by m and J_m, respectively, substituting $m = \infty$, $J_m = \infty$ and $h = 0$ in these two equations, assembling them into matrix form, and solving the matrix equation for the reflection matrix, the same reflection matrix for a clamped boundary as Eq. (9.33) is obtained.

- Pinned boundary when $m = \infty$ and $J_m = 0$

Dividing both sides of Eq. (9.44a) by m, assembling this equation and Eq. (9.44b) into matrix form, substituting $m = \infty$, $J_m = 0$, and $h = 0$ into the matrix equation, and solving the matrix equation for the reflection matrix, the same reflection matrix for a pinned boundary as Eq. (9.37) is obtained.

The reflection matrices at various boundaries obtained using the Rayleigh bending vibration theory are summarized in Table 9.3.

9.2.2 Shear and Timoshenko Bending Vibration Theories

The differences between the Timoshenko and Shear bending vibration theories in the governing differential equations and wavenumbers are listed in Tables 9.1 and 9.2. The Timoshenko bending vibration theory for $\omega < \omega_c$ (the most common situation in audio frequency applications) and the Shear bending vibration theory, however, share the following common expressions:

- The expressions for bending deflection $y(x)$ and bending slope $\psi(x)$, which are described by Eqs. (9.12a) and (9.12b);
- The expressions for magnitude ratios $\overline{a_{1,2}^{\pm}} / a_{1,2}^{\pm}$ between wave components of the bending slope $\psi(x)$ and those of bending deflection $y(x)$, in terms of coefficients P and N, which are described by Eq. (9.14);
- The expressions for coefficients P and N, which are given by Eq. (9.15).

Table 9.3 Reflection matrices at various boundaries by the Rayleigh bending vibration theory.

Boundary	Reflection Matrix
Free	$$\mathbf{r} = \begin{bmatrix} k_1^2 & -k_2^2 \\ ik_1^3 - ik_1\left(\dfrac{C_r}{C_b}\omega\right)^2 & -k_2^3 - k_2\left(\dfrac{C_r}{C_b}\omega\right)^2 \end{bmatrix}^{-1} \begin{bmatrix} -k_1^2 & k_2^2 \\ ik_1^3 - ik_1\left(\dfrac{C_r}{C_b}\omega\right)^2 & -k_2^3 - k_2\left(\dfrac{C_r}{C_b}\omega\right)^2 \end{bmatrix}$$
Clamped	$$\mathbf{r} = \begin{bmatrix} -1 & -1 \\ ik_1 & k_2 \end{bmatrix}^{-1} \begin{bmatrix} 1 & 1 \\ ik_1 & k_2 \end{bmatrix} = \frac{1}{k_2 - ik_1}\begin{bmatrix} -k_2 - ik_1 & -2k_2 \\ i2k_1 & k_2 + ik_1 \end{bmatrix}$$
Pinned	$$\mathbf{r} = \begin{bmatrix} -1 & 0 \\ 0 & -1 \end{bmatrix}$$
Spring	$$\mathbf{r} = \begin{bmatrix} ik_1^3 - ik_1\left(\dfrac{C_r}{C_b}\omega\right)^2 + \dfrac{\overline{K_T}}{EI} & -k_2^3 - k_2\left(\dfrac{C_r}{C_b}\omega\right)^2 + \dfrac{\overline{K_T}}{EI} \\ -k_1^2 + ik_1\dfrac{\overline{K_R}}{EI} & k_2^2 + k_2\dfrac{\overline{K_R}}{EI} \end{bmatrix}^{-1}$$ $$\begin{bmatrix} ik_1^3 - ik_1\left(\dfrac{C_r}{C_b}\omega\right)^2 - \dfrac{\overline{K_T}}{EI} & -k_2^3 - k_2\left(\dfrac{C_r}{C_b}\omega\right)^2 - \dfrac{\overline{K_T}}{EI} \\ k_1^2 + ik_1\dfrac{\overline{K_R}}{EI} & -k_2^2 + k_2\dfrac{\overline{K_R}}{EI} \end{bmatrix}$$
Rigid Mass Block	$$\mathbf{r} = \begin{bmatrix} ik_1^3 - \dfrac{m}{EI}\omega^2\left(1 + ik_1\dfrac{h}{2}\right) - ik_1\left(\dfrac{C_r}{C_b}\omega\right)^2 & -k_2^3 - \dfrac{m}{EI}\omega^2\left(1 + k_2\dfrac{h}{2}\right) - k_2\left(\dfrac{C_r}{C_b}\omega\right)^2 \\ -\left[ik_1^3 - ik_1\left(\dfrac{C_r}{C_b}\omega\right)^2\right]\dfrac{h}{2} - k_1^2 - ik_1\dfrac{J_m}{EI}\omega^2 & \left[k_2^3 + k_2\left(\dfrac{C_r}{C_b}\omega\right)^2\right]\dfrac{h}{2} + k_2^2 - k_2\dfrac{J_m}{EI}\omega^2 \end{bmatrix}^{-1}$$ $$\begin{bmatrix} ik_1^3 + \dfrac{m}{EI}\omega^2\left(1 - ik_1\dfrac{h}{2}\right) - ik_1\left(\dfrac{C_r}{C_b}\omega\right)^2 & -k_2^3 + \dfrac{m}{EI}\omega^2\left(1 - k_2\dfrac{h}{2}\right) - k_2\left(\dfrac{C_r}{C_b}\omega\right)^2 \\ -\left[ik_1^3 - ik_1\left(\dfrac{C_r}{C_b}\omega\right)^2\right]\dfrac{h}{2} + k_1^2 - ik_1\dfrac{J_m}{EI}\omega^2 & \left[k_2^3 + k_2\left(\dfrac{C_r}{C_b}\omega\right)^2\right]\dfrac{h}{2} - k_2^2 - k_2\dfrac{J_m}{EI}\omega^2 \end{bmatrix}$$

In summary, from Eqs. (9.12a), (9.12b), and (9.14), the bending deflection $y(x)$ and bending slope $\psi(x)$ based on both the Timoshenko (for $\omega < \omega_c$) and Shear bending vibration theories are given by the same expressions

$$y(x) = a_1^+ e^{-ik_1 x} + a_2^+ e^{-k_2 x} + a_1^- e^{ik_1 x} + a_2^- e^{k_2 x} \tag{9.47a}$$

$$\psi(x) = -iPa_1^+ e^{-ik_1 x} - Na_2^+ e^{-k_2 x} + iPa_1^- e^{ik_1 x} + Na_2^- e^{k_2 x} \tag{9.47b}$$

where $P = k_1\left(1 - \dfrac{\omega^2}{k_1^2 C_s^2}\right)$ and $N = k_2\left(1 + \dfrac{\omega^2}{k_2^2 C_s^2}\right)$, as described in Eq. (9.15).

In addition, based on both the Timoshenko (for $\omega < \omega_c$) and Shear bending vibration theories, the internal resistant shear force $V(x,t)$ and bending moment $M(x,t)$ are related to the bending deflection $y(x,t)$ and bending slope $\psi(x,t)$ by the same expressions,

$$V(x,t) = GA\kappa \left[\frac{\partial y(x,t)}{\partial x} - \psi(x,t) \right] \tag{9.48a}$$

$$M(x,t) = EI \frac{\partial \psi(x,t)}{\partial x} \tag{9.48b}$$

when following the sign convention defined in Figure 9.3.

As a result, boundary reflection relationships, based on both the Timoshenko (for $\omega < \omega_c$) and Shear bending vibration theories, are described by the same expressions. One must keep in mind, however, that the values will not be the same because of the differences in the governing differential equations and wavenumbers between the Timoshenko and Shear bending vibration theories, as shown in Table 9.1 and Eqs. (9.13), (9.21a), and (9.21b).

a) Free boundary

At a free boundary, the equilibrium conditions are

$$V(x,t) = 0, \ M(x,t) = 0 \tag{9.49}$$

From Eqs. (9.47a), (9.47b), (9.48a), (9.48b), and (9.49),

$$\begin{aligned} i(P - k_1)a_1^+ e^{-ik_1 x} + (N - k_2)a_2^+ e^{-k_2 x} - i(P - k_1)a_1^- e^{ik_1 x} - (N - k_2)a_2^- e^{k_2 x} = 0 \\ -Pk_1 a_1^+ e^{-ik_1 x} + Nk_2 a_2^+ e^{-k_2 x} - Pk_1 a_1^- e^{ik_1 x} + Nk_2 a_2^- e^{k_2 x} = 0 \end{aligned} \tag{9.50}$$

Choosing the origin at the boundary and putting Eq. (9.50) into matrix form,

$$\begin{bmatrix} i(P - k_1) & N - k_2 \\ -Pk_1 & Nk_2 \end{bmatrix} \begin{bmatrix} a_1^+ \\ a_2^+ \end{bmatrix} - \begin{bmatrix} i(P - k_1) & N - k_2 \\ Pk_1 & -Nk_2 \end{bmatrix} \begin{bmatrix} a_1^- \\ a_2^- \end{bmatrix} = 0 \tag{9.51}$$

From Eqs. (9.6), (9.24), and (9.51), the reflection matrix at a free boundary by the Timoshenko and Shear bending vibration theories is

$$\mathbf{r} = \begin{bmatrix} i(P - k_1) & N - k_2 \\ Pk_1 & -Nk_2 \end{bmatrix}^{-1} \begin{bmatrix} i(P - k_1) & N - k_2 \\ -Pk_1 & Nk_2 \end{bmatrix} \tag{9.52a}$$

or

$$\mathbf{r} = \begin{vmatrix} \dfrac{-Pk_1(k_2 - N) + ik_2 N(k_1 - P)}{Pk_1(k_2 - N) + ik_2 N(k_1 - P)} & \dfrac{2Nk_2(k_2 - N)}{Pk_1(k_2 - N) + ik_2 N(k_1 - P)} \\[2ex] \dfrac{i2Pk_1(k_1 - P)}{Pk_1(k_2 - N) + ik_2 N(k_1 - P)} & \dfrac{Pk_1(k_2 - N) - ik_2 N(k_1 - P)}{Pk_1(k_2 - N) + ik_2 N(k_1 - P)} \end{vmatrix} \tag{9.52b}$$

b) Clamped boundary

The boundary conditions at a clamped boundary are

$$y(x,t) = 0, \ \psi(x,t) = 0 \tag{9.53}$$

From Eqs. (9.47a), (9.47b), and (9.53),

$$\begin{aligned} a_1^+ e^{-ik_1 x} + a_2^+ e^{-k_2 x} + a_1^- e^{ik_1 x} + a_2^- e^{k_2 x} = 0 \\ -iPa_1^+ e^{-ik_1 x} - Na_2^+ e^{-k_2 x} + iPa_1^- e^{ik_1 x} + Na_2^- e^{k_2 x} = 0 \end{aligned} \tag{9.54}$$

Choosing the origin at the boundary and putting Eq. (9.54) into matrix form,

$$
\begin{bmatrix} 1 & 1 \\ iP & N \end{bmatrix} \begin{bmatrix} a_1^+ \\ a_2^+ \end{bmatrix} - \begin{bmatrix} -1 & -1 \\ iP & N \end{bmatrix} \begin{bmatrix} a_1^- \\ a_2^- \end{bmatrix} = 0
\tag{9.55}
$$

From Eqs. (9.6), (9.24), and (9.55), the reflection matrix at a clamped boundary by the Timoshenko and Shear bending vibration theories is

$$
\mathbf{r} = \begin{bmatrix} -1 & -1 \\ iP & N \end{bmatrix}^{-1} \begin{bmatrix} 1 & 1 \\ iP & N \end{bmatrix}
\tag{9.56a}
$$

or

$$
\mathbf{r} = \begin{bmatrix} \dfrac{P-iN}{P+iN} & \dfrac{-i2N}{P+iN} \\[2ex] \dfrac{-2P}{P+iN} & -\dfrac{P-iN}{P+iN} \end{bmatrix}
\tag{9.56b}
$$

c) Pinned boundary

The boundary conditions at a pinned boundary are

$$
y(x,t) = 0, \; M(x,t) = 0
\tag{9.57}
$$

From Eqs. (9.47a), (9.47b), (9.48a), (9.48b), and (9.57),

$$
\begin{aligned}
a_1^+ e^{-ik_1 x} + a_2^+ e^{-k_2 x} + a_1^- e^{ik_1 x} + a_2^- e^{k_2 x} &= 0 \\
-Pk_1 a_1^+ e^{-ik_1 x} + Nk_2 a_2^+ e^{-k_2 x} - Pk_1 a_1^- e^{ik_1 x} + Nk_2 a_2^- e^{k_2 x} &= 0
\end{aligned}
\tag{9.58}
$$

Choosing the origin at the boundary and putting Eq. (9.58) into matrix form,

$$
\begin{bmatrix} 1 & 1 \\ -Pk_1 & Nk_2 \end{bmatrix} \begin{bmatrix} a_1^+ \\ a_2^+ \end{bmatrix} - \begin{bmatrix} -1 & -1 \\ Pk_1 & -Nk_2 \end{bmatrix} \begin{bmatrix} a_1^- \\ a_2^- \end{bmatrix} = 0
\tag{9.59}
$$

From Eqs. (9.6), (9.24), and (9.59), the reflection matrix at a pinned boundary by the Timoshenko and Shear bending vibration theories is

$$
\mathbf{r} = \begin{bmatrix} -1 & 0 \\ 0 & -1 \end{bmatrix}
\tag{9.60}
$$

d) Boundary with spring attachments

For a boundary with spring attachments, the boundary support exerts both translational and rotational constraints on the beam. These are described by the translational and rotational dynamic stiffnesses $\overline{K_T}$ and $\overline{K_R}$, as shown in Figure 9.4. When bending vibrations are modeled by the Timoshenko or Shear bending vibration theories, there may be coupling between translation and rotation because of the coupled motions in the bending deflection and bending slope. As a result, the boundary support may induce a moment in response to a translation and a force in response to a rotation. These are described by the transfer dynamic stiffnesses $\overline{K_{TR}}$ and $\overline{K_{RT}}$. These dynamic stiffnesses are, in general, complex and frequency dependent, and can include the effects of mass, stiffness and damping of the springs.

From the free body diagram of Figure 9.4, taking into account the resistant forces and moments due to the transfer dynamic stiffnesses $\overline{K_{TR}}$ and $\overline{K_{RT}}$, the equilibrium equations are

$$
\begin{aligned}
-V(x,t) - \overline{K_T} y(x,t) - \overline{K_{TR}} \psi(x,t) &= 0, \\
-M(x,t) - \overline{K_R} \psi(x,t) - \overline{K_{RT}} y(x,t) &= 0
\end{aligned}
\tag{9.61}
$$

Substituting Eqs. (9.48a) and (9.48b), into (9.61), the equilibrium equations become

$$\left[\frac{\partial y(x,t)}{\partial x} - \psi(x,t)\right] + K_T y(x,t) + K_{TR}\psi(x,t) = 0, \quad \frac{\partial \psi(x,t)}{\partial x} + K_R \psi(x,t) + K_{RT} y(x,t) = 0 \tag{9.62}$$

where the normalized spring stiffnesses are $K_T = \dfrac{\overline{K_T}}{GA\kappa}$, $K_{TR} = \dfrac{\overline{K_{TR}}}{GA\kappa}$, $K_R = \dfrac{\overline{K_R}}{EI}$, and $K_{RT} = \dfrac{\overline{K_{RT}}}{EI}$.

From Eqs. (9.47a), (9.47b), and (9.62),

$$\left[iP\left(-1+K_{TR}\right)+ik_1 - K_T\right]a_1^+ e^{-ik_1 x} + \left[N\left(-1+K_{TR}\right)+k_2 - K_T\right]a_2^+ e^{-k_2 x}$$
$$+\left[iP\left(1-K_{TR}\right)-ik_1 - K_T\right]a_1^- e^{ik_1 x} + \left[N\left(1-K_{TR}\right)-k_2 - K_T\right]a_2^- e^{k_2 x} = 0 \tag{9.63a}$$

$$\left[iP\left(-ik_1+K_R\right)-K_{RT}\right]a_1^+ e^{-ik_1 x} + \left[N\left(-k_2+K_R\right)-K_{RT}\right]a_2^+ e^{-k_2 x}$$
$$+\left[iP\left(-ik_1-K_R\right)-K_{RT}\right]a_1^- e^{ik_1 x} + \left[N\left(-k_2-K_R\right)-K_{RT}\right]a_2^- e^{k_2 x} = 0 \tag{9.63b}$$

Choosing the origin at the boundary and putting Eqs. (9.63a) and (9.63b) into matrix form,

$$\begin{bmatrix} iP\left(-1+K_{TR}\right)+ik_1 - K_T & N\left(-1+K_{TR}\right)+k_2 - K_T \\ iP\left(-ik_1+K_R\right)-K_{RT} & N\left(-k_2+K_R\right)-K_{RT} \end{bmatrix}\begin{bmatrix} a_1^+ \\ a_2^+ \end{bmatrix}$$
$$+\begin{bmatrix} iP\left(1-K_{TR}\right)-ik_1 - K_T & N\left(1-K_{TR}\right)-k_2 - K_T \\ iP\left(-ik_1-K_R\right)-K_{RT} & N\left(-k_2-K_R\right)-K_{RT} \end{bmatrix}\begin{bmatrix} a_1^- \\ a_2^- \end{bmatrix} = 0 \tag{9.64}$$

From Eqs. (9.6), (9.24), and (9.64), the reflection matrix at a boundary with spring attachments by the Timoshenko and Shear bending vibration theories is

$$\mathbf{r} = -\begin{bmatrix} iP\left(1-K_{TR}\right)+ik_1 + K_T & -N\left(1-K_{TR}\right)+k_2 + K_T \\ iP\left(ik_1+K_R\right)+K_{RT} & N\left(k_2+K_R\right)+K_{RT} \end{bmatrix}^{-1}$$
$$\begin{bmatrix} iP\left(-1+K_{TR}\right)+ik_1 - K_T & N\left(-1+K_{TR}\right)+k_2 - K_T \\ iP\left(-ik_1+K_R\right)-K_{RT} & N\left(-k_2+K_R\right)-K_{RT} \end{bmatrix} \tag{9.65}$$

The reflection matrices of the three classical boundaries, namely the free, clamped, and pinned boundaries, can be obtained as special situations by setting both K_{TR} and K_{RT} to 0, and letting K_T and K_R be either zero or infinite.

- Free boundary when $K_T = 0$ and $K_R = 0$

Substituting $K_T = 0$ and $K_R = 0$ into Eq. (9.65), the same reflection matrix for a free boundary as Eq. (9.52a) is obtained.

- Clamped boundary when $K_T = \infty$ and $K_R = \infty$

Dividing both sides of Eq. (9.63a) and Eq. (9.63b) by K_T and K_R, respectively, substituting $K_T = \infty$ and $K_R = \infty$ in these two equations, assembling them into matrix form, and solving the matrix equation for the reflection matrix, the same reflection matrix for a clamped boundary as Eq. (9.56a), is obtained.

- Pinned boundary when $K_T = \infty$ and $K_R = 0$

Dividing both sides of Eq. (9.63a) by K_T, assembling this equation and Eq. (9.63b) into matrix form, substituting $K_T = \infty$ and $K_R = 0$ into the matrix equation, and solving the matrix equation for the reflection matrix, the same reflection matrix for a pinned boundary as Eq. (9.60) is obtained.

e) Boundary with a mass attachment

For a boundary with a rigid mass block attachment shown in Figure 9.5, the equation of motion of the boundary are obtained in Eq. (9.42). The transverse displacement and rotation of the rigid mass block, namely, y_m and ψ_m, are related to the bending deflection and bending slope of the beam at the point where the mass is attached by Eq. (9.43).

From Eqs. (9.42), (9.43), (9.47a), (9.47b), (9.48a), and (9.48b), with the mass attached point chosen as the origin,

$$\left[iGA\kappa(k_1 - P) + m\omega^2\left(1 - iP\frac{h}{2}\right)\right]a_1^+ + \left[GA\kappa(k_2 - N) + m\omega^2\left(1 - N\frac{h}{2}\right)\right]a_2^+$$
$$+ \left[-iGA\kappa(k_1 - P) + m\omega^2\left(1 + iP\frac{h}{2}\right)\right]a_1^- + \left[-GA\kappa(k_2 - N) + m\omega^2\left(1 + N\frac{h}{2}\right)\right]a_2^- = 0 \tag{9.66a}$$

$$\left[EIPk_1 - iGA\kappa(k_1 - P)\frac{h}{2} - iPJ_m\omega^2\right]a_1^+ + \left[-EINk_2 - GA\kappa(k_2 - N)\frac{h}{2} - NJ_m\omega^2\right]a_2^+$$
$$+ \left[EIPk_1 + iGA\kappa(k_1 - P)\frac{h}{2} + iPJ_m\omega^2\right]a_1^- + \left[-EINk_2 + GA\kappa(k_2 - N)\frac{h}{2} + NJ_m\omega^2\right]a_2^- = 0 \tag{9.66b}$$

Putting Eqs. (9.66a) and (9.66b) into matrix form,

$$\begin{bmatrix} iGA\kappa(k_1 - P) + m\omega^2\left(1 - iP\dfrac{h}{2}\right) & GA\kappa(k_2 - N) + m\omega^2\left(1 - N\dfrac{h}{2}\right) \\ EIPk_1 - iGA\kappa(k_1 - P)\dfrac{h}{2} - iPJ_m\omega^2 & -EINk_2 - GA\kappa(k_2 - N)\dfrac{h}{2} - NJ_m\omega^2 \end{bmatrix}\begin{bmatrix} a_1^+ \\ a_2^+ \end{bmatrix}$$
$$+ \begin{bmatrix} -iGA\kappa(k_1 - P) + m\omega^2\left(1 + iP\dfrac{h}{2}\right) & -GA\kappa(k_2 - N) + m\omega^2\left(1 + N\dfrac{h}{2}\right) \\ EIPk_1 + iGA\kappa(k_1 - P)\dfrac{h}{2} + iPJ_m\omega^2 & -EINk_2 + GA\kappa(k_2 - N)\dfrac{h}{2} + NJ_m\omega^2 \end{bmatrix}\begin{bmatrix} a_1^- \\ a_2^- \end{bmatrix} = 0 \tag{9.67}$$

From Eqs. (9.6), (9.24), and (9.67), the reflection matrix at a mass block attached boundary by the Timoshenko and Shear bending vibration theories is

$$\mathbf{r} = \begin{bmatrix} iGA\kappa(k_1 - P) - m\omega^2\left(1 + iP\dfrac{h}{2}\right) & GA\kappa(k_2 - N) - m\omega^2\left(1 + N\dfrac{h}{2}\right) \\ -EIPk_1 - iGA\kappa(k_1 - P)\dfrac{h}{2} - iPJ_m\omega^2 & EINk_2 - GA\kappa(k_2 - N)\dfrac{h}{2} - NJ_m\omega^2 \end{bmatrix}^{-1}$$
$$\begin{bmatrix} iGA\kappa(k_1 - P) + m\omega^2\left(1 - iP\dfrac{h}{2}\right) & GA\kappa(k_2 - N) + m\omega^2\left(1 - N\dfrac{h}{2}\right) \\ EIPk_1 - iGA\kappa(k_1 - P)\dfrac{h}{2} - iPJ_m\omega^2 & -EINk_2 - GA\kappa(k_2 - N)\dfrac{h}{2} - NJ_m\omega^2 \end{bmatrix} \tag{9.68}$$

Under the special situation where $h = 0$, and m and J_m approach zero or infinity, the mass attached end is physically equivalent to a free, clamped, or pinned boundary.

- Free boundary when $m = 0$ and $J_m = 0$

Substituting $h = 0$, $m = 0$, and $J_m = 0$ into Eq. (9.68), the same reflection matrix for a free boundary as Eq. (9.52a) is obtained.

- Clamped boundary when $m = \infty$ and $J_m = \infty$

Dividing both sides of Eq. (9.66a) and Eq. (9.66b) by m and J_m, respectively, substituting $h = 0$, $m = \infty$, and $J_m = \infty$ in these two equations, assembling them into matrix form, and solving the matrix equation for the reflection matrix, the same reflection matrix for a clamped boundary as Eq. (9.56a) is obtained.

- Pinned boundary when $m = \infty$ and $J_m = 0$

Dividing both sides of Eq. (9.66a) by m, assembling this equation and Eq. (9.66b) into matrix form, substituting $h = 0$, $m = \infty$, and $J_m = 0$ into the matrix equation, and solving the matrix equation for the reflection matrix, the same reflection matrix for a pinned boundary as Eq. (9.60) is obtained.

The Timoshenko bending vibration theory related expressions obtained in this section are for $\omega < \omega_c$, which is the most common situation in audio frequency applications. For $\omega > \omega_c$, k_2 in the related expressions by the Timoshenko bending

Table 9.4 Reflection matrices at various boundaries by the Timoshenko and Shear bending vibration theories.

Boundary	Reflection Matrix
Free	$\mathbf{r} = \begin{bmatrix} \dfrac{-Pk_1(k_2-N)+ik_2N(k_1-P)}{Pk_1(k_2-N)+ik_2N(k_1-P)} & \dfrac{2Nk_2(k_2-N)}{Pk_1(k_2-N)+ik_2N(k_1-P)} \\[2ex] \dfrac{i2Pk_1(k_1-P)}{Pk_1(k_2-N)+ik_2N(k_1-P)} & \dfrac{Pk_1(k_2-N)-ik_2N(k_1-P)}{Pk_1(k_2-N)+ik_2N(k_1-P)} \end{bmatrix}$
Clamped	$\mathbf{r} = \begin{bmatrix} \dfrac{P-iN}{P+iN} & \dfrac{-i2N}{P+iN} \\[2ex] \dfrac{-2P}{P+iN} & -\dfrac{P-iN}{P+iN} \end{bmatrix}$
Pinned	$\mathbf{r} = \begin{bmatrix} -1 & 0 \\ 0 & -1 \end{bmatrix}$
Spring	$\mathbf{r} = \begin{bmatrix} -iP(1-K_{TR})+ik_1+K_T & -N(1-K_{TR})+k_2+K_T \\ iP(ik_1+K_R)+K_{RT} & N(k_2+K_R)+K_{RT} \end{bmatrix}^{-1}$ $\begin{bmatrix} iP(-1+K_{TR})+ik_1-K_T & N(-1+K_{TR})+k_2-K_T \\ iP(-ik_1+K_R)-K_{RT} & N(-k_2+K_R)-K_{RT} \end{bmatrix}$
Lumped Rigid Mass	$\mathbf{r} = \begin{bmatrix} iGA\kappa(k_1-P)-m\omega^2\left(1+iP\dfrac{h}{2}\right) & GA\kappa(k_2-N)-m\omega^2\left(1+N\dfrac{h}{2}\right) \\ -EIPk_1-iGA\kappa(k_1-P)\dfrac{h}{2}-iPJ_m\omega^2 & EINk_2-GA\kappa(k_2-N)\dfrac{h}{2}-NJ_m\omega^2 \end{bmatrix}^{-1}$ $\begin{bmatrix} iGA\kappa(k_1-P)+m\omega^2\left(1-iP\dfrac{h}{2}\right) & GA\kappa(k_2-N)+m\omega^2\left(1-N\dfrac{h}{2}\right) \\ EIPk_1-iGA\kappa(k_1-P)\dfrac{h}{2}-iPJ_m\omega^2 & -EINk_2-GA\kappa(k_2-N)\dfrac{h}{2}-NJ_m\omega^2 \end{bmatrix}$

vibration theory needs to be replaced with ik_2, because of the transition of wave mode from a pair of decaying waves below the cut-off frequency to a pair of propagating waves above the cut-off frequency in k_2 related wave components.

The reflection matrices at various boundaries obtained based on the Timoshenko and Shear bending vibration theories are listed in Table 9.4.

MATLAB symbolic scripts for reflection matrix at a rigid mass block attached boundary by the Timoshenko and Shear bending vibration theories are available in Section 9.7.

9.3 Waves Generated by Externally Applied Point Force and Moment on the Span

As discussed in Chapter 3, an externally applied transverse force or bending moment has the effect of injecting waves into a continuous structure. Figure 9.6, which is the same as Figure 3.6, illustrates a point transverse force $\bar{Q}$, a bending moment $\bar{M}$, and waves $\mathbf{a}^{\pm}$ and $\mathbf{b}^{\pm}$ at the point where the force and moment are applied. These waves are injected by and related to the externally applied force and moment. The physical parameters on the left and right side of the point where the external excitations are applied are denoted using subscripts $-$ and $+$, respectively.

The bending deflections on the left and right side of the point where the force and moment are applied are the same,

$$y_- = y_+, \ \psi_- = \psi_+ \tag{9.69}$$

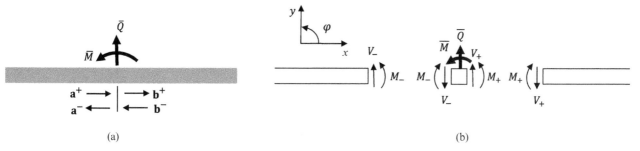

Figure 9.6 (a) Waves generated by external transverse force and bending moment applied on the span and (b) free body diagram.

From the free body diagram of Figure 9.6, the externally applied transverse force and bending moment are related to the internal resistant forces and moments on the left and right side of point where the external force and moment are applied,

$$\bar{Q} = V_- - V_+, \quad \bar{M} = M_- - M_+ \tag{9.70}$$

The relationships described in Eqs. (9.69) and (9.70) are called the continuity and equilibrium equations, respectively.

9.3.1 Rayleigh Bending Vibration Theory

From Eq. (9.3) and Figure 9.6a,

$$y_- = a_1^+ e^{-ik_1 x} + a_2^+ e^{-k_2 x} + a_1^- e^{ik_1 x} + a_2^- e^{k_2 x} \tag{9.71a}$$

$$y_+ = b_1^+ e^{-ik_1 x} + b_2^+ e^{-k_2 x} + b_1^- e^{ik_1 x} + b_2^- e^{k_2 x} \tag{9.71b}$$

Because the Rayleigh bending vibration theory neglects shear deformation by assuming infinite shear rigidity, bending slope and bending deflection are related by $\psi_\pm = dy_\pm / dx$, from Eqs. (9.71a) and (9.71b),

$$\psi_- = -ik_1 a_1^+ e^{-ik_1 x} - k_2 a_2^+ e^{-k_2 x} + ik_1 a_1^- e^{ik_1 x} + k_2 a_2^- e^{k_2 x} \tag{9.72a}$$

$$\psi_+ = -ik_1 b_1^+ e^{-ik_1 x} - k_2 b_2^+ e^{-k_2 x} + ik_1 b_1^- e^{ik_1 x} + k_2 b_2^- e^{k_2 x} \tag{9.72b}$$

From Eqs. (9.23a), (9.23b), and (9.70), the equilibrium conditions become

$$\bar{Q} = \left(-EI \frac{\partial^3 y_-(x,t)}{\partial x^3} + \rho I \frac{\partial^3 y_-(x,t)}{\partial x \partial t^2} \right) - \left(-EI \frac{\partial^3 y_+(x,t)}{\partial x^3} + \rho I \frac{\partial^3 y_+(x,t)}{\partial x \partial t^2} \right)$$

$$\bar{M} = EI \frac{\partial^2 y_-(x,t)}{\partial x^2} - EI \frac{\partial^2 y_+(x,t)}{\partial x^2} \tag{9.73}$$

With the origin chosen at the point where the external force and moment are applied, from Eqs. (9.69), (9.71a), (9.71b), (9.72a), and (9.72b), the continuity conditions become

$$a_1^+ + a_2^+ + a_1^- + a_2^- = b_1^+ + b_2^+ + b_1^- + b_2^- \tag{9.74a}$$

$$-ik_1 a_1^+ - k_2 a_2^+ + ik_1 a_1^- + k_2 a_2^- = -ik_1 b_1^+ - k_2 b_2^+ + ik_1 b_1^- + k_2 b_2^- \tag{9.74b}$$

Putting Eqs. (9.74a) and (9.74b) into matrix form,

$$\begin{bmatrix} 1 & 1 \\ -ik_1 & -k_2 \end{bmatrix} (\mathbf{a}^+ - \mathbf{b}^+) + \begin{bmatrix} 1 & 1 \\ ik_1 & k_2 \end{bmatrix} (\mathbf{a}^- - \mathbf{b}^-) = 0 \tag{9.75}$$

With the origin chosen at the point where the external force and moment are applied, from (9.71a), (9.71b), and (9.73), the equilibrium conditions become

$$\bar{M} = EI\left[\left(-k_1^2 a_1^+ + k_2^2 a_2^+ - k_1^2 a_1^- + k_2^2 a_2^-\right) - \left(-k_1^2 b_1^+ + k_2^2 b_2^+ - k_1^2 b_1^- + k_2^2 b_2^-\right)\right]$$

$$\bar{Q} = \left[\left(-ik_1^3 EI + ik_1 \omega^2 \rho I\right) a_1^+ + \left(k_2^3 EI + k_2 \omega^2 \rho I\right) a_2^+ + \left(ik_1^3 EI - ik_1 \omega^2 \rho I\right) a_1^- + \left(-k_2^3 EI - k_2 \omega^2 \rho I\right) a_2^-\right]$$

$$- \left[\left(-ik_1^3 EI + ik_1 \omega^2 \rho I\right) b_1^+ + \left(k_2^3 EI + k_2 \omega^2 \rho I\right) b_2^+ + \left(ik_1^3 EI - ik_1 \omega^2 \rho I\right) b_1^- + \left(-k_2^3 EI - k_2 \omega^2 \rho I\right) b_2^-\right]$$

(9.76)

Putting Eq. (9.76) into matrix form,

$$\begin{bmatrix} -ik_1^3 EI + ik_1 \omega^2 \rho I & k_2^3 EI + k_2 \omega^2 \rho I \\ -k_1^2 & k_2^2 \end{bmatrix}\left(\mathbf{a}^+ - \mathbf{b}^+\right) + \begin{bmatrix} ik_1^3 EI - ik_1 \omega^2 \rho I & -k_2^3 EI - k_2 \omega^2 \rho I \\ -k_1^2 & k_2^2 \end{bmatrix}\left(\mathbf{a}^- - \mathbf{b}^-\right) = \begin{bmatrix} \bar{Q} \\ \bar{M} \end{bmatrix}$$

(9.77)

From Eqs. (9.75) and (9.77), the relationships between the injected bending waves and the externally applied transverse force and bending moment are

$$\mathbf{b}^+ - \mathbf{a}^+ = \mathbf{q} + \mathbf{m}, \quad \mathbf{a}^- - \mathbf{b}^- = \mathbf{q} - \mathbf{m}$$

(9.78a)

where the load vectors are

$$\mathbf{q} = \begin{bmatrix} -i/k_1 \\ -1/k_2 \end{bmatrix}\frac{\bar{Q}}{2EI\left(k_1^2 + k_2^2\right)}, \quad \mathbf{m} = \begin{bmatrix} 1 \\ -1 \end{bmatrix}\frac{\bar{M}}{2EI\left(k_1^2 + k_2^2\right)}$$

(9.78b)

Let $k_1 = k_2 = k$, which holds when the rotary inertia is assumed negligibly small, the force generated wave expressions by the Euler–Bernoulli bending vibration theory are obtained.

9.3.2 Shear and Timoshenko Bending Vibration Theories

For both the Timoshenko (for $\omega < \omega_c$) and Shear bending vibration theories, the bending deflection $y(x)$ and bending slope $\psi(x)$ are described by the same expressions of Eqs. (9.47a) and (9.47b), and the internal resistant shear force $V(x,t)$ and bending moment $M(x,t)$ are related to the bending deflection $y(x,t)$ and bending slope $\psi(x,t)$ by the same expressions of Eqs. (9.48a) and (9.48b) when following the sign convention defined in Figure 9.3. Consequently, the force generated wave expressions, based on both the Timoshenko (for $\omega < \omega_c$) and Shear bending vibration theories, are described by the same expressions.

Again, one must keep in mind that the values will not be the same because of the differences in the governing differential equations and wavenumbers between the Timoshenko and Shear bending vibration theories.

The continuity and equilibrium equations are obtained in Eqs. (9.69) and (9.70), respectively.

From Figure 9.6a and Eqs. (9.47a) and (9.47b),

$$y_- = a_1^+ e^{-ik_1 x} + a_2^+ e^{-k_2 x} + a_1^- e^{ik_1 x} + a_2^- e^{k_2 x}$$

(9.79a)

$$y_+ = b_1^+ e^{-ik_1 x} + b_2^+ e^{-k_2 x} + b_1^- e^{ik_1 x} + b_2^- e^{k_2 x}$$

(9.79b)

and

$$\psi_- = -iP a_1^+ e^{-ik_1 x} - N a_2^+ e^{-k_2 x} + iP a_1^- e^{ik_1 x} + N a_2^- e^{k_2 x}$$

(9.80a)

$$\psi_+ = -iP b_1^+ e^{-ik_1 x} - N b_2^+ e^{-k_2 x} + iP b_1^- e^{ik_1 x} + N b_2^- e^{k_2 x}$$

(9.80b)

From Eqs. (9.48a), (9.48b) and (9.70),

$$\bar{Q} = GA\kappa\left[\frac{\partial y_-(x,t)}{\partial x} - \psi_-(x,t)\right] - GA\kappa\left[\frac{\partial y_+(x,t)}{\partial x} - \psi_+(x,t)\right]$$

(9.81a)

$$\bar{M} = EI \frac{\partial \psi_-(x,t)}{\partial x} - EI \frac{\partial \psi_+(x,t)}{\partial x} \tag{9.81b}$$

With the origin chosen at the point where the external force and moment are applied, from Eqs. (9.69), (9.79a), (9.79b), (9.80a), and (9.80b), the continuity conditions become

$$a_1^+ + a_2^+ + a_1^- + a_2^- = b_1^+ + b_2^+ + b_1^- + b_2^- \tag{9.82a}$$

$$-iPa_1^+ - Na_2^+ + iPa_1^- + Na_2^- = -iPb_1^+ - Nb_2^+ + iPb_1^- + Nb_2^- \tag{9.82b}$$

Putting Eqs. (9.82a) and (9.82b) into matrix form,

$$\begin{bmatrix} 1 & 1 \\ -iP & -N \end{bmatrix} (\mathbf{a}^+ - \mathbf{b}^+) + \begin{bmatrix} 1 & 1 \\ iP & N \end{bmatrix} (\mathbf{a}^- - \mathbf{b}^-) = 0 \tag{9.83}$$

With the origin chosen at the point where the external force and moment are applied, from Eqs. (9.79a) to (9.81b), the equilibrium conditions become

$$i(P - k_1)a_1^+ + (N - k_2)a_2^+ - i(P - k_1)a_1^- - (N - k_2)a_2^- -$$
$$\left[i(P - k_1)b_1^+ + (N - k_2)b_2^+ - i(P - k_1)b_1^- - (N - k_2)b_2^- \right] = \frac{\bar{Q}}{GA\kappa} \tag{9.84a}$$

$$-k_1 Pa_1^+ + k_2 Na_2^+ - k_1 Pa_1^- + k_2 Na_2^- - (-k_1 Pb_1^+ + k_2 Nb_2^+ - k_1 Pb_1^- + k_2 Nb_2^-) = \frac{\bar{M}}{EI} \tag{9.84b}$$

Putting Eqs. (9.84a) and (9.84b) into matrix form,

$$\begin{bmatrix} -ik_1 + iP & -k_2 + N \\ -k_1 P & k_2 N \end{bmatrix} (\mathbf{a}^+ - \mathbf{b}^+) + \begin{bmatrix} ik_1 - iP & k_2 - N \\ -k_1 P & k_2 N \end{bmatrix} (\mathbf{a}^- - \mathbf{b}^-) = \begin{bmatrix} \bar{Q}/(GA\kappa) \\ \bar{M}/(EI) \end{bmatrix} \tag{9.85}$$

From Eqs. (9.83) and (9.85), the relationships between the injected bending waves and the externally applied transverse force and bending moment are

$$\mathbf{b}^+ - \mathbf{a}^+ = \mathbf{q} + \mathbf{m}, \ \mathbf{a}^- - \mathbf{b}^- = \mathbf{q} - \mathbf{m} \tag{9.86a}$$

where the load vectors are

$$\mathbf{q} = \begin{bmatrix} -iN \\ -P \end{bmatrix} \frac{\bar{Q}}{2GA\kappa(k_1 N - k_2 P)}, \ \mathbf{m} = \begin{bmatrix} 1 \\ -1 \end{bmatrix} \frac{\bar{M}}{2EI(k_1 P + k_2 N)} \tag{9.86b}$$

The Timoshenko bending vibration theory related expressions presented in this section are for $\omega < \omega_c$, which is the most common situation in audio frequency applications. For $\omega > \omega_c$, k_2 in the related expressions by the Timoshenko bending vibration theory needs to be replaced with ik_2, because of the transition of wave mode from a pair of decaying waves below the cut-off frequency to a pair of propagating waves above the cut-off frequency in k_2 related wave components.

9.4 Waves Generated by Externally Applied Point Force and Moment at a Free End

An external force and moment may be applied at a free end of a beam, as shown in Figure 9.7.

From the free body diagram of Figure 9.7, the externally applied force and moment are related to the internal resistant force and moment by the following equilibrium equations

$$\bar{Q} = V, \ \bar{M} = M \tag{9.87}$$

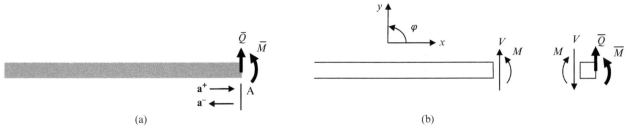

Figure 9.7 (a) Waves generated by external transverse force and bending moment applied at a free end and (b) free body diagram.

9.4.1 Rayleigh Bending Vibration Theory

Following the Rayleigh bending vibration theory and the sign convention defined in Figure 9.3, from Eqs. (9.23a), (9.23b), and (9.87),

$$\bar{Q} = -EI\frac{\partial^3 y(x,t)}{\partial x^3} + \rho I \frac{\partial^3 y(x,t)}{\partial x \partial t^2} \tag{9.88a}$$

$$\bar{M} = EI\frac{\partial^2 y(x,t)}{\partial x^2} \tag{9.88b}$$

Choosing the origin at the external force and moment applied end point, from Figure 9.7a and Eqs. (9.3), (9.4), (9.88a), and (9.88b),

$$\bar{Q} = \left(-ik_1^3 EI + ik_1\omega^2\rho I\right)a_1^+ + \left(k_2^3 EI + k_2\omega^2\rho I\right)a_2^+ + \left(ik_1^3 EI - ik_1\omega^2\rho I\right)a_1^- + \left(-k_2^3 EI - k_2\omega^2\rho I\right)a_2^-,$$
$$\bar{M} = EI\left(-k_1^2 a_1^+ + k_2^2 a_2^+ - k_1^2 a_1^- + k_2^2 a_2^-\right) \tag{9.89}$$

Putting Eq. (9.89) into matrix form, the end excitations generated waves by the Rayleigh bending vibration theory are

$$\mathbf{y}_E^+ \mathbf{a}^+ + \mathbf{y}_E^- \mathbf{a}^- = \mathbf{q}_F \tag{9.90a}$$

where $\mathbf{y}_E^+$ and $\mathbf{y}_E^-$ are matrices of the coefficients of the positive- and negative-going waves and $\mathbf{q}_F$ is the load vector, they are

$$\mathbf{y}_E^+ = \begin{bmatrix} -ik_1^3 EI + ik_1\omega^2\rho I & k_2^3 EI + k_2\omega^2\rho I \\ -k_1^2 EI & k_2^2 EI \end{bmatrix}, \mathbf{y}_E^- = \begin{bmatrix} ik_1^3 EI - ik_1\omega^2\rho I & -k_2^3 EI - k_2\omega^2\rho I \\ -k_1^2 EI & k_2^2 EI \end{bmatrix}, \text{ and}$$
$$\mathbf{q}_F = \begin{bmatrix} \bar{Q} \\ \bar{M} \end{bmatrix} \tag{9.90b}$$

or equivalently are

$$\mathbf{y}_E^+ = \begin{bmatrix} -ik_1^3 + ik_1\left(\dfrac{C_r}{C_b}\omega\right)^2 & k_2^3 + k_2\left(\dfrac{C_r}{C_b}\omega\right)^2 \\ -k_1^2 & k_2^2 \end{bmatrix}, \mathbf{y}_E^- = \begin{bmatrix} ik_1^3 - ik_1\left(\dfrac{C_r}{C_b}\omega\right)^2 & -k_2^3 - k_2\left(\dfrac{C_r}{C_b}\omega\right)^2 \\ -k_1^2 & k_2^2 \end{bmatrix}, \text{ and } \mathbf{q}_F = \begin{bmatrix} \bar{Q}/(EI) \\ \bar{M}/(EI) \end{bmatrix} \tag{9.90c}$$

Setting either $\rho I = 0$ in Eq. (9.90b) or equivalently $C_r = 0$ in Eq. (9.90c), the expressions of end force generated waves by the Euler–Bernoulli bending vibration theory are obtained. Recall that $k_1 = k_2$ by the Euler–Bernoulli bending vibration theory.

Under the special situation of $\bar{Q} = 0$ and $\bar{M} = 0$, which represents a free boundary, from Eq. (9.90c),

$$\begin{bmatrix} -ik_1^3 + ik_1\left(\dfrac{C_r}{C_b}\omega\right)^2 & k_2^3 + k_2\left(\dfrac{C_r}{C_b}\omega\right)^2 \\ -k_1^2 & k_2^2 \end{bmatrix}\mathbf{a}^+ + \begin{bmatrix} ik_1^3 - ik_1\left(\dfrac{C_r}{C_b}\omega\right)^2 & -k_2^3 - k_2\left(\dfrac{C_r}{C_b}\omega\right)^2 \\ -k_1^2 & k_2^2 \end{bmatrix}\mathbf{a}^- = \begin{bmatrix} 0 \\ 0 \end{bmatrix} \tag{9.91}$$

Because $\mathbf{a}^-$ and $\mathbf{a}^+$ are related by reflection matrix $\mathbf{r}$, that is, $\mathbf{a}^- = \mathbf{r}\mathbf{a}^+$, the reflection matrix is

$$\mathbf{r} = -\begin{bmatrix} ik_1^3 - ik_1\left(\dfrac{C_r}{C_b}\omega\right)^2 & -k_2^3 - k_2\left(\dfrac{C_r}{C_b}\omega\right)^2 \\ -k_1^2 & k_2^2 \end{bmatrix}^{-1}\begin{bmatrix} -ik_1^3 + ik_1\left(\dfrac{C_r}{C_b}\omega\right)^2 & k_2^3 + k_2\left(\dfrac{C_r}{C_b}\omega\right)^2 \\ -k_1^2 & k_2^2 \end{bmatrix} \tag{9.92}$$

which is equivalent to the reflection matrix of a free boundary by the Rayleigh bending vibration theory obtained in Eq. (9.29).

9.4.2 Shear and Timoshenko Bending Vibration Theories

Following the Shear and Timoshenko bending vibration theories and the sign convention defined in Figure 9.3, from Eqs. (9.48a), (9.48b), and (9.87),

$$\bar{Q} = GA\kappa\left[\frac{\partial y(x,t)}{\partial x} - \psi(x,t)\right] \tag{9.93a}$$

$$\bar{M} = EI\frac{\partial \psi(x,t)}{\partial x} \tag{9.93b}$$

From Figure 9.7a and Eqs. (9.47a) and (9.47b), the bending deflection and shear deformation by the Timoshenko (for $\omega < \omega_c$) and Shear bending vibration theories are

$$y(x) = a_1^+ e^{-ik_1 x} + a_2^+ e^{-k_2 x} + a_1^- e^{ik_1 x} + a_2^- e^{k_2 x} \tag{9.93c}$$

$$\psi(x) = -iPa_1^+ e^{-ik_1 x} - Na_2^+ e^{-k_2 x} + iPa_1^- e^{ik_1 x} + Na_2^- e^{k_2 x} \tag{9.93d}$$

where $P = k_1\left(1 - \dfrac{\omega^2}{k_1^2 C_s^2}\right)$ and $N = k_2\left(1 + \dfrac{\omega^2}{k_2^2 C_s^2}\right)$, as described in Eq. (9.15).

With the origin chosen at the point where the external force and moment are applied, from Eqs. (9.93a-d),

$$i(P - k_1)a_1^+ + (N - k_2)a_2^+ - i(P - k_1)a_1^- - (N - k_2)a_2^- = \frac{\bar{Q}}{GA\kappa} \tag{9.94a}$$

$$-k_1 Pa_1^+ + k_2 Na_2^+ - k_1 Pa_1^- + k_2 Na_2^- = \frac{\bar{M}}{EI} \tag{9.94b}$$

Putting Eqs. (9.94a) and (9.94b) into matrix form, the end excitation generated waves by the Timoshenko (for $\omega < \omega_c$) and Shear bending vibration theories are

$$\mathbf{y}_E^+\mathbf{a}^+ + \mathbf{y}_E^-\mathbf{a}^- = \mathbf{q}_F \tag{9.95a}$$

where $\mathbf{y}_E^+$ and $\mathbf{y}_E^-$ are the coefficient matrices of positive- and negative-going waves and $\mathbf{q}_F$ is the load vector, they are

$$\mathbf{y}_E^+ = \begin{bmatrix} -ik_1 + iP & -k_2 + N \\ -k_1 P & k_2 N \end{bmatrix}, \quad \mathbf{y}_E^- = \begin{bmatrix} ik_1 - iP & k_2 - N \\ -k_1 P & k_2 N \end{bmatrix}, \quad \text{and } \mathbf{q}_F = \begin{bmatrix} \bar{Q}/(GA\kappa) \\ \bar{M}/(EI) \end{bmatrix} \tag{9.95b}$$

Under the special situation of $\bar{Q} = 0$ and $\bar{M} = 0$, which represents a free boundary, from Eqs. (9.95a) and (9.95b)

$$\begin{bmatrix} -ik_1 + iP & -k_2 + N \\ -k_1 P & k_2 N \end{bmatrix}\mathbf{a}^+ + \begin{bmatrix} ik_1 - iP & k_2 - N \\ -k_1 P & k_2 N \end{bmatrix}\mathbf{a}^- = \begin{bmatrix} 0 \\ 0 \end{bmatrix} \tag{9.96}$$

Because $\mathbf{a}^-$ and $\mathbf{a}^+$ are related by reflection matrix $\mathbf{r}$, that is, $\mathbf{a}^- = \mathbf{r}\mathbf{a}^+$, the reflection matrix is

$$\mathbf{r} = -\begin{bmatrix} ik_1 - iP & k_2 - N \\ -k_1 P & k_2 N \end{bmatrix}^{-1} \begin{bmatrix} -ik_1 + iP & -k_2 + N \\ -k_1 P & k_2 N \end{bmatrix} \qquad (9.97)$$

which, as expected, is equivalent to the reflection matrix of a free boundary by the Timoshenko and Shear bending vibration theories obtained in Eq. (9.52a).

The Timoshenko bending vibration theory related expressions presented in this section are for $\omega < \omega_c$, which is the most common situation in audio frequency applications. When $\omega > \omega_c$, k_2 in the related expressions by the Timoshenko bending vibration theory needs to be replaced with ik_2, because of the transition of wave mode from a pair of decaying waves below the cut-off frequency to a pair of propagating waves above the cut-off frequency in k_2 related wave components.

9.5 Free and Forced Vibration Analysis

With the availability of the propagation relationships, boundary reflections, and force generated waves, both free and forced vibration analysis become concise and systematic, regardless of the vibration theories adopted.

9.5.1 Free Vibration Analysis

Figure 9.8 shows a uniform beam with Boundaries A and B. Denoting the incident and reflected waves at Boundaries A and B using $\mathbf{a}^-$, $\mathbf{a}^+$, $\mathbf{b}^+$, and $\mathbf{b}^-$, the following reflection and propagation relationships exist:

$$\mathbf{a}^+ = \mathbf{r}_A \mathbf{a}^-, \ \mathbf{b}^- = \mathbf{r}_B \mathbf{b}^+, \ \mathbf{a}^- = \mathbf{f}(L)\mathbf{b}^-, \ \mathbf{b}^+ = \mathbf{f}(L)\mathbf{a}^+ \qquad (9.98)$$

where $\mathbf{r}_A$ and $\mathbf{r}_B$ are the reflection matrices at Boundaries A and B, respectively, and $\mathbf{f}(L)$ is the propagation matrix for a distance L.

Free vibration problem may be solved by assembling the propagation and reflection relationships described in Eq. (9.98) into a matrix form, as described in Chapter 3. Alternatively, free vibration responses may be obtained by solving Eq. (9.98) directly,

$$\left[\mathbf{r}_A \mathbf{f}(L) \mathbf{r}_B \mathbf{f}(L) - \mathbf{I} \right] \mathbf{a}^+ = 0 \qquad (9.99)$$

where $\mathbf{I}$ denotes an identity matrix. For non-trivial solution, the determinant of the coefficient matrix of Eq. (9.99) equals zero,

$$\left| \mathbf{r}_A \mathbf{f}(L) \mathbf{r}_B \mathbf{f}(L) - \mathbf{I} \right| = 0 \qquad (9.100)$$

Equation (9.100) is the characteristic equation, from which the natural frequencies of the beam can be found.

9.5.2 Forced Vibration Analysis

Figure 9.9 shows a uniform beam with a point force and a moment applied at an arbitrary point C. The external excitation divides the structure into two regions, namely Region 1 and 2. The incoming and outgoing waves at the external excitation applied point C are denoted using $\mathbf{c}_1^+$, $\mathbf{c}_1^-$, $\mathbf{c}_2^+$, and $\mathbf{c}_2^-$.

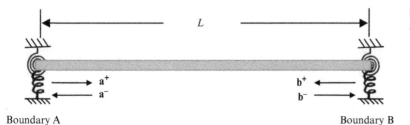

Figure 9.8 Wave propagation and reflection in a uniform beam.

Boundary A   Boundary B

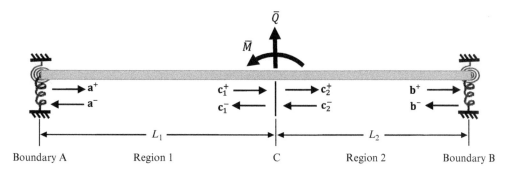

Figure 9.9 Waves in a uniform beam with external transverse force and bending moment applied on the span.

At Boundaries A and B,

$$\mathbf{a}^+ = \mathbf{r}_A\mathbf{a}^-, \ \mathbf{b}^- = \mathbf{r}_B\mathbf{b}^+ \tag{9.101}$$

where $\mathbf{r}_A$ and $\mathbf{r}_B$ are the reflection matrices at the boundaries.
The wave propagation relationships are

$$\mathbf{c}_1^+ = \mathbf{f}(L_1)\,\mathbf{a}^+, \ \mathbf{a}^- = \mathbf{f}(L_1)\mathbf{c}_1^-, \ \mathbf{b}^+ = \mathbf{f}(L_2)\mathbf{c}_2^+, \ \mathbf{c}_2^- = \mathbf{f}(L_2)\,\mathbf{b}^- \tag{9.102}$$

where $\mathbf{f}(L_1)$ and $\mathbf{f}(L_2)$ are propagation matrices for a distance L_1 and L_2, respectively.

From Eq. (9.78a) or (9.86a), the waves at the point where the external force and moment are applied are related to the external force and moment,

$$\mathbf{c}_2^+ - \mathbf{c}_1^+ = \mathbf{q} + \mathbf{m}, \ \ \mathbf{c}_1^- - \mathbf{c}_2^- = \mathbf{q} - \mathbf{m} \tag{9.103}$$

where the load vectors $\mathbf{q}$ and $\mathbf{m}$ by the Rayleigh bending vibration theory are given in Eq. (9.78b), and by the Timoshenko (for $\omega < \omega_c$) and Shear bending vibration theories are given in Eq. (9.86b).

In addition, $\mathbf{c}_1^+$ is related to $\mathbf{c}_1^-$ by reflection at Boundary A and propagation through Region 1. Similarly, $\mathbf{c}_2^+$ is related to $\mathbf{c}_2^-$ by reflection at Boundary B and propagation through Region 2. Matrices ρ_1 and ρ_2 are defined to relate the amplitudes of the incoming and outgoing waves on the left and right side of the point where the external force and moment are applied,

$$\mathbf{c}_1^+ = \rho_1\mathbf{c}_1^-, \ \mathbf{c}_2^- = \rho_2\mathbf{c}_2^+ \tag{9.104}$$

From Eq. (9.101) to (9.104),

$$\rho_1 = \mathbf{f}(L_1)\mathbf{r}_A\mathbf{f}(L_1), \ \rho_2 = \mathbf{f}(L_2)\mathbf{r}_B\mathbf{f}(L_2) \tag{9.105}$$

Forced vibrations can be solved by assembling the wave relationships from Eqs. (9.101) to (9.103) into matrix form, as described in Chapter 3. Alternatively, forced vibrations can be found by solving for the wave components at Point C, where the external excitations are applied, from Eqs. (9.103) and (9.104),

$$\mathbf{c}_1^+ = \rho_1\left(\mathbf{I} - \rho_1\rho_2\right)^{-1}\!\left[(\rho_2 + \mathbf{I})\mathbf{q} + (\rho_2 - \mathbf{I})\mathbf{m}\right], \ \mathbf{c}_1^- = \left(\mathbf{I} - \rho_1\rho_2\right)^{-1}\!\left[(\rho_2 + \mathbf{I})\mathbf{q} + (\rho_2 - \mathbf{I})\mathbf{m}\right]$$
$$\mathbf{c}_2^+ = \left(\mathbf{I} - \rho_1\rho_2\right)^{-1}\!\left[(\rho_1 + \mathbf{I})\mathbf{q} + (\mathbf{I} - \rho_1)\mathbf{m}\right], \ \mathbf{c}_2^- = \rho_2\left(\mathbf{I} - \rho_1\rho_2\right)^{-1}\!\left[(\rho_1 + \mathbf{I})\mathbf{q} + (\mathbf{I} - \rho_1)\mathbf{m}\right] \tag{9.106}$$

The deflection of any point on the beam can be found. For example, the deflection of a point in Region 1 that is a distance x from Point C is

$$y = \begin{bmatrix} 1 & 1 \end{bmatrix}\mathbf{f}(x)\mathbf{c}_1^- + \begin{bmatrix} 1 & 1 \end{bmatrix}\mathbf{f}(-x)\mathbf{c}_1^+ \tag{9.107}$$

Similarly, the deflection of a point in Region 2 that is a distance x from Point C is

$$y = \begin{bmatrix} 1 & 1 \end{bmatrix}\mathbf{f}(x)\mathbf{c}_2^+ + \begin{bmatrix} 1 & 1 \end{bmatrix}\mathbf{f}(-x)\mathbf{c}_2^- \tag{9.108}$$

Figure 9.10 Waves in a uniform beam with external transverse force and bending moment applied at a free end.

When external forces and moments are applied at a free end of a beam, as illustrated in Figure 9.10, the wave relationships are as follows.

At Boundary A,

$$\mathbf{a}^+ = \mathbf{r}_A \, \mathbf{a}^- \qquad (9.109)$$

where $\mathbf{r}_A$ is the reflection matrix at Boundary A.

The propagation relationships are

$$\mathbf{b}^+ = \mathbf{f}(L) \, \mathbf{a}^+, \; \mathbf{a}^- = \mathbf{f}(L) \, \mathbf{b}^- \qquad (9.110)$$

where $\mathbf{f}(L)$ is the propagation matrix for a distance L.

At the external excitations applied boundary B, from Eq. (9.90a) or (9.95a),

$$\mathbf{y}_E^+ \mathbf{b}^+ + \mathbf{y}_E^- \mathbf{b}^- = \mathbf{q}_F \qquad (9.111)$$

where the coefficient matrices $\mathbf{y}_E^+$ and $\mathbf{y}_E^-$, and load vector $\mathbf{q}_F$ by the Rayleigh bending vibration theory are given in Eq. (9.90b) or (9.90c), and by the Timoshenko (for $\omega < \omega_c$) and Shear bending vibration theories are given in Eq. (9.95b).

In addition, $\mathbf{b}^+$ is related to $\mathbf{b}^-$ by reflection at Boundary A and propagation over the length of the beam. Matrix ρ is defined to relate the incoming and outgoing waves at the end point where the external force and moment are applied,

$$\mathbf{b}^+ = \rho \mathbf{b}^- \qquad (9.112)$$

From Eqs. (9.109), (9.110), and (9.112),

$$\rho = \mathbf{f}(L)\mathbf{r}_A\mathbf{f}(L) \qquad (9.113)$$

Forced vibrations can be solved by assembling the wave relationships from Eqs. (9.109) to (9.111) into matrix form, as described in Chapter 3. Alternatively, forced vibrations can be found by solving for the wave components at Point B from Eqs. (9.111) and (9.112),

$$\mathbf{b}^+ = \rho \left(\mathbf{y}_E^+ \rho + \mathbf{y}_E^- \right)^{-1} \mathbf{q}_F,$$
$$\mathbf{b}^- = \left(\mathbf{y}_E^+ \rho + \mathbf{y}_E^- \right)^{-1} \mathbf{q}_F \qquad (9.114)$$

With the availability of the responses of the wave components, the deflection of any point on the beam can be obtained. For example, the deflection of a point located a distance x from Boundary B in Figure 9.10 is

$$y = \begin{bmatrix} 1 & 1 \end{bmatrix} \mathbf{f}(x)\mathbf{b}^- + \begin{bmatrix} 1 & 1 \end{bmatrix} \mathbf{f}(-x)\mathbf{b}^+ \qquad (9.115)$$

9.6 Numerical Examples and Experimental Studies

Free and forced bending vibrations of the same uniform steel beam with free-free boundaries studied in Chapters 2 and 3 are analyzed based on the various bending vibration theories. This allows analytical results to be compared with the experimental results. The material and geometrical properties of the steel beam are: the Young's modulus E is $198.87\,GN/m^2$, shear modulus G is $77.5\,GN/m^2$, Poisson's ratio ν is 0.30, and volume mass density ρ is $7664.5\,kg/m^3$. The shear coefficient κ

is related to the Poisson's ratio ν by $\kappa = 10(1+\nu)/(12+11\nu)$. The cross section of the beam is a rectangular shape whose dimensions are 25.4 *mm* by 12.7*mm* (or 1.0 *in* by 0.5 *in*). The length of the beam is 914.4 *mm* (or 3.0 *ft*).

Recall that two tests were conducted, with an impact force applied on the beam using a Brüel & Kjær Type 8202 impact hammer, along the direction normal to the plane formed by the beam and the wires. In the first test, the thickness of the beam is 12.7 *mm* (or 0.5 *in*); while in the second test, the thickness of the beam is 25.4 *mm* (or 1.0 *in*), that is, by rotating the beam 90°. Vibrations were measured using a Brüel & Kjær Type 4397 accelerometer. The impact force was applied at 489.0 *mm* measured from a free end, and the accelerometer is placed at 593.9 *mm* measured from the same free end. In the experiments, the frequency span and frequency resolution were 6400 *Hz* and 1 *Hz*, respectively. The inertance frequency responses were obtained using a PULSE data acquisition system, as described in Section 3.5.

Figures 9.11 to 9.13 present the responses with the width and thickness of the beam being 25.4 *mm* and 12.7 *mm*, respectively. MATLAB scripts for free and forced vibration analysis by the Timoshenko bending vibration theory are available in Section 9.7.

Figures 9.11 shows the magnitude, real part, and imaginary part of the two wavenumbers predicted by the Euler–Bernoulli, Rayleigh, Shear, and Timoshenko bending vibration theories. The positive and negative wavenumber values represent the fact that waves in the beam travel in both the positive and negative directions. Regardless of the bending vibration theory

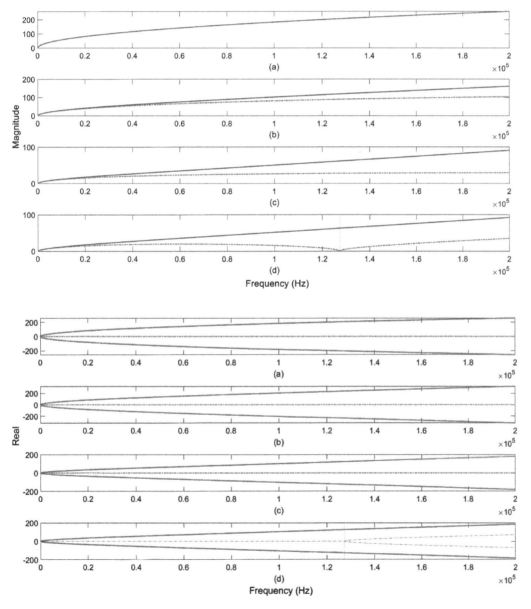

Figure 9.11 Magnitude, real part, and imaginary part of the two wavenumbers by the (a) Euler-Bernoulli, (b) Rayleigh, (c) Shear, and (d) Timoshenko bending vibration theories, with width and thickness being 25.4 *mm* and 12.7 *mm*, respectively.

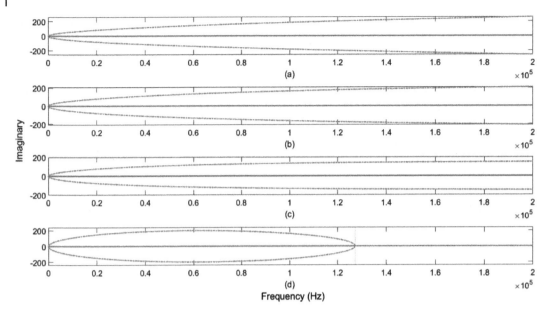

Figure 9.11 (Cont'd)

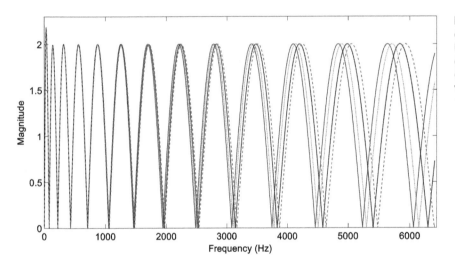

Figure 9.12 Magnitudes of the characteristic polynomial of the beam based on the Timoshenko (___), Shear (-.-.), Rayleigh (...), and Euler–Bernoulli (---) bending vibration theories, with width and thickness being 25.4 *mm* and 12.7 *mm*, respectively.

adopted, the bending waves are dispersive, in which the phase velocity ω/k is frequency dependent. Figure 9.11 shows that, in the absence of damping, the Euler–Bernoulli, Rayleigh, Shear bending vibration theories predict that one pair of wavenumbers is always real and the other pair is always imaginary. The pair of real wavenumbers corresponds to positive- and negative-going propagating waves, and the pair of imaginary wavenumbers corresponds to positive- and negative-going decaying waves.

The Timoshenko bending vibration theory also predicts that one pair of wavenumbers is always real. However, the second pair can be imaginary or real according to the Timoshenko bending vibration theory. A wave mode transition exists at a cut-off frequency ω_c given by C_s/C_r, which is marked on the plot using a vertical line. Below the cut-off frequency ω_c, the second pair of wavenumbers is imaginary; while above the cut-off frequency ω_c, this pair of wavenumbers becomes real. This indicates that below the cut-off frequency ω_c, a pair of positive- and negative-going propagating waves and a pair of positive- and negative-going decaying waves exist in the beam, while above the cut-off frequency ω_c, there are two pairs of propagating waves. In audio frequency applications, the former case is overwhelmingly the most common because the cut-off frequency is usually quite high, as shown in the example plots.

Figure 9.12 shows the dB magnitudes of the characteristic polynomial of the beam based on the Euler–Bernoulli, Rayleigh, Shear, and Timoshenko bending vibration theories. The local minima correspond to the natural frequencies. At lower frequencies, the natural frequencies predicted by the four bending vibration theories agree well with each other. However, discrepancies start to appear and become more significant as frequency moves higher.

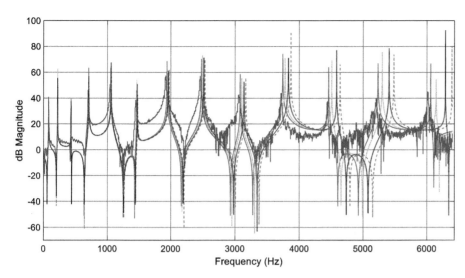

Figure 9.13 Overlaid experimental and analytical inertance frequency responses by the Timoshenko (___), Shear (-·-·), Rayleigh (...), and Euler–Bernoulli (---) bending vibration theories, with width and thickness being 25.4 *mm* and 12.7 *mm,* respectively.

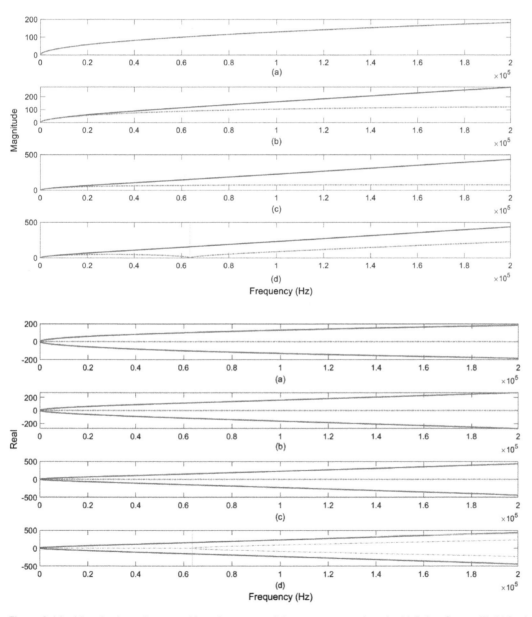

Figure 9.14 Magnitude, real part, and imaginary part of the two wavenumbers by (a) Euler–Bernoulli, (b) Rayleigh, (c) Shear, and (d) Timoshenko bending vibration theories, with width and thickness being 12.7 *mm* and 25.4 *mm,* respectively.

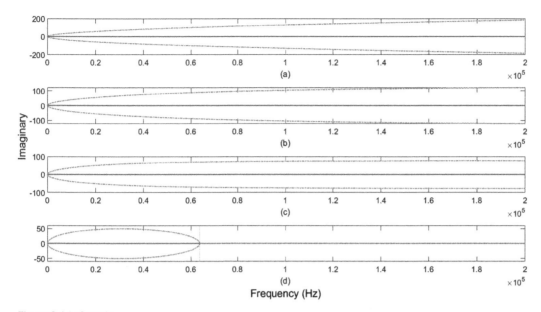

Figure 9.14 (Cont'd)

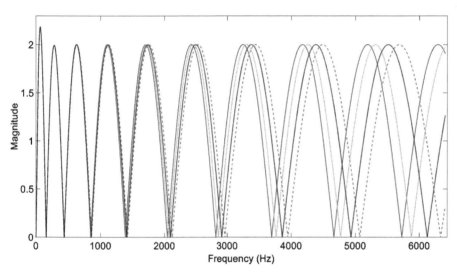

Figure 9.15 Magnitudes of the characteristic polynomial of the beam based on the Timoshenko (___), Shear (-.-.), Rayleigh (...), and Euler–Bernoulli (---) bending vibration theories, with width and thickness being 12.7 mm and 25.4 mm, respectively.

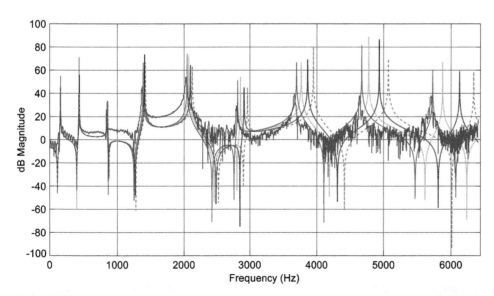

Figure 9.16 Overlaid experimental and analytical inertance frequency responses by the Timoshenko (___), Shear (-.-.), Rayleigh (...), and Euler–Bernoulli (---) bending vibration theories, with width and thickness being 12.7 mm and 25.4 mm, respectively.

Figure 9.13 shows the overlaid experimental and analytical inertance frequency responses (Acceleration/Force) by the Euler–Bernoulli, Rayleigh, Shear, and Timoshenko bending vibration theories. The experimental and analytical results from all four bending vibration theories agree well in the lower frequency range. As the frequency becomes higher, the Timoshenko bending vibration theory is the most accurate among the four bending vibration theories. The other three bending vibration theories, namely, the Euler–Bernoulli, Rayleigh, and Shear bending vibration theories, are shown to overpredict the natural frequencies at higher frequencies. This highlights the importance of taking into account both the rotary inertia and shear deformation at higher frequencies. The discrepancies in magnitudes between the experimental results and the responses obtained by various bending vibration theories are because of the structural damping being neglected in the analytical study.

Similar analytical and experimental responses are obtained by rotating the steel beam 90 degrees about its longitudinal axis, so that the width and thickness of the beam are $12.7\,mm$ and $25.4\,mm$, respectively. The responses are presented in Figures 9.14 to 9.16, which correspond to Figures 9.11 to 9.13, respectively. The same discussion and observations from above apply, in terms of wavenumbers, propagating/decaying nature of waves, and the accuracy of the bending vibration theories in predicting the natural frequencies of a structure. In summary, at lower frequencies, all four bending vibration theories work well. The Euler–Bernoulli bending vibration theory is often chosen because of its simplicity. However, at higher frequencies, the effects of rotary inertia and shear deformation become dominant, and the Timoshenko bending vibration theory must be adopted.

9.7 MATLAB Scripts

Script 1: Symbolic scripts for reflection matrix at a rigid mass block attached boundary by the Timoshenko and Shear bending vibration theories

```
clc, close all, clear all

syms P N k1 k2 k GAk EI h m Jm omega a1plus a2plus a1minus a2minus x real

y=a1plus*exp(-i*k1*x)+a2plus*exp(-k2*x)+a1minus*exp(i*k1*x)+a2minus*exp(k2*x)

phi=-i*P*a1plus*exp(-i*k1*x)-N*a2plus*exp(-k2*x)+i*P*a1minus*exp(i*k1*x)+N*a2minus*exp(k2*x)

y1=diff(y,x)
y2=diff(y,x,2)
y3=diff(y,x,3)

phi1=diff(phi,x)

phim=phi
ym=y+(h/2)*phim
V=GAk*(y1-phi)
M=EI*phi1

Expression1x=-V-m*(-omega^2)*ym
Expression2x=-M+V*(h/2)-Jm*(-omega^2)*phim

Expression1=subs(Expression1x,x,0)
Expression2=subs(Expression2x,x,0)

a1plusCoeff1=subs(Expression1, {a1plus a2plus a1minus a2minus}, {1 0 0 0})
a2plusCoeff1=subs(Expression1, {a1plus a2plus a1minus a2minus}, {0 1 0 0})
a1minusCoeff1=subs(Expression1, {a1plus a2plus a1minus a2minus}, {0 0 1 0})
a2minusCoeff1=subs(Expression1, {a1plus a2plus a1minus a2minus}, {0 0 0 1})
```

```
a1plusCoeff2=subs(Expression2, {a1plus a2plus a1minus a2minus}, {1 0 0 0})
a2plusCoeff2=subs(Expression2, {a1plus a2plus a1minus a2minus}, {0 1 0 0})
a1minusCoeff2=subs(Expression2, {a1plus a2plus a1minus a2minus}, {0 0 1 0})
a2minusCoeff2=subs(Expression2, {a1plus a2plus a1minus a2minus}, {0 0 0 1})

AminusCoeff=[a1minusCoeff1, a2minusCoeff1;a1minusCoeff2,a2minusCoeff2]
AplusCoeff=[a1plusCoeff1, a2plusCoeff1;a1plusCoeff2,a2plusCoeff2]
rEndMass=simplify(-inv(AminusCoeff)*(AplusCoeff))
```

Script 2: MATLAB scripts for free and forced vibration analysis by the Timoshenko bending vibration theory

```
clc; clear; close all;

%EITHER "in plane"
h=2.54/100        %Beam thickness
b=0.5*2.54/100    %Beam width

%OR "out of plane"
% h=0.5*2.54/100   %Beam thickness
% b=2.54/100       %Beam width

%Beam length
L=36.00*2.54/100;

%External excitation applied point
L1=19.25*2.54/100
L2=L-L1;

%Sensor to view responses (Pay attention to relative locations between sensor and external force)
x1sensor=23.38*2.54/100;   %Experimental observation point

%Material properties of the steel beam
E=198.87*10^9;
p=7664.5
v=0.3;
K=10*(1+v)/(12+11*v);
G=77.5*10^9;

A=b*h;
I=b*h^3/12;
EA=E*A;
EI=E*I;
KAG=K*A*G

Cb=sqrt(E*I/p/A);
Cs=sqrt(G*K*A/p/A);
Cr=sqrt(p*I/p/A);

%The cutoff frequency
wc=Cs/Cr
fc=wc/(2*pi)

%Frequency range in Hz
```

```matlab
f1 = 0;        %Starting frequency in Hz
f2 = 6400;     %Ending frequency in Hz
stepsize=0.1

freq = f1:stepsize:f2;

for n = 1:length(freq)

w=freq(n)*2*pi;
k1=abs(sqrt(0.5*((1/Cs)^2+(Cr/Cb)^2)*w^2+sqrt(w^2/Cb^2+0.25*((1/Cs)^2-(Cr/Cb)^2)^2*w^4)));
k2=abs(sqrt(0.5*((1/Cs)^2+(Cr/Cb)^2)*w^2-sqrt(w^2/Cb^2+0.25*((1/Cs)^2-(Cr/Cb)^2)^2*w^4)));

%%%%%%%%%%%%%%%%
%When w<wc, k2=k2
%When w>wc, k2=ik2
%%%%%%%%%%%%%%%%%%

if w<wc
k2=k2;
else
k2=i*k2;
end

P=-(w^2-Cs^2*k1^2)/(k1*Cs^2);
N=(w^2+Cs^2*k2^2)/(k2*Cs^2);

%The propagation matrices
fL1 = diag([exp(-i*k1*L1) exp(-k2*L1)]);
fL2 = diag([exp(-i*k1*L2) exp(-k2*L2)]);
fL = diag([exp(-i*k1*L) exp(-k2*L)]);
fx1FromG2=diag([exp(-i*k1*(x1sensor-L1)) exp(-k1*(x1sensor-L1))]);

%The reflection matrices

%At a pinned (or simply-supported) end
xs=[1,1;k1*P,-k2*N];
ys=[-1,-1;-k1*P,k2*N];
rsimple=inv(xs)*ys;

%At a clamped end
xc=[1,1;-i*P,-N];
yc=[-1,-1;-i*P,-N];
rclamped=inv(xc)*yc;

%At a free end
xf=[k1*P,-k2*N;i*k1-i*P,k2-N];
yf=[-k1*P,k2*N;i*k1-i*P,k2-N];
rfree=inv(xf)*yf;

%Boundaries of the Timoshenko beam
rA=rfree;
rB=rfree;
```

%Forming the characteristic matrix

BigA=rA*fL*rB*fL-eye(2);

%Force generated waves by the Timoshenko bending vibration theory
ShearForce=1;
Moment=0;
qforce=(ShearForce/(2*KAG*(k2*P-k1*N)))*[i*N;P];
mforce=(Moment/(2*EI*(k1*P+k2*N)))*[1;-1];

%Forced response by solving the wave components at force applied point directly

rho1=fL1*rA*fL1;
rho2=fL2*rB*fL2;

g1plus=rho1*inv(eye(2)-rho1*rho2)*((rho2+eye(2))*qforce+(rho2-eye(2))*mforce);
g1minus=inv(eye(2)-rho1*rho2)*((rho2+eye(2))*qforce+(rho2-eye(2))*mforce);
g2plus=inv(eye(2)-rho1*rho2)*((rho1+eye(2))*qforce+(eye(2)-rho1)*mforce);
g2minus=rho2*inv(eye(2)-rho1*rho2)*((rho1+eye(2))*qforce+(eye(2)-rho1)*mforce);

Y1sensor(n)=[1 1]*fx1FromG2*g2plus+[1 1]*inv(fx1FromG2)*g2minus;

Delta0(n)=det(BigA);

end

figure(1),plot(freq,abs(Delta0),':b','LineWidth',2)
xlabel('Frequency (Hz)'),ylabel('Linear Magnitude'),hold on

figure(2), plot(freq,real(Delta0),':b','LineWidth',2),hold on
plot(freq,imag(Delta0),'r','LineWidth',2),hold on
yline(0),hold on
xlabel('Frequency (Hz)'),ylabel('Amplitude')
legend('Real part','Imaginary part')

figure(3), plot(freq,20*log10(((2*pi*freq).^2).*abs(Y1sensor)),'g','LineWidth',2); hold on
xlabel('Frequency (Hz)')
ylabel('dB Magnitude of Inertance Frequency Response')

figure(4), plot(freq,20*log10(abs(Y1sensor)),'g','LineWidth',2); hold on
xlabel('Frequency (Hz)')
ylabel('dB Magnitude of Receptance Frequency Response')

References

Mei C. and Mace B.R. Wave Reflection and Transmission in Timoshenko Beams and Wave Analysis of Timoshenko Beam Structures, *ASME Journal of Vibration and Acoustics*, 127(4) 382–394 (2005).
Meirovitch L. *Principles and Techniques of Vibrations*, Prentice Hall, Englewood Cliffs, New Jersey (1997).

Homework Project

Free and Forced Wave Analysis of In-Plane Bending Vibrations in a Uniform Cantilever Beam with a Rigid Body Mass Attachment at Its Free End.

Consider in-plane bending vibration, that is, bending vibrations in the x–y plane, of the steel beam shown in the figure below. This uniform steel beam has one end clamped and the other end attached to a rigid mass block whose mass is $m = 1.7\,kg$, width is $h = 80\,mm$, and mass moment of inertia about an axis normal to the x–y plane and passing its center of mass is $J_m = 0.001\,kgm^2$.

  The material and geometrical properties of the steel beam are: the Young's modulus is $E = 198.87\,GN/m^2$ and mass density is $\rho = 7664.5\,kg/m^3$. The cross section of the beam is a rectangular shape whose thickness is $t = 12.7\,mm$ and depth is $d = 25.4\,mm$. The length of the beam is $L = 914.4\,mm$.

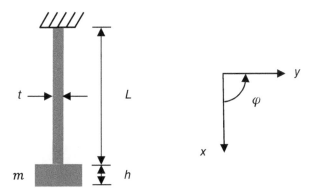

This structure is the same as that of homework project from Chapter 3, which has been analyzed using the Euler–Bernoulli bending vibration theory. Analyze the in-plane bending vibrations of the above structure using the advanced Timoshenko bending vibration theory. The shear modulus of the beam is $G = 77.5\,GN/m^2$, the Poisson's ratio is $\nu = 0.30$, and the shear coefficient κ is related to the Poisson's ratio ν by $\kappa = 10(1+\nu)/(12+11\nu)$.

Complete the following tasks based on the Timoshenko bending vibration theory:

i) Plot the magnitude of the characteristic polynomial, read the lowest five natural frequencies from the plot, and compare the natural frequency values with those obtained from Chapter 3 based on the Euler–Bernoulli bending vibration theory;

ii) Present the receptance frequency response of the beam to include the lowest five modes, with an external excitation force applied at 400 mm and with the response observed at 700 mm, both measured from the clamped end. Compare this forced response with that obtained from Chapter 3 based on the Euler–Bernoulli bending vibration theory;

iii) Are there discrepancies in the free and forced responses obtained from the Timoshenko and Euler–Bernoulli bending vibration theories? Why?

10

Longitudinal Waves in Beams Based on Various Vibration Theories

There exist four main engineering theories for longitudinally vibrating beams: the Elementary, Love, Mindlin–Herrmann, and Three-mode theories. The governing equations of motion vary from one partial differential equation to three coupled partial differential equations.

The Elementary longitudinal vibration theory, covered in Chapter 2, assumes that the axial deformations are the same at all points on a cross section and that there are no transverse deflections. The Love longitudinal vibration theory assumes that each material point of the beam has a transverse velocity. As a result, while the strain energy is the same as that of the Elementary longitudinal vibration theory, the kinetic energy contains additional terms. The Mindlin–Herrmann longitudinal vibration theory takes into account the lateral displacements by considering the Poisson effect between the axial and transverse deformations. It assumes uniform distribution of the axial displacement in the cross section of the beam. The Three-mode longitudinal vibration theory assumes a parabolic distribution of the longitudinal displacement along the cross section. It models the longitudinal displacement, transverse contraction, and parabolic distribution of the axial displacement along the cross section.

In this chapter, these various longitudinal vibration theories are introduced, and free and forced vibrations are analyzed from the wave vibration standpoint (Mei 2015).

10.1 The Governing Equations and the Propagation Relationships

10.1.1 Love Longitudinal Vibration Theory

The equation of motion for free longitudinal vibrations of a uniform beam according to the Love theory is (Doyle 1989)

$$EA\frac{\partial^2 u(x,t)}{\partial x^2}+\nu^2\rho J\frac{\partial^2}{\partial x^2}\left(\frac{\partial^2 u(x,t)}{\partial t^2}\right)-\rho A\frac{\partial^2 u(x,t)}{\partial t^2}=0 \tag{10.1}$$

where $u(x,t)$ is the longitudinal deflection of the centerline of the beam, x is the position along the beam axis, and t is the time. ν is the Poisson's ratio, E is the Young's modulus, ρ is the volume mass density, J is the polar moment of inertia of the cross section, and A is the cross-sectional area of the beam.

The middle term on the left side of Eq. (10.1) comes from the additional term in the kinetic energy based on the assumption of the Love longitudinal vibration theory, which states that each material point of the beam has a transverse velocity. The Elementary longitudinal vibration theory is recovered by neglecting this term.

Assuming time harmonic motion and using separation of variables, the solution to Eq. (10.1) can be written in the form $u(x,t)=u_0 e^{-ikx}e^{i\omega t}$, where u_0 is the amplitude of the longitudinal deflection of the centerline of the beam, ω is the circular frequency, k is the wavenumber, and i is the imaginary unit. Substituting this expression for $u(x,t)$ into Eq. (10.1) gives the dispersion equation, from which the longitudinal wavenumber k is obtained,

$$k=\omega\sqrt{\frac{\rho A}{EA-\nu^2\rho J\omega^2}} \tag{10.2}$$

Mechanical Wave Vibrations: Analysis and Control, First Edition. Chunhui Mei.
© 2023 Chunhui Mei. Published 2023 by John Wiley & Sons Ltd.
Companion Website: www.wiley.com/go/Mei/MechanicalWaveVibrations

which is a function of frequency ω, as well as the material and geometrical properties of the beam. Note that a critical frequency ω_{cl} exists at which the denominator in Eq. (10.2) becomes zero

$$\omega_{cl} = \sqrt{\frac{EA}{\nu^2 \rho J}} \tag{10.3}$$

With time dependence $e^{i\omega t}$ suppressed, the solution to Eq. (10.1) can be written as

$$u(x) = a^+ e^{-ikx} + a^- e^{ikx} \tag{10.4}$$

where superscripts $+$ and $-$ in wave amplitude denote positive- and negative-going waves, respectively.

Consider two points A and B on a uniform beam that are located a distance x apart, as shown in Figure 10.1. Waves propagate from one point to the other, with the propagation relationships determined by the appropriate wavenumber. The positive- and negative-going wave components at Points A and B are related by

$$a^- = f(x)b^-, \text{ and } b^+ = f(x)a^+ \tag{10.5}$$

where

$$f(x) = e^{-ikx} \tag{10.6}$$

is the propagation coefficient for a distance x.

10.1.2 Mindlin–Herrmann Longitudinal Vibration Theory

The equations of motion for free longitudinal vibrations of a uniform beam according to the Mindlin–Herrmann theory are (Krawczuk et. al. 2006)

$$(2\mu + \lambda)A \frac{\partial^2 u(x,t)}{\partial x^2} + \lambda A \frac{\partial \psi(x,t)}{\partial x} = \rho A \frac{\partial^2 u(x,t)}{\partial t^2},$$

$$\mu I K_1 \frac{\partial^2 \psi(x,t)}{\partial x^2} - (2\mu + \lambda)A\psi(x,t) - \lambda A \frac{\partial u(x,t)}{\partial x} = \rho I K_2 \frac{\partial^2 \psi(x,t)}{\partial t^2} \tag{10.7}$$

where $u(x,t)$ is the longitudinal deflection of the centerline of the beam, and $\psi(x,t)$ and I are the transverse contraction and area moment inertia of the cross section, respectively. μ and λ are Lame constants, which are related to the Poisson's ratio ν and Young's modulus E by $\mu = \dfrac{E}{2(1+\nu)}$ and $\lambda = \dfrac{\nu E}{(1+\nu)(1-2\nu)}$. K_1 and K_2 are adjusting coefficients related to the shear rigidity and rotary inertia, and are related by the Poisson's ratio ν as $K_1 = K_2 \left(\dfrac{0.87 + 1.12\nu}{1+\nu}\right)^2$. The remaining parameters are defined in Eq. (10.1).

Assuming time harmonic motion and using separation of variables, the solutions to Eq. (10.7) can be written in the form $u(x,t) = u_0 e^{-ikx} e^{i\omega t}$ and $\psi(x,t) = \psi_0 e^{-ikx} e^{i\omega t}$, where ω is the circular frequency, k is the wavenumber, and i is the imaginary unit. u_0 and ψ_0 are the amplitudes of the longitudinal deflection of the centerline of the beam and the transverse contraction of the cross section, respectively. Substituting these expressions for $u(x,t)$ and $\psi(x,t)$ into Eq. (10.7) and putting the equations into matrix form,

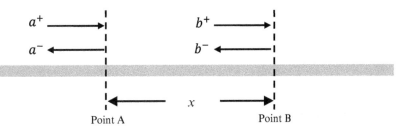

Figure 10.1 Wave propagation relationships.

$$\begin{bmatrix} -k^2(2\mu+\lambda)A + \rho A\omega^2 & -ik\lambda A \\ ik\lambda A & -k^2\mu IK_1 - (2\mu+\lambda)A + \rho IK_2\omega^2 \end{bmatrix}\begin{bmatrix} u_0 \\ \psi_0 \end{bmatrix} = \begin{bmatrix} 0 \\ 0 \end{bmatrix} \tag{10.8}$$

Setting the determinant of the coefficient matrix of Eq. (10.8) to zero gives the dispersion equation,

$$Z_2 k^4 + Z_1 k^2 + Z_0 = 0 \tag{10.9a}$$

where

$$Z_2 = (2\mu+\lambda)A\mu IK_1,$$

$$Z_1 = 4\mu(\mu+\lambda)A^2 - \rho I K_2 \omega^2(2\mu+\lambda)A - \rho A\omega^2 \mu IK_1, \tag{10.9b}$$

$$Z_0 = -\rho A\omega^2\left[(2\mu+\lambda)A - \rho IK_2\omega^2\right]$$

Equation (10.9a) is a quadratic equation in terms of k^2. As a result, the Mindlin–Herrmann longitudinal vibration theory (unlike the Elementary and Love longitudinal vibration theories) predicts two wave modes, with wavenumbers k_1 and k_2.

A cut-off frequency exists that can be solved by setting k in the dispersion equation obtained in Eq. (10.9a) to zero,

$$\omega_c = \sqrt{\frac{(2\mu+\lambda)A}{\rho IK_2}} \tag{10.10}$$

With time dependence $e^{i\omega t}$ suppressed, the solutions to Eq. (10.7) can be written as

$$u(x) = a_1^+ e^{-ik_1 x} + a_2^+ e^{-ik_2 x} + a_1^- e^{ik_1 x} + a_2^- e^{ik_2 x}$$

$$\psi(x) = \overline{a_1^+} e^{-ik_1 x} + \overline{a_2^+} e^{-ik_2 x} + \overline{a_1^-} e^{ik_1 x} + \overline{a_2^-} e^{ik_2 x} \tag{10.11}$$

The wave amplitudes a of $u(x)$ and $\bar{a}$ of $\psi(x)$ are related to each other, and the relationships are found from Eq. (10.8),

$$\frac{\overline{a_1^+}}{a_1^+} = -iP_1 \; ; \; \frac{\overline{a_1^-}}{a_1^-} = +iP_1 \; ; \; \frac{\overline{a_2^+}}{a_2^+} = -iP_2 \; ; \; \frac{\overline{a_2^-}}{a_2^-} = +iP_2 \tag{10.12a}$$

where

$$P_1 = \frac{-k_1^2(2\mu+\lambda) + \rho\omega^2}{k_1\lambda} \quad \text{and} \quad P_2 = \frac{-k_2^2(2\mu+\lambda) + \rho\omega^2}{k_2\lambda} \tag{10.12b}$$

From Eqs. (10.11) and (10.12a),

$$u(x) = a_1^+ e^{-ik_1 x} + a_2^+ e^{-ik_2 x} + a_1^- e^{ik_1 x} + a_2^- e^{ik_2 x},$$

$$\psi(x) = -iP_1 a_1^+ e^{-ik_1 x} - iP_2 a_2^+ e^{-ik_2 x} + iP_1 a_1^- e^{ik_1 x} + iP_2 a_2^- e^{ik_2 x} \tag{10.13}$$

The wave propagation relationships are similar to Eq. (10.5). However, because two wave components exist, $\mathbf{a}^\pm$ and $\mathbf{b}^\pm$ are 2 by 1 vectors, and $\mathbf{f}(x)$ becomes a 2 by 2 propagation matrix,

$$\mathbf{a}^\pm = \begin{bmatrix} a_1^\pm \\ a_2^\pm \end{bmatrix}, \quad \mathbf{b}^\pm = \begin{bmatrix} b_1^\pm \\ b_2^\pm \end{bmatrix}, \quad \text{and} \quad \mathbf{f}(x) = \begin{bmatrix} e^{-ik_1 x} & 0 \\ 0 & e^{-ik_2 x} \end{bmatrix} \tag{10.14}$$

10.1.3 Three-mode Longitudinal Vibration Theory

The equations of motion for free longitudinal vibrations of a uniform beam according to the Three-mode theory are (Doyle 1989)

$$(2\mu+\lambda)A\frac{\partial^2 u(x,t)}{\partial x^2}+\lambda A\frac{\partial \psi(x,t)}{\partial x}=\rho A\frac{\partial^2 u(x,t)}{\partial t^2},$$

$$\mu I\frac{\partial^2 \psi(x,t)}{\partial x^2}-(2\mu+\lambda)A\psi(x,t)-\lambda A\frac{\partial u(x,t)}{\partial x}-2\mu Ah\frac{\partial \phi(x,t)}{\partial x}=\rho I\frac{\partial^2 \psi(x,t)}{\partial t^2}, \tag{10.15}$$

$$(2\mu+\lambda)I\frac{\partial^2 \phi(x,t)}{\partial x^2}-5\mu A\phi(x,t)+\frac{10}{48}\mu Ah\frac{\partial \psi(x,t)}{\partial x}=\rho I\frac{\partial^2 \phi(x,t)}{\partial t^2}$$

where $u(x,t)$ is the longitudinal deflection of the centerline of the beam, $\psi(x,t)$ is the transverse contraction of the cross section, and $\phi(x,t)$ is the parabolic distribution function of the axial displacement along the thickness. h is the thickness of the beam. The remaining parameters are defined in Eqs. (10.1) and (10.7).

Assuming time harmonic motion and using separation of variables, the solutions to Eq. (10.15) can be written in the form $u(x,t)=u_0 e^{-ikx}e^{i\omega t}$, $\psi(x,t)=\psi_0 e^{-ikx}e^{i\omega t}$, and $\phi(x,t)=\phi_0 e^{-ikx}e^{i\omega t}$, where ω is the circular frequency, k is the wavenumber, and i is the imaginary unit. u_0, ψ_0, and ϕ_0 are the amplitudes of the longitudinal deflection of the centerline of the beam, the transverse contraction of the cross section, and the parabolic distribution function of the axial displacement along the thickness, repectively. Substituting these expressions for $u(x,t)$, $\psi(x,t)$, and $\phi(x,t)$ into Eq. (10.15) and putting the equations into matrix form,

$$\begin{vmatrix} -(2\mu+\lambda)Ak^2+\rho A\omega^2 & -ik\lambda A & 0 \\ ik\lambda A & -\mu Ik^2-(2\mu+\lambda)A+\rho I\omega^2 & i2k\mu Ah \\ 0 & -i\frac{10}{48}k\mu Ah & -(2\mu+\lambda)Ik^2-5\mu A+\rho I\omega^2 \end{vmatrix}\begin{bmatrix} u_0 \\ \psi_0 \\ \phi_0 \end{bmatrix}=\begin{bmatrix} 0 \\ 0 \\ 0 \end{bmatrix} \tag{10.16}$$

Setting the determinant of the coefficient matrix of Eq. (10.16) to zero gives the dispersion equation,

$$z_3 k^6+z_2 k^4+z_1 k^2+z_0=0 \tag{10.17a}$$

where

$$z_3=-AI^2\mu(2\mu+\lambda)^2,$$

$$z_2=A(\lambda+2\mu)\left[\frac{5}{12}A^2 h^2\mu^2-AI\mu(9\mu+4\lambda)+\rho I^2\omega^2(4\mu+\lambda)\right],$$

$$z_1=-20A^3\mu^2(\mu+\lambda)-A^2\rho\omega^2\left[\mu^2\left(\frac{5}{12}Ah^2-23I\right)-I\lambda(\lambda+13\mu)\right]-AI^2\rho^2\omega^4(2\lambda+5\mu), \tag{10.17b}$$

$$z_0=\rho A\omega^2\left(5A\mu-\rho I\omega^2\right)\left[A(2\mu+\lambda)-\rho I\omega^2\right]$$

Equation (10.17a) is a cubic equation in terms of k^2. As a result, the Three-mode longitudinal vibration theory predicts three wave modes, with wavenumbers k_1, k_2, and k_3.

Setting k in the dispersion equation obtained in Eq. (10.17a) to zero gives two cut-off frequencies,

$$\omega_{c1}=\sqrt{\frac{(2\mu+\lambda)A}{\rho I}}, \text{ and } \omega_{c2}=\sqrt{\frac{5\mu A}{\rho I}} \tag{10.18}$$

When the adjusting coefficient K_2 in Eq. (10.10) is set to unity, the cut-off frequency of the Mindlin–Herrmann longitudinal vibration theory matches cut-off frequency ω_{c1} of the Three-mode longitudinal vibration theory.

With time dependence $e^{i\omega t}$ suppressed, the solutions to Eq. (10.15) can be written as

$$u(x)=a_1^+ e^{-ik_1 x}+a_2^+ e^{-ik_2 x}+a_3^+ e^{-ik_3 x}+a_1^- e^{ik_1 x}+a_2^- e^{ik_2 x}+a_3^- e^{ik_3 x}$$

$$\psi(x)=\overline{a_1^+}e^{-ik_1 x}+\overline{a_2^+}e^{-ik_2 x}+\overline{a_3^+}e^{-ik_3 x}+\overline{a_1^-}e^{ik_1 x}+\overline{a_2^-}e^{ik_2 x}+\overline{a_3^-}e^{ik_3 x} \tag{10.19}$$

$$\phi(x)=\overline{\overline{a_1^+}}e^{-ik_1 x}+\overline{\overline{a_2^+}}e^{-ik_2 x}+\overline{\overline{a_3^+}}e^{-ik_3 x}+\overline{\overline{a_1^-}}e^{ik_1 x}+\overline{\overline{a_2^-}}e^{ik_2 x}+\overline{\overline{a_3^-}}e^{ik_3 x}$$

The wave amplitudes a of $u(x)$, $\bar{a}$ of $\psi(x)$, and $\bar{\bar{a}}$ of $\phi(x)$ are related to each other, and the relationships are found from Eq. (10.16),

$$\frac{\overline{a_1^+}}{a_1^+}=-iP_1, \quad \frac{\overline{a_2^+}}{a_2^+}=-iP_2, \quad \frac{\overline{a_3^+}}{a_3^+}=-iP_3, \quad \frac{\overline{a_1^-}}{a_1^-}=iP_1, \quad \frac{\overline{a_2^-}}{a_2^-}=iP_2, \quad \frac{\overline{a_3^-}}{a_3^-}=iP_3,$$

$$\frac{\overline{\overline{a_1^+}}}{a_1^+}=\frac{\overline{\overline{a_1^-}}}{a_1^-}=N_1, \quad \frac{\overline{\overline{a_2^+}}}{a_2^+}=\frac{\overline{\overline{a_2^-}}}{a_2^-}=N_2, \quad \frac{\overline{\overline{a_3^+}}}{a_3^+}=\frac{\overline{\overline{a_3^-}}}{a_3^-}=N_3$$

(10.20)

where $P_n=\dfrac{\rho A\omega^2-(2\mu+\lambda)Ak_n^2}{k_n\lambda A}$, $N_n=\dfrac{5\mu h\left[\rho A\omega^2-(2\mu+\lambda)Ak_n^2\right]}{24\lambda\left[\rho I\omega^2-(2\mu+\lambda)Ik_n^2-5\mu A\right]}$, and $n=1,2,3$.

From Eqs. (10.19) and (10.20),

$$u(x)=a_1^+e^{-ik_1x}+a_2^+e^{-ik_2x}+a_3^+e^{-ik_3x}+a_1^-e^{ik_1x}+a_2^-e^{ik_2x}+a_3^-e^{ik_3x}$$
$$\psi(x)=-iP_1a_1^+e^{-ik_1x}-iP_2a_2^+e^{-ik_2x}-iP_3a_3^+e^{-ik_3x}+iP_1a_1^-e^{ik_1x}+iP_2a_2^-e^{ik_2x}+iP_3a_3^-e^{ik_3x}$$
$$\phi(x)=N_1a_1^+e^{-ik_1x}+N_2a_2^+e^{-ik_2x}+N_3a_3^+e^{-ik_3x}+N_1a_1^-e^{ik_1x}+N_2a_2^-e^{ik_2x}+N_3a_3^-e^{ik_3x}$$

(10.21)

The wave propagation relationships are similar to Eq. (10.5). However, because three wave components exist, $\mathbf{a}^\pm$ and $\mathbf{b}^\pm$ are 3 by 1 vectors, and $\mathbf{f}(x)$ becomes a 3 by 3 propagation matrix,

$$\mathbf{a}^\pm=\begin{bmatrix}a_1^\pm\\a_2^\pm\\a_3^\pm\end{bmatrix}, \quad \mathbf{b}^\pm=\begin{bmatrix}b_1^\pm\\b_2^\pm\\b_3^\pm\end{bmatrix}, \quad \text{and } \mathbf{f}(x)=\begin{bmatrix}e^{-ik_1x} & 0 & 0\\0 & e^{-ik_2x} & 0\\0 & 0 & e^{-ik_3x}\end{bmatrix}$$

(10.22)

10.2 Wave Reflection at Classical Boundaries

Figure 10.2 shows the sign convention adopted in the analysis that follows.

An incident wave is reflected at a boundary. As illustrated in Figure 10.3, the incident wave a^+ and reflected wave a^- are related by reflection coefficient r,

$$a^-=ra^+$$

(10.23a)

When higher order longitudinal vibration theories are adopted, the wave components become multiple and the reflection coefficient becomes reflection matrix,

$$\mathbf{a}^-=\mathbf{r}\mathbf{a}^+$$

(10.23b)

For a beam that undergoes longitudinal vibration, a classical boundary is either fixed or free. The reflection coefficient/matrix is determined by boundary condition(s).

10.2.1 Love Longitudinal Vibration Theory

Following the sign convention defined in Figure 10.2 and according to the Love longitudinal vibration theory, the internal resistant force and deflection are related by

Figure 10.2 Definition of positive internal resistant forces.

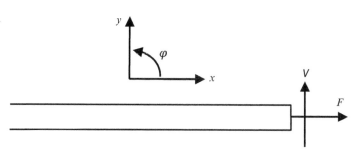

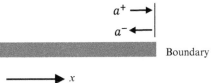

Figure 10.3 Wave reflection at a boundary.

$$F(x,t) = EA\frac{\partial u(x,t)}{\partial x} + \nu^2 \rho J \frac{\partial}{\partial x}\left(\frac{\partial^2 u(x,t)}{\partial t^2}\right) \tag{10.24}$$

At a free boundary, the equilibrium condition is

$$F(x,t) = 0 \tag{10.25}$$

From Eqs. (10.4), (10.24), and (10.25),

$$ik\left(-EA + \nu^2 \rho J \omega^2\right)a^+ e^{-ikx} + ik\left(EA - \nu^2 \rho J \omega^2\right)a^- e^{ikx} = 0 \tag{10.26}$$

Choosing the origin at the boundary, from Eqs. (10.23a) and (10.26), the reflection coefficient at a free boundary by the Love longitudinal vibration theory is

$$r = 1 \tag{10.27}$$

The boundary condition at a fixed boundary is

$$u(x,t) = 0 \tag{10.28}$$

From Eqs. (10.4) and (10.28),

$$a^+ e^{-ikx} + a^- e^{ikx} = 0 \tag{10.29}$$

Choosing the origin at the boundary, from Eqs. (10.23a) and (10.29), the reflection coefficient at a fixed boundary by the Love longitudinal vibration theory is

$$r = -1 \tag{10.30}$$

The reflection coefficients, at a free and a fixed boundary, are identical to those obtained from the Elementary longitudinal vibration theory in Chapter 2.

10.2.2 Mindlin–Herrmann Longitudinal Vibration Theory

Following the sign convention defined in Figure 10.2 and according to the Mindlin–Herrmann longitudinal vibration theory, the internal resistant forces and deflections are related as follows

$$F_u(x,t) = (2\mu + \lambda)A\frac{\partial u(x,t)}{\partial x} + \lambda A\psi(x,t), \text{ and } F_\psi(x,t) = \mu I K_1 \frac{\partial \psi(x,t)}{\partial x} \tag{10.31}$$

At a free boundary, the equilibrium conditions are

$$F_u(x,t) = 0, \text{ and } F_\psi(x,t) = 0 \tag{10.32}$$

From Eqs. (10.13), (10.31), and (10.32),

$$
\begin{aligned}
&-\left[(2\mu + \lambda)ik_1 + iP_1\lambda\right]a_1^+ e^{-ik_1 x} - \left[(2\mu + \lambda)ik_2 + iP_2\lambda\right]a_2^+ e^{-ik_2 x} \\
&+\left[(2\mu + \lambda)ik_1 + iP_1\lambda\right]a_1^- e^{ik_1 x} + \left[(2\mu + \lambda)ik_2 + iP_2\lambda\right]a_2^+ e^{ik_2 x} = 0, \\
&-P_1 k_1 a_1^+ e^{-ik_1 x} - P_2 k_2 a_2^+ e^{-ik_2 x} - P_1 k_1 a_1^- e^{ik_1 x} - P_2 k_2 a_2^- e^{ik_2 x} = 0
\end{aligned}
\tag{10.33}
$$

Choosing the origin at the boundary and putting Eq. (10.33) into matrix form,

$$\begin{bmatrix} -(2\mu+\lambda)ik_1 - iP_1\lambda & -(2\mu+\lambda)ik_2 - iP_2\lambda \\ -P_1k_1 & -P_2k_2 \end{bmatrix}\mathbf{a}^+ + \begin{bmatrix} (2\mu+\lambda)ik_1 + iP_1\lambda & (2\mu+\lambda)ik_2 + iP_2\lambda \\ -P_1k_1 & -P_2k_2 \end{bmatrix}\mathbf{a}^- = 0 \quad (10.34)$$

where $\mathbf{a}^+ = \begin{bmatrix} a_1^+ \\ a_2^+ \end{bmatrix}$ and $\mathbf{a}^- = \begin{bmatrix} a_1^- \\ a_2^- \end{bmatrix}$.

From Eqs. (10.23b) and (10.34), the reflection matrix at a free boundary by the Mindlin–Herrmann longitudinal vibration theory is

$$\mathbf{r} = \begin{bmatrix} (2\mu+\lambda)ik_1 + iP_1\lambda & (2\mu+\lambda)ik_2 + iP_2\lambda \\ -P_1k_1 & -P_2k_2 \end{bmatrix}^{-1} \begin{bmatrix} (2\mu+\lambda)ik_1 + iP_1\lambda & (2\mu+\lambda)ik_2 + iP_2\lambda \\ P_1k_1 & P_2k_2 \end{bmatrix} \quad (10.35)$$

The boundary conditions at a fixed boundary are

$$u(x,t) = 0, \text{ and } \psi(x,t) = 0 \quad (10.36)$$

From Eqs. (10.13) and (10.36),

$$a_1^+ e^{-ik_1x} + a_2^+ e^{-ik_2x} + a_1^- e^{ik_1x} + a_2^- e^{ik_2x} = 0,$$
$$-iP_1 a_1^+ e^{-ik_1x} - iP_2 a_2^+ e^{-ik_2x} + iP_1 a_1^- e^{ik_1x} + iP_2 a_2^- e^{ik_2x} = 0 \quad (10.37)$$

Choosing the origin at the boundary and putting Eq. (10.37) into matrix form,

$$\begin{bmatrix} 1 & 1 \\ -iP_1 & -iP_2 \end{bmatrix}\mathbf{a}^+ + \begin{bmatrix} 1 & 1 \\ iP_1 & iP_2 \end{bmatrix}\mathbf{a}^- = 0 \quad (10.38)$$

From Eqs. (10.23b) and (10.38), the reflection matrix at a fixed boundary by the Mindlin–Herrmann longitudinal vibration theory is

$$\mathbf{r} = \begin{bmatrix} 1 & 1 \\ iP_1 & iP_2 \end{bmatrix}^{-1} \begin{bmatrix} -1 & -1 \\ iP_1 & iP_2 \end{bmatrix} \quad (10.39)$$

The adjusting coefficients related to the shear rigidity and rotary inertia, namely, K_1 and K_2, do not play a role in the reflection matrices of classical boundaries.

10.2.3 Three-mode Longitudinal Vibration Theory

Following the sign convention defined in Figure 10.2 and according to the Three–mode longitudinal vibration theory, the internal resistant forces and deflections are related as follows

$$F_u(x,t) = (2\mu+\lambda)A\frac{\partial u(x,t)}{\partial x} + \lambda A\psi(x,t),$$
$$F_\psi(x,t) = \mu I\left(\frac{\partial \psi(x,t)}{\partial x} - 24\frac{\phi(x,t)}{h}\right),$$
$$F_\phi(x,t) = \frac{48}{5}(2\mu+\lambda)I\frac{\partial\phi(x,t)}{\partial x} \quad (10.40)$$

At a free boundary, the equilibrium conditions are

$$F_u(x,t) = 0, \ F_\psi(x,t) = 0, \text{ and } F_\phi(x,t) = 0 \quad (10.41)$$

From Eqs. (10.21), (10.40), and (10.41),

$$
\begin{aligned}
&-\left[(2\mu+\lambda)ik_1+iP_1\lambda\right]a_1^+e^{-ik_1x}-\left[(2\mu+\lambda)ik_2+iP_2\lambda\right]a_2^+e^{-ik_2x}-\left[(2\mu+\lambda)ik_3+iP_3\lambda\right]a_3^+e^{-ik_3x}\\
&+\left[(2\mu+\lambda)ik_1+iP_1\lambda\right]a_1^-e^{ik_1x}+\left[(2\mu+\lambda)ik_2+iP_2\lambda\right]a_2^-e^{ik_2x}+\left[(2\mu+\lambda)ik_3+iP_3\lambda\right]a_3^-e^{ik_3x}=0,\\
&\left(-P_1k_1-\frac{24}{h}N_1\right)a_1^+e^{-ik_1x}+\left(-P_2k_2-\frac{24}{h}N_2\right)a_2^+e^{-ik_2x}+\left(-P_3k_3-\frac{24}{h}N_3\right)a_3^+e^{-ik_3x}\\
&+\left(-P_1k_1-\frac{24}{h}N_1\right)a_1^-e^{ik_1x}+\left(-P_2k_2-\frac{24}{h}N_2\right)a_2^-e^{ik_2x}+\left(-P_3k_3-\frac{24}{h}N_3\right)a_3^-e^{ik_3x}=0,\\
&-ik_1N_1a_1^+e^{-ik_1x}-ik_2N_2a_2^+e^{-ik_2x}-ik_3N_3a_3^+e^{-ik_3x}+ik_1N_1a_1^-e^{ik_1x}+ik_2N_2a_2^-e^{ik_2x}+ik_3N_3a_3^-e^{ik_3x}=0
\end{aligned}
\tag{10.42}
$$

Choosing the origin at the boundary and putting Eq. (10.42) into matrix form,

$$
\begin{aligned}
&\begin{bmatrix}
-ik_1(2\mu+\lambda)-iP_1\lambda & -ik_2(2\mu+\lambda)-iP_2\lambda & -ik_3(2\mu+\lambda)-iP_3\lambda\\
-P_1k_1-\dfrac{24}{h}N_1 & -P_2k_2-\dfrac{24}{h}N_2 & -P_3k_3-\dfrac{24}{h}N_3\\
-ik_1N_1 & -ik_2N_2 & -ik_3N_3
\end{bmatrix}\mathbf{a}^+\\[2ex]
&+\begin{bmatrix}
ik_1(2\mu+\lambda)+iP_1\lambda & ik_2(2\mu+\lambda)+iP_2\lambda & ik_3(2\mu+\lambda)+iP_3\lambda\\
-P_1k_1-\dfrac{24}{h}N_1 & -P_2k_2-\dfrac{24}{h}N_2 & -P_3k_3-\dfrac{24}{h}N_3\\
ik_1N_1 & ik_2N_2 & ik_3N_3
\end{bmatrix}\mathbf{a}^-=0
\end{aligned}
\tag{10.43}
$$

where $\mathbf{a}^+=\begin{bmatrix}a_1^+\\a_2^+\\a_3^+\end{bmatrix}$ and $\mathbf{a}^-=\begin{bmatrix}a_1^-\\a_2^-\\a_3^-\end{bmatrix}$.

From Eqs. (10.23b) and (10.43), the reflection matrix at a free boundary by the Three–mode longitudinal vibration theory is

$$
\mathbf{r}=\begin{bmatrix}
ik_1(2\mu+\lambda)+iP_1\lambda & ik_2(2\mu+\lambda)+iP_2\lambda & ik_3(2\mu+\lambda)+iP_3\lambda\\
-P_1k_1-\dfrac{24}{h}N_1 & -P_2k_2-\dfrac{24}{h}N_2 & -P_3k_3-\dfrac{24}{h}N_3\\
ik_1N_1 & ik_2N_2 & ik_3N_3
\end{bmatrix}^{-1}
\begin{bmatrix}
ik_1(2\mu+\lambda)+iP_1\lambda & ik_2(2\mu+\lambda)+iP_2\lambda & ik_3(2\mu+\lambda)+iP_3\lambda\\
P_1k_1+\dfrac{24}{h}N_1 & P_2k_2+\dfrac{24}{h}N_2 & P_3k_3+\dfrac{24}{h}N_3\\
ik_1N_1 & ik_2N_2 & ik_3N_3
\end{bmatrix}
\tag{10.44}
$$

The boundary conditions at a fixed boundary are

$$
u(x,t)=0,\ \ \psi(x,t)=0,\ \text{and}\ \ \phi(x,t)=0
\tag{10.45}
$$

From Eqs. (10.21) and (10.45),

$$
\begin{aligned}
&a_1^+e^{-ik_1x}+a_2^+e^{-ik_2x}+a_3^+e^{-ik_3x}+a_1^-e^{ik_1x}+a_2^-e^{ik_2x}+a_3^-e^{ik_3x}=0,\\
&-iP_1a_1^+e^{-ik_1x}-iP_2a_2^+e^{-ik_2x}-iP_3a_3^+e^{-ik_3x}+iP_1a_1^-e^{ik_1x}+iP_2a_2^-e^{ik_2x}+iP_3a_3^-e^{ik_3x}=0,\\
&N_1a_1^+e^{-ik_1x}+N_2a_2^+e^{-ik_2x}+N_3a_3^+e^{-ik_3x}+N_1a_1^-e^{ik_1x}+N_2a_2^-e^{ik_2x}+N_3a_3^-e^{ik_3x}=0
\end{aligned}
\tag{10.46}
$$

Choosing the origin at the boundary and putting Eq. (10.46) into matrix form,

$$
\begin{bmatrix} 1 & 1 & 1 \\ -iP_1 & -iP_2 & -iP_3 \\ N_1 & N_2 & N_3 \end{bmatrix} \mathbf{a}^+ + \begin{bmatrix} 1 & 1 & 1 \\ iP_1 & iP_2 & iP_3 \\ N_1 & N_2 & N_3 \end{bmatrix} \mathbf{a}^- = 0 \tag{10.47}
$$

From Eqs. (10.23b) and (10.47), the reflection matrix at a fixed boundary by the Three–mode longitudinal vibration theory is

$$
\mathbf{r} = \begin{bmatrix} 1 & 1 & 1 \\ iP_1 & iP_2 & iP_3 \\ N_1 & N_2 & N_3 \end{bmatrix}^{-1} \begin{bmatrix} -1 & -1 & -1 \\ iP_1 & iP_2 & iP_3 \\ -N_1 & -N_2 & -N_3 \end{bmatrix} \tag{10.48}
$$

10.3 Waves Generated by External Excitations on the Span

An externally applied force has the effect of injecting waves into a continuous structure. Consider the axial waves injected by an external axial force $\bar{F}$, as shown in Figure 10.4. The relationships between an external excitation and its generated waves can be found from the continuity and equilibrium conditions at the point where the external excitation is applied. The detailed expressions corresponding to the various longitudinal vibration theories are derived as follows.

Note that when higher order longitudinal vibration theories are adopted, the scalar parameters for external excitation and wave components in Figure 10.4 become vectors.

10.3.1 Love Longitudinal Vibration Theory

The continuity and equilibrium conditions at the point where the external axial force is applied are

$$
u_- = u_+ \tag{10.49}
$$

and

$$
\bar{F} = F_- - F_+ \tag{10.50}
$$

where subscripts $-$ and $+$ denote physical parameters on the left and right side of the point where the external force is applied.

From Eq. (10.4) and Figure 10.4,

$$
u_- = a^+ e^{-ikx} + a^- e^{ikx} \tag{10.51a}
$$

$$
u_+ = b^+ e^{-ikx} + b^- e^{ikx} \tag{10.51b}
$$

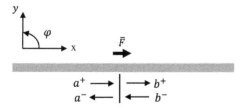

Figure 10.4 Waves generated by external excitation on the span.

From Eqs. (10.24) and (10.50),

$$\bar{F} = \left(EA\frac{\partial u_-(x,t)}{\partial x} + \nu^2\rho J\frac{\partial}{\partial x}\left(\frac{\partial^2 u_-(x,t)}{\partial t^2}\right)\right) - \left(EA\frac{\partial u_+(x,t)}{\partial x} + \nu^2\rho J\frac{\partial}{\partial x}\left(\frac{\partial^2 u_+(x,t)}{\partial t^2}\right)\right) \tag{10.52}$$

Choosing the origin at the point where the external force is applied, from Eqs. (10.49), (10.51a), and (10.51b), the continuity conditions become

$$a^+ + a^- = b^+ + b^- \tag{10.53a}$$

Equation (10.53a) can be written as

$$(a^+ - b^+) + (a^- - b^-) = 0 \tag{10.53b}$$

Noting that $\dfrac{\partial^2 u_\pm(x,t)}{\partial t^2} = -\omega^2 u_\pm(x,t)$ for time harmonic motion and choosing the origin at the point where the external force is applied, from Eqs. (10.51a), (10.51b), and (10.52), the equilibrium conditions become

$$\bar{F} = ik\left(-EA + \nu^2\rho J\omega^2\right)a^+ + ik\left(EA - \nu^2\rho J\omega^2\right)a^- + ik\left(EA - \nu^2\rho J\omega^2\right)b^+ + ik\left(-EA + \nu^2\rho J\omega^2\right)b^- \tag{10.54a}$$

Equation (10.54a) can be written as

$$(a^+ - b^+) - (a^- - b^-) = \frac{i\bar{F}}{k\left(EA - \nu^2\rho J\omega^2\right)} \tag{10.54b}$$

The relationships between an externally applied force and its generated waves can be solved from Eqs. (10.53b) and (10.54b),

$$b^+ - a^+ = -i\frac{\bar{F}}{2k\left(EA - \nu^2\rho J\omega^2\right)}, \text{ and } b^- - a^- = i\frac{\bar{F}}{2k\left(EA - \nu^2\rho J\omega^2\right)} \tag{10.55}$$

Alternatively, Eqs. (10.53b) and (10.54b) may be kept as they are for matrix assembly.

The relationships between the applied force and the generated waves by the Elementary longitudinal vibration theory are obtained when neglecting the kinetic energy related term, that is, the second term in the denominators of Eq. (10.55).

10.3.2 Mindlin–Herrmann Longitudinal Vibration Theory

The continuity conditions at the point where the external excitations are applied are

$$u_- = u_+, \text{ and } \psi_- = \psi_+ \tag{10.56}$$

The equilibrium conditions at the point where the external excitations are applied are

$$\overline{F_u} = F_{u-} - F_{u+}, \text{ and } \overline{F_\psi} = F_{\psi-} - F_{\psi+} \tag{10.57}$$

In Eqs. (10.56) and (10.57), subscripts $-$ and $+$ denote physical parameters on the left and right side of the point where the external excitations are applied.

From Figure 10.4 and Eq. (10.13),

$$\begin{aligned}
u_- &= a_1^+ e^{-ik_1 x} + a_2^+ e^{-ik_2 x} + a_1^- e^{ik_1 x} + a_2^- e^{ik_2 x}, \\
u_+ &= b_1^+ e^{-ik_1 x} + b_2^+ e^{-ik_2 x} + b_1^- e^{ik_1 x} + b_2^- e^{ik_2 x}, \\
\psi_- &= -iP_1 a_1^+ e^{-ik_1 x} - iP_2 a_2^+ e^{-ik_2 x} + iP_1 a_1^- e^{ik_1 x} + iP_2 a_2^- e^{ik_2 x}, \\
\psi_+ &= -iP_1 b_1^+ e^{-ik_1 x} - iP_2 b_2^+ e^{-ik_2 x} + iP_1 b_1^- e^{ik_1 x} + iP_2 b_2^- e^{ik_2 x}
\end{aligned} \tag{10.58}$$

From Eqs. (10.31) and (10.57),

$$\overline{F_u} = \left[(2\mu + \lambda)A\frac{\partial u_-(x,t)}{\partial x} + \lambda A\psi_-(x,t)\right] - \left[(2\mu + \lambda)A\frac{\partial u_+(x,t)}{\partial x} + \lambda A\psi_+(x,t)\right],$$

$$\overline{F_\psi} = \mu I K_1\frac{\partial \psi_-(x,t)}{\partial x} - \mu I K_1\frac{\partial \psi_+(x,t)}{\partial x} \tag{10.59}$$

Choosing the origin at the point where the external excitations are applied, from Eqs. (10.56) and (10.58), the continuity conditions become

$$a_1^+ + a_2^+ + a_1^- + a_2^- = b_1^+ + b_2^+ + b_1^- + b_2^-, \tag{10.60a}$$

$$-iP_1 a_1^+ - iP_2 a_2^+ + iP_1 a_1^- + iP_2 a_2^- = -iP_1 b_1^+ - iP_2 b_2^+ + iP_1 b_1^- + iP_2 b_2^- \tag{10.60b}$$

Putting Eqs. (10.60a) and (10.60b) into matrix form,

$$\mathbf{x}_C^+(\mathbf{a}^+ - \mathbf{b}^+) + \mathbf{x}_C^-(\mathbf{a}^- - \mathbf{b}^-) = 0 \tag{10.61a}$$

where $\mathbf{x}_C^+$ and $\mathbf{x}_C^-$ are the coefficient matrices of the positive- and negative-going waves of the continuity equation,

$$\mathbf{x}_C^+ = \begin{bmatrix} 1 & 1 \\ -iP_1 & -iP_2 \end{bmatrix}, \text{ and } \mathbf{x}_C^- = \begin{bmatrix} 1 & 1 \\ iP_1 & iP_2 \end{bmatrix} \tag{10.61b}$$

Choosing the origin at the point where the external excitations are applied, from Eqs. (10.58) and (10.59), the equilibrium conditions become

$$\begin{aligned} &\left[-ik_1(2\mu+\lambda)-iP_1\lambda\right]\left(a_1^+ - b_1^+\right) + \left[-ik_2(2\mu+\lambda)-iP_2\lambda\right]\left(a_2^+ - b_2^+\right) + \left[ik_1(2\mu+\lambda)+iP_1\lambda\right] \\ &\left(a_1^- - b_1^-\right) + \left[ik_2(2\mu+\lambda)+iP_2\lambda\right]\left(a_2^- - b_2^-\right) = \overline{F_u}/A, \\ &-k_1P_1\left(a_1^+ - b_1^+\right) - k_2P_2\left(a_2^+ - b_2^+\right) - k_1P_1\left(a_1^- - b_1^-\right) - k_2P_2\left(a_2^- - b_2^-\right) = \overline{F_\psi}/(\mu IK_1) \end{aligned} \tag{10.62}$$

Putting Eq. (10.62) into matrix form gives

$$\mathbf{x}_E^+(\mathbf{a}^+ - \mathbf{b}^+) + \mathbf{x}_E^-(\mathbf{a}^- - \mathbf{b}^-) = \mathbf{q}_E \tag{10.63a}$$

where $\mathbf{x}_E^+$ and $\mathbf{x}_E^-$ are the coefficient matrices of positive- and negative-going waves of the equilibrium equation and $\mathbf{q}_E$ is the load vector,

$$\begin{aligned} \mathbf{x}_E^+ &= \begin{bmatrix} -ik_1(2\mu+\lambda)-iP_1\lambda & -ik_2(2\mu+\lambda)-iP_2\lambda \\ -k_1P_1 & -k_2P_2 \end{bmatrix}, \\ \mathbf{x}_E^- &= \begin{bmatrix} ik_1(2\mu+\lambda)+iP_1\lambda & ik_2(2\mu+\lambda)+iP_2\lambda \\ -k_1P_1 & -k_2P_2 \end{bmatrix}, \text{ and } \mathbf{q}_E = \begin{bmatrix} \overline{F_u}/A \\ \overline{F_\psi}/(\mu IK_1) \end{bmatrix} \end{aligned} \tag{10.63b}$$

The relationships between the external excitations and the generated waves may be solved from Eqs. (10.61a) and (10.63a). Alternatively, Eqs. (10.61a) and (10.63a) may be kept as they are for matrix assembly.

10.3.3 Three-mode Longitudinal Vibration Theory

The continuity conditions at the point where the external excitations are applied are

$$u_- = u_+, \ \psi_- = \psi_+, \text{ and } \phi_- = \phi_+ \tag{10.64}$$

The equilibrium conditions at the point where the external excitations are applied are

$$\overline{F_u} = F_{u-} - F_{u+}, \ \overline{F_\psi} = F_{\psi-} - F_{\psi+}, \text{ and } \overline{F_\phi} = F_{\phi-} - F_{\phi+} \tag{10.65}$$

In Eqs. (10.64) and (10.65), subscripts − and + denote physical parameters on the left and right side of the point where the external excitations are applied.

From Figure 10.4 and Eq. (10.21),

$$
\begin{aligned}
u_- &= a_1^+ e^{-ik_1 x} + a_2^+ e^{-ik_2 x} + a_3^+ e^{-ik_3 x} + a_1^- e^{ik_1 x} + a_2^- e^{ik_2 x} + a_3^- e^{ik_3 x}, \\
u_+ &= b_1^+ e^{-ik_1 x} + b_2^+ e^{-ik_2 x} + b_3^+ e^{-ik_3 x} + b_1^- e^{ik_1 x} + b_2^- e^{ik_2 x} + b_3^- e^{ik_3 x}, \\
\psi_- &= -iP_1 a_1^+ e^{-ik_1 x} - iP_2 a_2^+ e^{-ik_2 x} - iP_3 a_3^+ e^{-ik_3 x} + iP_1 a_1^- e^{ik_1 x} + iP_2 a_2^- e^{ik_2 x} + iP_3 a_3^- e^{ik_3 x}, \\
\psi_+ &= -iP_1 b_1^+ e^{-ik_1 x} - iP_2 b_2^+ e^{-ik_2 x} - iP_3 b_3^+ e^{-ik_3 x} + iP_1 b_1^- e^{ik_1 x} + iP_2 b_2^- e^{ik_2 x} + iP_3 b_3^- e^{ik_3 x}, \\
\phi_- &= N_1 a_1^+ e^{-ik_1 x} + N_2 a_2^+ e^{-ik_2 x} + N_3 a_3^+ e^{-ik_3 x} + N_1 a_1^- e^{ik_1 x} + N_2 a_2^- e^{ik_2 x} + N_3 a_3^- e^{ik_3 x}, \\
\phi_+ &= N_1 b_1^+ e^{-ik_1 x} + N_2 b_2^+ e^{-ik_2 x} + N_3 b_3^+ e^{-ik_3 x} + N_1 b_1^- e^{ik_1 x} + N_2 b_2^- e^{ik_2 x} + N_3 b_3^- e^{ik_3 x}
\end{aligned}
\tag{10.66}
$$

From Eqs. (10.40) and (10.65),

$$
\begin{aligned}
\overline{F_u} &= (2\mu + \lambda) A \frac{\partial u_-(x,t)}{\partial x} + \lambda A \psi_-(x,t) - (2\mu + \lambda) A \frac{\partial u_+(x,t)}{\partial x} - \lambda A \psi_+(x,t), \\
\overline{F_\psi} &= \mu I \left[\frac{\partial \psi_-(x,t)}{\partial x} - 24 \frac{\phi_-(x,t)}{h} \right] - \mu I \left[\frac{\partial \psi_+(x,t)}{\partial x} - 24 \frac{\phi_+(x,t)}{h} \right], \\
\overline{F_\phi} &= \frac{48}{5}(2\mu + \lambda) I \frac{\partial \phi_-(x,t)}{\partial x} - \frac{48}{5}(2\mu + \lambda) I \frac{\partial \phi_+(x,t)}{\partial x}
\end{aligned}
\tag{10.67}
$$

Choosing the origin at the point where the external excitations are applied, from Eqs. (10.64) and (10.66), the continuity conditions become

$$
\begin{aligned}
a_1^+ + a_2^+ + a_3^+ + a_1^- + a_2^- + a_3^- &= b_1^+ + b_2^+ + b_3^+ + b_1^- + b_2^- + b_3^- \\
-iP_1 a_1^+ - iP_2 a_2^+ - iP_3 a_3^+ + iP_1 a_1^- + iP_2 a_2^- + iP_3 a_3^- &= -iP_1 b_1^+ - iP_2 b_2^+ - iP_3 b_3^+ + iP_1 b_1^- + iP_2 b_2^- + iP_3 b_3^- \\
N_1 a_1^+ + N_2 a_2^+ + N_3 a_3^+ + N_1 a_1^- + N_2 a_2^- + N_3 a_3^- &= N_1 b_1^+ + N_2 b_2^+ + N_3 b_3^+ + N_1 b_1^- + N_2 b_2^- + N_3 b_3^-
\end{aligned}
\tag{10.68}
$$

Putting Eq. (10.68) into matrix form gives

$$
\mathbf{x}_C^+ (\mathbf{a}^+ - \mathbf{b}^+) + \mathbf{x}_C^- (\mathbf{a}^- - \mathbf{b}^-) = 0
\tag{10.69a}
$$

where $\mathbf{x}_C^+$ and $\mathbf{x}_C^-$ are the coefficient matrices of the positive- and negative-going waves of the continuity equation,

$$
\mathbf{x}_C^+ =
\begin{vmatrix}
1 & 1 & 1 \\
-iP_1 & -iP_2 & -iP_3 \\
N_1 & N_2 & N_3
\end{vmatrix}, \text{ and } \mathbf{x}_C^- =
\begin{vmatrix}
1 & 1 & 1 \\
iP_1 & iP_2 & iP_3 \\
N_1 & N_2 & N_3
\end{vmatrix}
\tag{10.69b}
$$

Choosing the origin at the point where the external excitations are applied, from Eqs. (10.66) and (10.67), the equilibrium conditions become

$$
\begin{aligned}
&\left[-ik_1(2\mu + \lambda) - iP_1 \lambda \right]\left(a_1^+ - b_1^+\right) + \left[-ik_2(2\mu + \lambda) - iP_2 \lambda \right]\left(a_2^+ - b_2^+\right) + \left[-ik_3(2\mu + \lambda) - iP_3 \lambda \right]\left(a_3^+ - b_3^+\right) \\
&+ \left[ik_1(2\mu + \lambda) + iP_1 \lambda \right]\left(a_1^- - b_1^-\right) + \left[ik_2(2\mu + \lambda) + iP_2 \lambda \right]\left(a_2^- - b_2^-\right) + \left[ik_3(2\mu + \lambda) + iP_3 \lambda \right]\left(a_3^- - b_3^-\right) = \overline{F_u}/A, \\
&\left(-k_1 P_1 - \frac{24}{h} N_1 \right)\left(a_1^+ - b_1^+\right) + \left(-k_2 P_2 - \frac{24}{h} N_2 \right)\left(a_2^+ - b_2^+\right) + \left(-k_3 P_3 - \frac{24}{h} N_3 \right)\left(a_3^+ - b_3^+\right) \\
&+ \left(-k_1 P_1 - \frac{24}{h} N_1 \right)\left(a_1^- - b_1^-\right) + \left(-k_2 P_2 - \frac{24}{h} N_2 \right)\left(a_2^- - b_2^-\right) + \left(-k_3 P_3 - \frac{24}{h} N_3 \right)\left(a_3^- - b_3^-\right) = \overline{F_\psi}/(\mu I), \\
&-ik_1 N_1 \left(a_1^+ - b_1^+\right) - ik_2 N_2 \left(a_2^+ - b_2^+\right) - ik_3 N_3 \left(a_3^+ - b_3^+\right) + ik_1 N_1 \left(a_1^- - b_1^-\right) + ik_2 N_2 \left(a_2^- - b_2^-\right) \\
&+ ik_3 N_3 \left(a_3^- - b_3^-\right) = 5\overline{F_\phi}/\left(48(2\mu + \lambda) I\right)
\end{aligned}
\tag{10.70}
$$

Putting Eq. (10.70) into matrix form gives

$$
\mathbf{x}_E^+ (\mathbf{a}^+ - \mathbf{b}^+) + \mathbf{x}_E^- (\mathbf{a}^- - \mathbf{b}^-) = \mathbf{q}_E
\tag{10.71a}
$$

where $\mathbf{x}_E^+$ and $\mathbf{x}_E^-$ are the coefficient matrices of the positive- and negative-going waves of the equilibrium equation and $\mathbf{q}_E$ is the load vector,

$$
\mathbf{x}_E^+ = \begin{bmatrix} -ik_1\left(2\mu+\lambda\right)-iP_1\lambda & -ik_2\left(2\mu+\lambda\right)-iP_2\lambda & -ik_3\left(2\mu+\lambda\right)-iP_3\lambda \\ -k_1P_1-\dfrac{24}{h}N_1 & -k_2P_2-\dfrac{24}{h}N_2 & -k_3P_3-\dfrac{24}{h}N_3 \\ -ik_1N_1 & -ik_2N_2 & -ik_3N_3 \end{bmatrix},
$$

$$
\mathbf{x}_E^- = \begin{bmatrix} ik_1\left(2\mu+\lambda\right)+iP_1\lambda & ik_2\left(2\mu+\lambda\right)+iP_2\lambda & ik_3\left(2\mu+\lambda\right)+iP_3\lambda \\ -k_1P_1-\dfrac{24}{h}N_1 & -k_2P_2-\dfrac{24}{h}N_2 & -k_3P_3-\dfrac{24}{h}N_3 \\ ik_1N_1 & ik_2N_2 & ik_3N_3 \end{bmatrix}, \text{ and}
$$

$$
\mathbf{q}_E = \begin{bmatrix} \overline{F_u}/A \\ \overline{F_\psi}/\left(\mu I\right) \\ \overline{F_\phi}/\left(48(2\mu+\lambda)I/5\right) \end{bmatrix}
$$

(10.71b)

The relationships between the applied force and the generated waves may be solved from Eqs. (10.69a) and (10.71a). Alternatively, Eqs. (10.69a) and (10.71a) may be kept as they are for matrix assembly.

10.4 Waves Generated by External Excitations at a Free End

An external excitation may be applied at a free end of a beam, as shown in Figure 10.5.

The externally applied force $\overline{F}$ is related to the corresponding internal resistant force F at the point where the external force is applied by the following equilibrium equation

$$
\overline{F} = F
$$

(10.72)

Note that when higher order longitudinal vibration theories are adopted, the scalar parameters for external excitation and wave components in Figure 10.5 become vectors, and the scalar equation in Eq. (10.72) becomes a vector equation.

10.4.1 Love Longitudinal Vibration Theory

From Figure 10.5 and Eq. (10.4), the deflection at the point where the external force is applied is

$$
u = a^+e^{-ikx} + a^-e^{ikx}
$$

(10.73)

From Eqs. (10.24) and (10.72),

$$
\overline{F} = EA\frac{\partial u\left(x,t\right)}{\partial x} + \nu^2\rho J\frac{\partial}{\partial x}\left(\frac{\partial^2 u\left(x,t\right)}{\partial t^2}\right)
$$

(10.74)

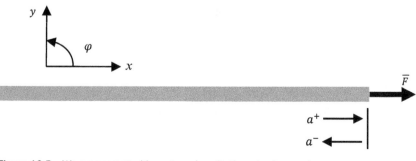

Figure 10.5 Waves generated by external excitation at a free end.

Recalling that $\dfrac{\partial^2 u(x,t)}{\partial t^2} = -\omega^2 u(x,t)$ for time harmonic motion and choosing the origin at the point where the external force is applied, from Eqs. (10.73) and (10.74),

$$\bar{F} = ik\left(-EA + \nu^2 \rho J\omega^2\right)a^+ + ik\left(EA - \nu^2 \rho J\omega^2\right)a^- \tag{10.75}$$

From Eq. (10.75), the relationship between the externally applied end excitation and its generated waves by the Love longitudinal vibration theory is

$$a^+ - a^- = q \tag{10.76}$$

where

$$q = i\frac{\bar{F}}{k\left(EA - \nu^2 \rho J\omega^2\right)} \tag{10.77}$$

The relationship between the applied force and the generated waves by the Elementary longitudinal vibration theory is obtained when neglecting the kinetic energy related term, that is, the second term in the denominator of Eq. (10.77).

10.4.2   Mindlin–Herrmann Longitudinal Vibration Theory

From Figure 10.5 and Eq. (10.13), the deflections at the point where the external excitations are applied are

$$\begin{aligned}
u &= a_1^+ e^{-ik_1 x} + a_2^+ e^{-ik_2 x} + a_1^- e^{ik_1 x} + a_2^- e^{ik_2 x} \\
\psi &= -iP_1 a_1^+ e^{-ik_1 x} - iP_2 a_2^+ e^{-ik_2 x} + iP_1 a_1^- e^{ik_1 x} + iP_2 a_2^- e^{ik_2 x}
\end{aligned} \tag{10.78}$$

From Eqs. (10.31) and (10.72),

$$\begin{aligned}
\overline{F_u} &= (2\mu + \lambda)A\frac{\partial u(x,t)}{\partial x} + \lambda A\psi(x,t) \\
\overline{F_\psi} &= \mu I K_1 \frac{\partial \psi(x,t)}{\partial x}
\end{aligned} \tag{10.79}$$

Choosing the origin at the point where the external excitations are applied, from Eqs. (10.78) and (10.79),

$$\begin{aligned}
&\left[-ik_1(2\mu + \lambda) - iP_1\lambda\right]a_1^+ + \left[-ik_2(2\mu + \lambda) - iP_2\lambda\right]a_2^+ + \left[ik_1(2\mu + \lambda) + iP_1\lambda\right]a_1^- + \left[ik_2(2\mu + \lambda) + iP_2\lambda\right]a_2^- = \overline{F_u}/A, \\
&-k_1 P_1 a_1^+ - k_2 P_2 a_2^+ - k_1 P_1 a_1^- - k_2 P_2 a_2^- = \overline{F_\psi}/(\mu I K_1)
\end{aligned} \tag{10.80}$$

Putting Eq. (10.80) into matrix form, the relationships between the externally applied end excitations and the generated waves by the Mindlin–Herrmann longitudinal vibration theory are

$$\mathbf{y}_E^+ \mathbf{a}^+ + \mathbf{y}_E^- \mathbf{a}^- = \mathbf{q}_F \tag{10.81a}$$

where $\mathbf{y}_E^+$ and $\mathbf{y}_E^-$ are the coefficient matrices of the positive- and negative-going waves of the equilibrium equation and $\mathbf{q}_F$ is the load vector,

$$\begin{aligned}
\mathbf{y}_E^+ &= \begin{bmatrix} -ik_1(2\mu + \lambda) - iP_1\lambda & -ik_2(2\mu + \lambda) - iP_2\lambda \\ -k_1 P_1 & -k_2 P_2 \end{bmatrix}, \\
\mathbf{y}_E^- &= \begin{bmatrix} ik_1(2\mu + \lambda) + iP_1\lambda & ik_2(2\mu + \lambda) + iP_2\lambda \\ -k_1 P_1 & -k_2 P_2 \end{bmatrix}, \text{ and } \mathbf{q}_F = \begin{bmatrix} \overline{F_u}/A \\ \overline{F_\psi}/(\mu I K_1) \end{bmatrix}
\end{aligned} \tag{10.81b}$$

10.4.3   Three-mode Longitudinal Vibration Theory

From Figure 10.5 and Eq. (10.21), the deflections at the point where the external excitations are applied are

$$\begin{aligned}
u &= a_1^+ e^{-ik_1 x} + a_2^+ e^{-ik_2 x} + a_3^+ e^{-ik_3 x} + a_1^- e^{ik_1 x} + a_2^- e^{ik_2 x} + a_3^- e^{ik_3 x} \\
\psi &= -iP_1 a_1^+ e^{-ik_1 x} - iP_2 a_2^+ e^{-ik_2 x} - iP_3 a_3^+ e^{-ik_3 x} + iP_1 a_1^- e^{ik_1 x} + iP_2 a_2^- e^{ik_2 x} + iP_3 a_3^- e^{ik_3 x} \\
\phi &= N_1 a_1^+ e^{-ik_1 x} + N_2 a_2^+ e^{-ik_2 x} + N_3 a_3^+ e^{-ik_3 x} + N_1 a_1^- e^{ik_1 x} + N_2 a_2^- e^{ik_2 x} + N_3 a_3^- e^{ik_3 x}
\end{aligned} \tag{10.82}$$

From Eqs. (10.40) and (10.72),

$$
\begin{aligned}
\overline{F_u} &= (2\mu + \lambda)A\frac{\partial u(x,t)}{\partial x} + \lambda A\psi(x,t), \\
\overline{F_\psi} &= \mu I\left(\frac{\partial \psi(x,t)}{\partial x} - 24\frac{\phi(x,t)}{h}\right), \\
\overline{F_\phi} &= \frac{48}{5}(2\mu + \lambda)I\frac{\partial \phi(x,t)}{\partial x}
\end{aligned}
\tag{10.83}
$$

Choosing the origin at the point where the external excitations are applied, from Eqs. (10.82) and (10.83),

$$
\begin{aligned}
&\left[-ik_1(2\mu+\lambda)-iP_1\lambda\right]a_1^+ + \left[-ik_2(2\mu+\lambda)-iP_2\lambda\right]a_2^+ + \left[-ik_3(2\mu+\lambda)-iP_3\lambda\right]a_3^+ \\
&+ \left[ik_1(2\mu+\lambda)+iP_1\lambda\right]a_1^- + \left[ik_2(2\mu+\lambda)+iP_2\lambda\right]a_2^- + \left[ik_3(2\mu+\lambda)+iP_3\lambda\right]a_3^- = \overline{F_u}/A, \\
&\left(-k_1P_1 - \frac{24}{h}N_1\right)a_1^+ + \left(-k_2P_2 - \frac{24}{h}N_2\right)a_2^+ + \left(-k_3P_3 - \frac{24}{h}N_3\right)a_3^+ \\
&+ \left(-k_1P_1 - \frac{24}{h}N_1\right)a_1^- + \left(-k_2P_2 - \frac{24}{h}N_2\right)a_2^- + \left(-k_3P_3 - \frac{24}{h}N_3\right)a_3^- = \overline{F_\psi}/(\mu I), \\
&-ik_1N_1a_1^+ - ik_2N_2a_2^+ - ik_3N_3a_3^+ + ik_1N_1a_1^- + ik_2N_2a_2^- + ik_3N_3a_3^- = 5\overline{F_\phi}/(48(2\mu+\lambda)I)
\end{aligned}
\tag{10.84}
$$

Putting Eq. (10.84) into matrix form, the relationships between externally applied end excitations and the generated waves by the Three-mode longitudinal vibration theory are

$$
\mathbf{y}_E^+ \mathbf{a}^+ + \mathbf{y}_E^- \mathbf{a}^- = \mathbf{q}_F
\tag{10.85a}
$$

where $\mathbf{y}_E^+$ and $\mathbf{y}_E^-$ are the coefficient matrices of positive- and negative-going waves of the equilibrium equation and $\mathbf{q}_F$ is the load vector,

$$
\begin{aligned}
\mathbf{y}_E^+ &= \begin{bmatrix}
-ik_1(2\mu+\lambda)-iP_1\lambda & -ik_2(2\mu+\lambda)-iP_2\lambda & -ik_3(2\mu+\lambda)-iP_3\lambda \\
-k_1P_1 - \dfrac{24}{h}N_1 & -k_2P_2 - \dfrac{24}{h}N_2 & -k_3P_3 - \dfrac{24}{h}N_3 \\
-ik_1N_1 & -ik_2N_2 & -ik_3N_3
\end{bmatrix}, \\
\mathbf{y}_E^- &= \begin{bmatrix}
ik_1(2\mu+\lambda)+iP_1\lambda & ik_2(2\mu+\lambda)+iP_2\lambda & ik_3(2\mu+\lambda)+iP_3\lambda \\
-k_1P_1 - \dfrac{24}{h}N_1 & -k_2P_2 - \dfrac{24}{h}N_2 & -k_3P_3 - \dfrac{24}{h}N_3 \\
ik_1N_1 & ik_2N_2 & ik_3N_3
\end{bmatrix}, \text{ and} \\
\mathbf{q}_F &= \begin{bmatrix}
\overline{F_u}/A \\
\overline{F_\psi}/(\mu I) \\
\overline{F_\phi}/(48(2\mu+\lambda)I/5)
\end{bmatrix}
\end{aligned}
\tag{10.85b}
$$

10.5 Free and Forced Vibration Analysis

With the availability of propagation relationships, boundary reflections, and force generated waves, both free and forced vibration analysis becomes concise and systematic, regardless of the vibration theories adopted.

In this section, the equations are presented in vector form, which are applicable to the analyses based on the Mindlin-Herrmann and Three-mode longitudinal vibration theories. For analysis based on the Love longitudinal vibration theory, the vector expressions become scalars, because only a single wave vibration mode is predicted by the Love longitudinal vibration theory.

Figure 10.6 Wave propagation and reflection in a uniform beam.

10.5.1   Free Vibration Analysis

Figure 10.6 illustrates waves in a uniform beam of length L. The following reflection and propagation relationships exist:

$$\mathbf{a}^+ = \mathbf{r}_A\, \mathbf{a}^-,\; \mathbf{b}^- = \mathbf{r}_B\, \mathbf{b}^+,\; \mathbf{b}^+ = \mathbf{f}(L)\, \mathbf{a}^+,\; \mathbf{a}^- = \mathbf{f}(L)\, \mathbf{b}^- \tag{10.86}$$

where $\mathbf{r}_A$ and $\mathbf{r}_B$ are the reflection matrices at Boundaries A and B, respectively. $\mathbf{f}(L)$ is the propagation matrix for a distance L.

Assembling the wave relationships in Eq. (10.86) into matrix form,

$$\begin{bmatrix} -\mathbf{I} & \mathbf{r}_A & \mathbf{0} & \mathbf{0} \\ \mathbf{0} & \mathbf{0} & \mathbf{r}_B & -\mathbf{I} \\ \mathbf{f}(L) & \mathbf{0} & -\mathbf{I} & \mathbf{0} \\ \mathbf{0} & -\mathbf{I} & \mathbf{0} & \mathbf{f}(L) \end{bmatrix} \begin{bmatrix} \mathbf{a}^+ \\ \mathbf{a}^- \\ \mathbf{b}^+ \\ \mathbf{b}^- \end{bmatrix} = \mathbf{0} \tag{10.87}$$

where $\mathbf{I}$ and $\mathbf{0}$ denote an identity and a zero matrix, respectively.

Equation (10.87) can be written in the form of,

$$\mathbf{A}_0 \mathbf{z}_0 = 0 \tag{10.88}$$

where $\mathbf{A}_0$ is the square coefficient matrix and $\mathbf{z}_0$ is the wave vector of Eq. (10.87).

Setting the determinant of coefficient matrix $\mathbf{A}_0$ to zero gives the characteristic equation. Natural frequencies of a structure are roots of its characteristic equation, which correspond to the non-trivial solutions of Eq. (10.87), or equivalently Eq. (10.88).

10.5.2   Forced Vibration Analysis

Figure 10.7 illustrates a beam with an external excitation force applied at an arbitrary point C on the beam. The following wave relationships exist:

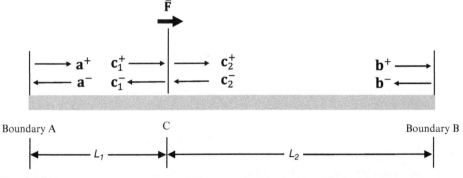

Figure 10.7 Waves in a uniform beam with external excitation(s) applied on the span.

- At Boundaries A and B,

$$\mathbf{a}^+ = \mathbf{r}_A\,\mathbf{a}^-,\ \mathbf{b}^- = \mathbf{r}_B\,\mathbf{b}^+ \tag{10.89}$$

where $\mathbf{r}_A$ and $\mathbf{r}_B$ are the reflection matrices at Boundaries A and B, respectively.

- At external force applied point C,

$$
\begin{aligned}
\mathbf{x}_C^+(\mathbf{c}_1^+ - \mathbf{c}_2^+) + \mathbf{x}_C^-(\mathbf{c}_1^- - \mathbf{c}_2^-) &= 0,\\
\mathbf{x}_E^+(\mathbf{c}_1^+ - \mathbf{c}_2^+) + \mathbf{x}_E^-(\mathbf{c}_1^- - \mathbf{c}_2^-) &= \mathbf{q}_E
\end{aligned}
\tag{10.90}
$$

where $\mathbf{x}_C^+$ and $\mathbf{x}_C^-$ are the coefficient matrices of the positive- and negative-going waves of the continuity equation, $\mathbf{x}_E^+$ and $\mathbf{x}_E^-$ are the coefficients of positive- and negative-going waves of the equilibrium equation, and $\mathbf{q}_E$ is the load vector.

- The propagation relationships are

$$\mathbf{c}_1^+ = \mathbf{f}(L_1)\,\mathbf{a}^+,\ \mathbf{a}^- = \mathbf{f}(L_1)\,\mathbf{c}_1^-,\ \mathbf{c}_2^- = \mathbf{f}(L_2)\,\mathbf{b}^-,\ \mathbf{b}^+ = \mathbf{f}(L_2)\,\mathbf{c}_2^+ \tag{10.91}$$

where $\mathbf{f}(L_1)$ and $\mathbf{f}(L_2)$ are the propagation matrices for a distance L_1 and L_2, respectively.

Assembling the wave relationships from Eqs. (10.89) to (10.91) into matrix form,

$$
\begin{bmatrix}
-\mathbf{I} & \mathbf{r}_A & \mathbf{0} & \mathbf{0} & \mathbf{0} & \mathbf{0} & \mathbf{0} & \mathbf{0} \\
\mathbf{0} & \mathbf{0} & \mathbf{r}_B & -\mathbf{I} & \mathbf{0} & \mathbf{0} & \mathbf{0} & \mathbf{0} \\
\mathbf{0} & \mathbf{0} & \mathbf{0} & \mathbf{0} & \mathbf{x}_C^+ & \mathbf{x}_C^- & -\mathbf{x}_C^+ & -\mathbf{x}_C^- \\
\mathbf{0} & \mathbf{0} & \mathbf{0} & \mathbf{0} & \mathbf{x}_E^+ & \mathbf{x}_E^- & -\mathbf{x}_E^+ & -\mathbf{x}_E^- \\
\mathbf{f}(L_1) & \mathbf{0} & \mathbf{0} & \mathbf{0} & -\mathbf{I} & \mathbf{0} & \mathbf{0} & \mathbf{0} \\
\mathbf{0} & -\mathbf{I} & \mathbf{0} & \mathbf{0} & \mathbf{0} & \mathbf{f}(L_1) & \mathbf{0} & \mathbf{0} \\
\mathbf{0} & \mathbf{0} & \mathbf{0} & \mathbf{f}(L_2) & \mathbf{0} & \mathbf{0} & \mathbf{0} & -\mathbf{I} \\
\mathbf{0} & \mathbf{0} & -\mathbf{I} & \mathbf{0} & \mathbf{0} & \mathbf{0} & \mathbf{f}(L_2) & \mathbf{0}
\end{bmatrix}
\begin{bmatrix}
\mathbf{a}^+ \\ \mathbf{a}^- \\ \mathbf{b}^+ \\ \mathbf{b}^- \\ \mathbf{c}_1^+ \\ \mathbf{c}_1^- \\ \mathbf{c}_2^+ \\ \mathbf{c}_2^-
\end{bmatrix}
=
\begin{bmatrix}
\mathbf{0} \\ \mathbf{0} \\ \mathbf{0} \\ \mathbf{q}_E \\ \mathbf{0} \\ \mathbf{0} \\ \mathbf{0} \\ \mathbf{0}
\end{bmatrix}
\tag{10.92}
$$

where $\mathbf{I}$ and $\mathbf{0}$ denote an identity and a zero matrix, respectively.

Equation (10.92) can be written in the form of,

$$\mathbf{A}_f\mathbf{z}_f = \mathbf{F} \tag{10.93}$$

where $\mathbf{A}_f$ is the square coefficient matrix, $\mathbf{z}_f$ is the wave vector, and $\mathbf{F}$ is the load vector of Eq. (10.92).

Solving for $\mathbf{z}_f$ from Eq. (10.93),

$$\mathbf{z}_f = \mathbf{A}_f^{-1}\mathbf{F} \tag{10.94}$$

Equation (10.94) provides the responses of all wave components, from which the deflection of any point on the beam can be obtained. For example, the deflection of a point located between Boundary A and excitation point C that is a distance x_1 from Point C in Figure 10.7, by the Mindlin-Herrmann longitudinal vibration theory, is

$$u_1 = [1 \quad 1]\mathbf{f}(x_1)\mathbf{c}_1^- + [1 \quad 1]\mathbf{f}(-x_1)\mathbf{c}_1^+ \tag{10.95a}$$

or equivalently is

$$u_1 = [1 \quad 1]\mathbf{f}(L_1 - x_1)\mathbf{a}^+ + [1 \quad 1]\mathbf{f}\big(-(L_1 - x_1)\big)\mathbf{a}^- \tag{10.95b}$$

Similarly, the deflection of a point located between Boundary B and the excitation point C that is a distance x_2 from Point C in Figure 10.7, by the Mindlin-Herrmann longitudinal vibration theory, is

$$u_2 = [1 \quad 1]\mathbf{f}(x_2)\mathbf{c}_2^+ + [1 \quad 1]\mathbf{f}(-x_2)\mathbf{c}_2^- \tag{10.96a}$$

or equivalently is

$$u_2 = [1 \quad 1]\mathbf{f}(L_2 - x_2)\mathbf{b}^- + [1 \quad 1]\mathbf{f}\big(-(L_2 - x_2)\big)\mathbf{b}^+ \tag{10.96b}$$

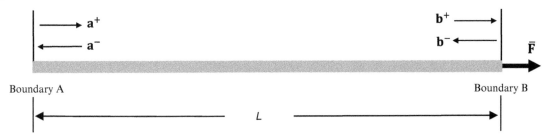

Figure 10.8 Waves in a uniform beam with external excitation(s) applied at a free end.

Figure 10.8 illustrates a beam with external excitation(s) applied at a free end. The following wave relationships exist:

- At Boundary A

$$\mathbf{a}^+ = \mathbf{r}_A \, \mathbf{a}^-$$

(10.97)

where $\mathbf{r}_A$ is the reflection matrix at A.

- At external excitation(s) applied end B

$$\mathbf{y}_E^+ \mathbf{b}^+ + \mathbf{y}_E^- \mathbf{b}^- = \mathbf{q}_F$$

(10.98)

where $\mathbf{y}_E^+$ and $\mathbf{y}_E^-$ are the coefficient matrices of the positive- and negative-going waves of the equilibrium equation and $\mathbf{q}_F$ is the load vector.

- The propagation relationships are

$$\mathbf{b}^+ = \mathbf{f}(L) \, \mathbf{a}^+, \quad \mathbf{a}^- = \mathbf{f}(L) \, \mathbf{b}^-$$

(10.99)

where $\mathbf{f}(L)$ is propagation matrix for a distance L.

Assembling the wave relationships from Eqs. (10.97) to (10.99) into matrix form,

$$\begin{bmatrix} -\mathbf{I} & \mathbf{r}_A & \mathbf{0} & \mathbf{0} \\ \mathbf{0} & \mathbf{0} & \mathbf{y}_E^+ & \mathbf{y}_E^- \\ \mathbf{f}(L) & \mathbf{0} & -\mathbf{I} & \mathbf{0} \\ \mathbf{0} & -\mathbf{I} & \mathbf{0} & \mathbf{f}(L) \end{bmatrix} \begin{bmatrix} \mathbf{a}^+ \\ \mathbf{a}^- \\ \mathbf{b}^+ \\ \mathbf{b}^- \end{bmatrix} = \begin{bmatrix} \mathbf{0} \\ \mathbf{q}_F \\ \mathbf{0} \\ \mathbf{0} \end{bmatrix}$$

(10.100)

where $\mathbf{I}$ and $\mathbf{0}$ denote an identity and a zero matrix, respectively.

Equation (10.100) can be written in the form of Eq. (10.93). The wave components can be solved from Eq. (10.94). With the availability of responses of the wave components, the deflection of any point on the beam can be obtained. For example, the deflection of a point that is a distance x away from Boundary A in Figure 10.8, by the Mindlin-Herrmann longitudinal vibration theory, is

$$u = \begin{bmatrix} 1 & 1 \end{bmatrix} \mathbf{f}(x)\mathbf{a}^+ + \begin{bmatrix} 1 & 1 \end{bmatrix} \mathbf{f}(-x)\mathbf{a}^-$$

(10.101a)

or equivalently is

$$u = \begin{bmatrix} 1 & 1 \end{bmatrix} \mathbf{f}(L-x)\mathbf{b}^- + \begin{bmatrix} 1 & 1 \end{bmatrix} \mathbf{f}(-(L-x))\mathbf{b}^+$$

(10.101b)

The equations presented in this section are in vector or matrix form, which can be applied directly to multiple wave modes related analysis. Note that Eqs. (10.95a), (10.95b), (10.96a), (10.96b), (10.101a), and (10.101b) are based on the Mindlin–Herrmann longitudinal vibration theory. For analysis based on the Three-mode longitudinal vibration theory, vector [1 1] in the above mentioned equations need to be replaced by vector [1 1 1]. For analysis based on the Love longitudinal vibration theory, which involves a single wave mode, the vectors become scalars. In particular, from Eq. (10.6), the propagation coefficient for a distance x is e^{-ikx}. From Eqs. (10.27) and (10.30), the reflection coefficients at a free and a fixed boundary are 1 and -1, respectively. From Eqs. (10.53b) and (10.54b), for waves generated by an external force on the span: $x_C^+ = 1$, $x_C^- = 1$, $x_E^+ = 1$, $x_E^- = -1$, and $q_E = \dfrac{i\bar{F}}{k\left(EA - \nu^2 \rho J \omega^2\right)}$. From Eqs. (10.76) and (10.77), for waves

generated by an external force at a free end: $y_E^+ = 1$, $y_E^- = -1$, and $q_F = \dfrac{i\bar{F}}{k\left(EA - \nu^2 \rho J \omega^2\right)}$.

10.6 Numerical Examples and Experimental Studies

Free and forced longitudinal vibrations of the same uniform steel beam with free-free boundaries studied in Chapters 2, 3, and 9 are analyzed based on the various longitudinal vibration theories. This allows the analytical results to be compared with the experimental results. The material and geometrical properties of the steel beam are: the Young's modulus is 198.87 GN/m^2 and volume mass density is 7664.5 kg/m^3. The cross section of the beam is a rectangular shape whose dimensions are 25.4 mm by 12.7 mm (or 1.0 in by 0.5 in). The length of the beam is 914.4 mm (or 3.0 ft).

Longitudinal vibration tests are conducted on the beam. The experimental set up is shown in Figure 2.15 and the experiments follow the same procedures described in Section 2.5. To have free-free boundary conditions, the metal beam is

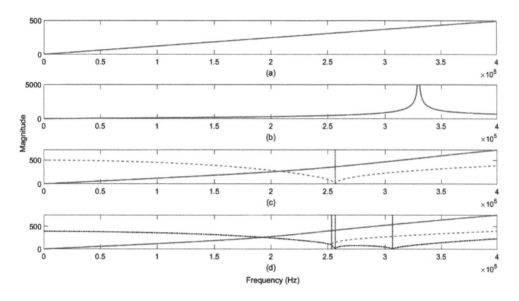

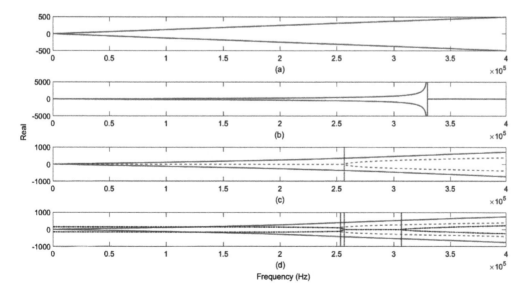

Figure 10.9 Magnitude, real part, and imaginary part of longitudinal wavenumbers by the (a) Elementary, (b) Love, (c) Mindlin–Herrmann, and (d) Three-mode theories.

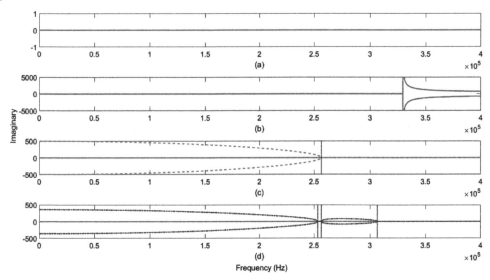

Figure 10.9 (Cont'd)

suspended by two thin wires. The test is conducted with an impact force applied at one end of the beam using a Brüel & Kjær Type 8202 impact hammer. Vibrations are picked up by a Brüel & Kjær Type 4397 accelerometer, which is mounted at the other end of the beam. Frequency responses are obtained using the PULSE unit from Brüel & Kjær. The frequency span and frequency resolution are $6400\,Hz$ and $1\,Hz$, respectively.

Figure 10.9 shows the magnitude, real part, and imaginary part of the longitudinal wavenumbers by the Elementary, Love, Mindlin–Herrmann, and Three-mode longitudinal vibration theories. The positive and negative wavenumber values represent positive- and negative-going waves in the beam. The longitudinal waves of a uniform beam predicted by the Elementary longitudinal vibration theory is non-dispersive, in that the phase velocity ω/k is of a constant value. The other three longitudinal vibration theories, namely, the Love, Mindlin–Herrmann, and Three-mode longitudinal vibration theories, predict dispersive longitudinal waves, in that the phase velocity ω/k is frequency dependent. The number of pairs of positive- and negative-going wave components varies by the longitudinal vibration theories. According to the Elementary and Love longitudinal vibration theories, one single pair of longitudinal waves exists. The Mindlin–Herrmann and Three-mode longitudinal vibration theories predict two and three pairs of wave components, respectively.

According to the Elementary longitudinal vibration theory, the wavenumber is always real, indicating that the pair of longitudinal vibration waves is always propagating. For the Love longitudinal vibration theory, there is a critical frequency at which the denominator of the wavenumber becomes zero. For the example steel beam, this critical frequency, calculated from Eq. (10.3), is $330\,kHz$. Below this critical frequency, the wavenumber is real, while above this critical frequency, the wavenumber becomes imaginary. This indicates that below the critical frequency, there is a pair of propagating waves; while above this critical frequency, there is a pair of decaying waves in an axially vibrating beam.

Between the two wavenumbers predicted by the Mindlin–Herrmann longitudinal vibration theory, one is always real. The root situation of the other is frequency dependent. Below the cut-off frequency, it is imaginary; while above the cut-off frequency, it becomes real. This indicates that below the cut-off frequency, a pair of propagating waves and a pair of decaying waves exist; while above the cut-off frequency, two pairs of propagating waves exist in an axially vibrating beam. The cut-off frequency can be calculated from Eq. (10.10), which is $257\,kHz$ for the example steel beam when the adjusting coefficient K_2 is set to unity.

The root situation of the three wavenumbers predicted by the Three-mode longitudinal vibration theory is complicated, as shown in Figure 10.9. To get a better view of the wave mode transition at the cut-off frequencies, the magnitude, real, and imaginary of each individual wavenumber by the Three-mode longitudinal vibration theory are presented in Figure 10.10.

From Eq. (10.18), there are two cut-off frequencies by the Three-mode longitudinal vibration theory, which correspond to two wave mode transitions. However, the wavenumber plots show three wave mode transitions. Where is this additional cut-off frequency from? Recall that the two cut-off frequencies in Eq. (10.18) are found from solving the roots of the dispersion equation obtained in Eq. (10.17a). Interestingly, this additional cut-off frequency is related to the discriminant polynomial Δ of the dispersion equation obtained in Eq. (10.17a), which is a cubic equation of k^2. Setting the discriminant polynomial Δ to zero and solving for the roots gives this additional cut-off frequency. This is because the sign of Δ determines the nature of the roots of the cubic polynomial equation of k^2:

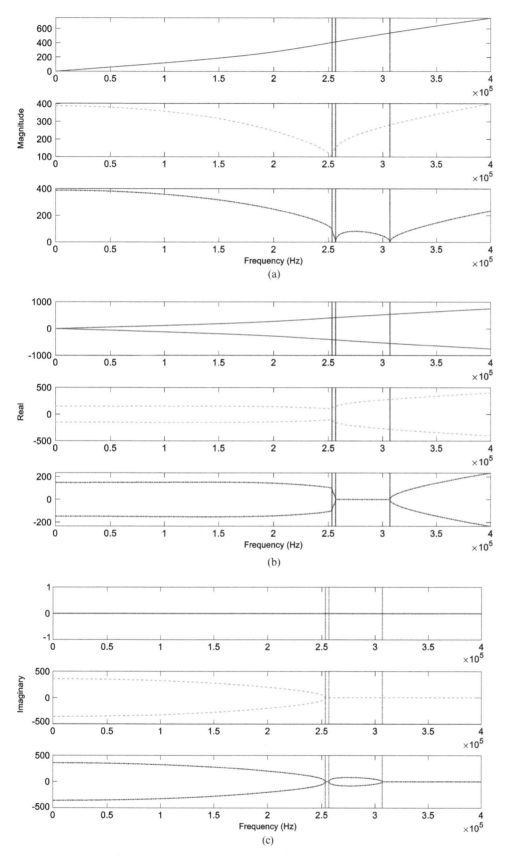

Figure 10.10 Longitudinal wavenumbers by the Three-mode theory: (a) magnitude, (b) real part, and (c) imaginary part.

$$\Delta = 18z_3 z_2 z_1 z_0 - 4z_2^3 z_0 + z_2^2 z_1^2 - 4z_3 z_1^3 - 27z_3^2 z_0^2 \tag{10.102}$$

where z_0, z_1, z_2, z_3 are given in Eq. (10.17b).

- When $\Delta < 0$, the equation has one real root and two complex conjugate roots;
- When $\Delta = 0$, the equation has a repeated root and all its roots are real;
- When $\Delta > 0$, the equation has three distinct real roots.

The frequency at which $\Delta = 0$ therefore corresponds to a change in root situation. Consequently there is a wave mode transition at this frequency. This additional cut-off frequency is named the pseudo cut-off frequency by the author.

The pseudo cut-off frequency, which is obtained from setting Eq. (10.102) to zero and solving for the roots, and the two cut-off frequencies calculated from Eq. (10.18) are marked using vertical lines in Figures 10.9 and 10.10. For the steel beam, the pseudo cut-off frequency calculated from setting Eq. (10.102) to zero is 253 kHz. The two cut-off frequencies calculated from Eq. (10.18) are 257 kHz and 307 kHz, which are greater than the pseudo cut-off frequency. Note that the lower cut-off frequency of the Three-mode longitudinal vibration theory calculated from Eq. (10.18) agrees with the cut-off frequency of the Mindlin–Herrmann longitudinal vibration theory, because the adjusting coefficient K_2 in Eq. (10.10) is set to unity.

Among the three wavenumbers predicted by the Three-mode longitudinal vibration theory, one is always real. The root situations of the other two are frequency dependent:

- Below the pseudo cut-off frequency, the two wavenumbers are complex conjugates;
- Between the pseudo and lower calculated cut-off frequency, the two wavenumbers are real;
- Between the two calculated cut-off frequencies, one wavenumber is real and the other wavenumber is imaginary;
- Above the higher calculated cut-off frequency, the two wavenumbers are real.

Consequently, in addition to one pair of waves that is always propagating, the other two pairs of waves experiences wave mode transitions as follows:

- Below the pseudo cut-off frequency, two pairs of propagating waves with decaying magnitudes;
- Between the pseudo and lower calculated cut-off frequencies, two pairs of propagating waves;
- Between the two calculated cut-off frequencies, one pair of propagating waves and one pair of decaying waves;
- Above the higher calculated cut-off frequency, two pairs of propagating waves.

Recall that the critical frequency of the steel beam predicted by the Love longitudinal vibration theory is 330 kHz, which is greater than all three cut-off frequencies of the Three-mode longitudinal vibration theory. According to the Three-mode longitudinal vibration theory, above the highest cut-off frequency of 307 kHz, all three pairs of waves are propagating waves.

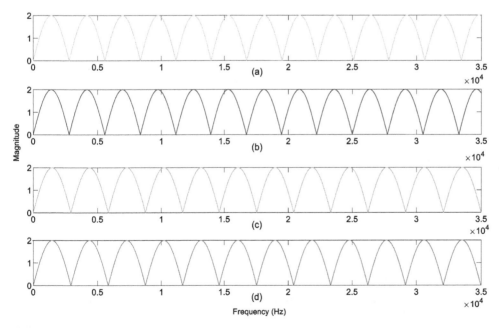

Figure 10.11 Magnitudes of the characteristic polynomials of the beam by the (a) Elementary, (b) Love, (c) Mindlin-Herrmann, and (d) Three-mode theories.

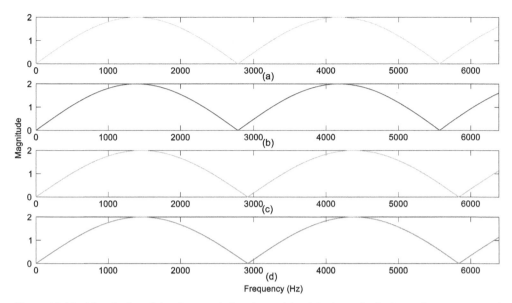

Figure 10.12 Magnitudes of the characteristic polynomials of the beam in the lower frequency range by the (a) Elementary, (b) Love, (c) Mindlin-Herrmann, and (d) Three-mode theories.

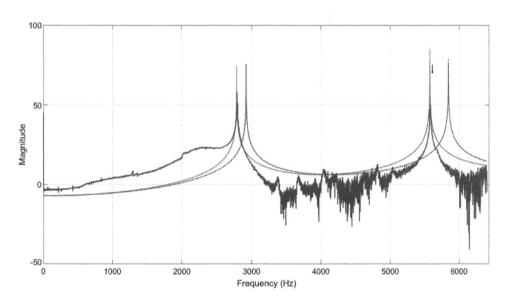

Figure 10.13 Overlaid experimental and analytical inertance frequency responses by the Elementary (__), Love (-·-·), Mindlin–Herrmann (...), and Three-mode (---) theories.

However, the Love longitudinal vibration theory predicts one single pair of decaying waves above the critical frequency $330\,kHz$, which contrasts the prediction from the Three-mode longitudinal vibration theory in the corresponding frequency range.

Figure 10.11 shows the magnitudes of the characteristic polynomials of the axially vibrating beam based on the Elementary, Love, Mindlin–Herrmann, and Three-mode longitudinal vibration theories. The frequencies at local minima correspond to the natural frequencies. The natural frequencies predicted by the Elementary and Love longitudinal vibration theories agree well, up to about $20\,kHz$. The natural frequencies predicted by the Mindlin–Herrmann and Three-mode longitudinal vibration theories agree well over the entire frequency range shown, up to about $30\,kHz$. However, the natural frequencies from the two groups do not seem to agree with each other.

Figure 10.12 shows the magnitude responses in the lower frequency range up to $6400\,Hz$, which is the frequency range for experimental study. Clearly these two groups do not at all agree at low frequencies. This contrasts the predicted bending vibrations, studied in Chapter 9, by the four bending vibration theories, namely, the Euler–Bernoulli, Rayleigh, Shear, and Timoshenko bending vibration theories, where all four theories agree at low frequencies.

Figure 10.13 shows the overlaid experimental and analytical inertance frequency (Acceleration/Force) responses by the Elementary, Love, Mindlin–Herrmann, and Three-mode longitudinal vibration theories. Because the external excitation force is applied at a free end, the analytical results are obtained following the analysis described in Eqs. (10.97) to (10.101b). The experimental results agree well with the analytical results from the Elementary and Love longitudinal vibration theories. The Mindlin–Herrmann and Three-mode longitudinal vibration theories overpredict the natural frequencies at low frequency range.

From the comparisons of experimental results with analytical predictions of bending vibrations based on the various bending vibration theories presented in Figures (9.13) and (9.16) of Chapter 9, and the comparisons of experimental results with analytical predictions of longitudinal vibrations based on the various longitudinal vibration theories shown in Figure 10.13 in this chapter, it seems that the Timoshenko bending vibration theory and the Elementary longitudinal vibration theory are all that are needed for structural vibration analysis.

References

Doyle J.F. *Wave Propagation in Structures*, Spring-Verlag, New York (1989).

Krawczuk M., Grabowska J., and Palacz M. Longitudinal Wave Propagation Part I --- Comparison of Rod Theories, *Journal of Sound and Vibration*, 295, 461–478 (2006)

Mei C. Comparison of the Four Rod Theories of Longitudinally Vibrating Rods, *Journal of Vibration and Control*, 21(8), 1639–1656 (2015).

11

Bending and Longitudinal Waves in Built-up Planar Frames

A built-up planar frame often comprises an angle joint, such as that in an L-shaped, T-shaped, or H-shaped frame. When vibrations propagate in a frame, an angle joint in general causes wave mode conversion. As a result, bending and longitudinal waves co-exist in a frame that undergoes in-plane vibrations.

In-Plane vibrations in built-up planar frame structures are analyzed from the wave standpoint. In the study that follows, longitudinal vibrations are modeled using the Elementary vibration theory, and bending vibrations are modeled using the Euler–Bernoulli vibration theory as well as the Timoshenko vibration theory.

11.1 The Governing Equations and the Propagation Relationships

The governing equations of motion for free vibration of a uniform beam element in a planar frame, with bending vibrations modeled by the Euler–Bernoulli vibration theory and longitudinal vibrations modeled using the Elementary vibration theory, are

$$EI\frac{\partial^4 y(x,t)}{\partial x^4} + \rho A\frac{\partial^2 y(x,t)}{\partial t^2} = 0$$

$$\rho A\frac{\partial^2 u(x,t)}{\partial t^2} - EA\frac{\partial^2 u(x,t)}{\partial x^2} = 0 \tag{11.1a}$$

The governing equations of motion for free vibration of a uniform beam element in a planar frame, with bending vibrations modeled by the Timoshenko vibration theory and longitudinal vibrations modeled using the Elementary vibration theory, are

$$GA\kappa\left[\frac{\partial \psi(x,t)}{\partial x} - \frac{\partial^2 y(x,t)}{\partial x^2}\right] + \rho A\frac{\partial^2 y(x,t)}{\partial t^2} = 0$$

$$EI\frac{\partial^2 \psi(x,t)}{\partial x^2} + GA\kappa\left[\frac{\partial y(x,t)}{\partial x} - \psi(x,t)\right] - \rho I\frac{\partial^2 \psi(x,t)}{\partial t^2} = 0$$

$$\rho A\frac{\partial^2 u(x,t)}{\partial t^2} - EA\frac{\partial^2 u(x,t)}{\partial x^2} = 0 \tag{11.1b}$$

where x is the position along the beam axis, t is the time, A is the cross-sectional area, I is the area moment of inertia of the cross section, E is the Young's modulus, and ρ is the volume mass density of the beam. $y(x,t)$ is the bending deflection of the centerline of the beam, $u(x,t)$ is the longitudinal deflection of the centerline of the beam, $\psi(x,t)$ is the slope due to bending, $\partial y(x,t)/\partial x$ is the slope of the centerline of the beam, and $\partial y(x,t)/\partial x - \psi(x,t)$ is the shear angle. G and κ are the shear modulus and shear coefficient, respectively. According to the Euler–Bernoulli bending vibration theory, the shear deformation is zero, that is, $\frac{\partial y(x,t)}{\partial x} - \psi(x,t) = 0$. As a result, $\psi(x,t) = \frac{\partial y(x,t)}{\partial x}$ by the Euler–Bernoulli bending vibration theory.

Mechanical Wave Vibrations: Analysis and Control, First Edition. Chunhui Mei.
© 2023 Chunhui Mei. Published 2023 by John Wiley & Sons Ltd.
Companion Website: www.wiley.com/go/Mei/MechanicalWaveVibrations

The wavenumbers for longitudinal and bending vibrations have been obtained in Chapters 2, 3, and 9, by assuming time harmonic motion and using separation of variables. The bending wavenumbers k_1 and k_2 based on the Euler–Bernoulli bending vibration theory are

$$k_1 = k_2 = \sqrt[4]{\rho A \omega^2 / EI} \tag{11.2a}$$

and based on the Timoshenko bending vibration theory are

$$k_1 = \left[\frac{1}{2}\left(\frac{1}{C_s}\right)^2 \omega^2 + \sqrt{\frac{\omega^2}{C_b^2} + \frac{1}{4}\left(\frac{1}{C_s}\right)^4 \omega^4} \right]^{\frac{1}{2}}$$

$$k_2 = \left[\left| \frac{1}{2}\left(\frac{1}{C_s}\right)^2 \omega^2 - \sqrt{\frac{\omega^2}{C_b^2} + \frac{1}{4}\left(\frac{1}{C_s}\right)^4 \omega^4} \right| \right]^{\frac{1}{2}} \tag{11.2b}$$

The longitudinal wavenumber based on the Elementary vibration theory is

$$k_3 = \sqrt{\frac{\rho}{E}} \omega \tag{11.2c}$$

where ω is the circular frequency, $C_b = \sqrt{\dfrac{EI}{\rho A}}$, and $C_s = \sqrt{\dfrac{GA\kappa}{\rho A}}$.

Recall that the Timoshenko bending vibration theory predicts a wave mode transition at a cut-off frequency ω_c, which is given by

$$\omega_c = \frac{C_s}{C_r} \tag{11.3}$$

where $C_r = \sqrt{\dfrac{\rho I}{\rho A}} = \sqrt{\dfrac{I}{A}}$. Below the cut-off frequency ω_c, a pair of propagating waves and a pair of decaying waves exist in the beam, while above the cut-off frequency ω_c, there are two pairs of propagating waves. In audio frequency applications, the former case is overwhelmingly the most common. Consequently, all expressions based on the Timoshenko bending vibration theory in this chapter are for $\omega < \omega_c$, which assumes a pair of propagating and a pair of decaying bending waves in the beam. When $\omega > \omega_c$, both pairs of bending waves become propagating waves, k_2 needs to be replaced by ik_2 in these expressions to reflect the wave mode transition at the cut-off frequency ω_c, where i is the imaginary unit.

With time dependence $e^{i\omega t}$ suppressed, the solutions to Eq. (11.1a) can be written as

$$y(x) = a_1^+ e^{-ik_1 x} + a_2^+ e^{-k_2 x} + a_1^- e^{ik_1 x} + a_2^- e^{k_2 x}$$

$$\psi(x) = \frac{\partial y(x)}{\partial x} = -ik_1 a_1^+ e^{-ik_1 x} - k_2 a_2^+ e^{-k_2 x} + ik_1 a_1^- e^{ik_1 x} + k_2 a_2^- e^{k_2 x}$$

$$u(x) = c^+ e^{-ik_3 x} + c^- e^{ik_3 x} \tag{11.4a}$$

and the solutions to Eq. (11.1b) are

$$y(x) = a_1^+ e^{-ik_1 x} + a_2^+ e^{-k_2 x} + a_1^- e^{ik_1 x} + a_2^- e^{k_2 x}$$

$$\psi(x) = -iPa_1^+ e^{-ik_1 x} - Na_2^+ e^{-k_2 x} + iPa_1^- e^{ik_1 x} + Na_2^- e^{k_2 x}$$

$$u(x) = c^+ e^{-ik_3 x} + c^- e^{ik_3 x} \tag{11.4b}$$

where i is the imaginary unit, and $P = k_1\left(1 - \dfrac{\omega^2}{k_1^2 C_s^2}\right)$ and $N = k_2\left(1 + \dfrac{\omega^2}{k_2^2 C_s^2}\right)$ from Eq. (9.15). Superscripts $+$ and $-$ in wave amplitude denote positive- and negative-going bending or longitudinal waves. Subscripts 1 and 2 in $a_{1,2}^{\pm}$ refer to the two bending wave components.

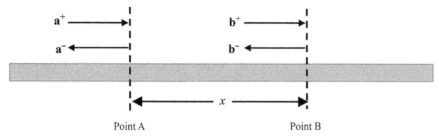

Figure 11.1 Wave propagation relationships.

Consider two points A and B that are a distance x apart on a uniform beam, as shown in Figure 11.1. Waves propagate from one point to the other, with the propagation relationships determined by the appropriate wavenumber. The positive- and negative-going waves at Points A and B are related by

$$\mathbf{b}^+ = \mathbf{f}(x)\mathbf{a}^+, \mathbf{a}^- = \mathbf{f}(x)\mathbf{b}^- \tag{11.5a}$$

where

$$\mathbf{a}^+ = \begin{bmatrix} a_1^+ \\ a_2^+ \\ c^+ \end{bmatrix}, \mathbf{a}^- = \begin{bmatrix} a_1^- \\ a_2^- \\ c^- \end{bmatrix}, \mathbf{b}^+ = \begin{bmatrix} b_1^+ \\ b_2^+ \\ d^+ \end{bmatrix}, \mathbf{b}^- = \begin{bmatrix} b_1^- \\ b_2^- \\ d^- \end{bmatrix} \tag{11.5b}$$

are the wave vectors, in which $a_1^\pm$, $a_2^\pm$, $b_1^\pm$, and $b_2^\pm$ are bending wave components, and $c^\pm$ and $d^\pm$ are longitudinal wave components, and

$$\mathbf{f}(x) = \begin{bmatrix} e^{-ik_1 x} & 0 & 0 \\ 0 & e^{-k_2 x} & 0 \\ 0 & 0 & e^{-ik_3 x} \end{bmatrix} \tag{11.5c}$$

is the propagation matrix for a distance x.

11.2 Wave Reflection at Classical Boundaries

Figure 11.2 shows the sign convention adopted in the analysis that follows.

Following the sign convention defined in Figure 11.2, and according to the Elementary longitudinal vibration theory, the internal resistant axial force $F(x,t)$ and the axial deflection $u(x,t)$ are related by

$$F(x,t) = EA \frac{\partial u(x,t)}{\partial x} \tag{11.6a}$$

The internal resistant shear force $V(x,t)$ and bending moment $M(x,t)$ are related to the transverse deflection $y(x,t)$ and bending slope $\psi(x,t)$ by the Euler–Bernoulli bending vibration theory as

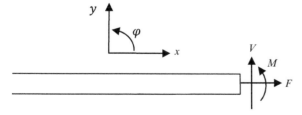

Figure 11.2 Definition of positive internal resistant axial force, shear force, and bending moment.

$$V(x,t) = -EI \frac{\partial^3 y(x,t)}{\partial x^3}, \ M(x,t) = EI \frac{\partial \psi(x,t)}{\partial x} = EI \frac{\partial^2 y(x,t)}{\partial x^2} \tag{11.6b}$$

where $\psi(x,t) = \dfrac{\partial y(x,t)}{\partial x}$ according to the Euler–Bernoulli bending vibration theory.

By the Timoshenko bending vibration theory, the internal resistant shear force $V(x,t)$ and bending moment $M(x,t)$ are related to the transverse deflection $y(x,t)$ and bending slope $\psi(x,t)$ by

$$V(x,t) = GA\kappa \left[\frac{\partial y(x,t)}{\partial x} - \psi(x,t) \right], \ M(x,t) = EI \frac{\partial \psi(x,t)}{\partial x} \tag{11.6c}$$

As illustrated in Figure 11.3, incident waves are reflected at a boundary. The reflected waves $\mathbf{a}^-$ and incident waves $\mathbf{a}^+$ are related by reflection matrix $\mathbf{r}$,

$$\mathbf{a}^- = \mathbf{r}\mathbf{a}^+ \tag{11.7}$$

where the reflection matrix is determined by boundary conditions.

The equilibrium conditions at a free boundary are

$$M(x,t) = 0, \ V(x,t) = 0, \text{ and } F(x,t) = 0 \tag{11.8a}$$

At a clamped boundary, the boundary conditions are

$$y(x,t) = 0, \ \psi(x,t) = 0, \text{ and } u(x,t) = 0 \tag{11.8b}$$

The boundary conditions at a pinned boundary are

$$y(x,t) = 0, \ M(x,t) = 0, \text{ and } F(x,t) = 0 \tag{11.8c}$$

The reflection matrices at the classical free, clamped, and pinned boundaries are a combination of the corresponding reflection matrices for longitudinal and bending waves obtained in Chapters 2, 3, and 9. The reflection matrices at classical boundaries, by the Euler–Bernoulli bending vibration theory and Elementary longitudinal vibration theory; and by the Timoshenko bending vibration theory and Elementary longitudinal vibration theory, respectively, are

- At a free boundary

$$\mathbf{r} = \begin{vmatrix} -i & 1+i & 0 \\ 1-i & i & 0 \\ 0 & 0 & 1 \end{vmatrix} \tag{11.9a}$$

$$\mathbf{r} = \begin{vmatrix} \dfrac{-Pk_1(k_2 - N) + ik_2 N(k_1 - p)}{Pk_1(k_2 - N) + ik_2 N(k_1 - p)} & \dfrac{2Nk_2(k_2 - N)}{Pk_1(k_2 - N) + ik_2 N(k_1 - P)} & 0 \\ \dfrac{i2Pk_1(k_1 - P)}{Pk_1(k_2 - N) + ik_2 N(k_1 - P)} & \dfrac{Pk_1(k_2 - N) + ik_2 N(k_1 - p)}{Pk_1(k_2 - N) + ik_2 N(k_1 - p)} & 0 \\ 0 & 0 & 1 \end{vmatrix} \tag{11.9b}$$

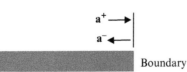

Figure 11.3 Wave reflection at a boundary.

- At a clamped boundary

$$\mathbf{r} = \begin{bmatrix} -i & -1-i & 0 \\ -1+i & i & 0 \\ 0 & 0 & -1 \end{bmatrix} \tag{11.10a}$$

$$\mathbf{r} = \begin{bmatrix} \dfrac{P-iN}{P+iN} & \dfrac{-i2N}{P+iN} & 0 \\ \dfrac{-2P}{P+iN} & -\dfrac{P-iN}{P+iN} & 0 \\ 0 & 0 & -1 \end{bmatrix} \tag{11.10b}$$

- At a pinned boundary

$$\mathbf{r} = \begin{bmatrix} -1 & 0 & 0 \\ 0 & -1 & 0 \\ 0 & 0 & -1 \end{bmatrix} \tag{11.11a}$$

$$\mathbf{r} = \begin{bmatrix} -1 & 0 & 0 \\ 0 & -1 & 0 \\ 0 & 0 & -1 \end{bmatrix} \tag{11.11b}$$

where i is the imaginary unit. The reflection matrices at a pinned boundary obtained from the two bending vibration theories are identical.

11.3 Force Generated Waves

Externally applied force and moment have the effect of injecting waves into a continuous structure, as shown in Figure 11.4. Waves $\mathbf{a}^{\pm}$ and $\mathbf{b}^{\pm}$ are injected by and related to the externally applied axial force $\bar{F}$, transverse force $\bar{Q}$, and bending moment $\bar{M}$.

Since the axial and bending deflections on the left and right side of the point where the external force and moment are applied are the same, denoting physical parameters on the left and right side of the point where the force and moment are applied using subscripts − and +, respectively,

$$u_- = u_+, \, y_- = y_+, \, \psi_- = \psi_+ \tag{11.12}$$

From the free body diagram of Figure 11.4, the externally applied axial force, transverse force, and bending moment are related to the internal resistant forces and moments on the left and right side of the point where the external excitations are applied by

$$\bar{F} = F_- - F_+, \, \bar{Q} = V_- - V_+, \, \bar{M} = M_- - M_+ \tag{11.13}$$

The relationships described in Eqs. (11.12) and (11.13) are called the continuity and equilibrium equations, respectively.

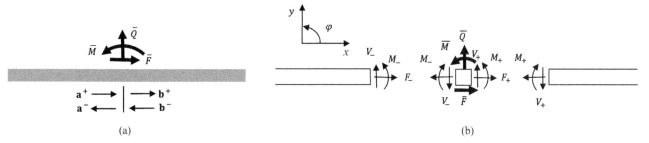

Figure 11.4   (a) Waves generated by externally applied force and moment and (b) free body diagram.

The relationships between injected waves and externally applied axial force, transverse force, and bending moment are found to be a combination of the corresponding relationships obtained in Chapters 2, 3, and 9. The relationships between injected waves and externally applied force and moment are

$$\mathbf{b}^+ - \mathbf{a}^+ = \mathbf{q} + \mathbf{m} + \mathbf{f}$$

$$\mathbf{a}^- - \mathbf{b}^- = \mathbf{q} - \mathbf{m} + \mathbf{f} \tag{11.14}$$

where the load vectors by the Euler–Bernoulli bending vibration theory and Elementary longitudinal vibration theory are

$$\mathbf{q} = \begin{bmatrix} -i \\ -1 \\ 0 \end{bmatrix} \frac{\bar{Q}}{4EIk_1^3} = \begin{bmatrix} -i \\ -1 \\ 0 \end{bmatrix} \frac{\bar{Q}}{4EIk_2^3}, \ \mathbf{m} = \begin{bmatrix} 1 \\ -1 \\ 0 \end{bmatrix} \frac{\bar{M}}{4EIk_1^2} = \begin{bmatrix} 1 \\ -1 \\ 0 \end{bmatrix} \frac{\bar{M}}{4EIk_2^2}, \ \mathbf{f} = \begin{bmatrix} 0 \\ 0 \\ -i \end{bmatrix} \frac{\bar{F}}{2EAk_3} \tag{11.15a}$$

and the load vectors by the Timoshenko bending vibration theory and Elementary longitudinal vibration theory are

$$\mathbf{q} = \begin{bmatrix} -iN \\ -P \\ 0 \end{bmatrix} \frac{\bar{Q}}{2GA\kappa(k_1 N - k_2 P)}, \ \mathbf{m} = \begin{bmatrix} 1 \\ -1 \\ 0 \end{bmatrix} \frac{\bar{M}}{2EI(k_1 P + k_2 N)}, \ \mathbf{f} = \begin{bmatrix} 0 \\ 0 \\ -i \end{bmatrix} \frac{\bar{F}}{2EAk_3} \tag{11.15b}$$

11.4   Free and Forced Vibration Analysis of a Multi-story Multi-bay Planar Frame

From the wave standpoint, vibrations travel along uniform structural waveguides and are reflected and transmitted at structural discontinuities.

Figure 11.5 shows waves in a general planar frame of n stories and m bays. Superscripts + and − in wave components denote positive- and negative-going waves, respectively. On the horizontal beam elements in Figure 11.5, waves move rightwards are defined positive going, and waves move leftwards are defined negative going. On the vertical beam elements in Figure 11.5, waves move upwards are defined positive going, and waves move downwards are defined negative going. Each wave component on Figure 11.5 is associated with a structural discontinuity, either a joint or a boundary, the location of the related joint or boundary is denoted in the subscript of a wave component.

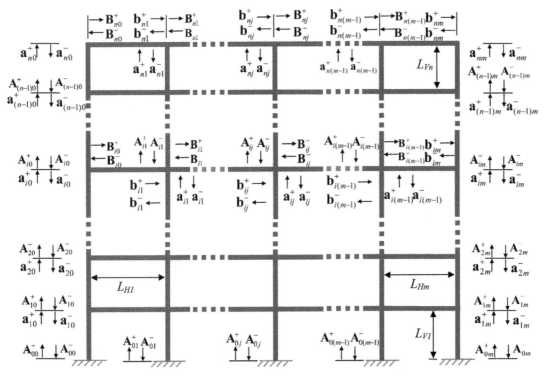

Figure 11.5 Waves in a general planar frame.

Because an n-story m-bay planar frame comprises $(2m + 1)n$ uniform beam elements, $(2m + 1)n$ pairs of propagation relationships exist. At the boundary of the frame, $(m + 1)$ reflection relationships exist. There are three types of joints in an n-story m-bay planar frame, namely, the L, T, and "+" joints. There is one L joint on each corner of the top side. There are $(n-1)$ T joints on the left and right side of the planar frame, respectively, and $(m-1)$ T joints on the top side. The "+" joints are all located inside the planar frame, $(n-1)(m-1)$ in total counts. Each L, T, and "+" joint contains two, three, and four equations, respectively, in describing the relationships among the incoming and outgoing waves from the beam elements joined at the corresponding joint.

In total, there are $(2m + 1)n \times 2 + (m + 1) + 2 \times 2 + [2(n\text{-}1) + (m\text{-}1)] \times 3 + (n\text{-}1)(m\text{-}1) \times 4$ matrix equations in describing the propagation, reflection, and transmission relationships of vibration waves in an n-story m-bay planar frame:

- There are $(2m + 1)n$ pairs of propagation relationships in total, among them, $(2m + 1)n$ are along the vertical beam elements and $(2m + 1)n$ along the horizontal beam elements,

$$\mathbf{a}_{ij}^+ = \mathbf{f}(L_{Vi})\mathbf{A}_{(i-1)j}^+, \quad \mathbf{A}_{(i-1)j}^- = \mathbf{f}(L_{Vi})\mathbf{a}_{ij}^-, \quad \text{where } i = 1, 2, \dots, n; \ j = 0, 1, 2, \dots, m,$$

$$\mathbf{b}_{ij}^+ = \mathbf{f}(L_{Hj})\mathbf{B}_{i(j-1)}^+, \quad \mathbf{B}_{i(j-1)}^- = \mathbf{f}(L_{Hj})\mathbf{b}_{ij}^-, \quad \text{where } i = 1, 2, \dots, n; \ j = 1, 2, \dots, m \tag{11.16}$$

where $\mathbf{f}(x)$ is the propagation matrix for a distance x described in Eq. (11.5c), and subscripts H and V indicate that the propagations are along the horizontal and vertical beam elements, respectively.

- There are $(m + 1)$ reflection relationships at the boundaries,

$$\mathbf{A}_{0j}^+ = \mathbf{r}_{0j}\mathbf{A}_{0j}^-, \quad \text{where } j = 0, 1, 2, \dots, m \tag{11.17}$$

where $\mathbf{r}$ is the reflection matrix at a boundary, whose location is denoted in the subscript.

- The two sets of reflection and transmission relationships at the L joints on the left and right corner of the top side are

$$\mathbf{B}_{n0}^+ = \mathbf{r}_{22}^{n0}\mathbf{B}_{n0}^- + \mathbf{t}_{12}^{n0}\mathbf{a}_{n0}^+, \quad \mathbf{a}_{n0}^- = \mathbf{r}_{11}^{n0}\mathbf{a}_{n0}^+ + \mathbf{t}_{21}^{n0}\mathbf{B}_{n0}^-;$$

$$\mathbf{b}_{nm}^- = \mathbf{r}_{22}^{nm}\mathbf{b}_{nm}^+ + \mathbf{t}_{12}^{nm}\mathbf{a}_{nm}^+, \quad \mathbf{a}_{nm}^- = \mathbf{r}_{11}^{nm}\mathbf{a}_{nm}^+ + \mathbf{t}_{21}^{nm}\mathbf{b}_{nm}^+ \tag{11.18}$$

- The $2(n-1)+(m-1)$ sets of reflection and transmission relationships at the T joints on the left side, right side, and top side, respectively, are

$$\mathbf{a}_{i0}^- = \mathbf{r}_{11}^{i0}\mathbf{a}_{i0}^+ + \mathbf{t}_{21}^{i0}\mathbf{B}_{i0}^- + \mathbf{t}_{31}^{i0}\mathbf{A}_{i0}^-,$$
$$\mathbf{B}_{i0}^+ = \mathbf{r}_{22}^{i0}\mathbf{B}_{i0}^- + \mathbf{t}_{12}^{i0}\mathbf{a}_{i0}^+ + \mathbf{t}_{32}^{i0}\mathbf{A}_{i0}^-,$$
$$\mathbf{A}_{i0}^+ = \mathbf{r}_{33}^{i0}\mathbf{A}_{i0}^- + \mathbf{t}_{23}^{i0}\mathbf{B}_{i0}^- + \mathbf{t}_{13}^{i0}\mathbf{a}_{i0}^+;$$

where $i = 1, 2, \dots, n\text{-}1$

$$\mathbf{a}_{im}^- = \mathbf{r}_{11}^{im}\mathbf{a}_{im}^+ + \mathbf{t}_{21}^{im}\mathbf{b}_{im}^+ + \mathbf{t}_{31}^{im}\mathbf{A}_{im}^-,$$
$$\mathbf{b}_{im}^- = \mathbf{r}_{22}^{im}\mathbf{b}_{im}^+ + \mathbf{t}_{12}^{im}\mathbf{a}_{im}^+ + \mathbf{t}_{32}^{im}\mathbf{A}_{im}^-,$$
$$\mathbf{A}_{im}^+ = \mathbf{r}_{33}^{im}\mathbf{A}_{im}^- + \mathbf{t}_{23}^{im}\mathbf{b}_{im}^+ + \mathbf{t}_{13}^{im}\mathbf{a}_{im}^+;$$

where $i = 1, 2, \dots, n\text{-}1$

$$\mathbf{a}_{nj}^- = \mathbf{r}_{11}^{nj}\mathbf{a}_{nj}^+ + \mathbf{t}_{21}^{nj}\mathbf{b}_{nj}^+ + \mathbf{t}_{31}^{nj}\mathbf{B}_{nj}^-,$$
$$\mathbf{b}_{nj}^- = \mathbf{r}_{22}^{nj}\mathbf{b}_{nj}^+ + \mathbf{t}_{12}^{nj}\mathbf{a}_{nj}^+ + \mathbf{t}_{32}^{nj}\mathbf{B}_{nj}^-, \tag{11.19}$$
$$\mathbf{B}_{nj}^+ = \mathbf{r}_{33}^{nj}\mathbf{B}_{nj}^- + \mathbf{t}_{23}^{nj}\mathbf{b}_{nj}^+ + \mathbf{t}_{13}^{nj}\mathbf{a}_{nj}^+,$$

where $j = 1, 2, \dots, m\text{-}1$

- The $(n-1)(m-1)$ sets of reflection and transmission relationships at the "+" joints located inside the planar frame are

$$
\begin{aligned}
\mathbf{a}_{ij}^{-} &= \mathbf{r}_{11}^{ij}\mathbf{a}_{ij}^{+} + \mathbf{t}_{21}^{ij}\mathbf{b}_{ij}^{+} + \mathbf{t}_{31}^{ij}\mathbf{A}_{ij}^{-} + \mathbf{t}_{41}^{ij}\mathbf{B}_{ij}^{-}, \\
\mathbf{b}_{ij}^{-} &= \mathbf{r}_{22}^{ij}\mathbf{b}_{ij}^{+} + \mathbf{t}_{12}^{ij}\mathbf{a}_{ij}^{+} + \mathbf{t}_{32}^{ij}\mathbf{A}_{ij}^{-} + \mathbf{t}_{42}^{ij}\mathbf{B}_{ij}^{-}, \\
\mathbf{A}_{ij}^{+} &= \mathbf{r}_{33}^{ij}\mathbf{A}_{ij}^{-} + \mathbf{t}_{13}^{ij}\mathbf{a}_{ij}^{+} + \mathbf{t}_{23}^{ij}\mathbf{b}_{ij}^{+} + \mathbf{t}_{43}^{ij}\mathbf{B}_{ij}^{-}, \\
\mathbf{B}_{ij}^{+} &= \mathbf{r}_{44}^{ij}\mathbf{B}_{ij}^{-} + \mathbf{t}_{14}^{ij}\mathbf{a}_{ij}^{+} + \mathbf{t}_{24}^{ij}\mathbf{b}_{ij}^{+} + \mathbf{t}_{34}^{ij}\mathbf{A}_{ij}^{-};
\end{aligned}
\tag{11.20}
$$

where $i = 1, 2, ..., n\text{-}1, j = 1, 2, ..., m\text{-}1$

In Eqs. (11.18) to (11.20), $\mathbf{r}$ and $\mathbf{t}$ are the reflection and transmission matrices, respectively. The superscripts in $\mathbf{r}$ and $\mathbf{t}$ denote the location of a joint, and the two numerical digits in the subscripts refer to the beam elements connected at the joint. The leading digit in a subscript identifies the beam element carrying the incident waves.

Free vibration responses can be obtained by assembling Eqs. (11.16) to (11.20) into matrix form,

$$
\mathbf{A}\mathbf{z} = 0
\tag{11.21}
$$

where $\mathbf{A}$ is a square coefficient matrix of size $\{(2m + 1)n \times 2 + (m + 1) + 2 \times 2 + [2(n-1) + (m-1)] \times 3 + (n-1)(m\text{-}1) \times 4\} \times 3$, and $\mathbf{z}$ a wave vector of size $\{(2m +)n \times 2 + (m + 1) + 2 \times 2 + [2(n-1) + (m\text{-}1)] \times 3 + (n-1)(m\text{-}1) \times 4\} \times 3$. The natural frequencies of the frame are obtained by setting the determinant of the coefficient matrix $\mathbf{A}$ to zero.

To obtain forced responses, Eq. (11.14), the relationships between injected waves and externally applied force and moment, needs to be added into the matrix assembly. In addition, the propagation relationships along the external excitation applied structural element(s) need to be modified. These wave relationships in matrix form are

$$
\mathbf{A}_f\mathbf{z}_f = \mathbf{F}
\tag{11.22}
$$

where $\mathbf{A}_f$ is a square coefficient matrix, $\mathbf{z}_f$ is a wave vector, and $\mathbf{F}$ is a force vector.

From Eq. (11.22), the wave components are solved,

$$
\mathbf{z}_f = \mathbf{A}_f^{-1}\mathbf{F}
\tag{11.23}
$$

From Eq. (11.23), the forced responses at any location on the structure can be obtained.

To obtain the reflection and transmission relationships at a structural discontinuity, a free body diagram needs to be drawn and analyzed for establishing the continuity relationships and the equations of motion at each discontinuity. A local coordinate must be defined for each beam element at a joint and for the joint itself. In selecting a local coordinate, the guidelines listed below need to be followed:

a) A two-dimensional Cartesian coordinate system is chosen to be in the plane of the frame. The x-axis needs to be along the longitudinal axis of a beam element, as is the case that the equations of motion of a beam are derived. The positive direction of the x-axis for any portion of a beam element must be consistently defined.
b) The positive direction of rotation angle φ adopted in deriving the governing equations of motion of a beam needs to be followed. In this book, it is the direction of rotating from the x-axis to the y-axis.
c) The same sign conventions adopted in deriving the equations of motion of a beam must be followed in defining positive internal resistant force and moment in each beam element with regard to its local coordinate. In this chapter, the sign convention defined in Figure 11.2 is followed, that is, an axial force is positive when it stretches the element, a shear force is positive when it rotates its element along the positive direction of angle φ, and a bending moment is positive when it bends its element concave towards the positive y-axis.

Other than the above rules, much freedom is allowed in selecting local coordinate systems for deriving the reflection and transmission relationships at a discontinuity.

Without loss of generality, a three-story two-bay planar frame shown in Figure 11.6 is considered, to better understand the selection of local coordinates for each waveguide with consistency. Ensuring that the positive direction of x-axis for any portion of a beam element is consistently defined is critical in describing the wave propagation relationships in the uniform beam element, as well as in finding the reflection and transmission relationships at a joint.

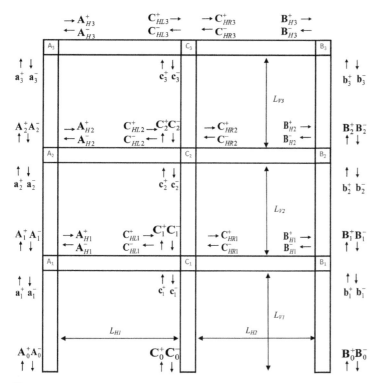

Figure 11.6 Waves in an example three-story two-bay frame.

In this three-story two-bay planar frame, there are fifteen uniform beam elements, hence fifteen waveguides. There are nine intermediate structural discontinuities in the frame, namely the two "L" shaped, five "T" shaped, and two "+" shaped joints. In addition, there are three boundary discontinuities. Incoming vibration waves are reflected and transmitted at an intermediate structural discontinuity, and are reflected at a boundary discontinuity.

The propagation relationships of the waves in the three-story two-bay frame shown in Figure 11.6 are

$$\mathbf{a}_i^+ = \mathbf{f}(L_{Vi})\mathbf{A}_{i-1}^+, \ \mathbf{A}_{i-1}^- = \mathbf{f}(L_{Vi})\mathbf{a}_i^-$$

$$\mathbf{c}_i^+ = \mathbf{f}(L_{Vi})\mathbf{C}_{i-1}^+, \ \mathbf{C}_{i-1}^- = \mathbf{f}(L_{Vi})\mathbf{c}_i^-$$

$$\mathbf{b}_i^+ = \mathbf{f}(L_{Vi})\mathbf{B}_{i-1}^+, \ \mathbf{B}_{i-1}^- = \mathbf{f}(L_{Vi})\mathbf{b}_i^-$$

$$\mathbf{C}_{HLi}^+ = \mathbf{f}(L_{H1})\mathbf{A}_{Hi}^+, \ \mathbf{A}_{Hi}^- = \mathbf{f}(L_{H1})\mathbf{C}_{HLi}^-$$

$$\mathbf{B}_{Hi}^+ = \mathbf{f}(L_{H2})\mathbf{C}_{HRi}^+, \ \mathbf{C}_{HRi}^- = \mathbf{f}(L_{H2})\mathbf{B}_{Hi}^- \tag{11.24}$$

where $i = 1, 2,$ and 3.

At each of the discontinuities, namely the boundaries and the L, T, and "+" joints, waves are related by the corresponding reflection and transmission relationships. Figure 11.7 shows a map of beam sequence numbers with respect to an L, T, or "+" joint, in which the numerical digit, following the joint type letter or symbol in the subscripts, refers to the sequence number of a beam element joined at the corresponding joint. Note that apart from the three vertical beam elements at the bottom, each of the remaining beam elements carries two sequence numbers because each of these beam elements is associated with two joints, one at each end of the beam element.

The reflection and transmission relationships of vibration waves at the boundaries and joints of Figure 11.6 are:

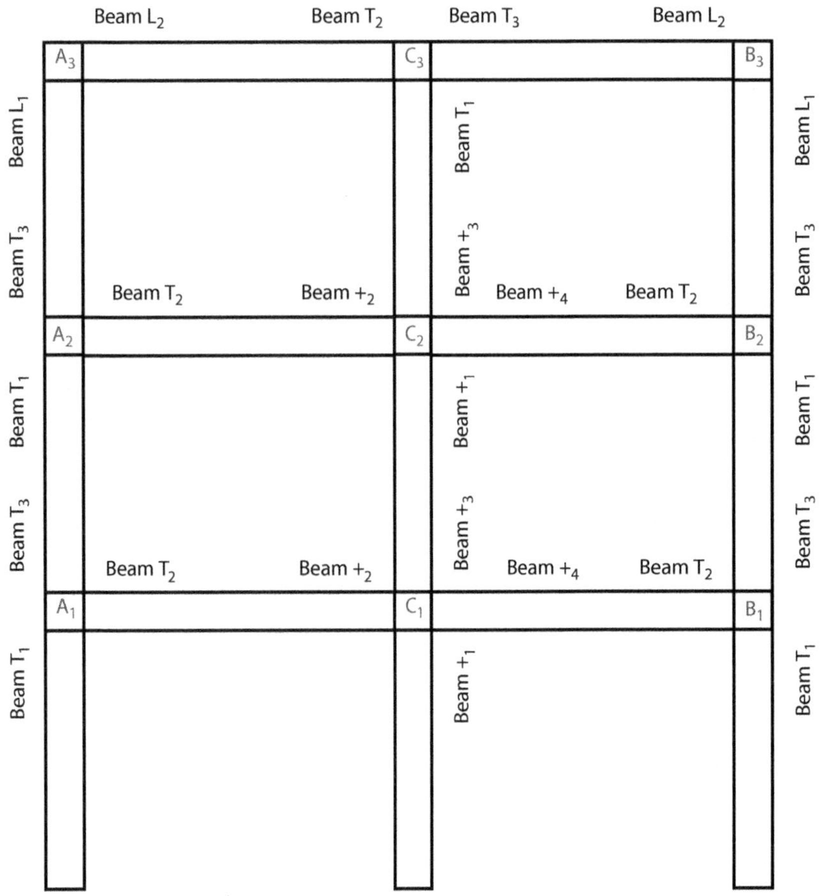

Figure 11.7 Map of beam sequence number by its joint(s).

- At the boundaries,

$$\mathbf{A}_0^+ = \mathbf{r}_A \mathbf{A}_0^-, \ \mathbf{C}_0^+ = \mathbf{r}_C \mathbf{C}_0^-, \ \mathbf{B}_0^+ = \mathbf{r}_B \mathbf{B}_0^- \tag{11.25}$$

where $\mathbf{r}$ is the reflection matrix at a boundary, whose location is denoted in the subscript.

- At the L joints on the left and right corner of the top side,

$$\mathbf{A}_{H3}^+ = \mathbf{r}_{22}^{A3} \mathbf{A}_{H3}^- + \mathbf{t}_{12}^{A3} \mathbf{a}_3^+, \ \mathbf{a}_3^- = \mathbf{r}_{11}^{A3} \mathbf{a}_3^+ + \mathbf{t}_{21}^{A3} \mathbf{A}_{H3}^-;$$

$$\mathbf{B}_{H3}^- = \mathbf{r}_{22}^{B3} \mathbf{B}_{H3}^+ + \mathbf{t}_{12}^{B3} \mathbf{b}_3^+, \ \mathbf{b}_3^- = \mathbf{r}_{11}^{B3} \mathbf{b}_3^+ + \mathbf{t}_{21}^{B3} \mathbf{B}_{H3}^+ \tag{11.26}$$

- At the T joints on the left side, right side, and top side,

$$\mathbf{a}_i^- = \mathbf{r}_{11}^{Ai} \mathbf{a}_i^+ + \mathbf{t}_{21}^{Ai} \mathbf{A}_{Hi}^- + \mathbf{t}_{31}^{Ai} \mathbf{A}_i^-,$$

$$\mathbf{A}_{Hi}^+ = \mathbf{r}_{22}^{Ai} \mathbf{A}_{Hi}^- + \mathbf{t}_{12}^{Ai} \mathbf{a}_i^+ + \mathbf{t}_{32}^{Ai} \mathbf{A}_i^-,$$

$$\mathbf{A}_i^+ = \mathbf{r}_{33}^{Ai} \mathbf{A}_i^- + \mathbf{t}_{23}^{Ai} \mathbf{A}_{Hi}^- + \mathbf{t}_{13}^{Ai} \mathbf{a}_i^+;$$

$$\mathbf{b}_i^- = \mathbf{r}_{11}^{Bi} \mathbf{b}_i^+ + \mathbf{t}_{21}^{Bi} \mathbf{B}_{Hi}^+ + \mathbf{t}_{31}^{Bi} \mathbf{B}_i^-,$$

$$\mathbf{B}_{Hi}^- = \mathbf{r}_{22}^{Bi}\mathbf{B}_{Hi}^+ + \mathbf{t}_{12}^{Bi}\mathbf{b}_i^+ + \mathbf{t}_{32}^{Bi}\mathbf{B}_i^-,$$

$$\mathbf{B}_i^+ = \mathbf{r}_{33}^{Bi}\mathbf{B}_i^- + \mathbf{t}_{23}^{Bi}\mathbf{B}_{Hi}^+ + \mathbf{t}_{13}^{Bi}\mathbf{b}_i^+;$$

where $i = 1, 2$, and

$$\mathbf{c}_3^- = \mathbf{r}_{11}^{C3}\mathbf{c}_3^+ + \mathbf{t}_{21}^{C3}\mathbf{C}_{HL3}^+ + \mathbf{t}_{31}^{C3}\mathbf{C}_{HR3}^-,$$

$$\mathbf{C}_{HL3}^- = \mathbf{r}_{22}^{C3}\mathbf{C}_{HL3}^+ + \mathbf{t}_{12}^{C3}\mathbf{c}_3^+ + \mathbf{t}_{32}^{C3}\mathbf{C}_{HR3}^-,$$

$$\mathbf{C}_{HR3}^+ = \mathbf{r}_{33}^{C3}\mathbf{C}_{HR3}^- + \mathbf{t}_{23}^{C3}\mathbf{C}_{HL3}^+ + \mathbf{t}_{13}^{C3}\mathbf{c}_3^+ \tag{11.27}$$

- At the "+" joints located inside the planar frame,

$$\mathbf{c}_i^- = \mathbf{r}_{11}^{Ci}\mathbf{c}_i^+ + \mathbf{t}_{21}^{Ci}\mathbf{C}_{HLi}^+ + \mathbf{t}_{31}^{Ci}\mathbf{C}_i^- + \mathbf{t}_{41}^{Ci}\mathbf{C}_{HRi}^-,$$

$$\mathbf{C}_{HLi}^- = \mathbf{r}_{22}^{Ci}\mathbf{C}_{HLi}^+ + \mathbf{t}_{12}^{Ci}\mathbf{c}_i^+ + \mathbf{t}_{32}^{Ci}\mathbf{C}_i^- + \mathbf{t}_{42}^{Ci}\mathbf{C}_{HRi}^-,$$

$$\mathbf{C}_i^+ = \mathbf{r}_{33}^{Ci}\mathbf{C}_i^- + \mathbf{t}_{13}^{Ci}\mathbf{c}_i^+ + \mathbf{t}_{23}^{Ci}\mathbf{C}_{HLi}^+ + \mathbf{t}_{43}^{Ci}\mathbf{C}_{HRi}^-,$$

$$\mathbf{C}_{HRi}^+ = \mathbf{r}_{44}^{Ci}\mathbf{C}_{HRi}^- + \mathbf{t}_{14}^{Ci}\mathbf{c}_i^+ + \mathbf{t}_{24}^{Ci}\mathbf{C}_{HLi}^+ + \mathbf{t}_{34}^{Ci}\mathbf{C}_i^- \tag{11.28}$$

where $i = 1, 2$.

In Eqs. (11.26) to (11.28), $\mathbf{r}$ and $\mathbf{t}$ are the reflection and transmission matrices, respectively. The superscripts in $\mathbf{r}$ and $\mathbf{t}$ denote the location of a joint, and the two numerical digits in the subscripts refer to the beam elements connected at the joint. The leading digit in a subscript identifies the beam element carrying the incident waves.

Free vibration responses of this example three-story two-bay frame can be obtained by assembling the wave relationships from Eqs. (11.24) to (11.28) into the matrix form of Eq. (11.21). Forced responses can be obtained from Eqs. (11.22) and (11.23), by considering the relationships between the injected waves and externally applied force and moment described in Eq. (11.14), and by modifying the propagation relationships along the structural element(s) where external excitations are applied.

Two example local coordinate systems for the three-story two-bay planar frame are described. In both examples, the local coordinate systems for the joints are kept the same, that is, rightward and upward being positive for the *x*- and *y*-axis, respectively. The local coordinate systems for the beam elements are different. In the first example, rightward and upward directions are positive, as shown in the free body diagram of the three-story two-bay planar frame in Figure 11.8. Because the coordinate systems for the same type of joints with different orientation are not related, this example coordinate system is named the "varying" coordinate system.

In the second example, the local coordinates for the beam elements are determined in such a way that beams connected to the same type of joints are related by a rotational relationship. This example coordinate system is called the "rotating" coordinate system. This is achieved by first defining a set of base coordinates for beams joined at each type of joint. An example set of base coordinates for beams joined at the L joints and T joints are shown in Figure 11.9. Because the beam elements connected to the "+" joints are also beam elements of the surrounding T joints, the coordinates of the beam elements of the "+" joint in the middle of the planar frame cannot be randomly selected, they are mostly predetermined by the surrounding T joints. Recall that the positive direction of the *x*-axis for any portion of a beam element must be consistently defined. Likewise, for those beam elements that are connected to a T joint at one end and to an L joint at the other end, one must make sure that the positive directions of the *x*-axis on the same beam element with respect to both end joints do not have conflict.

Figure 11.10 shows the free body diagram of the three-story two-bay planar frame with its local beam coordinates related by a rotational relationship to the base coordinates defined in Figure 11.9.

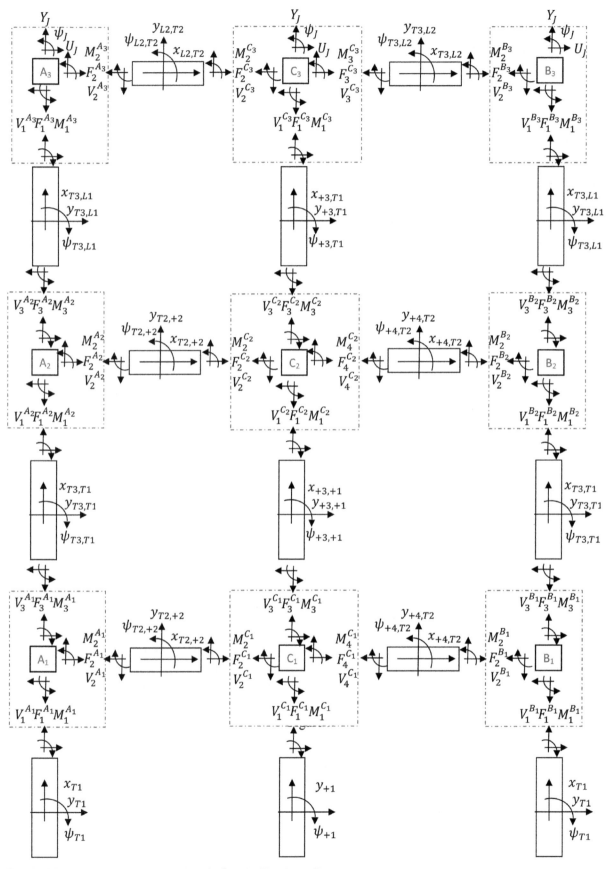

Figure 11.8 Free body diagram based on the "varying" local coordinate systems.

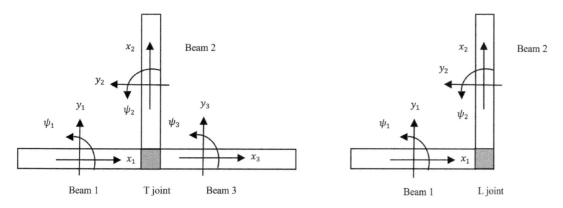

Figure 11.9 Definition of base coordinates for the "rotating" local coordinate systems at T and L joints.

The equations of motion and continuity equations at each joint by the "varying" local coordinate system are obtained from the free body diagram illustrated in Figure 11.8 as follows:

- At the right side L joint B_3

$$-V_1 - F_2 = m_J \ddot{U}_J, \quad -V_2 - F_1 = m_J \ddot{Y}_J, \quad M_1 - M_2 - V_1 \frac{h_2}{2} + V_2 \frac{h_1}{2} = J \ddot{\psi}_J \qquad (11.29a)$$

$$u_1 = Y_J, \quad u_2 = U_J$$

$$y_1 = U_J + \frac{h_2}{2}\psi_J, \quad y_2 = Y_J - \frac{h_1}{2}\psi_J$$

$$\psi_1 = -\psi_J, \quad \psi_2 = \psi_J \qquad (11.29b)$$

- At the left side L joint A_3

$$F_2 - V_1 = m_J \ddot{U}_J, \quad V_2 - F_1 = m_J \ddot{Y}_J, \quad M_1 + M_2 - V_1 \frac{h_2}{2} + V_2 \frac{h_1}{2} = J \ddot{\psi}_J \qquad (11.30a)$$

$$u_1 = Y_J, \quad u_2 = U_J$$

$$y_1 = U_J + \frac{h_2}{2}\psi_J, \quad y_2 = Y_J + \frac{h_1}{2}\psi_J$$

$$\psi_1 = -\psi_J, \quad \psi_2 = \psi_J \qquad (11.30b)$$

- At the right side T joint B_1 or B_2

$$-V_1 - F_2 + V_3 = m_J \ddot{U}_J, \quad -V_2 - F_1 + F_3 = m_J \ddot{Y}_J$$

$$M_1 - M_2 - M_3 - V_1 \frac{h_2}{2} + V_2 \frac{h_1}{2} - V_3 \frac{h_2}{2} = J \ddot{\psi}_J \qquad (11.31a)$$

$$u_1 = Y_J, \quad u_2 = U_J, \quad u_3 = Y_J$$

$$y_1 = U_J + \frac{h_2}{2}\psi_J, \quad y_2 = Y_J - \frac{h_1}{2}\psi_J, \quad y_3 = U_J - \frac{h_2}{2}\psi_J$$

$$\psi_1 = -\psi_J, \quad \psi_2 = \psi_J, \quad \psi_3 = -\psi_J \qquad (11.31b)$$

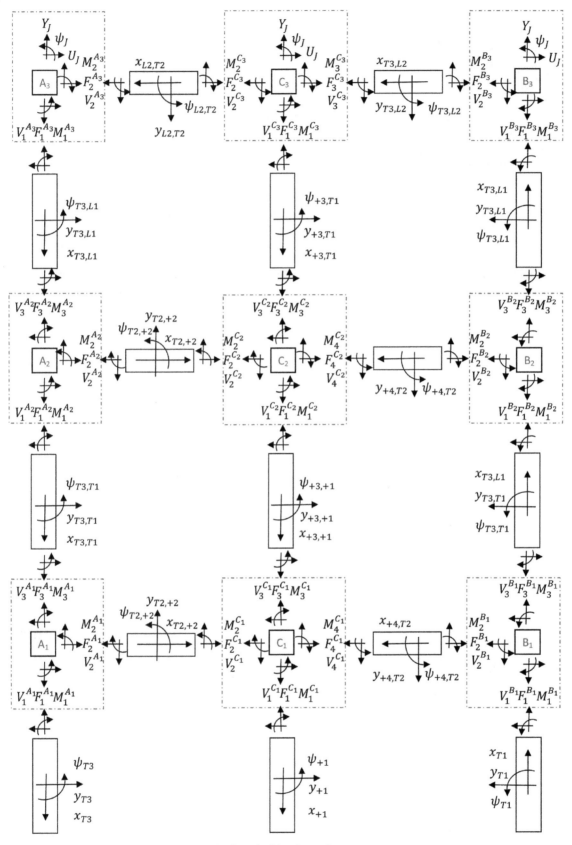

Figure 11.10 Free body diagram based on the "rotating" local coordinate systems.

- At the left side T joint A_1 or A_2

$$F_2 + V_3 - V_1 = m_J \ddot{U}_J, \ V_2 + F_3 - F_1 = m_J \ddot{Y}_J$$

$$M_1 + M_2 - M_3 - V_1 \frac{h_2}{2} + V_2 \frac{h_1}{2} - V_3 \frac{h_2}{2} = J \ddot{\psi}_J \tag{11.32a}$$

$$u_1 = Y_J, \ u_2 = U_J, \ u_3 = Y_J$$

$$y_1 = U_J + \frac{h_2}{2} \psi_J, \ y_2 = Y_J + \frac{h_1}{2} \psi_J, \ y_3 = U_J - \frac{h_2}{2} \psi_J$$

$$\psi_1 = -\psi_J, \ \psi_2 = \psi_J, \ \psi_3 = -\psi_J \tag{11.32b}$$

- At the top side T joint C_3

$$-V_1 - F_2 + F_3 = m_J \ddot{U}_J, \ -V_2 - F_1 + V_3 = m_J \ddot{Y}_J$$

$$M_1 - M_2 + M_3 - V_1 \frac{h_2}{2} + V_2 \frac{h_1}{2} + V_3 \frac{h_1}{2} = J \ddot{\psi}_J \tag{11.33a}$$

$$u_1 = Y_J, \ u_2 = U_J, \ u_3 = U_J$$

$$y_1 = U_J + \frac{h_2}{2} \psi_J, \ y_2 = Y_J - \frac{h_1}{2} \psi_J, \ y_3 = Y_J + \frac{h_1}{2} \psi_J$$

$$\psi_1 = -\psi_J, \ \psi_2 = \psi_J, \ \psi_3 = \psi_J \tag{11.33b}$$

- At the "+" joint C_1 or C_2

$$-V_1 - F_2 + V_3 + F_4 = m_J \ddot{U}_J$$

$$-V_2 - F_1 + F_3 + V_4 = m_J \ddot{Y}_J$$

$$M_1 - M_2 - M_3 + M_4 - V_1 \frac{h_2}{2} + V_2 \frac{h_1}{2} - V_3 \frac{h_2}{2} + V_4 \frac{h_1}{2} = J \ddot{\psi}_J \tag{11.34a}$$

$$u_1 = Y_J, \ u_2 = U_J, \ u_3 = Y_J, \ u_4 = U_J$$

$$y_1 = U_J + \frac{h_2}{2} \psi_J, \ y_2 = Y_J - \frac{h_1}{2} \psi_J, \ y_3 = U_J - \frac{h_2}{2} \psi_J, \ y_4 = Y_J + \frac{h_1}{2} \psi_J$$

$$\psi_1 = -\psi_J, \ \psi_2 = \psi_J, \ \psi_3 = -\psi_J, \ \psi_4 = \psi_J \tag{11.34b}$$

The equations of motion and continuity equations at each joint by the "rotating" local coordinate system are obtained from the free body diagram of Figure 11.10 and by applying Newton's second law for translational and rotational motion,

- At the right side L joint B_3

$$V_1 - F_2 = m_J \ddot{U}_J, \ -V_2 - F_1 = m_J \ddot{Y}_J, \ -M_1 + M_2 + V_1 \frac{h_2}{2} + V_2 \frac{h_1}{2} = J \ddot{\psi}_J \tag{11.35a}$$

$$u_1 = Y_J, \ u_2 = -U_J$$

$$y_1 = -U_J - \frac{h_2}{2}\psi_J, \ y_2 = -Y_J + \frac{h_1}{2}\psi_J$$

$$\psi_1 = \psi_J, \ \psi_2 = \psi_J \tag{11.35b}$$

- At the left side L joint A_3

$$F_2 + V_1 = m_J\ddot{U}_J, \ V_2 - F_1 = m_J\ddot{Y}_J, \ M_1 - M_2 + V_1\frac{h_2}{2} + V_2\frac{h_1}{2} = J\ddot{\psi}_J \tag{11.36a}$$

$$u_1 = -Y_J, \ u_2 = -U_J$$

$$y_1 = \ U_J + \frac{h_2}{2}\psi_J, \ y_2 = -Y_J - \frac{h_1}{2}\psi_J$$

$$\psi_1 = \psi_J, \ \psi_2 = \psi_J \tag{11.36b}$$

- At the right side T joint B_1 or B_2

$$V_1 - F_2 - V_3 = m_J\ddot{U}_J, \ -V_2 - F_1 + F_3 = \ m_J\ddot{Y}_J$$

$$-M_1 + M_2 + M_3 + V_1\frac{h_2}{2} + V_2\frac{h_1}{2} + V_3\frac{h_2}{2} = J\ddot{\psi}_J \tag{11.37a}$$

$$u_1 = Y_J, \ u_2 = -U_J, \ u_3 = Y_J$$

$$y_1 = -U_J - \frac{h_2}{2}\psi_J, \ y_2 = -Y_J + \frac{h_1}{2}\psi_J, \ y_3 = -U_J + \frac{h_2}{2}\psi_J$$

$$\psi_1 = \psi_J, \ \psi_2 = \ \psi_J, \ \psi_3 = \psi_J \tag{11.37b}$$

- At the left side T joint A_1 or A_2

$$F_2 - V_3 + V_1 = m_J\ddot{U}_J, \ V_2 + F_3 - F_1 = \ m_J\ddot{Y}_J$$

$$M_1 + M_2 - M_3 + V_1\frac{h_2}{2} + V_2\frac{h_1}{2} + V_3\frac{h_2}{2} = J\ddot{\psi}_J \tag{11.38a}$$

$$u_1 = -Y_J, \ u_2 = \ U_J, \ u_3 = -Y_J$$

$$y_1 = \ U_J + \frac{h_2}{2}\psi_J, \ y_2 = \ Y_J + \frac{h_1}{2}\psi_J, \ y_3 = U_J - \frac{h_2}{2}\psi_J$$

$$\psi_1 = \ \psi_J, \ \psi_2 = \ \psi_J, \ \psi_3 = \ \psi_J \tag{11.38b}$$

- At the top side T joint C_3

$$V_1 - F_2 + F_3 = m_J\ddot{U}_J, \ -V_2 - F_1 + V_3 = m_J\ddot{Y}_J$$

$$M_1 + M_2 - M_3 + V_1\frac{h_2}{2} + V_2\frac{h_1}{2} + V_3\frac{h_1}{2} = J\ddot{\psi}_J \tag{11.39a}$$

$$u_1 = -Y_J, \ u_2 = -U_J, \ u_3 = -U_J$$

$$y_1 = \ U_J + \frac{h_2}{2}\psi_J, \ y_2 = \ -Y_J + \frac{h_1}{2}\psi_J, \ y_3 = -Y_J - \frac{h_1}{2}\psi_J$$

$$\psi_1 = \psi_J, \; \psi_2 = \psi_J, \; \psi_3 = \psi_J \tag{11.39b}$$

- At the "+" joint C_1 or C_2

$$V_1 - F_2 - V_3 + F_4 = m_J \ddot{U}_J$$

$$-V_2 - F_1 + F_3 + V_4 = m_J \ddot{Y}_J$$

$$M_1 - M_2 - M_3 - M_4 + V_1 \frac{h_2}{2} + V_2 \frac{h_1}{2} + V_3 \frac{h_2}{2} + V_4 \frac{h_1}{2} = J \ddot{\psi}_J \tag{11.40a}$$

$$u_1 = -Y_J, \; u_2 = U_J, \; u_3 = -Y_J, \; u_4 = -U_J$$

$$y_1 = U_J + \frac{h_2}{2}\psi_J, \; y_2 = Y_J - \frac{h_1}{2}\psi_J, \; y_3 = U_J - \frac{h_2}{2}\psi_J, \; y_4 = -Y_J - \frac{h_1}{2}\psi_J$$

$$\psi_1 = \psi_J, \; \psi_2 = \psi_J, \; \psi_3 = \psi_J, \; \psi_4 = \psi_J \tag{11.40b}$$

In Eqs. (11.29) to (11.40), a joint is modeled as a rigid body. m_J is the mass of the joint. J and ψ_J are the mass moment inertia and rotation of the joint about the centroidal axis that is normal to the the plane of the frame, respectively. U_J and Y_J denote the translational motion of the rigid joint along the x and y direction, respectively. h is the thickness of a beam element. $V(x, t)$, $M(x, t)$, and $F(x, t)$ are the internal resistant shear force, bending moment, and longitudinal force at the section of a beam, respectively. Numerical subscripts 1, 2, 3, and 4 denote parameters relating to beam 1, beam 2, beam 3, and beam 4, respectively.

The sign differences in the equations of the same joint are because of the difference in the two sets of local coordinate systems, namely the "varying" and "rotating" coordinate systems. The above equations allow the reflection and transmission relationships at each joint to be solved based on the two sets of local coordinate systems illustrated in Figures 11.8 and 11.10, respectively.

The set of local coordinate systems illustrated in Figure 11.10, whose local coordinates for L and T joints are obtained through rotating the coordinates of the Base L and T joints of Figure 11.9, offers more convenience compared to that of Figure 11.8. When the "rotating" coordinate systems are adopted, there is generally no need to derive the reflection and transmission relationships multiple times for the same type of joints. As a result, the "rotating" local coordinate system may be preferable to the "varying" local coordinate system.

Let us look at how the local coordinates of the beam elements at the two L joints B_3 and A_3 in Figure 11.10 are related to those of the Base L joint defined in Figure 10.9. It is not difficult to see that the local coordinates of the beam elements at joints B_3 and A_3 can be viewed as the Base L joint being rotated counterclockwise by a 90° and 180° angle, respectively.

There are five T joints in the planar frame, among them two on the right side (B_1 and B_2), one on the top side (C_3), and two on the left side (A_1 and A_2). Similarly, in the "rotating" coordinate systems shown in Figure 11.10, the local coordinates of beam elements at the T joints on the right, top, and left sides can be viewed as the Base T joint defined in Figure 11.9 being rotated counterclockwise by a 90°, 180°, and 270° angle, respectively.

The local coordinates of the "+" joint in the middle of the planar frame, however, are mostly predetermined. The local coordinates of three out of the four beam elements joined at "+" joint C_2 are predetermined by their connection to the surrounding T joints, because the positive direction of x-axis for the same beam element must be consistently defined. There is freedom in defining the local coordinates of the lower vertical beam element C_1C_2 of joint C_2. However, because this beam element is also one of the beam elements of "+" joint C_1, its position to joint C_1 is equivalent to that of beam element C_2C_3 to joint C_2, its local coordinates are therefore chosen to be the same as those of C_2C_3. The lower vertical beam element of joint C_1 is chosen along the same line. By keeping this consistency in the two "+" joints C_1 and C_2, only one set of reflection and transmission relationships needs to be derived for the two cross joints in Figure 11.10.

Let us denote the reflection and transmission matrices $\mathbf{r}$ and $\mathbf{t}$ at the Base L joint as $\mathbf{r}_{L11}$, $\mathbf{t}_{L12}$, $\mathbf{r}_{L22}$, and $\mathbf{t}_{L21}$, and those at the Base T joints as $\mathbf{r}_{T11}$, $\mathbf{t}_{T12}$, $\mathbf{t}_{T13}$, $\mathbf{r}_{T22}$, $\mathbf{t}_{T21}$, $\mathbf{t}_{T23}$, $\mathbf{r}_{T33}$, $\mathbf{t}_{T31}$, and $\mathbf{t}_{T32}$. The letter in the subscripts names the joint, and the two numerical digits in the subscripts refer to the related beam elements. The leading digit in a subscript identifies the beam element carrying the incident waves.

Keeping in mind the rotational relationships between the beam elements of each of the L and T joints and the corresponding Base L and T joints, the following equivalencies exist in the planar frame shown in Figure 11.10:

- At L joint B_3 on the top right side of the planar frame

$$\mathbf{r}_{11}^{B_3} \leftrightarrow \mathbf{r}_{L11}, \mathbf{t}_{12}^{B_3} \leftrightarrow \mathbf{t}_{L12}, \mathbf{r}_{22}^{B_3} \leftrightarrow \mathbf{r}_{L22}, \mathbf{t}_{21}^{B_3} \leftrightarrow \mathbf{t}_{L21} \tag{11.41a}$$

- At L joint A_3 on the top left side of the planar frame

$$\mathbf{r}_{11}^{A_3} \leftrightarrow \mathbf{r}_{L22}, \mathbf{t}_{12}^{A_3} \leftrightarrow \mathbf{t}_{L21}, \mathbf{r}_{22}^{A_3} \leftrightarrow \mathbf{r}_{L11}, \mathbf{t}_{21}^{A_3} \leftrightarrow \mathbf{t}_{L12} \tag{11.41b}$$

- At T joints B_1 and B_2 on the right side of the planar frame

$$\mathbf{r}_{11}^{B_1, B_2} \leftrightarrow \mathbf{r}_{T11}, \mathbf{t}_{12}^{B_1, B_2} \leftrightarrow \mathbf{t}_{T12}, \mathbf{t}_{13}^{B_1, B_2} \leftrightarrow \mathbf{t}_{T13}$$

$$\mathbf{r}_{22}^{B_1, B_2} \leftrightarrow \mathbf{r}_{T22}, \mathbf{t}_{21}^{B_1, B_2} \leftrightarrow \mathbf{t}_{T21}, \mathbf{t}_{23}^{B_1, B_2} \leftrightarrow \mathbf{t}_{T23} \tag{11.41c}$$

$$\mathbf{r}_{33}^{B_1, B_2} \leftrightarrow \mathbf{r}_{T33}, \mathbf{t}_{31}^{B_1, B_2} \leftrightarrow \mathbf{t}_{T31}, \mathbf{t}_{32}^{B_1, B_2} \leftrightarrow \mathbf{t}_{T32}$$

- At T joint C_3 on the top side of the planar frame

$$\mathbf{r}_{11}^{C_3} \leftrightarrow \mathbf{r}_{T22}, \mathbf{t}_{12}^{C_3} \leftrightarrow \mathbf{t}_{T23}, \mathbf{t}_{13}^{C_3} \leftrightarrow \mathbf{t}_{T21}$$

$$\mathbf{r}_{22}^{C_3} \leftrightarrow \mathbf{r}_{T33}, \mathbf{t}_{21}^{C_3} \leftrightarrow \mathbf{t}_{T32}, \mathbf{t}_{23}^{C_3} \leftrightarrow \mathbf{t}_{T31} \tag{11.41d}$$

$$\mathbf{r}_{33}^{C_3} \leftrightarrow \mathbf{r}_{T11}, \mathbf{t}_{31}^{C_3} \leftrightarrow \mathbf{t}_{T12}, \mathbf{t}_{32}^{C_3} \leftrightarrow \mathbf{t}_{T13}$$

- At T joints A_1 and A_2 on the left side of the planar frame

$$\mathbf{r}_{11}^{A_1, A_2} \leftrightarrow \mathbf{r}_{T33}, \mathbf{t}_{12}^{A_1, A_2} \leftrightarrow \mathbf{t}_{T32}, \mathbf{t}_{13}^{A_1, A_2} \leftrightarrow \mathbf{t}_{T31}$$

$$\mathbf{r}_{22}^{A_1, A_2} \leftrightarrow \mathbf{r}_{T22}, \mathbf{t}_{21}^{A_1, A_2} \leftrightarrow \mathbf{t}_{T23}, \mathbf{t}_{23}^{A_1, A_2} \leftrightarrow \mathbf{t}_{T21} \tag{11.41e}$$

$$\mathbf{r}_{33}^{A_1, A_2} \leftrightarrow \mathbf{r}_{T11}, \mathbf{t}_{31}^{A_1, A_2} \leftrightarrow \mathbf{t}_{T13}, \mathbf{t}_{32}^{A_1, A_2} \leftrightarrow \mathbf{t}_{T12}$$

where symbol "$\leftrightarrow$" indicates that the two expressions on the left and right side of the symbol are equivalent to each other. The superscripts in $\mathbf{r}$ and $\mathbf{t}$ identify the related joint in the planar frame shown in Figure 11.10.

11.5 Reflection and Transmission of Waves in a Multi-story Multi-bay Planar Frame

The reflection and transmission at the L, T, and "+" joints corresponding to both sets of local coordinates can be obtained from the equations of motion and continuity equations.

As an example, the reflection and transmission of the L and T joints using the base coordinates of Figures 11.9, and the cross joint using the local coordinates shown in Figure 11.10, are discussed.

Although it is not practical to study all possible joints in planar engineering structures in this chapter, the wave reflection and transmission relationships of other joints, such as oblique angled joints, can be obtained following a similar procedure.

11.5.1 Wave Reflection and Transmission at an L-shaped Joint

Figure 11.11 shows the free body diagram of the Base L joint of Figure 11.9, from which the equations of motion of the L joint are obtained by applying Newton's second law for translational and rotational motion,

$$-V_1 + F_2 = m_J \ddot{Y}_J$$

$$-V_2 - F_1 = m_J \ddot{U}_J$$

$$-M_1 + M_2 + V_1 \frac{h_2}{2} + V_2 \frac{h_1}{2} = J \ddot{\psi}_J \qquad (11.42)$$

where h_1 and h_2 are the thickness of Beam 1 and Beam 2, respectively. Let us call Eq. (11.42) the equilibrium equations, because by the principle of d'Alembert, these equations represent dynamic equilibrium.

From Figure 11.11, the continuity conditions at the joint are

$$u_1 = U_J, \; u_2 = Y_J$$

$$y_1 = Y_J - \frac{h_2}{2} \psi_J, \; y_2 = -U_J + \frac{h_1}{2} \psi_J$$

$$\psi_1 = \psi_J, \; \psi_2 = \psi_J \qquad (11.43)$$

To find the reflection and transmission relationships of vibration waves in beam elements connected at a joint, the incident, reflected, and transmitted waves must be related directly. As a result, the rigid body joint related parameters $\{U_J, Y_J, \psi_J\}$ need to be eliminated from both the equilibrium and continuity equations, namely, Eqs. (11.42) and (11.43).

Eliminating the rigid body joint related parameters $\{U_J, Y_J, \psi_J\}$ from the continuity equations of Eq. (11.43),

$$\psi_1 = \psi_2 \qquad (11.44a)$$

$$y_1 = u_2 - \frac{h_2}{2} \psi_1, \text{ or equivalently } y_1 = u_2 - \frac{h_2}{2} \psi_2 \qquad (11.44b)$$

$$y_2 = -u_1 + \frac{h_1}{2} \psi_1, \text{ or equivalently } y_2 = -u_1 + \frac{h_1}{2} \psi_2 \qquad (11.44c)$$

From Eqs. (11.44a-c), there are four sets of equivalent continuity equations. Each set of continuity equations consists of three scalar equations by taking one scalar equation each from Eqs. (11.44a), (11.44b), and (11.44c).

From the continuity equations of Eq. (11.43), the rigid body joint related parameters U_J, Y_J, and ψ_J can be expressed in terms of the beam elements related parameters u_n, y_n, and ψ_n, where $n = 1, 2$, as follows

$$\psi_J = \psi_1, \text{ or equivalently } \psi_J = \psi_2;$$

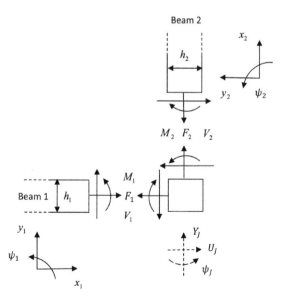

Figure 11.11 Free body diagram of the Base L joint of Figure 11.9 (Courtesy: (Mei, 2012)).

$$U_J = u_1, \text{ or equivalently } U_J = -y_2 + \frac{h_1}{2}\psi_1, \text{ or equivalently } U_J = -y_2 + \frac{h_1}{2}\psi_2;$$

$$Y_J = u_2, \text{ or equivalently } Y_J = y_1 + \frac{h_2}{2}\psi_1, \text{ or equivalently } Y_J = y_1 + \frac{h_2}{2}\psi_2 \tag{11.45}$$

From Eqs. (11.42) and (11.45),

$$-V_1 + F_2 = m_J\ddot{u}_2, \text{ or equivalently}$$

$$-V_1 + F_2 = m_J\left(\ddot{y}_1 + \frac{h_2}{2}\ddot{\psi}_1\right), \text{ or equivalently}$$

$$-V_1 + F_2 = m_J\left(\ddot{y}_1 + \frac{h_2}{2}\ddot{\psi}_2\right) \tag{11.46a}$$

$$-V_2 - F_1 = m_J\ddot{u}_1, \text{ or equivalently}$$

$$-V_2 - F_1 = m_J\left(-\ddot{y}_2 + \frac{h_1}{2}\ddot{\psi}_1\right), \text{ or equivalently}$$

$$-V_2 - F_1 = m_J\left(-\ddot{y}_2 + \frac{h_1}{2}\ddot{\psi}_2\right) \tag{11.46b}$$

$$-M_1 + M_2 + V_1\frac{h_2}{2} + V_2\frac{h_1}{2} = J\ddot{\psi}_1, \text{ or equivalently}$$

$$-M_1 + M_2 + V_1\frac{h_2}{2} + V_2\frac{h_1}{2} = J\ddot{\psi}_2 \tag{11.46c}$$

From Eqs. (11.46a-c), there are eighteen sets of equivalent equilibrium equations. Each set consists of three scalar equations of equilibrium by taking one scalar equation each from Eqs. (11.46a), (11.46b), and (11.46c).

The reflection and transmission relationships corresponding to waves incident upon a joint from a different beam element are in general different. Figures 11.12a and 11.12b show the reflection and transmission of waves incident upon an L joint from Beam 1 and Beam 2, respectively.

For waves incident from Beam 1, as shown in Figure 11.12a, the reflected waves $\mathbf{a}^-$ on Beam 1 and transmitted waves $\mathbf{b}^+$ to Beam 2 are related to the incident waves $\mathbf{a}^+$ by reflection matrix $\mathbf{r}_{11}$ and transmission matrix $\mathbf{t}_{12}$,

$$\mathbf{a}^- = \mathbf{r}_{11}\mathbf{a}^+, \ \mathbf{b}^+ = \mathbf{t}_{12}\mathbf{a}^+ \tag{11.47}$$

When incident waves are from Beam 2, as shown in Figure 11.12b, the reflected waves $\mathbf{a}^-$ on Beam 2 and transmitted waves $\mathbf{b}^+$ to Beam 1 are related to the incident waves $\mathbf{a}^+$ by reflection matrix $\mathbf{r}_{22}$ and transmission matrix $\mathbf{t}_{21}$,

$$\mathbf{a}^- = \mathbf{r}_{22}\mathbf{a}^+, \ \mathbf{b}^+ = \mathbf{t}_{21}\mathbf{a}^+ \tag{11.48}$$

The wave vectors in Eqs. (11.47) and (11.48) are defined in Eq. (11.5b).

First, find $\mathbf{r}_{11}$ and $\mathbf{t}_{12}$, that is, the reflection and transmission matrices at the L joint of Figure 11.12a, corresponding to incident waves from Beam 1.

Following the local coordinate systems of Figure 11.11, the axial and bending deflections of the beam elements are

$$u_1 = c^+ e^{-ik_{13}x_1} + c^- e^{ik_{13}x_1}$$

$$u_2 = d^+ e^{-ik_{23}x_2}$$

$$y_1 = a_1^+ e^{-ik_{11}x_1} + a_2^+ e^{-k_{12}x_1} + a_1^- e^{ik_{11}x_1} + a_2^- e^{k_{12}x_1}$$

$$y_2 = b_1^+ e^{-ik_{21}x_2} + b_2^+ e^{-k_{22}x_2} \tag{11.49a}$$

and the bending slopes by the Euler–Bernoulli bending vibration theory are

$$\psi_1 = -ik_{11}a_1^+ e^{-ik_{11}x_1} - k_{12}a_2^+ e^{-k_{12}x_1} + ik_{11}a_1^- e^{ik_{11}x_1} + k_{12}a_2^- e^{k_{12}x_1}$$

$$\psi_2 = -ik_{21}b_1^+ e^{-ik_{21}x_2} - k_{22}b_2^+ e^{-k_{22}x_2} \tag{11.49b}$$

and the bending slopes by the Timoshenko bending vibration theory are

$$\psi_1 = -iP_1 a_1^+ e^{-ik_{11}x_1} - N_1 a_2^+ e^{-k_{12}x_1} + iP_1 a_1^- e^{ik_{11}x_1} + N_1 a_2^- e^{k_{12}x_1}$$

$$\psi_2 = -iP_2 b_1^+ e^{-ik_{21}x_2} - N_2 b_2^+ e^{-k_{22}x_2} \tag{11.49c}$$

where i is the imaginary unit and subscripts of P and N identify the beam elements by numerical numbers as marked on Figures 11.11 and 11.12. Each wavenumber carries two numerical digits in its subscript, with the leading digit identifying a beam element and the other digit denoting the wavenumber sequence in the beam element.

Substituting Eqs. (11.49a) and (11.49b) into a set of continuity equations from Eqs. (11.44a-c), and choosing the origin at the joint, a set of three scalar equations of continuity is obtained based on the Euler–Bernoulli bending vibration theory. Similarly, substituting Eqs. (11.49a) and (11.49c) into a set of continuity equations from Eqs. (11.44a-c), and choosing the origin at the joint, a set of three scalar equations of continuity is obtained based on the Timoshenko bending vibration theory. This set of three scalar equations of continuity, either by the Euler–Bernoulli or Timoshenko bending vibration theory, can be placed in a vector equation, in terms of wave components $\mathbf{a}^+$, $\mathbf{a}^-$, and $\mathbf{b}^+$, which are defined in Eq. (11.5b). A vector equation from this group is named $\mathbf{L\_Beam1}_{C12}$, because it is a vector equation on the continuity between Beam 1 and Beam 2 at an L joint corresponding to incident waves from Beam 1.

Substituting Eqs. (11.49a) and (11.49b) into Eqs. (11.6a) and (11.6b) gives the functional relationships of the internal resistant forces and moment and the wave components, based on the Euler–Bernoulli bending vibration theory. Substituting these functional relationships into a set of equilibrium equations from Eqs. (11.46a-c), and choosing the origin at the joint, a set of three scalar equations of equilibrium is obtained based on the Euler–Bernoulli bending vibration theory. Similarly, Substituting Eqs. (11.49a) and (11.49c) into Eqs. (11.6a) and (11.6c) gives the functional relationships of the internal resistant forces and moment and the wave components, based on the Timoshenko bending vibration theory. Substituting these functional relationships into a set of equilibrium equations from Eqs. (11.46a-c), and choosing the origin at the joint, a set of three scalar equations of equilibrium is obtained based on the Timoshenko bending vibration theory. This set of three scalar equations of equilibrium, either by the Euler–Bernoulli or Timoshenko bending vibration theory, can be placed in a vector equation, in terms of wave components $\mathbf{a}^+$, $\mathbf{a}^-$, and $\mathbf{b}^+$, which are defined in Eq. (11.5b). A vector equation from this group is named $\mathbf{L\_Beam1}_E$, because it is a vector equation on equilibrium at an L joint corresponding to incident waves from Beam 1.

The reflection and transmission matrices $\mathbf{r}_{11}$ and $\mathbf{t}_{12}$ are solved from Eq. (11.47) and the vector equations of continuity and equilibrium, namely, $\mathbf{L\_Beam1}_{C12}$ and $\mathbf{L\_Beam1}_E$.

Next, find $\mathbf{r}_{22}$ and $\mathbf{t}_{21}$, that is, the reflection and transmission matrices at the L joint of Figure 11.12b, corresponding to incident waves from Beam 2.

Following the local coordinate systems of Figure 11.11, the axial and bending deflections of the beam elements are

$$u_2 = c^+ e^{ik_{23}x_2} + c^- e^{-ik_{23}x_2}$$

$$u_1 = d^+ e^{ik_{13}x_1}$$

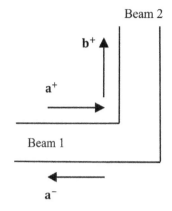

(a)

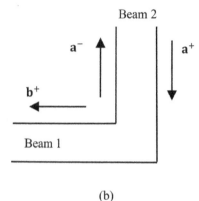

(b)

Figure 11.12 Wave reflection and transmission at an L joint with (a) incident waves from Beam 1 and (b) incident waves from Beam 2.

$$y_2 = a_1^+ e^{ik_{21}x_2} + a_2^+ e^{k_{22}x_2} + a_1^- e^{-ik_{21}x_2} + a_2^- e^{-k_{22}x_2}$$

$$y_1 = b_1^+ e^{ik_{11}x_1} + b_2^+ e^{k_{12}x_1} \tag{11.50a}$$

and the bending slopes by the Euler–Bernoulli bending vibration theory are

$$\psi_2 = ik_{21}a_1^+ e^{ik_{21}x_2} + k_{22}a_2^+ e^{k_{22}x_2} - ik_{21}a_1^- e^{-ik_{21}x_2} - k_{22}a_2^- e^{-k_{22}x_2}$$

$$\psi_1 = ik_{11}b_1^+ e^{ik_{11}x_1} + k_{12}b_2^+ e^{k_{12}x_1} \tag{11.50b}$$

and the bending slopes by the Timoshenko bending vibration theory are

$$\psi_2 = iP_2 a_1^+ e^{ik_{21}x_2} + N_2 a_2^+ e^{k_{22}x_2} - iP_2 a_1^- e^{-ik_{21}x_2} - N_2 a_2^- e^{-k_{22}x_2}$$

$$\psi_1 = iP_1 b_1^+ e^{ik_{11}x_1} + N_1 b_2^+ e^{k_{12}x_1} \tag{11.50c}$$

where i is the imaginary unit and subscripts of P and N identify the beam elements by numerical numbers as marked on Figures 11.11 and 11.12. Each wavenumber carries two numerical digits in its subscript, with the leading digit identifying a beam element and the other digit denoting the wavenumber sequence in the beam element.

Substituting Eqs. (11.50a) and (11.50b) into a set of continuity equations from Eqs. (11.44a-c), and choosing the origin at the joint, a set of three scalar equations of continuity is obtained based on the Euler–Bernoulli bending vibration theory. Similarly, substituting Eqs. (11.50a) and (11.50c) into a set of continuity equations from Eqs. (11.44a-c), and choosing the origin at the joint, a set of three scalar equations of continuity is obtained based on the Timoshenko bending vibration theory. This set of three scalar equations of continuity, either by the Euler–Bernoulli or Timoshenko bending vibration theory, can be placed in a vector equation, in terms of wave components $\mathbf{a}^+$, $\mathbf{a}^-$, and $\mathbf{b}^+$, which are defined in Eq. (11.5b). A vector equation from this group is named $\mathbf{L\_Beam2}_{C21}$, because it is a vector equation on the continuity between Beam 2 and Beam 1 at an L joint corresponding to incident waves from Beam 2.

Substituting Eqs. (11.50a) and (11.50b) into Eqs. (11.6a) and (11.6b) gives the functional relationships of the internal resistant forces and moment and the wave components, based on the Euler–Bernoulli bending vibration theory. Substituting these functional relationships into a set of equilibrium equations from Eqs. (11.46a-c), and choosing the origin at the joint, a set of three scalar equations of equilibrium is obtained based on the Euler–Bernoulli bending vibration theory. Similarly, Substituting Eqs. (11.50a) and (11.50c) into Eqs. (11.6a) and (11.6c) gives the functional relationships of the internal resistant forces and moment and the wave components, based on the Timoshenko bending vibration theory. Substituting these functional relationships into a set of equilibrium equations from Eqs. (11.46a-c), and choosing the origin at the joint, a set of three scalar equations of equilibrium is obtained based on the Timoshenko bending vibration theory. This set of three scalar equations of equilibrium, either by the Euler–Bernoulli or Timoshenko bending vibration theory, can be placed in a vector equation, in terms of wave components $\mathbf{a}^+$, $\mathbf{a}^-$, and $\mathbf{b}^+$, which are defined in Eq. (11.5b). A vector equation from this group is named $\mathbf{L\_Beam2}_E$, because it is a vector equation on equilibrium at an L joint corresponding to incident waves from Beam 2.

The reflection and transmission matrices $\mathbf{r}_{22}$ $\mathbf{t}_{21}$ are solved from Eq. (11.48) and the vector equations of continuity and equilibrium, namely, $\mathbf{L\_Beam2}_{C21}$ and $\mathbf{L\_Beam2}_E$.

Note that because of the equivalent sets of continuity and equilibrium equations, the reflection and transmission matrices, $\mathbf{r}_{11}$, $\mathbf{r}_{22}$, $\mathbf{t}_{12}$, and $\mathbf{t}_{21}$, may have various forms of equivalent expressions.

11.5.2 Wave Reflection and Transmission at a T-shaped Joint

Figure 11.13 shows the free body diagram of the Base T joint of Figure 11.9, from which the equations of motion of the T joint are obtained by applying Newton's second law for translational and rotational motion,

$$-F_1 + F_3 - V_2 = m_J \ddot{U}_J$$

$$F_2 - V_1 + V_3 = m_J \ddot{Y}_J$$

$$-M_1 + M_2 + M_3 + V_1 \frac{h_2}{2} + V_2 \frac{h_1}{2} + V_3 \frac{h_2}{2} = J \ddot{\psi}_J \tag{11.51}$$

where h denotes the thickness of the beam elements at the rigid body joint. The subscript in h identifies the beam by sequence number at the joint. The rigid body T joint is assumed a rectangular shape, that is, $h_3 = h_1$. Let us call Eq. (11.51) the equilibrium equations, because by the principle of d'Alembert, these equations represent dynamic equilibrium.

From Figure 11.13, the continuity conditions at the joint are

$$u_1 = U_J, \ u_2 = Y_J, \ u_3 = U_J,$$

$$y_1 = Y_J - \frac{h_2}{2}\psi_J, \ y_3 = Y_J + \frac{h_2}{2}\psi_J, \ y_2 = -U_J + \frac{h_1}{2}\psi_J,$$

$$\psi_1 = \psi_J, \ \psi_2 = \psi_J, \ \psi_3 = \psi_J \tag{11.52}$$

To find the reflection and transmission relationships of vibration waves in beam elements connected at a joint, the incident, reflected, and transmitted waves must be related directly. As a result, the rigid body joint related parameters $\{U_J, Y_J, \psi_J\}$ need to be eliminated from both the equilibrium and continuity equations, namely, Eqs. (11.51) and (11.52).

Eliminating the rigid body joint related parameters $\{U_J, Y_J, \psi_J\}$ from the continuity equations of Eq. (11.52),

$$u_1 = u_3$$

$$\psi_1 = \psi_2 = \psi_3$$

$$y_1 = u_2 - \frac{h_2}{2}\psi_{1,2,3}$$

$$y_2 = -u_{1,3} + \frac{h_1}{2}\psi_{1,2,3}$$

$$y_3 = u_2 + \frac{h_2}{2}\psi_{1,2,3} \tag{11.53}$$

where $\psi_{1,2,3} = \psi_1 = \psi_2 = \psi_3$ and $u_{1,3} = u_1 = u_3$.

The continuity equations of Eq. (11.53) need to be sorted into three groups. Each group consists of three scalar equations, relating waves in two beams at a time.

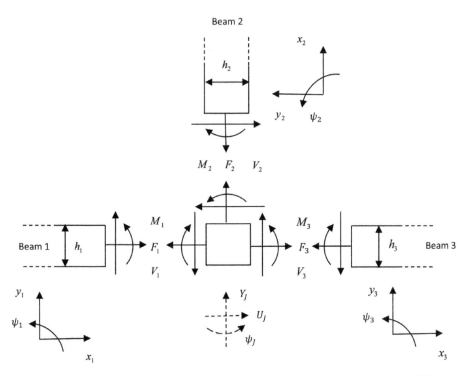

Figure 11.13 Free body diagram of the Base T joint of Figure 11.9 (Courtesy: (Mei 2008)).

The group of continuity equations relating Beam 1 and Beam 2, by sorting Eq. (11.53), is

$$\psi_1 = \psi_2 \tag{11.54a}$$

$$y_1 = u_2 - \frac{h_2}{2}\psi_{1,2} \tag{11.54b}$$

$$y_2 = -u_1 + \frac{h_1}{2}\psi_{1,2} \tag{11.54c}$$

where $\psi_{1,2} = \psi_1 = \psi_2$.

From Eqs. (11.54a–c), in the group of continuity equations relating Beam 1 and Beam 2, there are four sets of equivalent continuity equations. Each set consists of three scalar equations of continuity by taking Eq. (11.54a), and one equation each from Eqs. (11.54b) and (11.54c).

The group of continuity equations relating Beam 1 and Beam 3, by sorting Eq. (11.53), is

$$\psi_1 = \psi_3 \tag{11.55a}$$

$$u_1 = u_3 \tag{11.55b}$$

$$y_1 - y_3 = -h_2\psi_{1,3} \tag{11.55c}$$

where $\psi_{1,3} = \psi_1 = \psi_3$.

From Eqs. (11.55a–c), in the group of continuity equations relating Beam 1 and Beam 3, there are two sets of equivalent continuity equations. Each set consists of three scalar equations of continuity by taking Eqs. (11.55a) and (11.55b), and one equation from Eq. (11.55c).

The group of continuity equations relating Beam 2 and Beam 3, by sorting the continuity equations of Eq. (11.53), is

$$\psi_2 = \psi_3 \tag{11.56a}$$

$$y_2 = -u_3 + \frac{h_1}{2}\psi_{2,3} \tag{11.56b}$$

$$y_3 = u_2 + \frac{h_2}{2}\psi_{2,3} \tag{11.56c}$$

where $\psi_{2,3} = \psi_2 = \psi_3$.

From Eqs. (11.56a–c), in the group of continuity equations relating Beam 2 and Beam 3, there are four sets of equivalent continuity equations. Each set consists of three scalar equations of continuity by taking Eq. (11.56a), and one equation each from Eqs. (11.56b) and (11.56c).

From the continuity equations of Eq. (11.52), the rigid body joint related parameters U_J, Y_J, and ψ_J can be expressed in terms of the beam elements related parameters u_n, y_n, and ψ_n, where $n = 1, 2, 3$, as follows

$$\psi_J = \psi_{1,2,3};$$

$$U_J = u_{1,3}, \text{ or equivalently } U_J = -y_2 + \frac{h_1}{2}\psi_{1,2,3};$$

$$Y_J = u_2, \text{ or equivalently } Y_J = y_1 + \frac{h_2}{2}\psi_{1,2,3}, \text{ or equivalently } Y_J = y_3 - \frac{h_2}{2}\psi_{1,2,3} \tag{11.57}$$

From Eqs. (11.51) and (11.57),

$$-F_1 + F_3 - V_2 = m_J\ddot{u}_{1,3}, \text{ or equivalently}$$

$$-F_1 + F_3 - V_2 = m_J\left(-\ddot{y}_2 + \frac{h_1}{2}\ddot{\psi}_{1,2,3}\right) \tag{11.58a}$$

$$F_2 - V_1 + V_3 = m_J\ddot{u}_2, \text{ or equivalently}$$

$$F_2 - V_1 + V_3 = m_J \left(\ddot{y}_1 + \frac{h_2}{2} \ddot{\psi}_{1,2,3} \right), \text{ or equivalently}$$

$$F_2 - V_1 + V_3 = m_J \left(\ddot{y}_3 - \frac{h_2}{2} \ddot{\psi}_{1,2,3} \right) \tag{11.58b}$$

$$-M_1 + M_2 + M_3 + V_1 \frac{h_2}{2} + V_2 \frac{h_1}{2} + V_3 \frac{h_2}{2} = J\ddot{\psi}_{1,2,3} \tag{11.58c}$$

where $\ddot{u}_{1,3} = \ddot{u}_1 = \ddot{u}_3$ and $\ddot{\psi}_{1,2,3} = \ddot{\psi}_1 = \ddot{\psi}_2 = \ddot{\psi}_3$.

From Eqs. (11.58a-c), there are numerous sets of equivalent equilibrium equations. Each set consists of three scalar equations of equilibrium by taking one scalar equation each from Eqs. (11.58a), (11.58b), and (11.58c).

The reflection and transmission relationships corresponding to waves incident upon a joint from a different beam element are in general different. Figures 11.14 a-c show the reflection and transmission of waves incident upon a T joint from Beam 1, Beam 3, and Beam 2, respectively.

For waves incident from Beam 1, as shown in Figure 11.14a, the reflected waves $\mathbf{a}^-$ on Beam 1, the transmitted waves $\mathbf{b}^+$ to Beam 2, and the transmitted waves $\mathbf{e}^+$ to Beam 3 are related to the incident waves $\mathbf{a}^+$ by reflection matrix $\mathbf{r}_{11}$, transmission matrix $\mathbf{t}_{12}$, and transmission matrix $\mathbf{t}_{13}$,

$$\mathbf{a}^- = \mathbf{r}_{11}\mathbf{a}^+, \ \mathbf{b}^+ = \mathbf{t}_{12}\mathbf{a}^+, \ \mathbf{e}^+ = \mathbf{t}_{13}\mathbf{a}^+ \tag{11.59}$$

For waves incident from Beam 3, as shown in Figure 11.14b, the reflected waves $\mathbf{a}^-$ on Beam 3, the transmitted waves $\mathbf{b}^+$ to Beam 2, and the transmitted waves $\mathbf{e}^+$ to Beam 1 are related to the incident waves $\mathbf{a}^+$ by reflection matrix $\mathbf{r}_{33}$, transmission matrix $\mathbf{t}_{32}$, and transmission matrix $\mathbf{t}_{31}$,

$$\mathbf{a}^- = \mathbf{r}_{33}\mathbf{a}^+, \ \mathbf{b}^+ = \mathbf{t}_{32}\mathbf{a}^+, \ \mathbf{e}^+ = \mathbf{t}_{31}\mathbf{a}^+ \tag{11.60}$$

For waves incident from Beam 2, as shown in Figure 11.14c, the reflected waves $\mathbf{a}^-$ on Beam 2, the transmitted waves $\mathbf{b}^+$ to Beam 1, and the transmitted waves $\mathbf{e}^+$ to Beam 3 are related to the incident waves $\mathbf{a}^+$ by reflection matrix $\mathbf{r}_{22}$, transmission matrix $\mathbf{t}_{21}$, and transmission matrix $\mathbf{t}_{23}$,

$$\mathbf{a}^- = \mathbf{r}_{22}\mathbf{a}^+, \ \mathbf{b}^+ = \mathbf{t}_{21}\mathbf{a}^+, \ \mathbf{e}^+ = \mathbf{t}_{23}\mathbf{a}^+ \tag{11.61}$$

The wave vectors in Eqs. (11.59) to (11.61) are

$$\mathbf{a}^\pm = \begin{bmatrix} a_1^\pm \\ a_2^\pm \\ c^\pm \end{bmatrix}, \ \mathbf{b}^\pm = \begin{bmatrix} b_1^\pm \\ b_2^\pm \\ d^\pm \end{bmatrix}, \ \mathbf{e}^\pm = \begin{bmatrix} e_1^\pm \\ e_2^\pm \\ g^\pm \end{bmatrix} \tag{11.62}$$

where in each wave vector, the first two elements are for bending waves, and the third element is for longitudinal wave.

First, solve for $\mathbf{r}_{11}$, $\mathbf{t}_{12}$, and $\mathbf{t}_{13}$, that is, the reflection and transmission matrices at the T joint of Figure 11.14a corresponding to incident waves from Beam 1.

Following the local coordinate systems of Figure 11.13, the axial and bending deflections of the beam elements are

$$u_1 = c^+ e^{-ik_{13}x_1} + c^- e^{ik_{13}x_1}$$

$$u_2 = d^+ e^{-ik_{23}x_2}$$

$$u_3 = g^+ e^{-ik_{33}x_3}$$

$$y_1 = a_1^+ e^{-ik_{11}x_1} + a_2^+ e^{-k_{12}x_1} + a_1^- e^{ik_{11}x_1} + a_2^- e^{k_{12}x_1}$$

$$y_2 = b_1^+ e^{-ik_{21}x_2} + b_2^+ e^{-k_{22}x_2}$$

$$y_3 = e_1^+ e^{-ik_{31}x_3} + e_2^+ e^{-k_{32}x_3} \tag{11.63a}$$

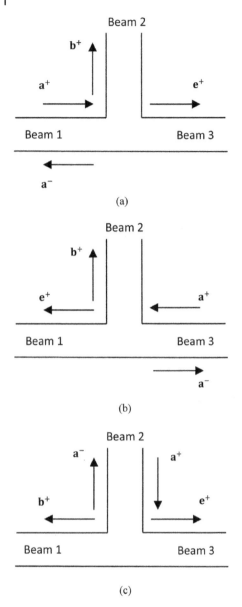

Figure 11.14 Wave reflection and transmission at a T joint with (a) incident waves from Beam 1, (b) incident waves from Beam 3, and (c) incident waves from Beam 2.

and the bending slopes by the Euler–Bernoulli bending vibration theory are

$$\psi_1 = -ik_{11}a_1^+e^{-ik_{11}x_1} - k_{12}a_2^+e^{-k_{12}x_1} + ik_{11}a_1^-e^{ik_{11}x_1} + k_{12}a_2^-e^{k_{12}x_1}$$

$$\psi_2 = -ik_{21}b_1^+e^{-ik_{21}x_2} - k_{22}b_2^+e^{-k_{22}x_2}$$

$$\psi_3 = -ik_{31}e_1^+e^{-ik_{31}x_3} - k_{32}e_2^+e^{-k_{32}x_3} \tag{11.63b}$$

and the bending slopes by the Timoshenko bending vibration theory are

$$\psi_1 = -iP_1a_1^+e^{-ik_{11}x_1} - N_1a_2^+e^{-k_{12}x_1} + iP_1a_1^-e^{ik_{11}x_1} + N_1a_2^-e^{k_{12}x_1}$$

$$\psi_2 = -iP_2b_1^+e^{-ik_{21}x_2} - N_2b_2^+e^{-k_{22}x_2}$$

$$\psi_3 = -iP_3e_1^+e^{-ik_{31}x_3} - N_3e_2^+e^{-k_{32}x_3} \tag{11.63c}$$

where i is the imaginary unit and subscripts of P and N identify the beam elements by numerical numbers as marked on Figures 11.13 and 11.14. Each wavenumber carries two numerical digits in its subscript, with the leading digit identifying a beam element and the other digit denoting the wavenumber sequence in the beam element.

Substituting Eqs. (11.63a) and (11.63b) into a set of continuity equations from Eqs. (11.54a-c), and choosing the origin at the joint, a set of three scalar equations of continuity is obtained based on the Euler–Bernoulli bending vibration theory. Similarly, substituting Eqs. (11.63a) and (11.63c) into a set of continuity equations from Eqs. (11.54a-c), and choosing the origin at the joint, a set of three scalar equations of continuity is obtained based on the Timoshenko bending vibration theory. This set of three scalar equations of continuity, either by the Euler–Bernoulli or Timoshenko bending vibration theory, can be placed in a vector equation, in terms of wave components $\mathbf{a}^+$, $\mathbf{a}^-$, and $\mathbf{b}^+$, which are defined in Eq. (11.62). A vector equation from this group is named $\mathbf{T\_Beam1}_{C12}$, because it is a vector equation on the continuity between Beam 1 and Beam 2 at a T joint corresponding to incident waves from Beam 1.

Substituting Eqs. (11.63a) and (11.63b) into a set of continuity equations from Eqs. (11.55a-c), and choosing the origin at the joint, a set of three scalar equations of continuity is obtained based on the Euler–Bernoulli bending vibration theory. Similarly, substituting Eqs. (11.63a) and (11.63c) into a set of continuity equations from Eqs. (11.55a-c), and choosing the origin at the joint, a set of three scalar equations of continuity is obtained based on the Timoshenko bending vibration theory. This set of three scalar equations of continuity, either by the Euler–Bernoulli or Timoshenko bending vibration theory, can be placed in a vector equation, in terms of wave components $\mathbf{a}^+$, $\mathbf{a}^-$, and $\mathbf{e}^+$, which are defined in Eq. (11.62). A vector equation from this group is named $\mathbf{T\_Beam1}_{C13}$, because it is a vector equation on the continuity between Beam 1 and Beam 3 at a T joint corresponding to incident waves from Beam 1.

Substituting Eqs. (11.63a) and (11.63b) into Eqs. (11.6a) and (11.6b) gives the functional relationships of the internal resistant forces and moment and the wave components, based on the Euler–Bernoulli bending vibration theory. Substituting these functional relationships into a set of equilibrium equations from Eqs. (11.58a-c), and choosing the origin at the joint, a set of three scalar equations of equilibrium is obtained based on the Euler–Bernoulli bending vibration theory. Similarly, Substituting Eqs. (11.63a) and (11.63c) into Eqs. (11.6a) and (11.6c) gives the functional relationships of the internal resistant forces and moment and the wave components, based on the Timoshenko bending vibration theory. Substituting these functional relationships into a set of equilibrium equations from Eqs. (11.58a-c), and choosing the origin at the joint, a set of three scalar equations of equilibrium is obtained based on the Timoshenko bending vibration theory. This set of three scalar equations of equilibrium, either by the Euler–Bernoulli or Timoshenko bending vibration theory, can be placed in a vector equation, in terms of wave components $\mathbf{a}^+$, $\mathbf{a}^-$, $\mathbf{b}^+$, and $\mathbf{e}^+$, which are defined in Eq. (11.62). A vector equation from this group is named $\mathbf{T\_Beam1}_E$, because it is a vector equation on equilibrium at a T joint corresponding to incident waves from Beam 1.

The reflection and transmission matrices, $\mathbf{r}_{11}$, $\mathbf{t}_{12}$, and $\mathbf{t}_{13}$, are found from Eq. (11.59) and the three vector equations of continuity and equilibrium, namely, $\mathbf{T\_Beam1}_{C12}$, $\mathbf{T\_Beam1}_{C13}$, and $\mathbf{T\_Beam1}_E$.

Next, solve for $\mathbf{r}_{33}$, $\mathbf{t}_{32}$, and $\mathbf{t}_{31}$, that is, the reflection and transmission matrices at the T joint of Figure 11.14b, corresponding to incident waves from Beam 3.

Following the local coordinate systems of Figure 11.13, the axial and bending deflections of the beam elements are

$$u_1 = g^+ e^{ik_{13}x_1}$$

$$u_2 = d^+ e^{-ik_{23}x_2}$$

$$u_3 = c^+ e^{ik_{33}x_3} + c^- e^{-ik_{33}x_3}$$

$$y_1 = e_1^+ e^{ik_{11}x_1} + e_2^+ e^{k_{12}x_1}$$

$$y_2 = b_1^+ e^{-ik_{21}x_2} + b_2^+ e^{-k_{22}x_2}$$

$$y_3 = a_1^+ e^{ik_{31}x_3} + a_2^+ e^{k_{32}x_3} + a_1^- e^{-ik_{31}x_3} + a_2^- e^{-k_{32}x_3} \tag{11.64a}$$

and the bending slopes by the Euler–Bernoulli bending vibration theory are

$$\psi_1 = ik_{11}e_1^+ e^{ik_{11}x_1} + k_{12}e_2^+ e^{k_{12}x_1}$$

$$\psi_2 = -ik_{21}b_1^+ e^{-ik_{21}x_2} - k_{22}b_2^+ e^{-k_{22}x_2}$$

$$\psi_3 = ik_{31}a_1^+ e^{ik_{31}x_3} + k_{32}a_2^+ e^{k_{32}x_3} - ik_{31}a_1^- e^{-ik_{31}x_3} - k_{32}a_2^- e^{-k_{32}x_3} \tag{11.64b}$$

and the bending slopes by the Timoshenko bending vibration theory are

$$\psi_1 = iP_1 e_1^+ e^{ik_{11}x_1} + N_1 e_2^+ e^{k_{12}x_1}$$

$$\psi_2 = -iP_2 b_1^+ e^{-ik_{21}x_2} - N_2 b_2^+ e^{-k_{22}x_2}$$

$$\psi_3 = iP_3 a_1^+ e^{ik_{31}x_3} + N_3 a_2^+ e^{k_{32}x_3} - iP_3 a_1^- e^{-ik_{31}x_3} - N_3 a_2^- e^{-k_{32}x_3} \tag{11.64c}$$

where i is the imaginary unit and subscripts of P and N identify the beam elements by numerical numbers as marked on Figures 11.13 and 11.14. Each wavenumber carries two numerical digits in its subscript, with the leading digit identifying a beam element and the other digit denoting the wavenumber sequence in the beam element.

Substituting Eqs. (11.64a) and (11.64b) into a set of continuity equations from Eqs. (11.55a-c), and choosing the origin at the joint, a set of three scalar equations of continuity is obtained based on the Euler–Bernoulli bending vibration theory. Similarly, substituting Eqs. (11.64a) and (11.64c) into a set of continuity equations from Eqs. (11.55a-c), and choosing the origin at the joint, a set of three scalar equations of continuity is obtained based on the Timoshenko bending vibration theory. This set of three scalar equations of continuity, either by the Euler–Bernoulli or Timoshenko bending vibration theory, can be placed in a vector equation, in terms of wave components $\mathbf{a}^+$, $\mathbf{a}^-$, and $\mathbf{e}^+$, which are defined in Eq. (11.62). A vector equation from this group is named $\mathbf{T\_Beam3}_{C31}$, because it is a vector equation on the continuity between Beam 3 and Beam 1 at a T joint corresponding to incident waves from Beam 3.

Substituting Eqs. (11.64a) and (11.64b) into a set of continuity equations from Eqs. (11.56a-c), and choosing the origin at the joint, a set of three scalar equations of continuity is obtained based on the Euler–Bernoulli bending vibration theory. Similarly, substituting Eqs. (11.64a) and (11.64c) into a set of continuity equations from Eqs. (11.56a-c), and choosing the origin at the joint, a set of three scalar equations of continuity is obtained based on the Timoshenko bending vibration theory. This set of three scalar equations of continuity, either by the Euler–Bernoulli or Timoshenko bending vibration theory, can be placed in a vector equation, in terms of wave components $\mathbf{a}^+$, $\mathbf{a}^-$, and $\mathbf{b}^+$, which are defined in Eq. (11.62). A vector equation from this group is named $\mathbf{T\_Beam3}_{C32}$, because it is a vector equation on the continuity between Beam 3 and Beam 2 at a T joint corresponding to incident waves from Beam 3.

Substituting Eqs. (11.64a) and (11.64b) into Eqs. (11.6a) and (11.6b) gives the functional relationships of the internal resistant forces and moment and the wave components, based on the Euler–Bernoulli bending vibration theory. Substituting these functional relationships into a set of equilibrium equations from Eqs. (11.58a-c), and choosing the origin at the joint, a set of three scalar equations of equilibrium is obtained based on the Euler–Bernoulli bending vibration theory. Similarly, Substituting Eqs. (11.64a) and (11.64c) into Eqs. (11.6a) and (11.6c) gives the functional relationships of the internal resistant forces and moment and the wave components, based on the Timoshenko bending vibration theory. Substituting these functional relationships into a set of equilibrium equations from Eqs. (11.58a-c), and choosing the origin at the joint, a set of three scalar equations of equilibrium is obtained based on the Timoshenko bending vibration theory. This set of three scalar equations of equilibrium, either by the Euler–Bernoulli or Timoshenko bending vibration theory, can be placed in a vector equation, in terms of wave components $\mathbf{a}^+$, $\mathbf{a}^-$, $\mathbf{b}^+$, and $\mathbf{e}^+$, which are defined in Eq. (11.62). A vector equation from this group is named $\mathbf{T\_Beam3}_E$, because it is a vector equation on equilibrium at a T joint corresponding to incident waves from Beam 3.

The reflection and transmission matrices, $\mathbf{r}_{33}$, $\mathbf{t}_{32}$, and $\mathbf{t}_{31}$, are found from Eq. (11.60) and the three vector equations of continuity and equilibrium, namely, $\mathbf{T\_Beam3}_{C31}$, $\mathbf{T\_Beam3}_{C32}$, and $\mathbf{T\_Beam3}_E$.

Last, solve for $\mathbf{r}_{22}$, $\mathbf{t}_{21}$, and $\mathbf{t}_{23}$, that is, the reflection and transmission matrices at the T joint of Figure 11.14c, corresponding to incident waves from Beam 2.

Following the local coordinate systems of Figure 11.13, the axial and bending deflections of the beam elements are

$$u_1 = d^+ e^{ik_{13}x_1}$$

$$u_2 = c^+ e^{ik_{23}x_2} + c^- e^{-ik_{23}x_2}$$

$$u_3 = g^+ e^{-ik_{33}x_3}$$

$$y_1 = b_1^+ e^{ik_{11}x_1} + b_2^+ e^{k_{12}x_1}$$

$$y_2 = a_1^+ e^{ik_{21}x_2} + a_2^+ e^{k_{22}x_2} + a_1^- e^{-ik_{21}x_2} + a_2^- e^{-k_{22}x_2}$$

$$y_3 = e_1^+ e^{-ik_{31}x_3} + e_2^+ e^{-k_{32}x_3} \tag{11.65a}$$

and the bending slopes by the Euler–Bernoulli bending vibration theory are

$$\psi_1 = ik_{11}b_1^+ e^{ik_{11}x_1} + k_{12}b_2^+ e^{k_{12}x_1}$$

$$\psi_2 = ik_{21}a_1^+ e^{ik_{21}x_2} + k_{22}a_2^+ e^{k_{22}x_2} - ik_{21}a_1^- e^{-ik_{21}x_2} - k_{22}a_2^- e^{-k_{22}x_2}$$

$$\psi_3 = -ik_{31}e_1^+ e^{-ik_{31}x_3} - k_{32}e_2^+ e^{-k_{32}x_3} \tag{11.65b}$$

and the bending slopes by the Timoshenko bending vibration theory are

$$\psi_1 = iP_1 b_1^+ e^{ik_{11}x_1} + N_1 b_2^+ e^{k_{12}x_1}$$

$$\psi_2 = iP_2 a_1^+ e^{ik_{21}x_2} + N_2 a_2^+ e^{k_{22}x_2} - iP_2 a_1^- e^{-ik_{21}x_2} - N_2 a_2^- e^{-k_{22}x_2}$$

$$\psi_3 = -iP_3 e_1^+ e^{-ik_{31}x_3} - N_3 e_2^+ e^{-k_{32}x_3} \tag{11.65c}$$

where i is the imaginary unit and subscripts of P and N identify the beam elements by numerical numbers as marked on Figures 11.13 and 11.14. Each wavenumber carries two numerical digits in its subscript, with the leading digit identifying a beam element and the other digit denoting the wavenumber sequence in the beam element.

Substituting Eqs. (11.65a) and (11.65b) into a set of continuity equations from Eqs. (11.54a-c), and choosing the origin at the joint, a set of three scalar equations of continuity is obtained based on the Euler–Bernoulli bending vibration theory. Similarly, substituting Eqs. (11.65a) and (11.65c) into a set of continuity equations from Eqs. (11.54a-c), and choosing the origin at the joint, a set of three scalar equations of continuity is obtained based on the Timoshenko bending vibration theory. This set of three scalar equations of continuity, either by the Euler–Bernoulli or Timoshenko bending vibration theory, can be placed in a vector equation, in terms of wave components $\mathbf{a}^+$, $\mathbf{a}^-$, and $\mathbf{b}^+$, which are defined in Eq. (11.62). A vector equation from this group is named $\mathbf{T\_Beam2}_{C21}$, because it is a vector equation on the continuity between Beam 2 and Beam 1 at a T joint corresponding to incident waves from Beam 2.

Substituting Eqs. (11.65a) and (11.65b) into a set of continuity equations from Eqs. (11.56a-c), and choosing the origin at the joint, a set of three scalar equations of continuity is obtained based on the Euler–Bernoulli bending vibration theory. Similarly, substituting Eqs. (11.65a) and (11.65c) into a set of continuity equations from Eqs. (11.56a-c), and choosing the origin at the joint, a set of three scalar equations of continuity is obtained based on the Timoshenko bending vibration theory. This set of three scalar equations of continuity, either by the Euler–Bernoulli or Timoshenko bending vibration theory, can be placed in a vector equation, in terms of wave components $\mathbf{a}^+$, $\mathbf{a}^-$, and $\mathbf{e}^+$, which are defined in Eq. (11.62). A vector equation from this group is named $\mathbf{T\_Beam2}_{C23}$, because it is a vector equation on the continuity between Beam 2 and Beam 3 at a T joint corresponding to incident waves from Beam 2.

Substituting Eqs. (11.65a) and (11.65b) into Eqs. (11.6a) and (11.6b) gives the functional relationships of the internal resistant forces and moment and the wave components, based on the Euler–Bernoulli bending vibration theory. Substituting these functional relationships into a set of equilibrium equations from Eqs. (11.58a-c), and choosing the origin at the joint, a set of three scalar equations of equilibrium is obtained based on the Euler–Bernoulli bending vibration theory. Similarly, Substituting Eqs. (11.65a) and (11.65c) into Eqs. (11.6a) and (11.6c) gives the functional relationships of the internal resistant forces and moment and the wave components, based on the Timoshenko bending vibration theory. Substituting these functional relationships into a set of equilibrium equations from Eqs. (11.58a-c), and choosing the origin at the joint, a set of three scalar equations of equilibrium is obtained based on the Timoshenko bending vibration theory. This set of three scalar equations of equilibrium, either by the Euler–Bernoulli or Timoshenko bending vibration theory, can be placed in a vector equation, in terms of wave components $\mathbf{a}^+$, $\mathbf{a}^-$, $\mathbf{b}^+$, and $\mathbf{e}^+$, which are defined in Eq. (11.62). A vector equation from this group is named $\mathbf{T\_Beam2}_E$, because it is a vector equation on equilibrium at a T joint corresponding to incident waves from Beam 2.

The reflection and transmission matrices, $\mathbf{r}_{22}$, $\mathbf{t}_{21}$, and $\mathbf{t}_{23}$, are found from Eq. (11.61) and the three vector equations of continuity and equilibrium, namely, $\mathbf{T\_Beam2}_{C21}$, $\mathbf{T\_Beam2}_{C23}$, and $\mathbf{T\_Beam2}_E$.

11.5.3 Wave Reflection and Transmission at a Cross Joint

Figure 11.15 shows the free body diagram of the cross joint following the local coordinates defined in Figure 11.10. Recall that the local coordinates at the "+" joints are mostly predetermined by the surrounding T joints.

From the free body diagram of Figure 11.15, the equations of motion of the cross joint are obtained by applying Newton's second law for translational and rotational motion,

$$V_1 - F_2 - V_3 + F_4 = m_J \ddot{U}_J$$

$$-V_2 - F_1 + F_3 + V_4 = m_J \ddot{Y}_J$$

$$M_1 - M_2 - M_3 - M_4 + V_1\frac{h_2}{2} + V_2\frac{h_1}{2} + V_3\frac{h_2}{2} + V_4\frac{h_1}{2} = J\ddot{\psi}_J \tag{11.66}$$

where h denotes the thickness of the beam elements at the rigid body joint. The subscript in h identifies the beam by sequence number at the joint. The rigid body "+" joint is assumed a rectangular shape, that is, $h_3 = h_1$, and $h_4 = h_2$. Let us call Eq. (11.66) the equilibrium equations, because by the principle of d'Alembert, these equations represent dynamic equilibrium.

From Figure 11.15, the continuity conditions at the cross joint are

$$u_1 = -Y_J, \; u_2 = U_J, \; u_3 = -Y_J, \; u_4 = -U_J$$

$$y_1 = U_J + \frac{h_2}{2}\psi_J, \; y_2 = Y_J - \frac{h_1}{2}\psi_J, \; y_3 = U_J - \frac{h_2}{2}\psi_J, \; y_4 = -Y_J - \frac{h_1}{2}\psi_J$$

$$\psi_1 = \psi_J, \; \psi_2 = \psi_J, \; \psi_3 = \psi_J, \; \psi_4 = \psi_J \tag{11.67}$$

To find the reflection and transmission relationships of vibration waves in beam elements connected at a joint, the incident, reflected, and transmitted waves must be related directly. As a result, the rigid body joint related parameters $\{U_J, Y_J, \psi_J\}$ need to be eliminated from both the equilibrium and continuity equations, namely, Eqs. (11.66) and (11.67).

Eliminating the rigid body joint related parameters $\{U_J, Y_J, \psi_J\}$ from the continuity equations of Eq. (11.67),

$$u_1 = u_3$$

$$u_2 = -u_4$$

$$\psi_1 = \psi_2 = \psi_3 = \psi_4$$

$$y_1 = u_2 + \frac{h_2}{2}\psi_{1,2,3,4}, \text{ or equivalently } y_1 = -u_4 + \frac{h_2}{2}\psi_{1,2,3,4}$$

$$y_2 = -u_{1,3} - \frac{h_1}{2}\psi_{1,2,3,4}$$

$$y_3 = u_2 - \frac{h_2}{2}\psi_{1,2,3,4}, \text{ or equivalently } y_3 = -u_4 - \frac{h_2}{2}\psi_{1,2,3,4}$$

$$y_4 = u_{1,3} - \frac{h_1}{2}\psi_{1,2,3,4} \tag{11.68}$$

where $\psi_{1,2,3,4} = \psi_1 = \psi_2 = \psi_3 = \psi_4$ and $u_{1,3} = u_1 = u_3$.

The continuity equations of Eq. (11.68) need to be sorted into six groups. Each group consists of three scalar equations, relating waves in two beam elements at a time.

The group of continuity equations relating Beam 1 and Beam 2, by sorting Eq. (11.68), is

$$\psi_1 = \psi_2 \tag{11.69a}$$

$$y_1 = u_2 + \frac{h_2}{2}\psi_{1,2} \tag{11.69b}$$

$$y_2 = -u_1 - \frac{h_1}{2}\psi_{1,2} \tag{11.69c}$$

where $\psi_{1,2} = \psi_1 = \psi_2$.

From Eqs. (11.69a-c), in the group of continuity equations relating Beam 1 and Beam 2, there are four sets of equivalent continuity equations. Each set consists of three scalar equations of continuity by taking Eq. (11.69a), and one equation each from Eqs. (11.69b) and (11.69c).

The group of continuity equations relating Beam 1 and Beam 3, by sorting Eq. (11.68), is

$$\psi_1 = \psi_3 \tag{11.70a}$$

$$u_1 = u_3 \tag{11.70b}$$

$$y_1 - y_3 = h_2\psi_{1,3} \tag{11.70c}$$

where $\psi_{1,3} = \psi_1 = \psi_3$.

From Eqs. (11.70a-c), in the group of continuity equations relating Beam 1 and Beam 3, there are two sets of equivalent continuity equations. Each set consists of three scalar equations of continuity by taking Eq. (11.70a), Eq. (11.70b), and one equation from Eq. (11.70c).

The group of continuity equations relating Beam 1 and Beam 4, by sorting Eq. (11.68), is

$$\psi_1 = \psi_4 \tag{11.71a}$$

$$y_1 = -u_4 + \frac{h_2}{2}\psi_{1,4} \tag{11.71b}$$

$$y_4 = u_1 - \frac{h_1}{2}\psi_{1,4} \tag{11.71c}$$

where $\psi_{1,4} = \psi_1 = \psi_4$.

From Eqs. (11.71a-c), in the group of continuity equations relating Beam 1 and Beam 4, there are four sets of equivalent continuity equations. Each set consists of three scalar equations of continuity by taking Eq. (11.71a), and one equation each from Eqs. (11.71b) and (11.71c).

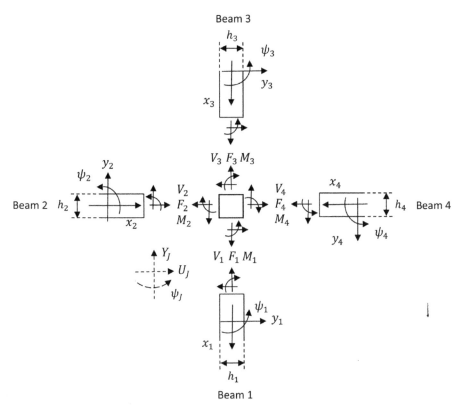

Figure 11.15 Free body diagram of the cross joints in Figure 11.10.

The group of continuity equations relating Beam 2 and Beam 3, by sorting Eqs. (11.68), is

$$\psi_2 = \psi_3 \tag{11.72a}$$

$$y_2 = -u_3 - \frac{h_1}{2}\psi_{2,3} \tag{11.72b}$$

$$y_3 = u_2 - \frac{h_2}{2}\psi_{2,3} \tag{11.72c}$$

where $\psi_{2,3} = \psi_2 = \psi_3$.

From Eqs. (11.72a-c), in the group of continuity equations relating Beam 2 and Beam 3, there are four sets of equivalent continuity equations. Each set consists of three scalar equations of continuity by taking Eq. (11.72a), and one equation each from Eqs. (11.72b) and (11.72c).

The group of continuity equations relating Beam 2 and Beam 4, by sorting Eq. (11.68), is

$$\psi_2 = \psi_4 \tag{11.73a}$$

$$u_2 = -u_4 \tag{11.73b}$$

$$y_2 + y_4 = -h_1\psi_{2,4} \tag{11.73c}$$

where $\psi_{2,4} = \psi_2 = \psi_4$.

From Eqs. (11.73a-c), in the group of continuity equations relating Beam 2 and Beam 4, there are two sets of equivalent continuity equations. Each set consists of three scalar equations of continuity by taking Eq. (11.73a), Eq. (11.73b), and one equation from Eq. (11.73c).

The group of continuity equations relating Beam 3 and Beam 4, by sorting Eq. (11.68), is

$$\psi_3 = \psi_4 \tag{11.74a}$$

$$y_3 = -u_4 - \frac{h_2}{2}\psi_{3,4} \tag{11.74b}$$

$$y_4 = u_3 - \frac{h_1}{2}\psi_{3,4} \tag{11.74c}$$

where $\psi_{3,4} = \psi_3 = \psi_4$.

From Eqs. (11.74a-c), in the group of continuity equations relating Beam 3 and Beam 4, there are four sets of equivalent continuity equations. Each set consists of three scalar equations of continuity by taking Eq. (11.74a), and one equation each from Eqs. (11.74b) and (11.74c).

From the continuity equations of Eq. (11.67), the rigid body joint related parameters U_J, Y_J, and ψ_J can be expressed in terms of the beam elements related parameters u_n, y_n, and ψ_n, where $n = 1, 2, 3, 4$, as follows

$$\psi_J = \psi_{1,2,3,4};$$

$U_J = u_2$, or equivalently $U_J = -u_4$, or equivalently

$$U_J = y_1 - \frac{h_2}{2}\psi_{1,2,3,4}, \text{ or equivalently } U_J = y_3 + \frac{h_2}{2}\psi_{1,2,3,4};$$

$$Y_J = -u_{1,3}, \text{ or equivalently } Y_J = y_2 + \frac{h_1}{2}\psi_{1,2,3,4}, \text{ or equivalently } Y_J = -y_4 - \frac{h_1}{2}\psi_{1,2,3,4} \tag{11.75}$$

From Eqs. (11.66) and (11.75),

$V_1 - F_2 - V_3 + F_4 = m_J \ddot{u}_2$, or equivalently

$V_1 - F_2 - V_3 + F_4 = -m_J \ddot{u}_4$, or equivalently

$$V_1 - F_2 - V_3 + F_4 = m_J\left(\ddot{y}_1 - \frac{h_2}{2}\ddot{\psi}_{1,2,3,4}\right), \text{ or equivalently}$$

$$V_1 - F_2 - V_3 + F_4 = m_J\left(\ddot{y}_3 + \frac{h_2}{2}\ddot{\psi}_{1,2,3,4}\right) \tag{11.76a}$$

$-V_2 - F_1 + F_3 + V_4 = -m_J \ddot{u}_{1,3}$, or equivalently

$$-V_2 - F_1 + F_3 + V_4 = m_J\left(\ddot{y}_2 + \frac{h_1}{2}\ddot{\psi}_{1,2,3,4}\right), \text{ or equivalently}$$

$$-V_2 - F_1 + F_3 + V_4 = m_J\left(-\ddot{y}_4 - \frac{h_1}{2}\ddot{\psi}_{1,2,3,4}\right) \tag{11.76b}$$

$$M_1 - M_2 - M_3 - M_4 + V_1\frac{h_2}{2} + V_2\frac{h_1}{2} + V_3\frac{h_2}{2} + V_4\frac{h_1}{2} = J\ddot{\psi}_{1,2,3,4} \tag{11.76c}$$

where $\ddot{u}_{1,3} = \ddot{u}_1 = \ddot{u}_3$, and $\ddot{\psi}_{1,2,3,4} = \ddot{\psi}_1 = \ddot{\psi}_2 = \ddot{\psi}_3 = \ddot{\psi}_4$.

From Eqs. (11.76a-c), there are numerous sets of equivalent equilibrium equations. Each set consists of three scalar equations of equilibrium by taking one scalar equation each from Eqs. (11.76a), (11.76b), and (11.76c).

The reflection and transmission relationships corresponding to waves incident upon a joint from a different beam element are in general different. Figures 11.16 a-d show the reflection and transmission of the waves incident upon a cross joint from Beam 1, Beam 2, Beam 3, and Beam 4, respectively.

For waves incident from Beam 1, as shown in Figure 11.16a, the reflected waves $\mathbf{a}^-$ on Beam 1, the transmitted waves $\mathbf{b}^+$ to Beam 2, the transmitted waves $\mathbf{e}^+$ to Beam 3, and the transmitted waves $\mathbf{p}^+$ to Beam 4 are related to the incident waves $\mathbf{a}^+$ by reflection matrix $\mathbf{r}_{11}$, transmission matrix $\mathbf{t}_{12}$, transmission matrix $\mathbf{t}_{13}$, and transmission matrix $\mathbf{t}_{14}$,

$$\mathbf{a}^- = \mathbf{r}_{11}\mathbf{a}^+, \ \mathbf{b}^+ = \mathbf{t}_{12}\mathbf{a}^+, \ \mathbf{e}^+ = \mathbf{t}_{13}\mathbf{a}^+, \ \mathbf{p}^+ = \mathbf{t}_{14}\mathbf{a}^+ \tag{11.77}$$

For waves incident from Beam 2, as shown in Figure 11.16b, the reflected waves $\mathbf{a}^-$ on Beam 2, the transmitted waves $\mathbf{b}^+$ to Beam 1, the transmitted waves $\mathbf{e}^+$ to Beam 3, and the transmitted waves $\mathbf{p}^+$ to Beam 4 are related to the incident waves $\mathbf{a}^+$ by reflection matrix $\mathbf{r}_{22}$, transmission matrix $\mathbf{t}_{21}$, transmission matrix $\mathbf{t}_{23}$, and transmission matrix $\mathbf{t}_{24}$,

$$\mathbf{a}^- = \mathbf{r}_{22}\mathbf{a}^+, \ \mathbf{b}^+ = \mathbf{t}_{21}\mathbf{a}^+, \ \mathbf{e}^+ = \mathbf{t}_{23}\mathbf{a}^+, \ \mathbf{p}^+ = \mathbf{t}_{24}\mathbf{a}^+ \tag{11.78}$$

For waves incident from Beam 3, as shown in Figure 11.16c, the reflected waves $\mathbf{a}^-$ on Beam 3, the transmitted waves $\mathbf{b}^+$ to Beam 2, the transmitted waves $\mathbf{e}^+$ to Beam 1, and the transmitted waves $\mathbf{p}^+$ to Beam 4 are related to the incident waves $\mathbf{a}^+$ by reflection matrix $\mathbf{r}_{33}$, transmission matrix $\mathbf{t}_{32}$, transmission matrix $\mathbf{t}_{31}$, and transmission matrix $\mathbf{t}_{34}$,

$$\mathbf{a}^- = \mathbf{r}_{33}\mathbf{a}^+, \ \mathbf{b}^+ = \mathbf{t}_{32}\mathbf{a}^+, \ \mathbf{e}^+ = \mathbf{t}_{31}\mathbf{a}^+, \ \mathbf{p}^+ = \mathbf{t}_{34}\mathbf{a}^+ \tag{11.79}$$

For waves incident from Beam 4, as shown in Figure 11.16d, the reflected waves $\mathbf{a}^-$ on Beam 4, the transmitted waves $\mathbf{b}^+$ to Beam 1, the transmitted waves $\mathbf{p}^+$ to Beam 2, and the transmitted waves $\mathbf{e}^+$ to Beam 3 are related to the incident waves $\mathbf{a}^+$ by reflection matrix $\mathbf{r}_{44}$, transmission matrix $\mathbf{t}_{41}$, transmission matrix $\mathbf{t}_{42}$, and transmission matrix $\mathbf{t}_{43}$,

$$\mathbf{a}^- = \mathbf{r}_{44}\mathbf{a}^+, \ \mathbf{b}^+ = \mathbf{t}_{41}\mathbf{a}^+, \ \mathbf{p}^+ = \mathbf{t}_{42}\mathbf{a}^+, \ \mathbf{e}^+ = \mathbf{t}_{43}\mathbf{a}^+ \tag{11.80}$$

The wave vectors in Eqs. (11.77) to (11.80) are

$$\mathbf{a}^\pm = \begin{bmatrix} a_1^\pm \\ a_2^\pm \\ c^\pm \end{bmatrix}, \ \mathbf{b}^\pm = \begin{bmatrix} b_1^\pm \\ b_2^\pm \\ d^\pm \end{bmatrix}, \ \mathbf{e}^\pm = \begin{bmatrix} e_1^\pm \\ e_2^\pm \\ g^\pm \end{bmatrix}, \ \mathbf{p}^\pm = \begin{bmatrix} p_1^\pm \\ p_2^\pm \\ q^\pm \end{bmatrix} \tag{11.81}$$

where in each wave vector, the first two elements are for bending waves, and the third element is for longitudinal wave.

There are four sets of reflection and transmission matrices to solve, corresponding to incident waves from each of the four beams joined at the cross joint, as described in Eqs. (11.77) to (11.80).

First, solve for $\mathbf{r}_{11}$, $\mathbf{t}_{12}$, $\mathbf{t}_{13}$, and $\mathbf{t}_{14}$, that is, the reflection and transmission matrices at the cross joint of Figure 11.16a, corresponding to incident waves from Beam 1.

Following the local coordinate systems of Figure 11.15, the axial and bending deflections of the beam elements are

$$u_1 = c^+ e^{ik_{13}x_1} + c^- e^{-ik_{13}x_1}$$

$$u_2 = d^+ e^{ik_{23}x_2}$$

$$u_3 = g^+ e^{ik_{33}x_3}$$

$$u_4 = q^+ e^{ik_{43}x_4}$$

$$y_1 = a_1^+ e^{ik_{11}x_1} + a_2^+ e^{k_{12}x_1} + a_1^- e^{-ik_{11}x_1} + a_2^- e^{-k_{12}x_1}$$

$$y_2 = b_1^+ e^{ik_{21}x_2} + b_2^+ e^{k_{22}x_2}$$

$$y_3 = e_1^+ e^{ik_{31}x_3} + e_2^+ e^{k_{32}x_3}$$

$$y_4 = p_1^+ e^{ik_{41}x_4} + p_2^+ e^{k_{42}x_4} \tag{11.82a}$$

and the bending slopes by the Euler–Bernoulli bending vibration theory are

$$\psi_1 = ik_{11}a_1^+ e^{ik_{11}x_1} + k_{12}a_2^+ e^{k_{12}x_1} - ik_{11}a_1^- e^{-ik_{11}x_1} - k_{12}a_2^- e^{-k_{12}x_1}$$

$$\psi_2 = ik_{21}b_1^+ e^{ik_{21}x_2} + k_{22}b_2^+ e^{k_{22}x_2}$$

$$\psi_3 = k_{31}e_1^+ e^{ik_{31}x_3} + k_{32}e_2^+ e^{k_{32}x_3}$$

$$\psi_4 = ik_{41}p_1^+ e^{ik_{41}x_4} + k_{42}p_2^+ e^{k_{42}x_4} \tag{11.82b}$$

and the bending slopes by the Timoshenko bending vibration theory are

$$\psi_1 = iP_1a_1^+ e^{ik_{11}x_1} + N_1a_2^+ e^{k_{12}x_1} - iP_1a_1^- e^{-ik_{11}x_1} - N_1a_2^- e^{-k_{12}x_1}$$

$$\psi_2 = iP_2b_1^+ e^{ik_{21}x_2} + N_2b_2^+ e^{k_{22}x_2}$$

$$\psi_3 = iP_3e_1^+ e^{ik_{31}x_3} + N_3e_2^+ e^{k_{32}x_3}$$

$$\psi_4 = iP_4p_1^+ e^{ik_{41}x_4} + N_4p_2^+ e^{k_{42}x_4} \tag{11.82c}$$

where i is the imaginary unit and subscripts of P and N identify the beam elements by numerical numbers as marked on Figures 11.15 and 11.16. Each wavenumber carries two numerical digits in its subscript, with the leading digit identifying a beam element and the other digit denoting the wavenumber sequence in the beam element.

Substituting Eqs. (11.82a) and (11.82b) into a set of continuity equations from Eqs. (11.69a-c), and choosing the origin at the joint, a set of three scalar equations of continuity is obtained based on the Euler–Bernoulli bending vibration theory. Similarly, substituting Eqs. (11.82a) and (11.82c) into a set of continuity equations from Eqs. (11.69a-c), and choosing the origin at the joint, a set of three scalar equations of continuity is obtained based on the Timoshenko bending vibration theory. This set of three scalar equations of continuity, either by the Euler–Bernoulli or Timoshenko bending vibration theory, can be placed in a vector equation, in terms of wave components $\mathbf{a}^+$, $\mathbf{a}^-$, and $\mathbf{b}^+$, which are defined in Eq. (11.81). A vector equation from this group is named $+\_\mathbf{Beam1}_{C12}$, because it is a vector equation on the continuity between Beam 1 and Beam 2 at a "+" joint corresponding to incident waves from Beam 1.

Substituting Eqs. (11.82a) and (11.82b) into a set of continuity equations from Eqs. (11.70a-c), and choosing the origin at the joint, a set of three scalar equations of continuity is obtained based on the Euler–Bernoulli bending vibration theory. Similarly, substituting Eqs. (11.82a) and (11.82c) into a set of continuity equations from Eqs. (11.70a-c), and choosing the origin at the joint, a set of three scalar equations of continuity is obtained based on the Timoshenko bending vibration theory. This set of three scalar equations of continuity, either by the Euler–Bernoulli or Timoshenko bending vibration theory, can be placed in a vector equation, in terms of wave components $\mathbf{a}^+$, $\mathbf{a}^-$, and $\mathbf{e}^+$, which are defined in Eq. (11.81). A vector equation from this group is named $+\_\mathbf{Beam1}_{C13}$, because it is a vector equation on the continuity between Beam 1 and Beam 3 at a "+" joint corresponding to incident waves from Beam 1.

Substituting Eqs. (11.82a) and (11.82b) into a set of continuity equations from Eqs. (11.71a-c), and choosing the origin at the joint, a set of three scalar equations of continuity is obtained based on the Euler–Bernoulli bending vibration theory. Similarly, substituting Eqs. (11.82a) and (11.82c) into a set of continuity equations from Eqs. (11.71a-c), and choosing the origin at the joint, a set of three scalar equations of continuity is obtained based on the Timoshenko bending vibration theory. This set of three scalar equations of continuity, either by the Euler–Bernoulli or Timoshenko bending vibration theory, can be placed in a vector equation, in terms of wave components $\mathbf{a}^+$, $\mathbf{a}^-$, and $\mathbf{p}^+$, which are defined in Eq. (11.81). A vector equation from this group is named $+\_\mathbf{Beam1}_{C14}$, because it is a vector equation on the continuity between Beam 1 and Beam 4 at a "+" joint corresponding to incident waves from Beam 1.

Substituting Eqs. (11.82a) and (11.82b) into Eqs. (11.6a) and (11.6b) gives the functional relationships of the internal resistant forces and moment and the wave components, based on the Euler–Bernoulli bending vibration theory. Substituting these functional relationships into a set of equilibrium equations from Eqs. (11.76a-c), and choosing the origin at the joint, a set of three scalar equations of equilibrium is obtained based on the Euler–Bernoulli bending vibration theory. Similarly, Substituting Eqs. (11.82a) and (11.82c) into Eqs. (11.6a) and (11.6c) gives the functional relationships of the internal resistant forces and moment and the wave components, based on the Timoshenko bending vibration theory. Substituting these functional relationships into a set of equilibrium equations from Eqs. (11.76a-c), and choosing the origin at the joint, a set of three scalar equations of equilibrium is obtained based on the Timoshenko bending vibration theory. This set of three scalar equations of equilibrium, either by the Euler–Bernoulli or Timoshenko bending vibration theory, can be placed in a vector equation, in terms of wave components $\mathbf{a}^+$, $\mathbf{a}^-$, $\mathbf{b}^+$, $\mathbf{e}^+$, and $\mathbf{p}^+$, which are defined in Eq. (11.81). A vector equation from this group is named $+\_\mathbf{Beam1}_E$, because it is a vector equation on equilibrium at a "+" joint corresponding to incident waves from Beam 1.

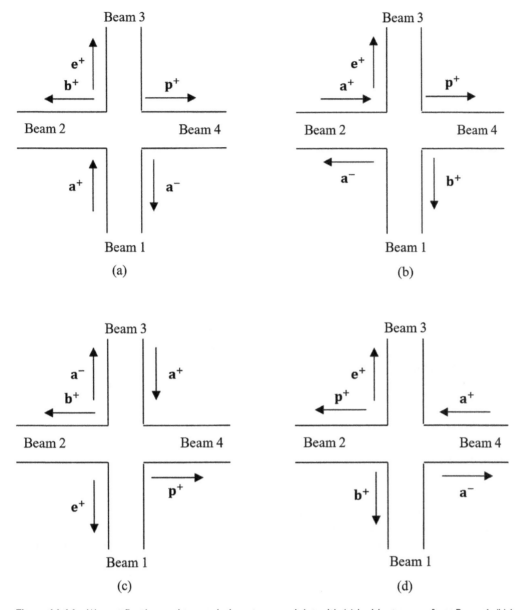

Figure 11.16 Wave reflection and transmission at a cross joint with (a) incident waves from Beam 1, (b) incident waves from Beam 2, (c) incident waves from Beam 3, and (d) incident waves from Beam 4.

The reflection and transmission matrices, $\mathbf{r}_{11}$, $\mathbf{t}_{12}$, $\mathbf{t}_{13}$, and $\mathbf{t}_{14}$, are found from Eq. (11.77) and the four vector equations of continuity and equilibrium, namely, $+\_\mathbf{Beam1}_{C12}$, $+\_\mathbf{Beam1}_{C13}$, $+\_\mathbf{Beam1}_{C14}$, and $+\_\mathbf{Beam1}_E$.

Next, solve for $\mathbf{r}_{22}$, $\mathbf{t}_{21}$, $\mathbf{t}_{23}$, and $\mathbf{t}_{24}$, that is, the reflection and transmission matrices at the cross joint of Figure 11.16b, corresponding to incident waves from Beam 2.

Following the local coordinate systems of Figure 11.15, the axial and bending deflections of the beam elements are

$$u_1 = d^+ e^{-ik_{13}x_1}$$

$$u_2 = c^+ e^{-ik_{23}x_2} + c^- e^{ik_{23}x_2}$$

$$u_3 = g^+ e^{ik_{33}x_3}$$

$$u_4 = q^+ e^{ik_{43}x_4}$$

$$y_1 = b_1^+ e^{-ik_{11}x_1} + b_2^+ e^{-k_{12}x_1}$$

$$y_2 = a_1^+ e^{-ik_{21}x_2} + a_2^+ e^{-k_{22}x_2} + a_1^- e^{ik_{21}x_2} + a_2^- e^{k_{22}x_2}$$

$$y_3 = e_1^+ e^{ik_{31}x_3} + e_2^+ e^{k_{32}x_3}$$

$$y_4 = p_1^+ e^{ik_{41}x_4} + p_2^+ e^{k_{42}x_4} \tag{11.83a}$$

and the bending slopes by the Euler–Bernoulli bending vibration theory are

$$\psi_1 = -ik_{11}b_1^+ e^{-ik_{11}x_1} - k_{12}b_2^+ e^{-k_{12}x_1}$$

$$\psi_2 = -ik_{21}a_1^+ e^{-ik_{21}x_2} - k_{22}a_2^+ e^{-k_{22}x_2} + ik_{21}a_1^- e^{ik_{21}x_2} + k_{22}a_2^- e^{k_{22}x_2}$$

$$\psi_3 = ik_{31}e_1^+ e^{ik_{31}x_3} + k_{32}e_2^+ e^{k_{32}x_3}$$

$$\psi_4 = ik_{41}p_1^+ e^{ik_{41}x_4} + k_{42}p_2^+ e^{k_{42}x_4} \tag{11.83b}$$

and the bending slopes by the Timoshenko bending vibration theory are

$$\psi_1 = -iP_1 b_1^+ e^{-ik_{11}x_1} - N_1 b_2^+ e^{-k_{12}x_1}$$

$$\psi_2 = -iP_2 a_1^+ e^{-ik_{21}x_2} - N_2 a_2^+ e^{-k_{22}x_2} + iP_2 a_1^- e^{ik_{21}x_2} + N_2 a_2^- e^{k_{22}x_2}$$

$$\psi_3 = iP_3 e_1^+ e^{ik_{31}x_3} + N_3 e_2^+ e^{k_{32}x_3}$$

$$\psi_4 = iP_4 p_1^+ e^{ik_{41}x_4} + N_4 p_2^+ e^{k_{42}x_4} \tag{11.83c}$$

where i is the imaginary unit and subscripts of P and N identify the beam elements by numerical numbers as marked on Figures 11.15 and 11.16. Each wavenumber carries two numerical digits in its subscript, with the leading digit identifying a beam element and the other digit denoting the wavenumber sequence in the beam element.

Substituting Eqs. (11.83a) and (11.83b) into a set of continuity equations from Eqs. (11.69a-c), and choosing the origin at the joint, a set of three scalar equations of continuity is obtained based on the Euler–Bernoulli bending vibration theory. Similarly, substituting Eqs. (11.83a) and (11.83c) into a set of continuity equations from Eqs. (11.69a-c), and choosing the origin at the joint, a set of three scalar equations of continuity is obtained based on the Timoshenko bending vibration theory. This set of three scalar equations of continuity, either by the Euler–Bernoulli or Timoshenko bending vibration theory, can be placed in a vector equation, in terms of wave components $\mathbf{a}^+$, $\mathbf{a}^-$, and $\mathbf{b}^+$, which are defined in Eq. (11.81). A vector equation from this group is named $+\_\mathbf{Beam2}_{C21}$, because it is a vector equation on the continuity between Beam 2 and Beam 1 at a "+" joint corresponding to incident waves from Beam 2.

Substituting Eqs. (11.83a) and (11.83b) into a set of continuity equations from Eqs. (11.72a-c), and choosing the origin at the joint, a set of three scalar equations of continuity is obtained based on the Euler–Bernoulli bending vibration theory. Similarly, substituting Eqs. (11.83a) and (11.83c) into a set of continuity equations from Eqs. (11.72a-c), and choosing the origin at the joint, a set of three scalar equations of continuity is obtained based on the Timoshenko bending vibration theory. This set of three scalar equations of continuity, either by the Euler–Bernoulli or Timoshenko bending vibration theory, can be placed in a vector equation, in terms of wave components $\mathbf{a}^+$, $\mathbf{a}^-$, and $\mathbf{e}^+$, which are defined in Eq. (11.81). A vector equation from this group is named $+\_\mathbf{Beam2}_{C23}$, because it is a vector equation on the continuity between Beam 2 and Beam 3 at a "+" joint corresponding to incident waves from Beam 2.

Substituting Eqs. (11.83a) and (11.83b) into a set of continuity equations from Eqs. (11.73a-c), and choosing the origin at the joint, a set of three scalar equations of continuity is obtained based on the Euler–Bernoulli bending vibration theory. Similarly, substituting Eqs. (11.83a) and (11.83c) into a set of continuity equations from Eqs. (11.73a-c), and choosing the origin at the joint, a set of three scalar equations of continuity is obtained based on the Timoshenko bending vibration theory. This set of three scalar equations of continuity, either by the Euler–Bernoulli or Timoshenko bending vibration theory, can be placed in a vector equation, in terms of wave components $\mathbf{a}^+$, $\mathbf{a}^-$, and $\mathbf{p}^+$, which are defined in Eq. (11.81). A vector equation from this group is named $+\_\mathbf{Beam2}_{C24}$, because it is a vector equation on the continuity between Beam 2 and Beam 4 at a "+" joint corresponding to incident waves from Beam 2.

Substituting Eqs. (11.83a) and (11.83b) into Eqs. (11.6a) and (11.6b) gives the functional relationships of the internal resistant forces and moment and the wave components, based on the Euler–Bernoulli bending vibration theory. Substituting these functional relationships into a set of equilibrium equations from Eqs. (11.76a-c), and choosing the origin at the joint, a set of three scalar equations of equilibrium is obtained based on the Euler–Bernoulli bending vibration theory. Similarly, Substituting Eqs. (11.83a) and (11.83c) into Eqs. (11.6a) and (11.6c) gives the functional relationships of the internal resistant forces and moment and the wave components, based on the Timoshenko bending vibration theory. Substituting these functional relationships into a set of equilibrium equations from Eqs. (11.76a-c), and choosing the origin at the joint, a set of three scalar equations of equilibrium is obtained based on the Timoshenko bending vibration theory. This set of three scalar equations of equilibrium, either by the Euler–Bernoulli or Timoshenko bending vibration theory, can be placed in a vector equation, in terms of wave components $\mathbf{a}^+$, $\mathbf{a}^-$, $\mathbf{b}^+$, $\mathbf{e}^+$, and $\mathbf{p}^+$, which are defined in Eq. (11.81). A vector equation from this group is named $+\_\mathbf{Beam2}_E$, because it is a vector equation on equilibrium at a "+" joint corresponding to incident waves from Beam 2.

The reflection and transmission matrices, $\mathbf{r}_{22}$, $\mathbf{t}_{21}$, $\mathbf{t}_{23}$, and $\mathbf{t}_{24}$, are found from Eq. (11.78) and the four vector equations of continuity and equilibrium, namely, $+\_\mathbf{Beam2}_{C21}$, $+\_\mathbf{Beam2}_{C23}$, $+\_\mathbf{Beam2}_{C24}$, and $+\_\mathbf{Beam2}_E$.

Then, solve for $\mathbf{r}_{33}$, $\mathbf{t}_{31}$, $\mathbf{t}_{32}$, and $\mathbf{t}_{34}$, that is, the reflection and transmission matrices at the cross joint of Figure 11.16c, corresponding to incident waves from Beam 3.

Following the local coordinate systems of Figure 11.15, the axial and bending deflections of the beam elements are

$$u_1 = g^+ e^{-ik_{13}x_1}$$

$$u_2 = d^+ e^{ik_{23}x_2}$$

$$u_3 = c^+ e^{-ik_{33}x_3} + c^- e^{ik_{33}x_3}$$

$$u_4 = q^+ e^{ik_{43}x_4}$$

$$y_1 = e_1^+ e^{-ik_{11}x_1} + e_2^+ e^{-k_{12}x_1}$$

$$y_2 = b_1^+ e^{ik_{21}x_2} + b_2^+ e^{k_{22}x_2}$$

$$y_3 = a_1^+ e^{-ik_{31}x_3} + a_2^+ e^{-k_{32}x_3} + a_1^- e^{ik_{31}x_3} + a_2^- e^{k_{32}x_3}$$

$$y_4 = p_1^+ e^{ik_{41}x_4} + p_2^+ e^{k_{42}x_4} \tag{11.84a}$$

and the bending slopes by the Euler–Bernoulli bending vibration theory are

$$\psi_1 = -ik_{11}e_1^+ e^{-ik_{11}x_1} - k_{12}e_2^+ e^{-k_{12}x_1}$$

$$\psi_2 = ik_{21}b_1^+ e^{ik_{21}x_2} + k_{22}b_2^+ e^{k_{22}x_2}$$

$$\psi_3 = -ik_{31}a_1^+ e^{-ik_{31}x_3} - k_{32}a_2^+ e^{-k_{32}x_3} + ik_{31}a_1^- e^{ik_{31}x_3} + k_{32}a_2^- e^{k_{32}x_3}$$

$$\psi_4 = ik_{41}p_1^+ e^{ik_{41}x_4} + k_{42}p_2^+ e^{k_{42}x_4} \tag{11.84b}$$

and the bending slopes by the Timoshenko bending vibration theory are

$$\psi_1 = -iP_1 e_1^+ e^{-ik_{11}x_1} - N_1 e_2^+ e^{-k_{12}x_1}$$

$$\psi_2 = iP_2 b_1^+ e^{ik_{21}x_2} + N_2 b_2^+ e^{k_{22}x_2}$$

$$\psi_3 = -iP_3 a_1^+ e^{-ik_{31}x_3} - N_3 a_2^+ e^{-k_{32}x_3} + iP_3 a_1^- e^{ik_{31}x_3} + N_3 a_2^- e^{k_{32}x_3}$$

$$\psi_4 = iP_4 p_1^+ e^{ik_{41}x_4} + N_4 p_2^+ e^{k_{42}x_4} \tag{11.84c}$$

where i is the imaginary unit and subscripts of P and N identify the beam elements by numerical numbers as marked on Figures 11.15 and 11.16. with the leading digit identifying a beam element and the other digit denoting the wavenumber sequence in the beam element.

Substituting Eqs. (11.84a) and (11.84b) into a set of continuity equations from Eqs. (11.70a-c), and choosing the origin at the joint, a set of three scalar equations of continuity is obtained based on the Euler–Bernoulli bending vibration theory. Similarly, substituting Eqs. (11.84a) and (11.84c) into a set of continuity equations from Eqs. (11.70a-c), and choosing the origin at the joint, a set of three scalar equations of continuity is obtained based on the Timoshenko bending vibration theory. This set of three scalar equations of continuity, either by the Euler–Bernoulli or Timoshenko bending vibration theory, can be placed in a vector equation, in terms of wave components $\mathbf{a}^+$, $\mathbf{a}^-$, and $\mathbf{e}^+$, which are defined in Eq. (11.81). A vector equation from this group is named $+\_\mathbf{Beam3}_{C31}$, because it is a vector equation on the continuity between Beam 3 and Beam 1 at a "+" joint corresponding to incident waves from Beam 3.

Substituting Eqs. (11.84a) and (11.84b) into a set of continuity equations from Eqs. (11.72a-c), and choosing the origin at the joint, a set of three scalar equations of continuity is obtained based on the Euler–Bernoulli bending vibration theory. Similarly, substituting Eqs. (11.84a) and (11.84c) into a set of continuity equations from Eqs. (11.72a-c), and choosing the origin at the joint, a set of three scalar equations of continuity is obtained based on the Timoshenko bending vibration theory. This set of three scalar equations of continuity, either by the Euler–Bernoulli or Timoshenko bending vibration theory, can be placed in a vector equation, in terms of wave components $\mathbf{a}^+$, $\mathbf{a}^-$, and $\mathbf{b}^+$, which are defined in Eq. (11.81). A vector equation from this group is named $+\_\mathbf{Beam3}_{C32}$, because it is a vector equation on the continuity between Beam 3 and Beam 2 at a "+" joint corresponding to incident waves from Beam 3.

Substituting Eqs. (11.84a) and (11.84b) into a set of continuity equations from Eqs. (11.74a-c), and choosing the origin at the joint, a set of three scalar equations of continuity is obtained based on the Euler–Bernoulli bending vibration theory. Similarly, substituting Eqs. (11.84a) and (11.84c) into a set of continuity equations from Eqs. (11.74a-c), and choosing the origin at the joint, a set of three scalar equations of continuity is obtained based on the Timoshenko bending vibration theory. This set of three scalar equations of continuity, either by the Euler–Bernoulli or Timoshenko bending vibration theory, can be placed in a vector equation, in terms of wave components $\mathbf{a}^+$, $\mathbf{a}^-$, and $\mathbf{p}^+$, which are defined in Eq. (11.81). A vector equation from this group is named $+\_\mathbf{Beam3}_{C34}$, because it is a vector equation on the continuity between Beam 3 and Beam 4 at a "+" joint corresponding to incident waves from Beam 3.

Substituting Eqs. (11.84a) and (11.84b) into Eqs. (11.6a) and (11.6b) gives the functional relationships of the internal resistant forces and moment and the wave components, based on the Euler–Bernoulli bending vibration theory. Substituting these functional relationships into a set of equilibrium equations from Eqs. (11.76a-c), and choosing the origin at the joint, a set of three scalar equations of equilibrium is obtained based on the Euler–Bernoulli bending vibration theory. Similarly, Substituting Eqs. (11.84a) and (11.84c) into Eqs. (11.6a) and (11.6c) gives the functional relationships of the internal resistant forces and moment and the wave components, based on the Timoshenko bending vibration theory. Substituting these functional relationships into a set of equilibrium equations from Eqs. (11.76a-c), and choosing the origin at the joint, a set of three scalar equations of equilibrium is obtained based on the Timoshenko bending vibration theory. This set of three scalar equations of equilibrium, either by the Euler–Bernoulli or Timoshenko bending vibration theory, can be placed in a vector equation, in terms of wave components $\mathbf{a}^+$, $\mathbf{a}^-$, $\mathbf{b}^+$, $\mathbf{e}^+$, and $\mathbf{p}^+$, which are defined in Eq. (11.81). A vector equation from this group is named $+\_\mathbf{Beam3}_E$, because it is a vector equation on equilibrium at a "+" joint corresponding to incident waves from Beam 3.

The reflection and transmission matrices, $\mathbf{r}_{33}$, $\mathbf{t}_{31}$, $\mathbf{t}_{32}$, and $\mathbf{t}_{34}$, are found from Eq. (11.79) and the four vector equations of continuity and equilibrium, namely, $+\_\mathbf{Beam3}_{C31}$, $+\_\mathbf{Beam3}_{C32}$, $+\_\mathbf{Beam3}_{C34}$, and $+\_\mathbf{Beam3}_E$.

Last, solve for $\mathbf{r}_{44}$, $\mathbf{t}_{41}$, $\mathbf{t}_{42}$, and $\mathbf{t}_{43}$, that is, the reflection and transmission matrices at the cross joint of Figure 11.16d, corresponding to incident waves from Beam 4.

Following the local coordinate systems of Figure 11.15, the axial and bending deflections of the beam elements are

$$u_1 = d^+ e^{-ik_{13}x_1}$$

$$u_2 = q^+ e^{ik_{23}x_2}$$

$$u_3 = g^+ e^{ik_{33}x_3}$$

$$u_4 = c^+ e^{-ik_{43}x_4} + c^- e^{ik_{43}x_4}$$

$$y_1 = b_1^+ e^{-ik_{11}x_1} + b_2^+ e^{-k_{12}x_1}$$

$$y_2 = p_1^+ e^{ik_{21}x_2} + p_2^+ e^{k_{22}x_2}$$

$$y_3 = e_1^+ e^{ik_{31}x_3} + e_2^+ e^{k_{32}x_3}$$

$$y_4 = a_1^+ e^{-ik_{41}x_4} + a_2^+ e^{-k_{42}x_4} + a_1^- e^{ik_{41}x_4} + a_2^- e^{k_{42}x_4} \tag{11.85a}$$

and the bending slopes by the Euler–Bernoulli bending vibration theory are

$$\psi_1 = -ik_{11}b_1^+ e^{-ik_{11}x_1} - k_{12}b_2^+ e^{-k_{12}x_1}$$

$$\psi_2 = ik_{21}p_1^+ e^{ik_{21}x_2} + k_{22}p_2^+ e^{k_{22}x_2}$$

$$\psi_3 = ik_{31}e_1^+ e^{ik_{31}x_3} + k_{32}e_2^+ e^{k_{32}x_3}$$

$$\psi_4 = -ik_{41}a_1^+ e^{-ik_{41}x_4} - k_{42}a_2^+ e^{-k_{42}x_4} + ik_{41}a_1^- e^{ik_{41}x_4} + k_{42}a_2^- e^{k_{42}x_4} \tag{11.85b}$$

and the bending slopes by the Timoshenko bending vibration theory are

$$\psi_1 = -iP_1 b_1^+ e^{-ik_{11}x_1} - N_1 b_2^+ e^{-k_{12}x_1}$$

$$\psi_2 = iP_2 p_1^+ e^{ik_{21}x_2} + N_2 p_2^+ e^{k_{22}x_2}$$

$$\psi_3 = iP_3 e_1^+ e^{ik_{31}x_3} + N_3 e_2^+ e^{k_{32}x_3}$$

$$\psi_4 = -iP_4 a_1^+ e^{-ik_{41}x_4} - N_4 a_2^+ e^{-k_{42}x_4} + iP_4 a_1^- e^{ik_{41}x_4} + N_4 a_2^- e^{k_{42}x_4} \tag{11.85c}$$

where i is the imaginary unit and subscripts in P and N identify the beam elements by numerical numbers as marked on Figures 11.15 and 11.16. Each wavenumber carries two numerical digits in its subscript, with the leading digit identifying a beam element and the other digit denoting the wavenumber sequence in the beam element.

Substituting Eqs. (11.85a) and (11.85b) into a set of continuity equations from Eqs. (11.71a-c), and choosing the origin at the joint, a set of three scalar equations of continuity is obtained based on the Euler–Bernoulli bending vibration theory. Similarly, substituting Eqs. (11.85a) and (11.85c) into a set of continuity equations from Eqs. (11.71a-c), and choosing the origin at the joint, a set of three scalar equations of continuity is obtained based on the Timoshenko bending vibration theory. This set of three scalar equations of continuity, either by the Euler–Bernoulli or Timoshenko bending vibration theory, can be placed in a vector equation, in terms of wave components $\mathbf{a}^+$, $\mathbf{a}^-$, and $\mathbf{b}^+$, which are defined in Eq. (11.81). A vector equation from this group is named $+\_\mathbf{Beam4}_{C41}$, because it is a vector equation on the continuity between Beam 4 and Beam 1 at a "+" joint corresponding to incident waves from Beam 4.

Substituting Eqs. (11.85a) and (11.85b) into a set of continuity equations from Eqs. (11.73a-c), and choosing the origin at the joint, a set of three scalar equations of continuity is obtained based on the Euler–Bernoulli bending vibration theory. Similarly, substituting Eqs. (11.85a) and (11.85c) into a set of continuity equations from Eqs. (11.73a-c), and choosing the origin at the joint, a set of three scalar equations of continuity is obtained based on the Timoshenko bending vibration theory. This set of three scalar equations of continuity, either by the Euler–Bernoulli or Timoshenko bending vibration theory, can be placed in a vector equation, in terms of wave components $\mathbf{a}^+$, $\mathbf{a}^-$, and $\mathbf{p}^+$, which are defined in Eq. (11.81). A vector equation from this group is named $+\_\mathbf{Beam4}_{C42}$, because it is a vector equation on the continuity between Beam 4 and Beam 2 at a "+" joint corresponding to incident waves from Beam 4.

Substituting Eqs. (11.85a) and (11.85b) into a set of continuity equations from Eqs. (11.74a-c), and choosing the origin at the joint, a set of three scalar equations of continuity is obtained based on the Euler–Bernoulli bending vibration theory. Similarly, substituting Eqs. (11.85a) and (11.85c) into a set of continuity equations from Eqs. (11.74a-c), and choosing the origin at the joint, a set of three scalar equations of continuity is obtained based on the Timoshenko bending vibration theory. This set of three scalar equations of continuity, either by the Euler–Bernoulli or Timoshenko bending vibration theory, can be placed in a vector equation, in terms of wave components $\mathbf{a}^+$, $\mathbf{a}^-$, and $\mathbf{e}^+$, which are defined in Eq. (11.81). A vector equation from this group is named $+\_\mathbf{Beam4}_{C43}$, because it is a vector equation on the continuity between Beam 4 and Beam 3 at a "+" joint corresponding to incident waves from Beam 4.

Substituting Eqs. (11.85a) and (11.85b) into Eqs. (11.6a) and (11.6b) gives the functional relationships of the internal resistant forces and moment and the wave components, based on the Euler–Bernoulli bending vibration theory. Substituting these functional relationships into a set of equilibrium equations from Eqs. (11.76a-c), and choosing the origin at the joint, a set of three scalar equations of equilibrium is obtained based on the Euler–Bernoulli bending vibration theory. Similarly, Substituting Eqs. (11.85a) and (11.85c) into Eqs. (11.6a) and (11.6c) gives the functional relationships of the internal resistant forces and moment and the wave components, based on the Timoshenko bending vibration theory. Substituting these functional relationships into a set of equilibrium equations from Eqs. (11.76a-c), and choosing the origin at the joint, a set of three scalar equations of equilibrium is obtained based on the Timoshenko bending vibration vibration theory. This set of three scalar equations of equilibrium, either by the Euler–Bernoulli or Timoshenko bending vibration theory, can be placed in a vector equation, in terms of wave components $\mathbf{a}^+$, $\mathbf{a}^-$, $\mathbf{b}^+$, $\mathbf{e}^+$, and $\mathbf{p}^+$, which are defined in Eq. (11.81). A vector equation from this group is named $+\_\mathbf{Beam4}_E$, because it is a vector equation on equilibrium at a "+" joint corresponding to incident waves from Beam 4.

The reflection and transmission matrices, $\mathbf{r}_{44}$, $\mathbf{t}_{41}$, $\mathbf{t}_{42}$, and $\mathbf{t}_{43}$, are found from Eq. (11.80) and the four vector equations of continuity and equilibrium, namely, $+\_\mathbf{Beam4}_{C41}$, $+\_\mathbf{Beam4}_{C42}$, $+\_\mathbf{Beam4}_{C43}$, and $+\_\mathbf{Beam4}_E$.

With the derived matrices of wave propagation along a uniform waveguide, wave reflection at a boundary, wave reflection and transmission at a structural joint, as well as the availability of relationships between the injected waves and an external excitation, both free and forced vibrations in built up planar frames can be obtained analytically. It involves a systematic assembling process as described in Section 11.4. Numerical examples of wave vibration analysis of built up planar frames can be found in the references.

For a planar frame undergoes out-of-plane vibrations, bending and torsional vibration waves co-exist. Free and forced out-of-plane vibrations in planar frames can be analyzed following a similar procedure described in this chapter, due to analogy between longitudinal and torsional vibrations (Mei 2019).

References

Mei C. Wave Analysis of In-Plane Vibrations of H- and T-shaped Planar Frame Structures, *ASME Journal of Vibration and Acoustics*, 130, 061004-1-10 (2008).

Mei, C. Wave Analysis of In-Plane Vibrations of L-shaped and Portal Planar Frame Structures, *ASME Journal of Vibration and Acoustics*, 134, 021011-1-12 (2012).

Mei, C. Analysis of In- and Out-of-Plane Vibrations in a Rectangular Frame based on Two- and Three-Dimensional Structural Model, *Journal of Sound and Vibration*, 440, 412-438 (2019).

Homework Project

Wave Analysis of Free In-Plane Vibrations in an H-shaped Planar Frame

An H-shaped frame was built by welding three uniform beam elements together, as shown in the figure below. These beam elements are made of cold-rolled steel whose material and geometrical properties are: the Young's modulus is $198.87 \, GN/m^2$ and mass density is $7664.5 \, kg/m^3$. The cross section of the beam is a rectangular shape whose width and thickness are $25.4 \, mm$ and $12.7 \, mm$ (or $1.0 \, in$ and $0.5 \, in$), respectively. The length of the upper and lower vertical leg is $L = 880 \, mm$. The length of the horizontal beam is $L_H = 300 \, mm$.

The natural frequencies of the H frame were measured in the laboratory. In the experiment, the H frame was laid on a large piece of foam, and all four boundaries are free. An Endevco 2302 impact hammer was used to excite the structure and a PCB 353 B12 accelerometer was used to measure the accelerations. A four-channel Pulse data acquisition system from Brüel & Kjær was used to collect and analyze the measured data. The measured natural frequencies are listed in the table below.

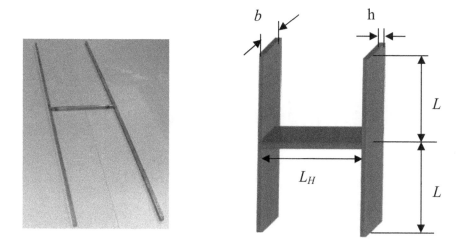

Mode Sequence	1	2	3	4	5	6	7	8	9	10	11	12
Natural frequency (*Hz*)	8.0	16.3	20.0	36.3	66.3	78.8	85.0	108.8	197.5	212.5	238.8	270.0

Perform free vibration analysis of this H-shaped frame based on the elementary vibration theories using the wave analysis approach to planar frame structures introduced in this chapter, plot the dB magnitude of the characteristic polynomial, read the natural frequencies from the plot, compare the analytical results with the tabulated experimental results, and identify the potential causes for discrepancy.

12

Bending, Longitudinal, and Torsional Waves in Built-up Space Frames

Space frame structures can be found in many fields of engineering, ranging from large scale high-rise buildings and towers to micro-frames used in modern electronic equipment. These frame structures are often subject to harmful vibrations. Vibration analysis of such structures is of practical importance.

A built-up space frame often consists of spatial angle joints such as Y-shaped or K-shaped joints. A spatial angle joint in general introduces wave mode conversion. Consequently, in a built-up space frame, there exist in- and out-of-plane bending, longitudinal, and torsional vibrations, which greatly challenges the conventional modal vibration analysis approach. The wave vibration analysis approach offers a concise assembly approach for systematically analyzing complex vibrations in built-up space frames. The in- and out-of-plane bending, longitudinal, and torsional vibration waves propagate along uniform structural elements and are reflected and transmitted at structural discontinuities such as joints and boundaries.

In this study, longitudinal and torsional vibrations are modeled using the Elementary vibration theories, and bending vibrations are modeled using the Euler–Bernoulli vibration theory as well as the Timoshenko vibration theory.

12.1 The Governing Equations and the Propagation Relationships

Consider a uniform beam element free of external forces and moments, lying along the x-axis, as shown in Figure 9.1. The equations of motion of the beam, with bending vibrations modeled by the Euler–Bernoulli vibration theory and longitudinal and torsional vibrations modeled using the Elementary vibration theories, are

$$EI_y \frac{\partial^4 y(x,t)}{\partial x^4} + \rho A \frac{\partial^2 y(x,t)}{\partial t^2} = 0$$

$$EI_z \frac{\partial^4 z(x,t)}{\partial x^4} + \rho A \frac{\partial^2 z(x,t)}{\partial t^2} = 0$$

$$\rho A \frac{\partial^2 u(x,t)}{\partial t^2} - EA \frac{\partial^2 u(x,t)}{\partial x^2} = 0$$

$$\rho J \frac{\partial^2 \theta(x,t)}{\partial t^2} - G\gamma \frac{\partial^2 \theta(x,t)}{\partial x^2} = 0 \tag{12.1a}$$

The equations of motion of the beam, with bending vibrations modeled by the Timoshenko vibration theory and longitudinal and torsional vibrations modeled using the Elementary vibration theories, are

$$GA\kappa \left[\frac{\partial \varphi_y(x,t)}{\partial x} - \frac{\partial^2 y(x,t)}{\partial x^2} \right] + \rho A \frac{\partial^2 y(x,t)}{\partial t^2} = 0$$

Mechanical Wave Vibrations: Analysis and Control, First Edition. Chunhui Mei.
© 2023 Chunhui Mei. Published 2023 by John Wiley & Sons Ltd.
Companion Website: www.wiley.com/go/Mei/MechanicalWaveVibrations

$$EI_y \frac{\partial^2 \varphi_y(x,t)}{\partial x^2} + GA\kappa \left[\frac{\partial y(x,t)}{\partial x} - \varphi_y(x,t) \right] - \rho I_y \frac{\partial^2 \varphi_y(x,t)}{\partial t^2} = 0$$

$$GA\kappa \left[\frac{\partial \varphi_z(x,t)}{\partial x} - \frac{\partial^2 z(x,t)}{\partial x^2} \right] + \rho A \frac{\partial^2 z(x,t)}{\partial t^2} = 0$$

$$EI_z \frac{\partial^2 \varphi_z(x,t)}{\partial x^2} + GA\kappa \left[\frac{\partial z(x,t)}{\partial x} - \varphi_z(x,t) \right] - \rho I_z \frac{\partial^2 \varphi_z(x,t)}{\partial t^2} = 0$$

$$\rho A \frac{\partial^2 u(x,t)}{\partial t^2} - EA \frac{\partial^2 u(x,t)}{\partial x^2} = 0$$

$$\rho J \frac{\partial^2 \theta(x,t)}{\partial t^2} - G\gamma \frac{\partial^2 \theta(x,t)}{\partial x^2} = 0 \tag{12.1b}$$

In the above equations, x is the position along the beam axis and t is the time. $y(x,t)$ and $z(x,t)$ are the transverse deflections in the x–y and x–z plane shown in Figure 9.1, $u(x,t)$ is the longitudinal deflection of the centerline of the beam, and $\theta(x,t)$ is the torsional deflection of the beam. $\varphi_y(x,t)$ is the slope due to bending, $\frac{\partial y(x,t)}{\partial x}$ is the slope of the centerline of the beam, and $\frac{\partial y(x,t)}{\partial x} - \varphi_y(x,t)$ is the shear angle in the x–y plane. $\varphi_z(x,t)$ is the slope due to bending, $\frac{\partial z(x,t)}{\partial x}$ is the slope of the centerline of the beam, and $\frac{\partial z(x,t)}{\partial x} - \varphi_z(x,t)$ is the shear angle in the x–z plane. The bending equations of motion are coupled through the slope due to bending and the transverse deflection of the beam. A, ρ, E, and G are the cross-sectional area, volume mass density, Young's modulus, and shear modulus, respectively. κ is the shear coefficient, which is related to the Poisson's ratio v. I_y is the area moment of inertia of the cross section for bending in the x-y plane, I_z is the area moment of inertia of the cross section for bending in the x-z plane, and J is the polar moment of inertia of the cross section, about the x-axis of Figure 9.1.

$G\gamma$ is the torsional rigidity of the beam. For a rotationally symmetric cross section, $G\gamma = GJ$. For a rotationally asymmetric cross section, such as a rectangular cross section with longer side l and shorter side s, $\gamma = \beta l s^3$, where β is an adjustment coefficient dependent upon the aspect ratio $l:s$. Table 12.1 lists a few β values of the rectangular cross section with various side ratios obtained from static elastic theory.

The wavenumbers for longitudinal and bending vibrations have been obtained in Chapters 2, 3, and 9, by assuming time harmonic motion and using separation of variables.

The bending wavenumbers k_1 and k_2 in the x-y plane and the bending wavenumbers k_3 and k_4 in the x-z plane, based on the Euler–Bernoulli bending vibration theory, are

$$k_1 = k_2 = \sqrt[4]{\frac{\rho A \omega^2}{EI_y}}$$

$$k_3 = k_4 = \sqrt[4]{\frac{\rho A \omega^2}{EI_z}} \tag{12.2a}$$

Table 12.1 Coefficient β for a rotationally asymmetric solid rectangular cross section (Cremer et. al. 1987).

$l:s$	1	1.5	2	3	6	$l \gg s$
β	0.141	0.196	0.229	0.263	0.298	1/3

and based on the Timoshenko bending vibration theory, the bending wavenumbers in the x–y and x–z planes are

$$
k_1 = \left[\frac{1}{2} \left(\frac{1}{C_s} \right)^2 \omega^2 + \sqrt{ \frac{\omega^2}{C_{by}^2} + \frac{1}{4} \left(\frac{1}{C_s} \right)^4 \omega^4 } \right]^{\frac{1}{2}}
$$

$$
k_2 = \left[\left| \frac{1}{2} \left(\frac{1}{C_s} \right)^2 \omega^2 - \sqrt{ \frac{\omega^2}{C_{by}^2} + \frac{1}{4} \left(\frac{1}{C_s} \right)^4 \omega^4 } \right| \right]^{\frac{1}{2}}
$$

$$
k_3 = \left[\frac{1}{2} \left(\frac{1}{C_s} \right)^2 \omega^2 + \sqrt{ \frac{\omega^2}{C_{bz}^2} + \frac{1}{4} \left(\frac{1}{C_s} \right)^4 \omega^4 } \right]^{\frac{1}{2}}
$$

$$
k_4 = \left[\left| \frac{1}{2} \left(\frac{1}{C_s} \right)^2 \omega^2 - \sqrt{ \frac{\omega^2}{C_{bz}^2} + \frac{1}{4} \left(\frac{1}{C_s} \right)^4 \omega^4 } \right| \right]^{\frac{1}{2}}
\tag{12.2b}
$$

The longitudinal wavenumber k_5 based on the Elementary vibration theory is

$$
k_5 = \sqrt{\frac{\rho}{E}} \, \omega
\tag{12.2c}
$$

From the analogy in the equations of motion of a beam in longitudinal and torsional vibrations derived in Chapter 1, the torsional wavenumber k_6 based on the Elementary vibration theory is

$$
k_6 = \sqrt{\frac{\rho J}{G \gamma}} \, \omega
\tag{12.2d}
$$

For a rotationally symmetric cross section, because $G\gamma = GJ$, $k_6 = \sqrt{\dfrac{\rho}{G}} \, \omega$.

In Eqs. (12.2a-d), ω is the circular frequency, $C_{by} = \sqrt{\dfrac{EI_y}{\rho A}}$, $C_{bz} = \sqrt{\dfrac{EI_z}{\rho A}}$, and $C_s = \sqrt{\dfrac{GA\kappa}{\rho A}}$. Recall that the Timoshenko bending vibration theory predicts a wave mode transition at a cut-off frequency ω_c. The cut-off frequencies for bending vibrations in the x-y and x-z planes are

$$
\omega_{cy} = \frac{C_s}{C_{ry}}
\tag{12.3a}
$$

$$
\omega_{cz} = \frac{C_s}{C_{rz}}
\tag{12.3b}
$$

where $C_{ry} = \sqrt{\dfrac{\rho I_y}{\rho A}} = \sqrt{\dfrac{I_y}{A}}$ and $C_{rz} = \sqrt{\dfrac{\rho I_z}{\rho A}} = \sqrt{\dfrac{I_z}{A}}$.

According to the Timoshenko bending vibration theory, below a cut-off frequency, there are a pair of propagating bending waves and a pair of decaying bending waves in the beam, while above a cut-off frequency, there are two pairs of propagating bending waves in the beam. Because in audio frequency applications, the former case is overwhelmingly the most common, expressions based on the Timoshenko bending vibration theory in this chapter are for frequency ω below the cut-off frequencies. When $\omega > \omega_{cy}$, k_2 needs to be replaced by ik_2, and when $\omega > \omega_{cz}$, k_4 needs to be replaced by ik_4, where i is the imaginary unit, in order to reflect the wave mode transition at the cut-off frequencies ω_{cy} and ω_{cz}, respectively.

With time dependence $e^{i\omega t}$ suppressed, the solutions to Eq. (12.1a) are

$$y(x) = a_1^+ e^{-ik_1 x} + a_2^+ e^{-k_2 x} + a_1^- e^{ik_1 x} + a_2^- e^{k_2 x}$$

$$\varphi_y(x) = \frac{\partial y(x)}{\partial x} = -ik_1 a_1^+ e^{-ik_1 x} - k_2 a_2^+ e^{-k_2 x} + ik_1 a_1^- e^{ik_1 x} + k_2 a_2^- e^{k_2 x}$$

$$z(x) = a_3^+ e^{-ik_3 x} + a_4^+ e^{-k_4 x} + a_3^- e^{ik_3 x} + a_4^- e^{k_4 x}$$

$$\varphi_z(x) = \frac{\partial z(x)}{\partial x} = -ik_3 a_3^+ e^{-ik_3 x} - k_4 a_4^+ e^{-k_4 x} + ik_3 a_3^- e^{ik_3 x} + k_4 a_4^- e^{k_4 x}$$

$$u(x) = c^+ e^{-ik_5 x} + c^- e^{ik_5 x}$$

$$\theta(x) = d^+ e^{-ik_6 x} + d^- e^{ik_6 x} \tag{12.4a}$$

and the solutions to Eq.(12.1b) are

$$y(x) = a_1^+ e^{-ik_1 x} + a_2^+ e^{-k_2 x} + a_1^- e^{ik_1 x} + a_2^- e^{k_2 x}$$

$$\varphi_y(x) = -iP_y a_1^+ e^{-ik_1 x} - N_y a_2^+ e^{-k_2 x} + iP_y a_1^- e^{ik_1 x} + N_y a_2^- e^{k_2 x}$$

$$z(x) = a_3^+ e^{-ik_3 x} + a_4^+ e^{-k_4 x} + a_3^- e^{ik_3 x} + a_4^- e^{k_4 x}$$

$$\varphi_z(x) = -iP_z a_3^+ e^{-ik_3 x} - N_z a_4^+ e^{-k_4 x} + iP_z a_3^- e^{ik_3 x} + N_z a_4^- e^{k_4 x}$$

$$u(x) = c^+ e^{-ik_5 x} + c^- e^{ik_5 x}$$

$$\theta(x) = d^+ e^{-ik_6 x} + d^- e^{ik_6 x} \tag{12.4b}$$

where i is the imaginary unit, superscripts $+$ and $-$ in wave amplitude denote positive- and negative-going bending, longitudinal, or torsional waves. Subscripts 1 to 4 in $a_{1,2}^\pm$ and $a_{3,4}^\pm$ refer to the two bending wave components in the x–y and x–z plane, respectively. From Eq. (9.15),

$$P_y = k_1 \left(1 - \frac{\omega^2}{k_1^2 C_S^2}\right), \quad N_y = k_2 \left(1 + \frac{\omega^2}{k_2^2 C_S^2}\right)$$

$$P_z = k_3 \left(1 - \frac{\omega^2}{k_3^2 C_S^2}\right), \quad N_z = k_4 \left(1 + \frac{\omega^2}{k_4^2 C_S^2}\right) \tag{12.5}$$

Vibration waves propagate along uniform structural elements, with the propagation relationships determined by the appropriate wavenumbers. Consider two points A and B that are a distance x apart on a uniform beam, the positive- and negative-going waves $\mathbf{a}^\pm$ and $\mathbf{b}^\pm$ at Points A and B are related,

$$\mathbf{a}^- = \mathbf{f}(x)\mathbf{b}^-; \quad \mathbf{b}^+ = \mathbf{f}(x)\mathbf{a}^+ \tag{12.6a}$$

where

$$\mathbf{a}^\pm = \begin{bmatrix} a_1^\pm \\ a_2^\pm \\ a_3^\pm \\ a_4^\pm \\ c^\pm \\ d^\pm \end{bmatrix}, \text{ and } \mathbf{b}^\pm = \begin{bmatrix} b_1^\pm \\ b_2^\pm \\ b_3^\pm \\ b_4^\pm \\ e^\pm \\ f^\pm \end{bmatrix} \tag{12.6b}$$

are the wave vectors, in which $a_1^{\pm}, a_2^{\pm}, a_3^{\pm}, a_4^{\pm}, b_1^{\pm}, b_2^{\pm}, b_3^{\pm}$, and $b_4^{\pm}$ are bending wave components, $c^{\pm}$ and $e^{\pm}$ are longitudinal wave components, and $d^{\pm}$ and $f^{\pm}$ are torsional wave components, and

$$
\mathbf{f}(x) = \begin{bmatrix}
e^{-ik_1 x} & 0 & 0 & & & \\
0 & e^{-k_2 x} & 0 & & 0 & \\
0 & 0 & e^{-ik_3 x} & & & \\
& & & e^{-k_4 x} & 0 & 0 \\
& 0 & & 0 & e^{-ik_5 x} & 0 \\
& & & 0 & 0 & e^{-ik_6 x}
\end{bmatrix}
\tag{12.6c}
$$

is the propagation matrix for a distance x.

12.2 Wave Reflection at Classical Boundaries

Figure 12.1 shows the sign convention adopted in the analysis that follows.

The internal resistant shear forces $V_y(x,t)$ and $V_z(x,t)$, and bending moments $M_y(x,t)$ and $M_z(x,t)$ at a section of the beam are related to the transverse deflections $y(x,t)$ and $z(x,t)$, and bending slopes $\varphi_y(x,t)$ and $\varphi_z(x,t)$. According to the Euler–Bernoulli bending vibration theory and recall that $\varphi_y(x,t) = \dfrac{\partial y(x,t)}{\partial x}$ and $\varphi_z(x,t) = \dfrac{\partial z(x,t)}{\partial x}$, the relationships are

$$
V_y(x,t) = -EI_y \frac{\partial^3 y(x,t)}{\partial x^3}
$$

$$
V_z(x,t) = -EI_z \frac{\partial^3 z(x,t)}{\partial x^3}
$$

$$
M_y(x,t) = EI_y \frac{\partial \varphi_y(x,t)}{\partial x} = EI_y \frac{\partial^2 y(x,t)}{\partial x^2}
$$

$$
M_z(x,t) = EI_z \frac{\partial \varphi_z(x,t)}{\partial x} = EI_z \frac{\partial^2 z(x,t)}{\partial x^2}
\tag{12.7a}
$$

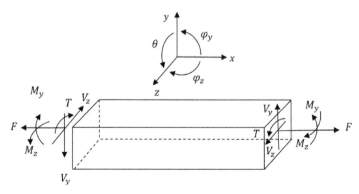

Figure 12.1 Definition of positive sign directions for shear forces, longitudinal force, bending moments, and torque.

and by the Timoshenko bending vibration theory the relationships are

$$V_y(x,t) = GA\kappa \left[\frac{\partial y(x,t)}{\partial x} - \varphi_y(x,t) \right]$$

$$V_z(x,t) = GA\kappa \left[\frac{\partial z(x,t)}{\partial x} - \varphi_z(x,t) \right]$$

$$M_y(x,t) = EI_y \frac{\partial \varphi_y(x,t)}{\partial x}$$

$$M_z(x,t) = EI_z \frac{\partial \varphi_z(x,t)}{\partial x} \tag{12.7b}$$

The internal resistant longitudinal force $F(x,t)$ and torque $T(x,t)$ are related to the longitudinal deflection $u(x,t)$ and torsional deflection $\theta(x,t)$ by

$$F(x,t) = EA \frac{\partial u(x,t)}{\partial x}$$

$$T(x,t) = G\gamma \frac{\partial \theta(x,t)}{\partial x} \tag{12.7c}$$

Incident waves $\mathbf{a}^+$ give rise to reflected waves $\mathbf{a}^-$ at a boundary, which are related by reflection matrix $\mathbf{r}$

$$\mathbf{a}^- = \mathbf{r}\,\mathbf{a}^+ \tag{12.8}$$

where $\mathbf{r}$ is determined by boundary conditions.

At a clamped boundary, the boundary conditions are

$$y(x,t) = 0,\ z(x,t) = 0,$$

$$\varphi_y(x,t) = 0,\ \varphi_z(x,t) = 0,$$

$$u(x,t) = 0,\ \theta(x,t) = 0 \tag{12.9}$$

At a free boundary, the equilibrium conditions are

$$V_y(x,t) = 0, V_z(x,t) = 0,$$

$$M_y(x,t) = 0,\ M_z(x,t) = 0,$$

$$F(x,t) = 0,\ T(x,t) = 0 \tag{12.10}$$

At a simply-supported boundary, the boundary conditions are

$$y(x,t) = 0,\ z(x,t) = 0,\ u(x,t) = 0,$$

$$M_y(x,t) = 0,\ M_z(x,t) = 0,\ T(x,t) = 0 \tag{12.11}$$

From Eqs. (12.4a), (12.8), and (12.9), with the origin chosen at the boundary, the reflection matrix at a clamped boundary by the Euler–Bernoulli bending vibration theory is

$$\mathbf{r} = \begin{bmatrix} -i & -1-i & 0 & 0 & 0 & 0 \\ -1+i & i & 0 & 0 & 0 & 0 \\ 0 & 0 & -i & -1-i & 0 & 0 \\ 0 & 0 & -1+i & i & 0 & 0 \\ 0 & 0 & 0 & 0 & -1 & 0 \\ 0 & 0 & 0 & 0 & 0 & -1 \end{bmatrix} \tag{12.12a}$$

From Eqs. (12.4b), (12.8), and (12.9), with the origin chosen at the boundary, the reflection matrix at a clamped boundary by the Timoshenko bending vibration theory is

$$\mathbf{r} = \begin{bmatrix} \dfrac{P_y - iN_y}{P_y + iN_y} & \dfrac{-2iN_y}{P_y + iN_y} & 0 & 0 & 0 & 0 \\[2ex] \dfrac{-2P_y}{P_y + iN_y} & -\dfrac{P_y - iN_y}{P_y + iN_y} & 0 & 0 & 0 & 0 \\[2ex] 0 & 0 & \dfrac{P_z - iN_z}{P_z + iN_z} & \dfrac{-2iN_z}{P_z + iN_z} & 0 & 0 \\[2ex] 0 & 0 & \dfrac{-2P_z}{P_z + iN_z} & -\dfrac{P_z - iN_z}{P_z + iN_z} & 0 & 0 \\[2ex] 0 & 0 & 0 & 0 & -1 & 0 \\[1ex] 0 & 0 & 0 & 0 & 0 & -1 \end{bmatrix} \tag{12.12b}$$

From Eqs. (12.4a), (12.7a), (12.8), and (12.10), with the origin chosen at the boundary, the reflection matrix at a free boundary by the Euler–Bernoulli bending vibration theory is

$$\mathbf{r} = \begin{bmatrix} -i & 1+i & 0 & 0 & 0 & 0 \\ 1-i & i & 0 & 0 & 0 & 0 \\ 0 & 0 & -i & 1+i & 0 & 0 \\ 0 & 0 & 1-i & i & 0 & 0 \\ 0 & 0 & 0 & 0 & 1 & 0 \\ 0 & 0 & 0 & 0 & 0 & 1 \end{bmatrix} \tag{12.13a}$$

From Eqs. (12.4b), (12.7b), (12.8), and (12.10), with the origin chosen at the boundary, the reflection matrix at a free boundary by the Timoshenko bending vibration theory is

$$\mathbf{r} = \begin{bmatrix} \dfrac{-P_y k_1(-N_y+k_2)+ik_2 N_y(k_1-P_y)}{P_y k_1(-N_y+k_2)+ik_2 N_y(k_1-P_y)} & \dfrac{2k_2 N_y(-N_y+k_2)}{P_y k_1(-N_y+k_2)+ik_2 N_y(k_1-P_y)} & 0 & 0 & 0 & 0 \\[2ex] \dfrac{2iP_y k_1(k_1-P_y)}{P_y k_1(-N_y+k_2)+ik_2 N_y(k_1-P_y)} & \dfrac{P_y k_1(-N_y+k_2)-ik_2 N_y(k_1-P_y)}{P_y k_1(-N_y+k_2)+ik_2 N_y(k_1-P_y)} & 0 & 0 & 0 & 0 \\[2ex] 0 & 0 & \dfrac{-P_z k_3(-N_z+k_4)+ik_4 N_z(k_3-P_z)}{P_z k_3(-N_z+k_4)+ik_4 N_z(k_3-P_z)} & \dfrac{2k_4 N_z(-N_z+k_4)}{P_z k_3(-N_z+k_4)+ik_4 N_z(k_3-P_z)} & 0 & 0 \\[2ex] 0 & 0 & \dfrac{2iP_z k_3(k_3-P_z)}{P_z k_3(-N_z+k_4)+ik_4 N_z(k_3-P_z)} & \dfrac{P_z k_3(-N_z+k_4)-ik_4 N_z(k_3-P_z)}{P_z k_3(-N_z+k_4)+ik_4 N_z(k_3-P_z)} & 0 & 0 \\[2ex] 0 & 0 & 0 & 0 & -1 & 0 \\[1ex] 0 & 0 & 0 & 0 & 0 & -1 \end{bmatrix}$$

$$\tag{12.13b}$$

From Eqs. (12.4a), (12.7a), (12.8), and (12.11), with the origin chosen at the boundary, the reflection matrix at a simply-supported boundary by the Euler–Bernoulli bending vibration theory is

$$\mathbf{r} = \begin{bmatrix} -1 & 0 & 0 & 0 & 0 & 0 \\ 0 & -1 & 0 & 0 & 0 & 0 \\ 0 & 0 & -1 & 0 & 0 & 0 \\ 0 & 0 & 0 & -1 & 0 & 0 \\ 0 & 0 & 0 & 0 & -1 & 0 \\ 0 & 0 & 0 & 0 & 0 & -1 \end{bmatrix} \tag{12.14}$$

From Eqs. (12.4b), (12.7b), (12.8), and (12.11), with the origin chosen at the boundary, the reflection matrix at a simply-supported boundary by the Timoshenko bending vibration theory is given by the same expression as Eq. (12.14).

12.3 Force Generated Waves

External forces and moments applied to a continuous structure have the effect of injecting waves into the structure. In Figure 12.2, $\overline{F}$, $\overline{Q_y}$, and $\overline{Q_z}$ denote the externally applied point forces along the x, y, and z directions, respectively. $\overline{M_y}$ and $\overline{M_z}$ are the externally applied moments in the x–y and x–z plane, respectively. $\overline{T}$ is the externally applied torque about the x-axis. The relationships between external excitations and the injected bending, longitudinal, as well as torsional vibration waves, can be found by considering the continuity and equilibrium conditions at the point where the external excitations are applied.

Because the bending, longitudinal, and torsional deflections, as well as the bending slope on the left and right side of the point where the external excitations are applied are the same, denoting physical parameters on the left and right side of the point where the external excitations are applied using subscripts − and +, respectively, the continuity conditions are

$$y_- = y_+, \; \varphi_{y-} = \varphi_{y+},$$

$$z_- = z_+, \; \varphi_{z-} = \varphi_{z+},$$

$$u_- = u_+, \; \theta_- = \theta_+ \tag{12.15}$$

The equilibrium conditions at the point where the external excitations are applied are

$$\overline{Q_y} = V_{y-} - V_{y+}, \; \overline{M_y} = M_{y-} - M_{y+}$$

$$\overline{Q_z} = V_{z-} - V_{z+}, \; \overline{M_z} = M_{z-} - M_{z+}$$

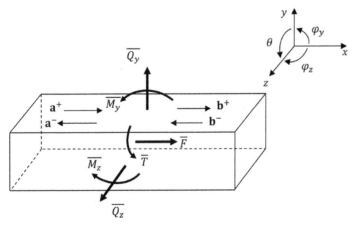

Figure 12.2 Waves generated by externally applied forces, moments, and torque.

$$\overline{F} = F_- - F_+, \quad \overline{T} = T_- - T_+ \tag{12.16}$$

The relationships between injected waves and externally applied excitations are found to be an expansion of the corresponding relationships obtained in Chapters 2, 3, and 9,

$$\mathbf{b}^+ - \mathbf{a}^+ = \mathbf{q}_y + \mathbf{q}_z + \mathbf{m}_y + \mathbf{m}_z + \mathbf{f} + \mathbf{t}$$

$$\mathbf{b}^- - \mathbf{a}^- = -\mathbf{q}_y - \mathbf{q}_z + \mathbf{m}_y + \mathbf{m}_z - \mathbf{f} - \mathbf{t} \tag{12.17}$$

where the load vectors by the Euler–Bernoulli bending vibration theory and the Elementary longitudinal and torsional theories are

$$\mathbf{q}_y = \begin{bmatrix} -i \\ -1 \\ 0 \\ 0 \\ 0 \\ 0 \end{bmatrix} \frac{\overline{Q_y}}{4EI_y k_1^3} = \begin{bmatrix} -i \\ -1 \\ 0 \\ 0 \\ 0 \\ 0 \end{bmatrix} \frac{\overline{Q_y}}{4EI_y k_2^3}, \quad \mathbf{q}_z = \begin{bmatrix} 0 \\ 0 \\ -i \\ -1 \\ 0 \\ 0 \end{bmatrix} \frac{\overline{Q_z}}{4EI_z k_3^3} = \begin{bmatrix} 0 \\ 0 \\ -i \\ -1 \\ 0 \\ 0 \end{bmatrix} \frac{\overline{Q_z}}{4EI_z k_4^3},$$

$$\mathbf{m}_y = \begin{bmatrix} 1 \\ -1 \\ 0 \\ 0 \\ 0 \\ 0 \end{bmatrix} \frac{\overline{M_y}}{4EI_y k_1^2} = \begin{bmatrix} 1 \\ -1 \\ 0 \\ 0 \\ 0 \\ 0 \end{bmatrix} \frac{\overline{M_y}}{4EI_y k_2^2}, \quad \mathbf{m}_z = \begin{bmatrix} 0 \\ 0 \\ 1 \\ -1 \\ 0 \\ 0 \end{bmatrix} \frac{\overline{M_z}}{4EI_z k_3^2} = \begin{bmatrix} 0 \\ 0 \\ 1 \\ -1 \\ 0 \\ 0 \end{bmatrix} \frac{\overline{M_z}}{4EI_z k_4^2},$$

$$\mathbf{f} = \begin{bmatrix} 0 \\ 0 \\ 0 \\ 0 \\ -i \\ 0 \end{bmatrix} \frac{\overline{F}}{2EAk_5}, \quad \mathbf{t} = \begin{bmatrix} 0 \\ 0 \\ 0 \\ 0 \\ 0 \\ -i \end{bmatrix} \frac{\overline{T}}{2G\gamma k_6} \tag{12.18a}$$

and the load vectors by the Timoshenko bending vibration theory and the Elementary longitudinal and torsional theories are

$$\mathbf{q}_y = \begin{bmatrix} -iN_y \\ -P_y \\ 0 \\ 0 \\ 0 \\ 0 \end{bmatrix} \frac{\overline{Q_y}}{2(N_y k_1 - P_y k_2)GA\kappa}, \quad \mathbf{q}_z = \begin{bmatrix} 0 \\ 0 \\ -iN_z \\ -P_z \\ 0 \\ 0 \end{bmatrix} \frac{\overline{Q_z}}{2(N_z k_3 - P_z k_4)GA\kappa}$$

$$\mathbf{m}_y = \begin{bmatrix} 1 \\ -1 \\ 0 \\ 0 \\ 0 \\ 0 \end{bmatrix} \frac{\overline{M_y}}{2(P_y k_1 + N_y k_2)EI_y}, \quad \mathbf{m}_z = \begin{bmatrix} 0 \\ 0 \\ 1 \\ -1 \\ 0 \\ 0 \end{bmatrix} \frac{\overline{M_z}}{2(P_z k_3 + N_z k_4)EI_z},$$

$$\mathbf{f} = \begin{bmatrix} 0 \\ 0 \\ 0 \\ 0 \\ -i \\ 0 \end{bmatrix} \frac{\bar{F}}{2EAk_5}, \; \mathbf{t} = \begin{bmatrix} 0 \\ 0 \\ 0 \\ 0 \\ 0 \\ -i \end{bmatrix} \frac{\bar{T}}{2G\gamma k_6} \tag{12.18b}$$

12.4 Free and Forced Vibration Analysis of a Multi-story Space Frame

Regardless of the complexity of a structure, from the wave standpoint, it consists of only two types of structural components, namely, structural elements and structural discontinuities. Vibrations propagate along uniform structural elements, and are reflected and transmitted at structural discontinuities. Assembling these propagation, reflection, and transmission relationships provides a systematic approach for free vibration analysis of a complex structure.

Figure 12.3 shows waves in an n-story space frame. Each level of the n-story space frame is labeled using a capital letter and the wave components on the level are denoted using the corresponding small letter. For example, on the top level "A", all wave components are denoted using small letter "a".

Each wave component is associated with a structural discontinuity, either a joint or a boundary. The joints and boundaries are numbered using numerical digits 1 to 4, starting from the front right side joint and following a counterclockwise rotation: Joints/Boundaries 1, 2, 3, and 4 refer to the front right side, back right side, back left side, and front left side joints/boundaries, respectively.

Subscripts in wave components that are associated with a joint comprise two numerical digits. The leading digit in a subscript is the sequence number of a joint, which varies from 1 to 4 because there are four joints on each level and four boundaries in the space frame. The other digit in the subscript is the sequence number of beam elements connected at the joint, which varies from 1 to 3 for a spatial Y joint and from 1 to 4 for a spatial K joint, because there are three and four beam elements connected at a spatial Y and K joint, respectively. Wave component associated with a boundary has a single numerical digit in the subscript, which is the sequence number of the boundary.

Superscripts $+$ and $-$ in the wave components denote positive- and negative-going waves, respectively. On the horizontal beam elements in Figure 12.3, waves traveling counterclockwise are defined positive going, and waves traveling clockwise are defined negative going. On the vertical beam elements in Figure 12.3, waves traveling downwards are defined positive going, and waves traveling upwards are defined negative going.

Because the n-story space frame in Figure 12.3 consists of $8n$ uniform beam elements, there are $8n$ pairs of propagation relationships. At the boundary of the space frame, there are four reflection relationships. There are four sets of reflection and transmission relationships corresponding to the four spatial Y joints on the top story, each set contains three equations describing the relationships among the incoming and outgoing waves from the three beam elements connected at each spatial Y joint. There are $4(n-1)$ sets of reflection and transmission relationships corresponding to the $(n-1)$ spatial K joints, each set contains four equations describing the relationships among the incoming and outgoing waves from the four beam elements connected at each spatial K joint.

In total, there are $8n \times 2 + 4 + 4 \times 3 + 4(n-1) \times 4$ matrix equations in describing the reflection, transmission, and propagation relationships of vibration waves in the n-story space frame.

There are $8n$ pairs of propagation relationships, shown below is an example set on the top story (Level A),

$$\mathbf{a}_{11}^+ = \mathbf{f}(L_x)\mathbf{a}_{42}^+, \; \mathbf{a}_{42}^- = \mathbf{f}(L_x)\mathbf{a}_{11}^-$$
$$\mathbf{a}_{12}^- = \mathbf{f}(L_y)\mathbf{a}_{21}^-, \; \mathbf{a}_{21}^+ = \mathbf{f}(L_y)\mathbf{a}_{12}^+$$
$$\mathbf{a}_{22}^- = \mathbf{f}(L_x)\mathbf{a}_{31}^-, \; \mathbf{a}_{31}^+ = \mathbf{f}(L_x)\mathbf{a}_{22}^+$$
$$\mathbf{a}_{32}^- = \mathbf{f}(L_y)\mathbf{a}_{41}^-, \; \mathbf{a}_{41}^+ = \mathbf{f}(L_y)\mathbf{a}_{32}^+$$
$$\mathbf{b}_{14}^+ = \mathbf{f}(L_{za})\mathbf{a}_{13}^+, \; \mathbf{a}_{13}^- = \mathbf{f}(L_{za})\mathbf{b}_{14}^- \tag{12.19}$$
$$\mathbf{b}_{24}^+ = \mathbf{f}(L_{za})\mathbf{a}_{23}^+, \; \mathbf{a}_{23}^- = \mathbf{f}(L_{za})\mathbf{b}_{24}^-$$
$$\mathbf{b}_{34}^+ = \mathbf{f}(L_{za})\mathbf{a}_{33}^+, \; \mathbf{a}_{33}^- = \mathbf{f}(L_{za})\mathbf{b}_{34}^-$$
$$\mathbf{b}_{44}^+ = \mathbf{f}(L_{za})\mathbf{a}_{43}^+, \; \mathbf{a}_{43}^- = \mathbf{f}(L_{za})\mathbf{b}_{44}^-$$

......

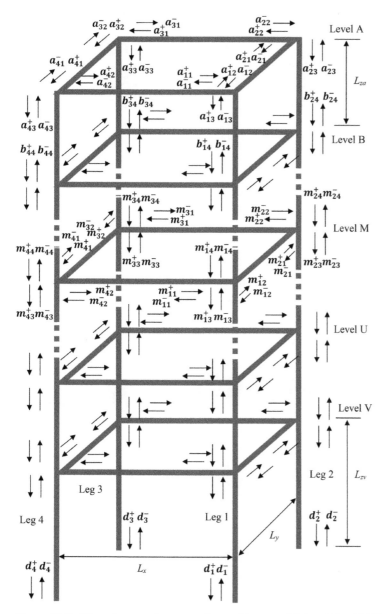

Figure 12.3 Waves in an *n*-story space frame. (Courtesy: (Mei and Sha, 2015)).

where $\mathbf{f}(L_x)$, $\mathbf{f}(L_y)$, and $\mathbf{f}(L_{za})$ are the propagation matrices for distances L_x, L_y, and L_{za}, respectively.

There are four reflection relationships at the boundaries,

$$\mathbf{d}_1^- = \mathbf{r}_{d1}\,\mathbf{d}_1^+, \ \mathbf{d}_2^- = \mathbf{r}_{d2}\,\mathbf{d}_2^+, \ \mathbf{d}_3^- = \mathbf{r}_{d3}\,\mathbf{d}_3^+, \ \mathbf{d}_4^- = \mathbf{r}_{d4}\,\mathbf{d}_4^+ \tag{12.20}$$

where $\mathbf{r}$ is the reflection matrix at a boundary, whose location is denoted in the subscript.

The four sets of reflection and transmission relationships at the spatial Y joint on the top story (Level A) are:

$$\mathbf{a}_{12}^+ = \mathbf{t}_{12}^{A1}\mathbf{a}_{11}^+ + \mathbf{r}_{22}^{A1}\mathbf{a}_{12}^- + \mathbf{t}_{32}^{A1}\mathbf{a}_{13}^-$$

$$\mathbf{a}_{11}^- = \mathbf{r}_{11}^{A1}\mathbf{a}_{11}^+ + \mathbf{t}_{21}^{A1}\mathbf{a}_{12}^- + \mathbf{t}_{31}^{A1}\mathbf{a}_{13}^-$$

$$\mathbf{a}_{13}^+ = \mathbf{t}_{13}^{A1}\mathbf{a}_{11}^+ + \mathbf{t}_{23}^{A1}\mathbf{a}_{12}^- + \mathbf{r}_{33}^{A1}\mathbf{a}_{13}^-$$

$$\mathbf{a}_{22}^{+} = \mathbf{t}_{12}^{A2}\mathbf{a}_{21}^{+} + \mathbf{r}_{22}^{A2}\mathbf{a}_{22}^{-} + \mathbf{t}_{32}^{A2}\mathbf{a}_{23}^{-}$$

$$\mathbf{a}_{21}^{-} = \mathbf{r}_{11}^{A2}\mathbf{a}_{21}^{+} + \mathbf{t}_{21}^{A2}\mathbf{a}_{22}^{-} + \mathbf{t}_{31}^{A2}\mathbf{a}_{23}^{-}$$

$$\mathbf{a}_{23}^{+} = \mathbf{t}_{13}^{A2}\mathbf{a}_{21}^{+} + \mathbf{t}_{23}^{A2}\mathbf{a}_{22}^{-} + \mathbf{r}_{33}^{A2}\mathbf{a}_{23}^{-}$$

$$\mathbf{a}_{32}^{+} = \mathbf{t}_{12}^{A3}\mathbf{a}_{31}^{+} + \mathbf{r}_{22}^{A3}\mathbf{a}_{32}^{-} + \mathbf{t}_{32}^{A3}\mathbf{a}_{33}^{-}$$

$$\mathbf{a}_{31}^{-} = \mathbf{r}_{11}^{A3}\mathbf{a}_{31}^{+} + \mathbf{t}_{21}^{A3}\mathbf{a}_{32}^{-} + \mathbf{t}_{31}^{A3}\mathbf{a}_{33}^{-}$$

$$\mathbf{a}_{33}^{+} = \mathbf{t}_{13}^{A3}\mathbf{a}_{31}^{+} + \mathbf{t}_{23}^{A3}\mathbf{a}_{32}^{-} + \mathbf{r}_{33}^{A3}\mathbf{a}_{33}^{-}$$

$$\mathbf{a}_{42}^{+} = \mathbf{t}_{12}^{A4}\mathbf{a}_{41}^{+} + \mathbf{r}_{22}^{A4}\mathbf{a}_{42}^{-} + \mathbf{t}_{32}^{A4}\mathbf{a}_{43}^{-}$$

$$\mathbf{a}_{41}^{-} = \mathbf{r}_{11}^{A4}\mathbf{a}_{41}^{+} + \mathbf{t}_{21}^{A4}\mathbf{a}_{42}^{-} + \mathbf{t}_{31}^{A4}\mathbf{a}_{43}^{-} \qquad (12.21)$$

$$\mathbf{a}_{43}^{+} = \mathbf{t}_{13}^{A4}\mathbf{a}_{41}^{+} + \mathbf{t}_{23}^{A4}\mathbf{a}_{42}^{-} + \mathbf{r}_{33}^{A4}\mathbf{a}_{43}^{-}$$

There is one set of 16 reflection and transmission equations corresponding to the four spatial K joints on each of the $(n-1)$ intermediate levels, an example set of reflection and transmission relationships at Level M is

$$\mathbf{m}_{12}^{+} = \mathbf{t}_{12}^{M1}\mathbf{m}_{11}^{+} + \mathbf{r}_{22}^{M1}\mathbf{m}_{12}^{-} + \mathbf{t}_{32}^{M1}\mathbf{m}_{13}^{-} + \mathbf{t}_{42}^{M1}\mathbf{m}_{14}^{+}$$

$$\mathbf{m}_{11}^{-} = \mathbf{r}_{11}^{M1}\mathbf{m}_{11}^{+} + \mathbf{t}_{21}^{M1}\mathbf{m}_{12}^{-} + \mathbf{t}_{31}^{M1}\mathbf{m}_{13}^{-} + \mathbf{t}_{41}^{M1}\mathbf{m}_{14}^{+}$$

$$\mathbf{m}_{13}^{+} = \mathbf{t}_{13}^{M1}\mathbf{m}_{11}^{+} + \mathbf{t}_{23}^{M1}\mathbf{m}_{12}^{-} + \mathbf{r}_{33}^{M1}\mathbf{m}_{13}^{-} + \mathbf{t}_{43}^{M1}\mathbf{m}_{14}^{+}$$

$$\mathbf{m}_{14}^{-} = \mathbf{t}_{14}^{M1}\mathbf{m}_{11}^{+} + \mathbf{t}_{24}^{M1}\mathbf{m}_{12}^{-} + \mathbf{t}_{34}^{M1}\mathbf{m}_{13}^{-} + \mathbf{r}_{44}^{M1}\mathbf{m}_{14}^{+}$$

$$\mathbf{m}_{22}^{+} = \mathbf{t}_{12}^{M2}\mathbf{m}_{21}^{+} + \mathbf{r}_{22}^{M2}\mathbf{m}_{22}^{-} + \mathbf{t}_{32}^{M2}\mathbf{m}_{23}^{-} + \mathbf{t}_{42}^{M2}\mathbf{m}_{24}^{+}$$

$$\mathbf{m}_{21}^{-} = \mathbf{r}_{11}^{M2}\mathbf{m}_{21}^{+} + \mathbf{t}_{21}^{M2}\mathbf{m}_{22}^{-} + \mathbf{t}_{31}^{M2}\mathbf{m}_{23}^{-} + \mathbf{t}_{41}^{M2}\mathbf{m}_{24}^{+}$$

$$\mathbf{m}_{23}^{+} = \mathbf{t}_{13}^{M2}\mathbf{m}_{21}^{+} + \mathbf{t}_{23}^{M2}\mathbf{m}_{22}^{-} + \mathbf{r}_{33}^{M2}\mathbf{m}_{23}^{-} + \mathbf{t}_{43}^{M2}\mathbf{m}_{24}^{+}$$

$$\mathbf{m}_{24}^{-} = \mathbf{t}_{14}^{M2}\mathbf{m}_{21}^{+} + \mathbf{t}_{24}^{M2}\mathbf{m}_{22}^{-} + \mathbf{t}_{34}^{M2}\mathbf{m}_{23}^{-} + \mathbf{r}_{44}^{M2}\mathbf{m}_{24}^{+}$$

$$\mathbf{m}_{32}^{+} = \mathbf{t}_{12}^{M3}\mathbf{m}_{31}^{+} + \mathbf{r}_{22}^{M3}\mathbf{m}_{32}^{-} + \mathbf{t}_{32}^{M3}\mathbf{m}_{33}^{-} + \mathbf{t}_{42}^{M3}\mathbf{m}_{34}^{+}$$

$$\mathbf{m}_{31}^{-} = \mathbf{r}_{11}^{M3}\mathbf{m}_{31}^{+} + \mathbf{t}_{21}^{M3}\mathbf{m}_{32}^{-} + \mathbf{t}_{31}^{M3}\mathbf{m}_{33}^{-} + \mathbf{t}_{41}^{M3}\mathbf{m}_{34}^{+}$$

$$\mathbf{m}_{33}^{+} = \mathbf{t}_{13}^{M3}\mathbf{m}_{31}^{+} + \mathbf{t}_{23}^{M3}\mathbf{m}_{32}^{-} + \mathbf{r}_{33}^{M3}\mathbf{m}_{33}^{-} + \mathbf{t}_{43}^{M3}\mathbf{m}_{34}^{+}$$

$$\mathbf{m}_{34}^{-} = \mathbf{t}_{14}^{M3}\mathbf{m}_{31}^{+} + \mathbf{t}_{24}^{M3}\mathbf{m}_{32}^{-} + \mathbf{t}_{34}^{M3}\mathbf{m}_{33}^{-} + \mathbf{r}_{44}^{M3}\mathbf{m}_{34}^{+}$$

$$\mathbf{m}_{42}^{+} = \mathbf{t}_{12}^{M4}\mathbf{m}_{41}^{+} + \mathbf{r}_{22}^{M4}\mathbf{m}_{42}^{-} + \mathbf{t}_{32}^{M4}\mathbf{m}_{43}^{-} + \mathbf{t}_{42}^{M4}\mathbf{m}_{44}^{+}$$

$$\mathbf{m}_{41}^{-} = \mathbf{r}_{11}^{M4}\mathbf{m}_{41}^{+} + \mathbf{t}_{21}^{M4}\mathbf{m}_{42}^{-} + \mathbf{t}_{31}^{M4}\mathbf{m}_{43}^{-} + \mathbf{t}_{41}^{M4}\mathbf{m}_{44}^{+}$$

$$\mathbf{m}_{43}^{+} = \mathbf{t}_{13}^{M4}\mathbf{m}_{41}^{+} + \mathbf{t}_{23}^{M4}\mathbf{m}_{42}^{-} + \mathbf{r}_{33}^{M4}\mathbf{m}_{43}^{-} + \mathbf{t}_{43}^{M4}\mathbf{m}_{44}^{+} \qquad (12.22)$$

$$\mathbf{m}_{44}^{-} = \mathbf{t}_{14}^{M4}\mathbf{m}_{41}^{+} + \mathbf{t}_{24}^{M4}\mathbf{m}_{42}^{-} + \mathbf{t}_{34}^{M4}\mathbf{m}_{43}^{-} + \mathbf{r}_{44}^{M4}\mathbf{m}_{44}^{+}$$

...

In Eqs. (12.21) and (12.22), $\mathbf{r}$ and $\mathbf{t}$ are the reflection and transmission matrices, respectively. The superscripts in $\mathbf{r}$ and $\mathbf{t}$ identify a joint by its level and sequence number, and the two numerical digits in the subscripts refer to the sequence number of beam elements connected at the joint. The leading digit in a subscript in both $\mathbf{r}$ and $\mathbf{t}$ identifies the beam element carrying the incident waves.

Free vibration responses can be obtained by assembling Eqs. (12.19) to (12.22) into matrix form,

$$\mathbf{A}\mathbf{z} = 0 \qquad (12.23)$$

where $\mathbf{A}$ is a square coefficient matrix of size $\{[\ 16n + 4 + 12 + 16(n-1)\ ] \times 6\}$, and $\mathbf{z}$ is a wave vector of size $\{[\ 16n + 4 + 12 + 16(n-1)\ ] \times 6\}$. The natural frequencies of the space frame are obtained by setting the determinant of the coefficient matrix $\mathbf{A}$ to zero.

To obtain forced responses, Eq. (12.17), the relationships between injected waves and externally applied forces and moments, needs to be added to the matrix assembly. In addition, the propagation relationships along the external excitation applied structural element(s) need to be modified. These wave relationships in matrix form are

$$\mathbf{A}_f \mathbf{z}_f = \mathbf{F} \tag{12.24}$$

where $\mathbf{A}_f$ is a square coefficient matrix, $\mathbf{z}$ is a wave vector, and $\mathbf{F}$ is a force vector.

From Eq. (12.24), the wave components are solved,

$$\mathbf{z}_f = \mathbf{A}_f^{-1} \mathbf{F} \tag{12.25}$$

With the availability of wave components from Eq. (12.25), the forced responses at any location on the structure can be obtained.

12.5 Reflection and Transmission of Waves in a Multi-story Space Frame

To obtain the reflection and transmission relationships at a structural discontinuity, a free body diagram needs to be drawn and analyzed for establishing the continuity relationships and equations of motion at each discontinuity. A local coordinate must be defined, for each beam element at a joint and for the joint itself. In selecting a local coordinate, the guidelines listed below need to be followed:

a) A three-dimensional Cartesian coordinate system is chosen. The x-axis needs to be along the longitudinal axis of a beam element, as is the case that the equations of motion of a beam are derived. The positive direction of the x-axis for any portion of a beam element must be consistently defined.
b) The positive directions of rotation angles φ_y and φ_z adopted in deriving the governing equations of motion of a beam need to be followed. In this book, they are the directions of rotating from the x-axis to the y- and z-axis, respectively.
c) The same sign conventions adopted in deriving the equations of motion of a beam must be followed in defining positive internal resistant torque, forces, and moments in each beam element with regard to its local coordinate. In this chapter, the sign convention defined in Figure 12.1 is followed, that is, an axial force F is positive when it stretches the element and a torque T is positive when it relates to the torsional deflection θ by the right hand rule. Shear forces V_y and V_z are positive when rotating an element along the positive directions of rotation angles φ_y and φ_z, and bending moments M_y and M_z are positive when bending an element concave towards the positive y- and z-axis, respectively.

Other than the above rules, much freedom is allowed in selecting local coordinate systems for deriving the reflection and transmission relationships at a discontinuity.

Without loss of generality, a two-story space frame shown in Figure 12.4 is considered to better understand the selection of local coordinate for each waveguide with consistency. Ensuring that the positive direction of the x-axis for any portion of a beam element is consistently defined is critical in describing wave propagation relationships in the uniform beam element, as well as in deriving the reflection and transmission relationships at a joint.

In this two-story space frame, there are 16 uniform beam elements, hence 16 waveguides. There are eight intermediate structural discontinuities in the space frame: the four spatial Y shaped spatial joints on the top story and the four spatial K shaped spatial joints in the middle of the space frame in Figure 12.4. In addition, there are four boundary discontinuities. Incoming vibration waves are reflected and transmitted at a joint, and are reflected at a boundary.

Figures 12.5 and 12.6 illustrate the front right side spatial Y joint and the front right side spatial K joint in Figure 12.4. These are Joint Y1 and Joint K1 by assigning numerical digit 1 to 4 to joints on the same level in the space frame from the front right side joint and following a counterclockwise rotation, as mentioned earlier. The beam elements joined at the spatial Y and K joints are numbered in Figures 12.5 and 12.6.

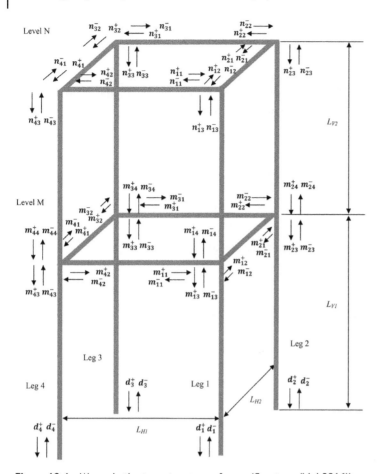

Figure 12.4 Waves in the two-story space frame. (Courtesy: (Mei 2016)).

From Figures 12.4 to 12.6, it can be seen that Beam 3 of Joint Y1 is Beam 4 of Joint K1. If a space frame is more than two stories tall, there may be two K joints on the same vertical leg of the space frame. On two adjacent levels, Beam 3 of an upper level K joint is Beam 4 of a lower level K joint. As a result, to guarantee the positive direction of the x-axis for any portion of a beam element is consistently defined, the positive x-axes on the vertical beam elements in both Figures 12.5 and 12.6 must be of the same direction. In this example set, the positive x-axes for vertical beam elements in the Y and K joints are chosen to point downwards.

In Figures 12.5 and 12.6, the local coordinate systems for the two horizontal beam elements in both the spatial Y and K joints are related by a rotational relationship: the local coordinate system of Beam 2 is obtained by rotating the the local coordinate system of Beam 1 counterclockwise 90 degrees about a vertical axis.

Because the waves traveling counterclockwise on the horizontal beam elements at each level of the two-story space frame in Figure 12.4 are defined positive going, by choosing the local coordinate systems for the horizontal beam elements in the spatial Y and K joints following a counterclockwise rotational relationship, and keeping the local coordinate systems for the vertical beam elements the same, the spatial Y and K joints in Figures 12.5 and 12.6 serve as the base joints for the remaining Y and K joints, respectively. Joints Y2, Y3, and Y4 in Figure 12.4 can be viewed as the Base Joint Y1 shown in Figure 12.5 being rotated counterclockwise about a vertical axis by a 90°, 180°, and 270° angle, respectively. Likewise, Joints K2, K3, and K4 in Figure 12.4 can be viewed as the Base Joint K1 shown in Figure 12.6 being rotated counterclockwise about a vertical axis by a 90°, 180°, and 270° angle, respectively.

When the rotating coordinate systems are adopted, there is no need to derive the reflection and transmission relationships multiple times for the same type of spatial joints. The reflection and transmission relationships obtained from the Base Y and K spatial joints can be applied to the remaining joints of the same type, by observing the rotational relationships in the beam elements. This is similar to the concept of base joint and "rotating" coordinate systems elaborated on wave vibration analysis of planar frames in Chapter 11.

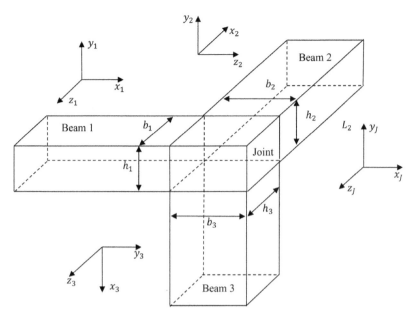

Figure 12.5 A spatial Y joint and its local coordinates. (Courtesy: (Mei 2016)).

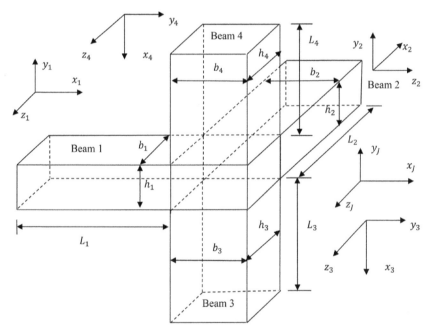

Figure 12.6 A spatial K joint and its local coordinates. (Courtesy: (Mei and Sha 2015)).

12.5.1 Wave Reflection and Transmission at a Y-shaped Spatial Joint

Figures 12.7a-c show the free body diagrams of the spatial Y joint of Figure 12.5, from which the equations of motion of the Y joint are obtained by applying Newton's second law for translational and rotational motion,

$$V_{z2} + V_{y3} - F_1 = m_J \ddot{x}_J, \quad V_{y2} - F_3 - V_{y1} = m_J \ddot{y}_J, \quad -F_2 - V_{z1} + V_{z3} = m_J \ddot{z}_J$$

$$-T_1 - M_{z3} + M_{y2} + V_{y2}\frac{h_3}{2} - V_{z3}\frac{h_2}{2} = J_{xJ}\ddot{\theta}_{xJ}$$

$$-T_3 - M_{z2} + M_{z1} - V_{z2}\frac{b_1}{2} - V_{z1}\frac{b_2}{2} = J_{yJ}\ddot{\theta}_{yJ}$$

$$-T_2 - M_{y1} + M_{y3} + V_{y1}\frac{b_3}{2} + V_{y3}\frac{h_1}{2} = J_{zJ}\ddot{\theta}_{zJ} \tag{12.26}$$

where m_J is the mass of the joint, which is modeled as a rigid body. J_{xJ}, J_{yJ}, and J_{zJ} are the mass moment inertia of the joint about the x, y, and z centroidal axis, respectively. x_J, y_J, and z_J refer to the translational movements of the rigid joint along the x, y, and z directions, respectively. θ_{xJ}, θ_{yJ}, and θ_{zJ} are the rotation of the rigid joint about the x-, y-, and z-axis, respectively. b and h are the width and thickness of the beam elements, respectively. Numerical subscripts 1, 2, and 3 in b and h refer to the width and thickness of Beam 1, Beam 2, and Beam 3, respectively. $V(x,t)$, $M(x,t)$, $F(x,t)$, and $T(x,t)$ are the internal resistant shear force, bending moment, longitudinal force, and torque at the section of a beam, respectively. Numerical subscripts 1, 2, and 3 in V, M, F, and T denote parameters related to Beam 1, Beam 2, and Beam 3, respectively. Letter subscripts y and z in $M(x,t)$ denote bending moments in the x–y and x–z plane, respectively. Letter subscripts y and z in $V(x,t)$ denote shear forces along the y- and z-axis direction, respectively. The shear force, bending moment, axial force, and torque are related to the corresponding deflections of each beam element by Eq. (12.7 a-c). Let us call Eq. (12.26) the equilibrium equations, because by the principle of d'Alembert, these equations represent dynamic equilibrium.

From Figures 12.7a-c, the continuity conditions at the spatial Y joint are

$$u_1 = x_J, \; u_2 = -z_J, \; u_3 = -y_J$$

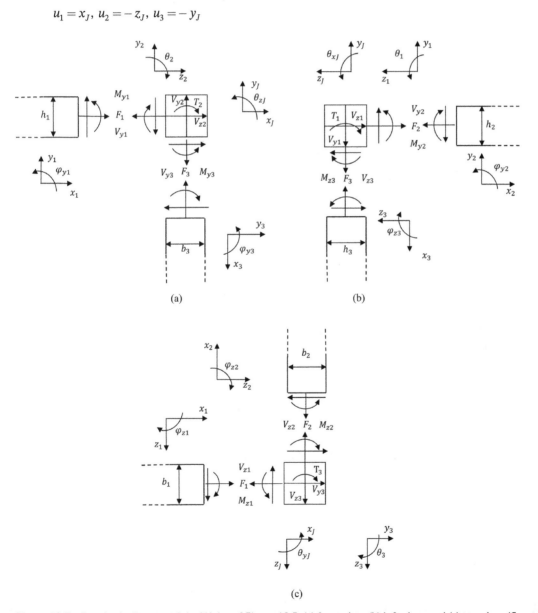

(a)   (b)

(c)

Figure 12.7 Free body diagram of the Y joint of Figure 12.5: (a) front view, (b) left view, and (c) top view. (Courtesy: (Mei 2016)).

$$y_1 = y_J - \theta_{zJ}\frac{b_3}{2}, \quad y_2 = y_J + \theta_{xJ}\frac{h_3}{2}, \quad y_3 = x_J + \theta_{zJ}\frac{h_1}{2}$$

$$z_1 = z_J + \theta_{yJ}\frac{b_2}{2}, \quad z_2 = x_J - \theta_{yJ}\frac{b_1}{2}, \quad z_3 = z_J - \theta_{xJ}\frac{h_2}{2}$$

$$\theta_1 = \theta_{xJ}, \quad \theta_2 = -\theta_{zJ}, \quad \theta_3 = -\theta_{yJ}$$

$$\varphi_{y1} = \theta_{zJ}, \quad \varphi_{y2} = \theta_{xJ}, \quad \varphi_{y3} = \theta_{zJ}$$

$$\varphi_{z1} = -\theta_{yJ}, \quad \varphi_{z2} = -\theta_{yJ}, \quad \varphi_{z3} = -\theta_{xJ} \tag{12.27}$$

where y and z denote the transverse deflections along the y-and z-axis direction, respectively. Numerical subscripts 1, 2, and 3 in Eq. (12.27) denote parameters relating to the correspondingly numbered beam at the joint. Letter subscripts y and z in φ denote the bending slope in the x–y and x–z plane, respectively.

To find the reflection and transmission relationships of vibration waves in beam elements connected at a joint, the incident, reflected, and transmitted waves must be related directly. As a result, the rigid body joint related parameters $\{x_J, y_J, z_J, \theta_{xJ}, \theta_{yJ}, \theta_{zJ}\}$ must be eliminated from both Eqs. (12.26) and (12.27).

Eliminating the rigid body joint related parameters $\{x_J, y_J, z_J, \theta_{xJ}, \theta_{yJ}, \theta_{zJ}\}$ from the continuity equations of Eq. (12.27),

$$\theta_1 = \varphi_{y2} = -\varphi_{z3}$$

$$\theta_2 = -\varphi_{y1} = -\varphi_{y3}$$

$$\theta_3 = \varphi_{z1} = \varphi_{z2}$$

$$y_1 = -u_3 + \theta_2\frac{b_3}{2}, \text{ or equivalently } y_1 = -u_3 - \varphi_{y1,y3}\frac{b_3}{2};$$

$$y_2 = -u_3 + \theta_1\frac{h_3}{2}, \text{ or equivalently } y_2 = -u_3 + \varphi_{y2}\frac{h_3}{2}, \text{ or equivalently } y_2 = -u_3 - \varphi_{z3}\frac{h_3}{2};$$

$$y_3 = u_1 - \theta_2\frac{h_1}{2}, \text{ or equivalently } y_3 = u_1 + \varphi_{y1,y3}\frac{h_1}{2}$$

$$z_1 = -u_2 - \theta_3\frac{b_2}{2}, \text{ or equivalently } z_1 = -u_2 - \varphi_{z1,z2}\frac{b_2}{2};$$

$$z_2 = u_1 + \theta_3\frac{b_1}{2}, \text{ or equivalently } z_2 = u_1 + \varphi_{z1,z2}\frac{b_1}{2};$$

$$z_3 = -u_2 - \theta_1\frac{h_2}{2}, \text{ or equivalently } z_3 = -u_2 - \varphi_{y2}\frac{h_2}{2}, \text{ or equivalently } z_3 = -u_2 + \varphi_{z3}\frac{h_2}{2} \tag{12.28}$$

where $\varphi_{y1,y3} = \varphi_{y1} = \varphi_{y3}$ and $\varphi_{z1,z2} = \varphi_{z1} = \varphi_{z2}$.

The continuity equations of Eq. (12.28) need to be sorted into three groups. Each group consists of six scalar equations relating waves in two beams at a time.

The group of continuity equations relating Beam 1 and Beam 2, by sorting Eq. (12.28), is

$$\theta_1 = \varphi_{y2}$$

$$\theta_2 = -\varphi_{y1}$$

$$\varphi_{z1} = \varphi_{z2}$$

$$z_1 = -u_2 - \varphi_{z1, z2}\frac{b_2}{2}$$

$$z_2 = u_1 + \varphi_{z1, z2}\frac{b_1}{2}$$

$$y_2 - y_1 = \theta_1\frac{h_3}{2} - \theta_2\frac{b_3}{2}, \text{ or equivalently } y_2 - y_1 = \varphi_{y2}\frac{h_3}{2} + \varphi_{y1}\frac{b_3}{2},$$

$$\text{or equivalently } y_2 - y_1 = \varphi_{y2}\frac{h_3}{2} - \theta_2\frac{b_3}{2}, \text{ or equivalently } y_2 - y_1 = \theta_1\frac{h_3}{2} + \varphi_{y1}\frac{b_3}{2} \tag{12.29}$$

where $\varphi_{z1,z2} = \varphi_{z1} = \varphi_{z2}$.

The group of continuity equations relating Beam 1 and Beam 3, by sorting Eq. (12.28), is

$$\theta_1 = -\varphi_{z3}$$

$$\varphi_{y1} = \varphi_{y3}$$

$$\theta_3 = \varphi_{z1}$$

$$y_1 = -u_3 - \varphi_{y1, y3}\frac{b_3}{2}$$

$$y_3 = u_1 + \varphi_{y1, y3}\frac{h_1}{2};$$

$$z_3 - z_1 = -\theta_1\frac{h_2}{2} + \theta_3\frac{b_2}{2}, \text{ or equivalently } z_3 - z_1 = \varphi_{z3}\frac{h_2}{2} + \varphi_{z1}\frac{b_2}{2},$$

$$\text{or equivalently } z_3 - z_1 = -\theta_1\frac{h_2}{2} + \varphi_{z1}\frac{b_2}{2}, \text{ or equivalently } z_3 - z_1 = \varphi_{z3}\frac{h_2}{2} + \theta_3\frac{b_2}{2} \tag{12.30}$$

where $\varphi_{y1,y3} = \varphi_{y1} = \varphi_{y3}$.

The group of continuity equations relating Beam 2 and Beam 3, by sorting Eq. (12.28), is

$$\varphi_{y2} = -\varphi_{z3}$$

$$\theta_2 = -\varphi_{y3}$$

$$\theta_3 = \varphi_{z2}$$

$$y_2 = -u_3 + \varphi_{y2}\frac{h_3}{2}, \text{ or equivalently } y_2 = -u_3 - \varphi_{z3}\frac{h_3}{2}$$

$$z_3 = -u_2 - \varphi_{y2}\frac{h_2}{2}, \text{ or equivalently } z_3 = -u_2 + \varphi_{z3}\frac{h_2}{2}$$

$$z_2 - y_3 = \theta_3\frac{b_1}{2} + \theta_2\frac{h_1}{2}, \text{ or equivalently } z_2 - y_3 = \varphi_{z2}\frac{b_1}{2} + \varphi_{y3}\frac{h_1}{2},$$

or equivalently $z_2 - y_3 = \theta_3 \dfrac{b_1}{2} - \varphi_{y3} \dfrac{h_1}{2}$, or equivalently $z_2 - y_3 = \varphi_{z2} \dfrac{b_1}{2} + \theta_2 \dfrac{h_1}{2}$ $\hspace{2cm}$ (12.31)

From Eqs. (12.29) to (12.31), the groups of continuity equations relating two of the beam elements joined at the spatial Y joint, it is clear that there are multiple sets of equivalent continuity equations in each group. Each set consists of six scalar equations of continuity.

From the continuity equations of Eq. (12.27), the rigid body joint related parameters $\{x_J, y_J, z_J, \theta_{xJ}, \theta_{yJ}, \theta_{zJ}\}$ can be expressed in terms of the beam elements related parameters as follows

$$\theta_{xJ} = \theta_1 = \varphi_{y2} = -\varphi_{z3};$$

$$\theta_{zJ} = -\theta_2 = \varphi_{y1} = \varphi_{y3};$$

$$\theta_{yJ} = -\theta_3 = -\varphi_{z1} = -\varphi_{z2} \hspace{2cm} (12.32a)$$

$$x_J = z_2 + \theta_{yJ} \frac{b_1}{2}, \text{ or equivalently}$$

$$x_J = y_3 - \theta_{zJ} \frac{h_1}{2};$$

$$y_J = y_1 + \theta_{zJ} \frac{b_3}{2}, \text{ or equivalently}$$

$$y_J = y_2 - \theta_{xJ} \frac{h_3}{2};$$

$$z_J = z_1 - \theta_{yJ} \frac{b_2}{2}, \text{ or equivalently}$$

$$z_J = z_3 + \theta_{xJ} \frac{h_2}{2} \hspace{2cm} (12.32b)$$

where θ_{xJ}, θ_{yJ}, and θ_{zJ} in Eq. (12.32b) can be replaced by the equivalent beam elements related parameters described in Eq. (12.32a).

From Eqs. (12.32a) and (12.32b), Eq. (12.26) can be expressed in numerous ways. In other words, there are numerous sets of equivalent equilibrium equations. Each set consists of six scalar equations of equilibrium that are obtained from Eqs. (12.32a), (12.32b), and (12.26) by eliminating the rigid body joint related parameters.

The reflection and transmission relationships corresponding to waves incident upon a joint from a different beam element are in general different. Figures 12.8 a-c show the reflection and transmission of waves incident upon a spatial Y joint from Beam 1, Beam 3, and Beam 2, respectively.

For waves incident from Beam 1, as shown in Figure 12.8a, the reflected waves $\mathbf{a}^-$ on Beam 1, the transmitted waves $\mathbf{b}^+$ to Beam 2, and the transmitted waves $\mathbf{g}^+$ to Beam 3 are related to the incident waves $\mathbf{a}^+$ by reflection matrix $\mathbf{r}_{11}$, transmission matrix $\mathbf{t}_{12}$, and transmission matrix $\mathbf{t}_{13}$ as follows

$$\mathbf{a}^- = \mathbf{r}_{11}\mathbf{a}^+, \ \mathbf{b}^+ = \mathbf{t}_{12}\mathbf{a}^+, \ \mathbf{g}^+ = \mathbf{t}_{13}\mathbf{a}^+ \hspace{2cm} (12.33)$$

For waves incident from Beam 3, as shown in Figure 12.8b, the reflected waves $\mathbf{a}^-$ on Beam 3, the transmitted waves $\mathbf{b}^+$ to Beam 2, and the transmitted waves $\mathbf{g}^+$ to Beam 1 are related to the incident waves $\mathbf{a}^+$ by reflection matrix $\mathbf{r}_{33}$, transmission matrix $\mathbf{t}_{32}$, and transmission matrix $\mathbf{t}_{31}$ as follows

$$\mathbf{a}^- = \mathbf{r}_{33}\mathbf{a}^+, \ \mathbf{b}^+ = \mathbf{t}_{32}\mathbf{a}^+, \ \mathbf{g}^+ = \mathbf{t}_{31}\mathbf{a}^+ \hspace{2cm} (12.34)$$

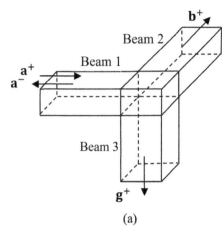

(a)

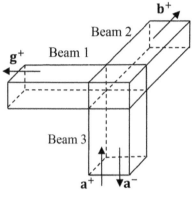

(b)

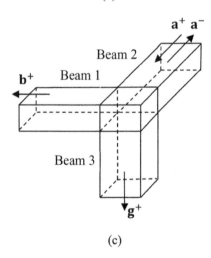

(c)

Figure 12.8 Wave transmission and reflection at the Y joint of Figure 12.5 with incident waves from (a) Beam 1, (b) Beam 3, and (c) Beam 2. (Courtesy: (Mei 2016)).

For waves incident from Beam 2, as shown in Figure 12.8c, the reflected waves $\mathbf{a}^-$ on Beam 2, the transmitted waves $\mathbf{b}^+$ to Beam 1, and the transmitted waves $\mathbf{g}^+$ to Beam 3 are related to the incident waves $\mathbf{a}^+$ by reflection matrix $\mathbf{r}_{22}$, transmission matrix $\mathbf{t}_{21}$, and transmission matrix $\mathbf{t}_{23}$ as follows

$$\mathbf{a}^- = \mathbf{r}_{22}\mathbf{a}^+, \ \mathbf{b}^+ = \mathbf{t}_{21}\mathbf{a}^+, \ \mathbf{g}^+ = \mathbf{t}_{23}\mathbf{a}^+ \tag{12.35}$$

In Eqs. (12.33) to (12.35), wave vectors $\mathbf{a}^\pm$ and $\mathbf{b}^+$ are defined in Eq. (12.6b) and wave vector $\mathbf{g}^+$ is

$$\mathbf{g}^+ = \begin{bmatrix} g_1^+ \\ g_2^+ \\ g_3^+ \\ g_4^+ \\ m^+ \\ n^+ \end{bmatrix} \tag{12.36}$$

In each wave vector, the first four elements are for bending waves in the x–y and x–z plane, and the last two elements are for longitudinal and torsional waves.

First, solve for $\mathbf{r}_{11}$, $\mathbf{t}_{12}$, and $\mathbf{t}_{13}$, that is, the reflection and transmission matrices at the spatial Y joint of Figure 12.8a, corresponding to incident waves from Beam 1.

Following the local coordinate systems of Figure 12.5, the longitudinal, torsional, and bending deflections of the beam elements are

$$u_1 = c^+ e^{-ik_{15}x_1} + c^- e^{ik_{15}x_1},$$

$$u_2 = e^+ e^{-ik_{25}x_2},$$

$$u_3 = m^+ e^{-ik_{35}x_3}$$

$$\theta_1 = d^+ e^{-ik_{16}x_1} + d^- e^{ik_{16}x_1},$$

$$\theta_2 = f^+ e^{-ik_{26}x_2},$$

$$\theta_3 = n^+ e^{-ik_{36}x_3}$$

$$y_1 = a_1^+ e^{-ik_{11}x_1} + a_2^+ e^{-k_{12}x_1} + a_1^- e^{ik_{11}x_1} + a_2^- e^{k_{12}x_1},$$

$$y_2 = b_1^+ e^{-ik_{21}x_2} + b_2^+ e^{-k_{22}x_2},$$

$$y_3 = g_1^+ e^{-ik_{31}x_3} + g_2^+ e^{-k_{32}x_3}$$

$$z_1 = a_3^+ e^{-ik_{13}x_1} + a_4^+ e^{-k_{14}x_1} + a_3^- e^{ik_{13}x_1} + a_4^- e^{k_{14}x_1},$$

$$z_2 = b_3^+ e^{-ik_{23}x_2} + b_4^+ e^{-k_{24}x_2},$$

$$z_3 = g_3^+ e^{-ik_{33}x_3} + g_4^+ e^{-k_{34}x_3} \tag{12.37a}$$

and the bending slopes by the Euler–Bernoulli bending vibration theory are

$$\varphi_{y1} = -ik_{11}a_1^+ e^{-ik_{11}x_1} - k_{12}a_2^+ e^{-k_{12}x_1} + ik_{11}a_1^- e^{ik_{11}x_1} + k_{12}a_2^- e^{k_{12}x_1},$$

$$\varphi_{y2} = -ik_{21}b_1^+ e^{-ik_{21}x_2} - k_{22}b_2^+ e^{-k_{22}x_2},$$

$$\varphi_{y3} = -ik_{31}g_1^+ e^{-ik_{31}x_3} - k_{32}g_2^+ e^{-k_{32}x_3},$$

$$\varphi_{z1} = -ik_{13}a_3^+ e^{-ik_{13}x_1} - k_{14}a_4^+ e^{-k_{14}x_1} + ik_{13}a_3^- e^{ik_{13}x_1} + k_{14}a_4^- e^{k_{14}x_1},$$

$$\varphi_{z2} = -ik_{23}b_3^+ e^{-ik_{23}x_2} - k_{24}b_4^+ e^{-k_{24}x_2},$$

$$\varphi_{z3} = -ik_{33}g_3^+ e^{-ik_{33}x_3} - k_{34}g_4^+ e^{-k_{34}x_3}; \tag{12.37b}$$

and the bending slopes by the Timoshenko bending vibration theory are

$$\varphi_{y1} = -iP_{1y}a_1^+ e^{-ik_{11}x_1} - N_{1y}a_2^+ e^{-k_{12}x_1} + iP_{1y}a_1^- e^{ik_{11}x_1} + N_{1y}a_2^- e^{k_{12}x_1},$$

$$\varphi_{y2} = -iP_{2y}b_1^+ e^{-ik_{21}x_2} - N_{2y}b_2^+ e^{-k_{22}x_2},$$

$$\varphi_{y3} = -iP_{3y}g_1^+ e^{-ik_{31}x_3} - N_{3y}g_2^+ e^{-k_{32}x_3}$$

$$\varphi_{z1} = -iP_{1z}a_3^+ e^{-ik_{13}x_1} - N_{1z}a_4^+ e^{-k_{14}x_1} + iP_{1z}a_3^- e^{ik_{13}x_1} + N_{1z}a_4^- e^{k_{14}x_1},$$

$$\varphi_{z2} = -iP_{2z}b_3^+ e^{-ik_{23}x_2} - N_{2z}b_4^+ e^{-k_{24}x_2},$$

$$\varphi_{z3} = -iP_{3z}g_3^+ e^{-ik_{33}x_3} - N_{3z}g_4^+ e^{-k_{34}x_3} \tag{12.37c}$$

where x_n denotes the position along the beam axis of the nth beam ($n = 1$, 2, 3), k_{nj} is the magnitude of the jth wavenumber (among them subscripts 1 and 2 are related to bending vibration in x–y plane, 3 and 4 are related to bending vibration in x–z plane, and 5 and 6 are for longitudinal and torsional vibrations, respectively) of the nth beam member. P_{ny} and N_{ny} are the coefficients relating bending slope φ_y to bending deflection y; P_{nz} and N_{nz} are the coefficients relating bending slope φ_z to bending deflection z of the nth beam, these coefficients are defined in Eq. (12.5).

Substituting Eqs. (12.37a) and (12.37b) into a set of continuity equations from Eq. (12.29), and choosing the origin at the joint, a set of six scalar equations of continuity is obtained based on the Euler–Bernoulli bending vibration theory. Similarly, substituting Eqs. (12.37a) and (12.37c) into a set of continuity equations from Eq. (12.29), and choosing the origin at the joint, a set of six scalar equations of continuity is obtained based on the Timoshenko bending vibration theory. This set of six scalar equations of continuity, either by the Euler–Bernoulli or Timoshenko bending vibration theory, can be placed in a vector equation, in terms of wave components $\mathbf{a}^+$, $\mathbf{a}^-$, and $\mathbf{b}^+$, which are defined in Eq. (12.6b). A vector equation from this group is named $\mathbf{Y\_Beam1}_{C12}$, because it is a vector equation on the continuity between Beam 1 and Beam 2 at a spatial Y joint corresponding to incident waves from Beam 1.

Substituting Eqs. (12.37a) and (12.37b) into a set of continuity equations from Eq. (12.30), and choosing the origin at the joint, a set of six scalar equations of continuity is obtained based on the Euler–Bernoulli bending vibration theory. Similarly, substituting Eqs. (12.37a) and (12.37c) into a set of continuity equations from Eq. (12.30), and choosing the origin at the joint, a set of six scalar equations of continuity is obtained based on the Timoshenko bending vibration theory. This set of six scalar equations of continuity, either by the Euler–Bernoulli or Timoshenko bending vibration theory, can be placed in a vector equation, in terms of wave components $\mathbf{a}^+$, $\mathbf{a}^-$, and $\mathbf{g}^+$, which are defined in Eqs. (12.6b) and (12.36). A vector equation from this group is named $\mathbf{Y\_Beam1}_{C13}$, because it is a vector equation on the continuity between Beam 1 and Beam 3 at a spatial Y joint corresponding to incident waves from Beam 1.

Eliminating the joint related parameters from Eq. (12.26) using Eqs. (12.32a) and (12.32b), substituting Eqs. (12.37a) and (12.37b) into Eqs. (12.7a), (12.7c), and (12.26) that is free of joint parameters, and choosing the origin at the joint, a set of six scalar equations of equilibrium is obtained based on the Euler–Bernoulli bending vibration theory. Similarly, eliminating the joint related parameters from Eq. (12.26) using Eqs. (12.32a) and (12.32b), substituting Eqs. (12.37a) and (12.37c) into Eqs. (12.7b), (12.7c), and (12.26) that is free of joint parameters, and choosing the origin at the joint, a set of six scalar equations of equilibrium based on the Timoshenko bending vibration theory. These six scalar equations of equilibrium can be placed in a vector equation in terms of $\mathbf{a}^+$, $\mathbf{a}^-$, $\mathbf{b}^+$, and $\mathbf{g}^+$, which are defined in Eqs. (12.6b) and (12.36). A vector equation from this group is named $\mathbf{Y\_Beam1}_E$, because it is a vector equation on equilibrium at a spatial Y joint corresponding to incident waves from Beam 1.

The reflection and transmission matrices, $\mathbf{r}_{11}$, $\mathbf{t}_{12}$, and $\mathbf{t}_{13}$, are found from Eq. (12.33) and the three vector equations of continuity and equilibrium, namely, $\mathbf{Y\_Beam1}_{C12}$, $\mathbf{Y\_Beam1}_{C13}$, and $\mathbf{Y\_Beam1}_E$.

Next, solve for $\mathbf{r}_{33}$, $\mathbf{t}_{32}$, and $\mathbf{t}_{31}$, that is, the reflection and transmission matrices at the spatial Y joint of Figure 12.8b, corresponding to incident waves from Beam 3.

Following the local coordinate systems of Figure 12.5, the longitudinal, torsional, and bending deflections of the beam elements are

$$u_1 = m^+ e^{ik_{15}x_1},$$

$$u_2 = e^+ e^{-ik_{25}x_2},$$

$$u_3 = c^+ e^{ik_{35}x_3} + c^- e^{-ik_{35}x_3}$$

$$\theta_1 = n^+ e^{ik_{16}x_1},$$

$$\theta_2 = f^+ e^{-ik_{26}x_2},$$

$$\theta_3 = d^+ e^{ik_{36}x_3} + d^- e^{-ik_{36}x_3}$$

$$y_1 = g_1^+ e^{ik_{11}x_1} + g_2^+ e^{k_{12}x_1},$$

$$y_2 = b_1^+ e^{-ik_{21}x_2} + b_2^+ e^{-k_{22}x_2},$$

$$y_3 = a_1^+ e^{ik_{31}x_3} + a_2^+ e^{k_{32}x_3} + a_1^- e^{-ik_{31}x_3} + a_2^- e^{-k_{32}x_3}$$

$$z_1 = g_3^+ e^{ik_{13}x_1} + g_4^+ e^{k_{14}x_1},$$

$$z_2 = b_3^+ e^{-ik_{23}x_2} + b_4^+ e^{-k_{24}x_2},$$

$$z_3 = a_3^+ e^{ik_{33}x_3} + a_4^+ e^{k_{34}x_3} + a_3^- e^{-ik_{33}x_3} + a_4^- e^{-k_{34}x_3} \tag{12.38a}$$

and the bending slopes by the Euler–Bernoulli bending vibration theory are

$$\varphi_{y1} = ik_{11}g_1^+ e^{ik_{11}x_1} + k_{12}g_2^+ e^{k_{12}x_1},$$

$$\varphi_{y2} = -ik_{21}b_1^+ e^{-ik_{21}x_2} - k_{22}b_2^+ e^{-k_{22}x_2},$$

$$\varphi_{y3} = ik_{31}a_1^+ e^{ik_{31}x_3} + k_{32}a_2^+ e^{k_{32}x_3} - ik_{31}a_1^- e^{-ik_{31}x_3} - k_{32}a_2^- e^{-k_{32}x_3}$$

$$\varphi_{z1} = ik_{13}g_3^+ e^{ik_{13}x_1} + k_{14}g_4^+ e^{k_{14}x_1},$$

$$\varphi_{z2} = -ik_{23}b_3^+ e^{-ik_{23}x_2} - k_{24}b_4^+ e^{-k_{24}x_2},$$

$$\varphi_{z3} = ik_{33}a_3^+ e^{ik_{33}x_3} + k_{34}a_4^+ e^{k_{34}x_3} - ik_{33}a_3^- e^{-ik_{33}x_3} - k_{34}a_4^- e^{-k_{34}x_3} \tag{12.38b}$$

and the bending slopes by the Timoshenko bending vibration theory are

$$\varphi_{y1} = iP_{1y}g_1^+ e^{ik_{11}x_1} + N_{1y}g_2^+ e^{k_{12}x_1},$$

$$\varphi_{y2} = -iP_{2y}b_1^+ e^{-ik_{21}x_2} - N_{2y}b_2^+ e^{-k_{22}x_2},$$

$$\varphi_{y3} = iP_{3y}a_1^+ e^{ik_{31}x_3} + N_{3y}a_2^+ e^{k_{32}x_3} - iP_{3y}a_1^- e^{-ik_{31}x_3} - N_{3y}a_2^- e^{-k_{32}x_3}$$

$$\varphi_{z1} = iP_{1z}g_3^+ e^{ik_{13}x_1} + N_{1z}g_4^+ e^{k_{14}x_1},$$

$$\varphi_{z2} = -iP_{2z}b_3^+ e^{-ik_{23}x_2} - N_{2z}b_4^+ e^{-k_{24}x_2},$$

$$\varphi_{z3} = iP_{3z}a_3^+ e^{ik_{33}x_3} + N_{3z}a_4^+ e^{k_{34}x_3} - iP_{3z}a_3^- e^{-ik_{33}x_3} - N_{3z}a_4^- e^{-k_{34}x_3} \tag{12.38c}$$

Substituting Eqs. (12.38a) and (12.38b) into a set of continuity equations from Eq. (12.30), and choosing the origin at the joint, a set of six scalar equations of continuity is obtained based on the Euler–Bernoulli bending vibration theory. Similarly, substituting Eqs. (12.38a) and (12.38c) into a set of continuity equations from Eq. (12.30), and choosing the origin at the joint, a set of six scalar equations of continuity is obtained based on the Timoshenko bending vibration theory. This set of six scalar equations of continuity, either by the Euler–Bernoulli or Timoshenko bending vibration theory, can be placed in a vector equation, in terms of wave components $\mathbf{a}^+$, $\mathbf{a}^-$, and $\mathbf{g}^+$, which are defined in Eqs. (12.6b) and (12.36). A vector equation from this group is named $\mathbf{Y\_Beam3}_{C31}$, because it is a vector equation on the continuity between Beam 3 and Beam 1 at a spatial Y joint corresponding to incident waves from Beam 3.

Substituting Eqs. (12.38a) and (12.38b) into a set of continuity equations from Eq. (12.31), and choosing the origin at the joint, a set of six scalar equations of continuity is obtained based on the Euler–Bernoulli bending vibration theory. Similarly, substituting Eqs. (12.38a) and (12.38c) into a set of continuity equations from Eq. (12.31), and choosing the origin at the joint, a set of six scalar equations of continuity is obtained based on the Timoshenko bending vibration theory. This set of six scalar equations of continuity, either by the Euler–Bernoulli or Timoshenko bending vibration theory, can be placed in a vector equation, in terms of wave components $\mathbf{a}^+$, $\mathbf{a}^-$, and $\mathbf{b}^+$, which are defined in Eq. (12.6b). A vector equation from this group is named $\mathbf{Y\_Beam3}_{C32}$, because it is a vector equation on the continuity between Beam 3 and Beam 2 at a spatial Y joint corresponding to incident waves from Beam 3.

Eliminating the joint related parameters from Eq. (12.26) using Eqs. (12.32a) and (12.32b), substituting Eqs. (12.38a) and (12.38b) into Eqs. (12.7a), (12.7c), and (12.26) that is free of joint parameters, and choosing the origin at the joint, a set of six scalar equations of equilibrium is obtained based on the Euler–Bernoulli bending vibration theory. Similarly, eliminating the joint related parameters from Eq. (12.26) using Eqs. (12.32a) and (12.32b), substituting Eqs. (12.38a) and (12.38c) into Eqs. (12.7b), (12.7c), and (12.26) that is free of joint parameters, and choosing the origin at the joint, a set of six scalar equations of equilibrium based on the Timoshenko bending vibration theory. These six scalar equations of equilibrium can be placed in a vector equation in terms of $\mathbf{a}^+$, $\mathbf{a}^-$, $\mathbf{b}^+$, and $\mathbf{g}^+$, which are defined in Eqs. (12.6b) and (12.36). A vector equation from this group is named $\mathbf{Y\_Beam3}_E$, because it is a vector equation on equilibrium at a spatial Y joint corresponding to incident waves from Beam 3.

The reflection and transmission matrices, $\mathbf{r}_{33}$, $\mathbf{t}_{32}$, and $\mathbf{t}_{31}$, are found from Eq. (12.34) and the three vector equations of continuity and equilibrium, namely, $\mathbf{Y\_Beam3}_{C31}$, $\mathbf{Y\_Beam3}_{C32}$, and $\mathbf{Y\_Beam3}_E$.

Last, solve for $\mathbf{r}_{22}$, $\mathbf{t}_{21}$, and $\mathbf{t}_{23}$, that is, the reflection and transmission matrices at the spatial Y joint of Figure 12.8c, corresponding to incident waves from Beam 2.

Following the local coordinate systems of Figure 12.5, the longitudinal, torsional, and bending deflections of the beam elements are

$$u_1 = e^+ e^{ik_{15}x_1},$$

$$u_2 = c^+ e^{ik_{25}x_2} + c^- e^{-ik_{25}x_2},$$

$$u_3 = m^+ e^{-ik_{35}x_3}$$

$$\theta_1 = f^+ e^{ik_{16}x_1},$$

$$\theta_2 = d^+ e^{ik_{26}x_2} + d^- e^{-ik_{26}x_2},$$

$$\theta_3 = n^+ e^{-ik_{36}x_3}$$

$$y_1 = b_1^+ e^{ik_{11}x_1} + b_2^+ e^{k_{12}x_1},$$

$$y_2 = a_1^+ e^{ik_{21}x_2} + a_2^+ e^{k_{22}x_2} + a_1^- e^{-ik_{21}x_2} + a_2^- e^{-k_{22}x_2},$$

$$y_3 = g_1^+ e^{-ik_{31}x_3} + g_2^+ e^{-k_{32}x_3}$$

$$z_1 = b_3^+ e^{ik_{13}x_1} + b_4^+ e^{k_{14}x_1},$$

$$z_2 = a_3^+ e^{ik_{23}x_2} + a_4^+ e^{k_{24}x_2} + a_3^- e^{-ik_{23}x_2} + a_4^- e^{-k_{24}x_2},$$

$$z_3 = g_3^+ e^{-ik_{33}x_3} + g_4^+ e^{-k_{34}x_3} \tag{12.39a}$$

and the bending slopes by the Euler–Bernoulli bending vibration theory are

$$\varphi_{y1} = ik_{11}b_1^+ e^{ik_{11}x_1} + k_{12}b_2^+ e^{k_{12}x_1},$$

$$\varphi_{y2} = ik_{21}a_1^+ e^{ik_{21}x_2} + k_{22}a_2^+ e^{k_{22}x_2} - ik_{21}a_1^- e^{-ik_{21}x_2} - k_{22}a_2^- e^{-k_{22}x_2},$$

$$\varphi_{y3} = -ik_{31}g_1^+ e^{-ik_{31}x_3} - k_{32}g_2^+ e^{-k_{32}x_3}$$

$$\varphi_{z1} = ik_{13}b_3^+ e^{ik_{13}x_1} + k_{14}b_4^+ e^{k_{14}x_1},$$

$$\varphi_{z2} = ik_{23}a_3^+ e^{ik_{23}x_2} + k_{24}a_4^+ e^{k_{24}x_2} - ik_{23}a_3^- e^{-ik_{23}x_2} - k_{24}a_4^- e^{-k_{24}x_2},$$

$$\varphi_{z3} = -ik_{33}g_3^+ e^{-ik_{33}x_3} - k_{34}g_4^+ e^{-k_{34}x_3} \tag{12.39b}$$

and the bending slopes by the Timoshenko bending vibration theory are

$$\varphi_{y1} = iP_{1y}b_1^+ e^{ik_{11}x_1} + N_{1y}b_2^+ e^{k_{12}x_1},$$

$$\varphi_{y2} = iP_{2y}a_1^+ e^{ik_{21}x_2} + N_{2y}a_2^+ e^{k_{22}x_2} - iP_{2y}a_1^- e^{-ik_{21}x_2} - N_{2y}a_2^- e^{-k_{22}x_2},$$

$$\varphi_{y3} = -iP_{3y}g_1^+ e^{-ik_{31}x_3} - N_{3y}g_2^+ e^{-k_{32}x_3}$$

$$\varphi_{z1} = iP_{1z}b_3^+ e^{ik_{13}x_1} + N_{1z}b_4^+ e^{k_{14}x_1},$$

$$\varphi_{z2} = iP_{2z}a_3^+ e^{ik_{23}x_2} + N_{2z}a_4^+ e^{k_{24}x_2} - iP_{2z}a_3^- e^{-ik_{23}x_2} - N_{2z}a_4^- e^{-k_{24}x_2},$$

$$\varphi_{z3} = -iP_{3z}g_3^+ e^{-ik_{33}x_3} - N_{3z}g_4^+ e^{-k_{34}x_3} \tag{12.39c}$$

Substituting Eqs. (12.39a) and (12.39b) into a set of continuity equations from Eq. (12.29), and choosing the origin at the joint, a set of six scalar equations of continuity is obtained based on the Euler–Bernoulli bending vibration theory. Similarly, substituting Eqs. (12.39a) and (12.39c) into a set of continuity equations from Eq. (12.29), and choosing the origin at the joint, a set of six scalar equations of continuity is obtained based on the Timoshenko bending vibration theory. This set of six scalar equations of continuity, either by the Euler–Bernoulli or Timoshenko bending vibration theory, can be placed in a vector equation, in terms of wave components $\mathbf{a}^+$, $\mathbf{a}^-$, and $\mathbf{b}^+$, which are defined in Eq. (12.6b). A vector equation from this group is named $\mathbf{Y\_Beam2}_{C21}$, because it is a vector equation on the continuity between Beam 2 and Beam 1 at a spatial Y joint corresponding to incident waves from Beam 2.

Substituting Eqs. (12.39a) and (12.39b) into a set of continuity equations from Eq. (12.31), and choosing the origin at the joint, a set of six scalar equations of continuity is obtained based on the Euler–Bernoulli bending vibration theory. Similarly, substituting Eqs. (12.39a) and (12.39c) into a set of continuity equations from Eq. (12.31), and choosing the origin at the joint, a set of six scalar equations of continuity is obtained based on the Timoshenko bending vibration theory. This set of six scalar equations of continuity, either by the Euler–Bernoulli or Timoshenko bending vibration theory, can be placed in a vector equation, in terms of wave components $\mathbf{a}^+$, $\mathbf{a}^-$, and $\mathbf{g}^+$, which are defined in Eqs. (12.6b) and (12.36). A vector equation from this group is named $\mathbf{Y\_Beam2}_{C23}$, because it is a vector equation on the continuity between Beam 2 and Beam 3 at a spatial Y joint corresponding to incident waves from Beam 2.

Eliminating the joint related parameters from Eq. (12.26) using Eqs. (12.32a) and (12.32b), substituting Eqs. (12.39a) and (12.39b) into Eqs. (12.7a), (12.7c), and (12.26) that is free of joint parameters, and choosing the origin at the joint, a set of six scalar equations of equilibrium is obtained based on the Euler–Bernoulli bending vibration theory. Similarly, eliminating the joint related parameters from Eq. (12.26) using Eqs. (12.32a) and (12.32b), substituting Eqs. (12.39a) and (12.39c) into Eqs. (12.7b), (12.7c), and (12.26) that is free of joint parameters, and choosing the origin at the joint, a set of six scalar equations of equilibrium based on the Timoshenko bending vibration theory. These six scalar equations of equilibrium can be placed in a vector equation in terms of $\mathbf{a}^+$, $\mathbf{a}^-$, $\mathbf{b}^+$, and $\mathbf{g}^+$, which are defined in Eqs. (12.6b) and (12.36). A vector equation from this group is named $\mathbf{Y\_Beam2}_E$, because it is a vector equation on equilibrium at a spatial Y joint corresponding to incident waves from Beam 2.

The reflection and transmission matrices, $\mathbf{r}_{22}$, $\mathbf{t}_{21}$, and $\mathbf{t}_{23}$, are found from Eq. (12.35) and the three vector equations of continuity and equilibrium, namely, $\mathbf{Y\_Beam2}_{C21}$, $\mathbf{Y\_Beam2}_{C23}$, and $\mathbf{Y\_Beam2}_E$.

12.5.2 Wave Reflection and Transmission at a K-shaped Spatial Joint

Figures 12.9a-c show the free body diagrams of the spatial K joint of Figure 12.6, from which the equations of motion of the K joint are obtained by applying Newton's second law for translational and rotational motion,

$$V_{z2} + V_{y3} - F_1 - V_{y4} = m_J \ddot{u}_J$$

$$V_{y2} - F_3 - V_{y1} + F_4 = m_J \ddot{y}_J$$

$$-F_2 - V_{z1} + V_{z3} - V_{z4} = m_J \ddot{z}_J$$

$$-T_1 - M_{z3} + M_{y2} + M_{z4} + V_{y2}\frac{h_3}{2} - V_{z3}\frac{h_2}{2} - V_{z4}\frac{h_2}{2} = J_{xJ}\ddot{\theta}_{xJ}$$

$$-T_3 - M_{z2} + M_{z1} + T_4 - V_{z2}\frac{b_1}{2} - V_{z1}\frac{b_2}{2} = J_{yJ}\ddot{\theta}_{yJ}$$

$$-T_2 - M_{y1} + M_{y3} - M_{y4} + V_{y1}\frac{b_3}{2} + V_{y3}\frac{h_1}{2} + V_{y4}\frac{h_1}{2} = J_{zJ}\ddot{\theta}_{zJ} \qquad (12.40)$$

where the parameters are defined in Eq. (12.26). Subscripts 1, 2, 3, 4, and J denote parameters related to Beam 1, Beam 2, Beam 3, Beam 4, and the joint, respectively. It is assumed that the vertical beam elements of Figure 12.4 are of the same cross section, that is, $h_3 = h_4$ and $b_3 = b_4$. The shear force, bending moment, axial force, and torque are related to the corresponding deflections in each beam element by Eq. (12.7 a-c). Let us call Eq. (12.40) the equilibrium equations, because by the principle of d'Alembert, these equations represent dynamic equilibrium.

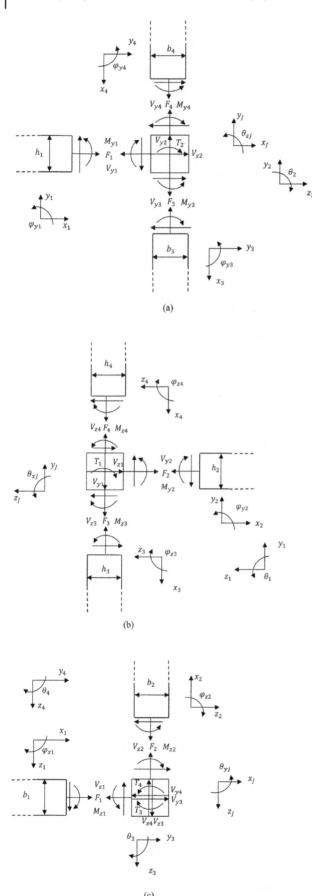

(a)

(b)

(c)

Figure 12.9 Free body diagram of the K joint of Figure 12.6: (a) front view, (b) left view, and (c) top view. (Courtesy: (Mei and Sha 2015)).

From Figures 12.9a–c, the continuity conditions at the spatial K joint are

$$u_1 = x_J, \ u_2 = -z_J, \ u_3 = -y_J, \ u_4 = -y_J$$

$$y_1 = y_J - \theta_{zJ}\frac{b_3}{2}, \ y_3 = x_J + \theta_{zJ}\frac{h_1}{2},$$

$$y_4 = x_J - \theta_{zJ}\frac{h_1}{2}, \ y_2 = y_J + \theta_{xJ}\frac{h_3}{2}$$

$$z_1 = z_J + \theta_{yJ}\frac{b_2}{2}, \ z_2 = x_J - \theta_{yJ}\frac{b_1}{2},$$

$$z_3 = z_J - \theta_{xJ}\frac{h_2}{2}, \ z_4 = z_J + \theta_{xJ}\frac{h_2}{2}$$

$$\theta_1 = \theta_{xJ}, \ \theta_2 = -\theta_{zJ}, \ \theta_3 = -\theta_{yJ}, \ \theta_4 = -\theta_{yJ}$$

$$\varphi_{y1} = \theta_{zJ}, \ \varphi_{y2} = \theta_{xJ}, \ \varphi_{y3} = \theta_{zJ}, \ \varphi_{y4} = \theta_{zJ}$$

$$\varphi_{z1} = -\theta_{yJ}, \ \varphi_{z2} = -\theta_{yJ},$$
$$\varphi_{z3} = -\theta_{xJ}, \ \varphi_{z4} = -\theta_{xJ} \tag{12.41}$$

where y and z denote the transverse deflections along the y- and z-axis direction, respectively. Numerical subscripts 1, 2, 3, and 4 in Eq. (12.41) denote parameters relating to the correspondingly numbered beam at the joint. Letter subscripts y and z in φ denote the bending slope in the x–y and x–z plane, respectively.

To find the reflection and transmission relationships of vibration waves in beam elements connected at a joint, the incident, reflected, and transmitted waves must be related directly. As a result, the rigid body joint related parameters $\{x_J, y_J, z_J, \theta_{xJ}, \theta_{yJ}, \theta_{zJ}\}$ need to be eliminated from both Eqs. (12.40) and (12.41).

Eliminating the rigid body joint related parameters $\{x_J, y_J, z_J, \theta_{xJ}, \theta_{yJ}, \theta_{zJ}\}$ from the continuity equations of Equation (12.41),

$$u_3 = u_4$$

$$\theta_1 = \varphi_{y2} = -\varphi_{z3} = -\varphi_{z4}$$

$$-\theta_2 = \varphi_{y1} = \varphi_{y3} = \varphi_{y4}$$

$$\varphi_{z1} = \varphi_{z2} = \theta_3 = \theta_4$$

$$y_1 = -u_{3,4} + \theta_2\frac{b_3}{2}, \text{ or equivalently}$$

$$y_1 = -u_{3,4} - \varphi_{y1,y3,y4}\frac{b_3}{2};$$

$$y_2 = -u_{3,4} + \theta_1\frac{h_3}{2}, \text{ or equivalently}$$

$$y_2 = -u_{3,4} + \varphi_{y2}\frac{h_3}{2}, \text{ or equivalently}$$

$$y_2 = -u_{3,4} - \varphi_{z3,z4}\frac{h_3}{2};$$

$$y_3 = u_1 + \varphi_{y1,y3,y4} \frac{h_1}{2}, \text{ or equivalently } y_3 = u_1 - \theta_2 \frac{h_1}{2};$$

$$y_4 = u_1 - \varphi_{y1,y3,y4} \frac{h_1}{2}, \text{ or equivalently } y_4 = u_1 + \theta_2 \frac{h_1}{2};$$

$$z_1 = -u_2 - \theta_{3,4} \frac{b_2}{2}, \text{ or equivalently } z_1 = -u_2 - \varphi_{z1,z2} \frac{b_2}{2};$$

$$z_2 = u_1 + \varphi_{z1,z2} \frac{b_1}{2}, \text{ or equivalently } z_2 = u_1 + \theta_{3,4} \frac{b_1}{2};$$

$$z_3 = -u_2 - \theta_1 \frac{h_2}{2}, \text{ or equivalently } z_3 = -u_2 - \varphi_{y2} \frac{h_2}{2}, \text{ or equivalently } z_3 = -u_2 + \varphi_{z3,z4} \frac{h_2}{2};$$

$$z_4 = -u_2 + \theta_1 \frac{h_2}{2}, \text{ or equivalently } z_4 = -u_2 + \varphi_{y2} \frac{h_2}{2}, \text{ or equivalently } z_4 = -u_2 - \varphi_{z3,z4} \frac{h_2}{2} \tag{12.42}$$

where $u_{3,4} = u_3 = u_4$, $\varphi_{y1,y3,y4} = \varphi_{y1} = \varphi_{y3} = \varphi_{y4}$, $\varphi_{z3,z4} = \varphi_{z3} = \varphi_{z4}$, $\varphi_{z1,z2} = \varphi_{z1} = \varphi_{z2}$, and $\theta_{3,4} = \theta_3 = \theta_4$.

The continuity equations of Eqs. (12.42) need to be sorted into six groups. Each group consists of six scalar equations relating waves in two beams at a time.

The group of continuity equations relating Beam 1 and Beam 2, by sorting Eq. (12.42), is

$$\varphi_{z1} = \varphi_{z2}; \; \theta_1 = \varphi_{y2}; \; \theta_2 = -\varphi_{y1};$$

$$y_2 - y_1 = \theta_1 \frac{h_3}{2} - \theta_2 \frac{b_3}{2}, \text{ or equivalently } y_2 - y_1 = \varphi_{y2} \frac{h_3}{2} + \varphi_{y1} \frac{b_3}{2},$$

$$y_2 - y_1 = \varphi_{y2} \frac{h_3}{2} - \theta_2 \frac{b_3}{2}, \text{ or equivalently } y_2 - y_1 = \theta_1 \frac{h_3}{2} + \varphi_{y1} \frac{b_3}{2};$$

$$z_1 = -u_2 - \varphi_{z1,z2} \frac{b_2}{2};$$

$$z_2 = u_1 + \varphi_{z1,z2} \frac{b_1}{2} \tag{12.43}$$

where $\varphi_{z1,z2} = \varphi_{z1} = \varphi_{z2}$.

The group of continuity equations relating Beam 1 and Beam 3, by sorting Eq. (12.42), is

$$\varphi_{y3} = \varphi_{y1}; \; \theta_1 = -\varphi_{z3}; \; \theta_3 = \varphi_{z1};$$

$$y_1 = -u_3 - \varphi_{y1,y3} \frac{b_3}{2};$$

$$y_3 = u_1 + \varphi_{y1,y3} \frac{h_1}{2};$$

$$z_3 - z_1 = -\theta_1 \frac{h_2}{2} + \theta_3 \frac{b_2}{2}, \text{ or equivalently } z_3 - z_1 = \varphi_{z3} \frac{h_2}{2} + \varphi_{z1} \frac{b_2}{2},$$

$$\text{or equivalently } z_3 - z_1 = -\theta_1 \frac{h_2}{2} + \varphi_{z1} \frac{b_2}{2}, \text{ or equivalently } z_3 - z_1 = \varphi_{z3} \frac{h_2}{2} + \theta_3 \frac{b_2}{2} \tag{12.44}$$

where $\varphi_{y1,y3} = \varphi_{y1} = \varphi_{y3}$.

The group of continuity equations relating Beam 1 and Beam 4, by sorting Eq. (12.42), is

$$\varphi_{y4} = \varphi_{y1}; \quad \theta_1 = -\varphi_{z4}; \quad \theta_4 = \varphi_{z1};$$

$$y_1 = -u_4 - \varphi_{y1,y4}\frac{b_3}{2};$$

$$y_4 = u_1 - \varphi_{y1,y4}\frac{h_1}{2};$$

$$z_4 - z_1 = \theta_1\frac{h_2}{2} + \theta_4\frac{b_2}{2}, \text{ or equivalently } z_4 - z_1 = -\varphi_{z4}\frac{h_2}{2} + \varphi_{z1}\frac{b_2}{2},$$

$$\text{or equivalently } z_4 - z_1 = \theta_1\frac{h_2}{2} + \varphi_{z1}\frac{b_2}{2}, \text{ or equivalently } z_4 - z_1 = -\varphi_{z4}\frac{h_2}{2} + \theta_4\frac{b_2}{2} \tag{12.45}$$

where $\varphi_{y1,y4} = \varphi_{y1} = \varphi_{y4}$.

The group of continuity equations relating Beam 2 and Beam 3, by sorting Eq. (12.42), is

$$\theta_2 = -\varphi_{y3}; \quad \theta_3 = \varphi_{z2}; \quad \varphi_{y2} = -\varphi_{z3};$$

$$y_2 = -u_3 + \varphi_{y2}\frac{h_3}{2}, \text{ or equivalently } y_2 = -u_3 - \varphi_{z3}\frac{h_3}{2};$$

$$z_3 = -u_2 - \varphi_{y2}\frac{h_2}{2}, \text{ or equivalently } z_3 = -u_2 + \varphi_{z3}\frac{h_2}{2};$$

$$y_3 - z_2 = \varphi_{y3}\frac{h_1}{2} - \varphi_{z2}\frac{b_1}{2}, \text{ or equivalently } y_3 - z_2 = -\theta_2\frac{h_1}{2} - \theta_3\frac{b_1}{2},$$

$$\text{or equivalently } y_3 - z_2 = \varphi_{y3}\frac{h_1}{2} - \theta_3\frac{b_1}{2}, \text{ or equivalently } y_3 - z_2 = -\theta_2\frac{h_1}{2} - \varphi_{z2}\frac{b_1}{2} \tag{12.46}$$

The group of continuity equations relating Beam 2 and Beam 4, by sorting Eq. (12.42), is

$$\theta_2 = -\varphi_{y4}; \quad \theta_4 = \varphi_{z2}; \quad \varphi_{z4} = -\varphi_{y2};$$

$$y_2 = -u_4 + \varphi_{y2}\frac{h_3}{2}, \text{ or equivalently } y_2 = -u_4 - \varphi_{z4}\frac{h_3}{2};$$

$$z_4 = -u_2 + \varphi_{y2}\frac{h_2}{2}, \text{ or equivalently } z_4 = -u_2 - \varphi_{z4}\frac{h_2}{2};$$

$$y_4 - z_2 = -\varphi_{y4}\frac{h_1}{2} - \varphi_{z2}\frac{b_1}{2}, \text{ or equivalently } y_4 - z_2 = \theta_2\frac{h_1}{2} - \theta_4\frac{b_1}{2},$$

$$\text{or equivalently } y_4 - z_2 = -\varphi_{y4}\frac{h_1}{2} - \theta_4\frac{b_1}{2}, \text{ or equivalently } y_4 - z_2 = \theta_2\frac{h_1}{2} - \varphi_{z2}\frac{b_1}{2} \tag{12.47}$$

The group of continuity equations relating Beam 3 and Beam 4, by sorting Eq. (12.42), is

$$u_3 = u_4; \quad \varphi_{z3} = \varphi_{z4}; \quad \varphi_{y3} = \varphi_{y4}; \quad \theta_3 = \theta_4;$$

$$y_4 - y_3 = -\varphi_{y3,y4}h_1;$$

$$z_4 - z_3 = -\varphi_{z3,z4}h_2 \tag{12.48}$$

where $\varphi_{y3,y4} = \varphi_{y3} = \varphi_{y4}$ and $\varphi_{z3,z4} = \varphi_{z3} = \varphi_{z4}$.

From Eqs. (12.43) to (12.48), the groups of continuity equations relating two of the beam elements joined at the spatial K joint, it is clear that there are multiple sets of equivalent continuity equations in each group. Each set consists of six scalar equations of continuity.

From the continuity equations of Eq. (12.41), the rigid body joint related parameters $\{x_J, y_J, z_J, \theta_{xJ}, \theta_{yJ}, \theta_{zJ}\}$ can be expressed in terms of the beam elements related parameters as follows

$$\theta_{xJ} = \theta_1 = \varphi_{y2} = -\varphi_{z3} = -\varphi_{z4}$$

$$\theta_{yJ} = -\varphi_{z1} = -\varphi_{z2} = -\theta_3 = -\theta_4$$

$$\theta_{zJ} = \varphi_{y1} = \varphi_{y3} = \varphi_{y4} = -\theta_2 \tag{12.49a}$$

$$x_J = u_1 = y_3 - \theta_{zJ}\frac{h_1}{2} = y_4 + \theta_{zJ}\frac{h_1}{2} = z_2 + \theta_{yJ}\frac{b_1}{2}$$

$$y_J = -u_3 = -u_4 = y_1 + \theta_{zJ}\frac{b_3}{2} = y_2 - \theta_{xJ}\frac{h_3}{2}$$

$$z_J = -u_2 = z_1 - \theta_{yJ}\frac{b_2}{2} = z_3 + \theta_{xJ}\frac{h_2}{2} = z_4 - \theta_{xJ}\frac{h_2}{2} \tag{12.49b}$$

where θ_{xJ}, θ_{yJ}, and θ_{zJ} in Eq. (12.49b) can be replaced by the equivalent beam elements related parameters described in Eq. (12.49a).

From Eqs. (12.49a) and (12.49b), Eq. (12.40) can be expressed in numerous ways. In other words, there are numerous sets of equivalent equilibrium equations. Each set consists of six scalar equations of equilibrium, which are obtained from Eqs. (12.49a), (12.49b), and (12.40) by eliminating the joint related parameters.

The reflection and transmission relationships corresponding to waves incident upon a joint from a different beam element are in general different. Figures 12.10 a-d show the reflection and transmission of waves incident upon a spatial K joint from Beam 1, Beam 2, Beam 3, and Beam 4, respectively.

For waves incident from Beam 1, as shown in Figure 12.10a, the reflected waves $\mathbf{a}^-$ on Beam 1, the transmitted waves $\mathbf{b}^+$ to Beam 2, the transmitted waves $\mathbf{g}^+$ to Beam 3, and the transmitted waves $\mathbf{p}^+$ to Beam 4 are related to the incident waves $\mathbf{a}^+$ by reflection matrix $\mathbf{r}_{11}$, transmission matrix $\mathbf{t}_{12}$, transmission matrix $\mathbf{t}_{13}$, and transmission matrix $\mathbf{t}_{14}$ as follows

$$\mathbf{a}^- = \mathbf{r}_{11}\mathbf{a}^+, \ \mathbf{b}^+ = \mathbf{t}_{12}\mathbf{a}^+, \ \mathbf{g}^+ = \mathbf{t}_{13}\mathbf{a}^+, \ \mathbf{p}^+ = \mathbf{t}_{14}\mathbf{a}^+ \tag{12.50}$$

For waves incident from Beam 2, as shown in Figure 12.10b, the reflected waves $\mathbf{a}^-$ on Beam 2, the transmitted waves $\mathbf{b}^+$ to Beam 1, the transmitted waves $\mathbf{g}^+$ to Beam 3, and the transmitted waves $\mathbf{p}^+$ to Beam 4 are related to the incident waves $\mathbf{a}^+$ by reflection matrix $\mathbf{r}_{22}$, transmission matrix $\mathbf{t}_{21}$, transmission matrix $\mathbf{t}_{23}$, and transmission matrix $\mathbf{t}_{24}$ as follows

$$\mathbf{a}^- = \mathbf{r}_{22}\mathbf{a}^+, \ \mathbf{b}^+ = \mathbf{t}_{21}\mathbf{a}^+, \ \mathbf{g}^+ = \mathbf{t}_{23}\mathbf{a}^+, \ \mathbf{p}^+ = \mathbf{t}_{24}\mathbf{a}^+ \tag{12.51}$$

For waves incident from Beam 3, as shown in Figure 12.10c, the reflected waves $\mathbf{a}^-$ on Beam 3, the transmitted waves $\mathbf{b}^+$ to Beam 2, the transmitted waves $\mathbf{g}^+$ to Beam 1, and the transmitted waves $\mathbf{p}^+$ to Beam 4 are related to the incident waves $\mathbf{a}^+$ by reflection matrix $\mathbf{r}_{33}$, transmission matrix $\mathbf{t}_{32}$, transmission matrix $\mathbf{t}_{31}$, and transmission matrix $\mathbf{t}_{34}$. as follows

$$\mathbf{a}^- = \mathbf{r}_{33}\mathbf{a}^+, \ \mathbf{b}^+ = \mathbf{t}_{32}\mathbf{a}^+, \ \mathbf{g}^+ = \mathbf{t}_{31}\mathbf{a}^+, \ \mathbf{p}^+ = \mathbf{t}_{34}\mathbf{a}^+ \tag{12.52}$$

For waves incident from Beam 4, as shown in Figure 12.10d, the reflected waves $\mathbf{a}^-$ on Beam 4, the transmitted waves $\mathbf{g}^+$ to Beam 1, the transmitted waves $\mathbf{b}^+$ to Beam 2, and the transmitted waves $\mathbf{p}^+$ to Beam 3 are related to the incident waves $\mathbf{a}^+$ by reflection matrix $\mathbf{r}_{44}$, transmission matrix $\mathbf{t}_{41}$, transmission matrix $\mathbf{t}_{42}$, and transmission matrix $\mathbf{t}_{43}$ as follows

$$\mathbf{a}^- = \mathbf{r}_{44}\mathbf{a}^+, \ \mathbf{g}^+ = \mathbf{t}_{41}\mathbf{a}^+, \ \mathbf{b}^+ = \mathbf{t}_{42}\mathbf{a}^+, \ \mathbf{p}^+ = \mathbf{t}_{43}\mathbf{a}^+ \tag{12.53}$$

In Eqs. (12.50) to (12.53), wave vectors $\mathbf{a}^{\pm}$, $\mathbf{b}^{+}$, $\mathbf{g}^{+}$, and $\mathbf{p}^{+}$ are defined as

$$\mathbf{a}^{\pm} = \begin{bmatrix} a_1^{\pm} \\ a_2^{\pm} \\ a_3^{\pm} \\ a_4^{\pm} \\ a_5^{\pm} \\ a_6^{\pm} \end{bmatrix}, \quad \mathbf{b}^{+} = \begin{bmatrix} b_1^{+} \\ b_2^{+} \\ b_3^{+} \\ b_4^{+} \\ b_5^{+} \\ b_6^{+} \end{bmatrix}, \quad \mathbf{g}^{+} = \begin{bmatrix} g_1^{+} \\ g_2^{+} \\ g_3^{+} \\ g_4^{+} \\ g_5^{+} \\ g_6^{+} \end{bmatrix}, \quad \mathbf{p}^{+} = \begin{bmatrix} p_1^{+} \\ p_2^{+} \\ p_3^{+} \\ p_4^{+} \\ p_5^{+} \\ p_6^{+} \end{bmatrix} \tag{12.54}$$

Although different symbols are adopted for longitudinal and torsional wave components in the wave vectors, the wave component sequence in each wave vector remains the same: the first four elements with subscripts 1 to 4 are for bending waves in the x–y and x–z plane, and the last two elements with subscripts 5 and 6 are for longitudinal and torsional waves.

First, solve for $\mathbf{r}_{11}$, $\mathbf{t}_{12}$, $\mathbf{t}_{13}$, and $\mathbf{t}_{14}$, that is, the reflection and transmission matrices at the spatial K joint of Figure 12.10a, corresponding to incident waves from Beam 1.

Following the local coordinate systems of Figure 12.6, the longitudinal, torsional, and bending deflections of the beam elements are

$$u_1 = a_5^{+} e^{-ik_{15}x_1} + a_5^{-} e^{ik_{15}x_1},$$

$$u_2 = b_5^{+} e^{-ik_{25}x_2},$$

$$u_3 = g_5^{+} e^{-ik_{35}x_3},$$

$$u_4 = p_5^{+} e^{ik_{45}x_4}$$

$$\theta_1 = a_6^{+} e^{-ik_{16}x_1} + a_6^{-} e^{ik_{16}x_1},$$

$$\theta_2 = b_6^{+} e^{-ik_{26}x_2},$$

$$\theta_3 = g_6^{+} e^{-ik_{36}x_3},$$

$$\theta_4 = p_6^{+} e^{ik_{46}x_4}$$

$$y_1 = a_1^{+} e^{-ik_{11}x_1} + a_2^{+} e^{-k_{12}x_1} + a_1^{-} e^{ik_{11}x_1} + a_2^{-} e^{k_{12}x_1},$$

$$y_2 = b_1^{+} e^{-ik_{21}x_2} + b_2^{+} e^{-k_{22}x_2},$$

$$y_3 = g_1^{+} e^{-ik_{31}x_3} + g_2^{+} e^{-k_{32}x_3},$$

$$y_4 = p_1^{+} e^{ik_{41}x_4} + p_2^{+} e^{k_{42}x_4}$$

$$z_1 = a_3^{+} e^{-ik_{13}x_1} + a_4^{+} e^{-k_{14}x_1} + a_3^{-} e^{ik_{13}x_1} + a_4^{-} e^{k_{14}x_1},$$

$$z_2 = b_3^{+} e^{-ik_{23}x_2} + b_4^{+} e^{-k_{24}x_2},$$

$$z_3 = g_3^{+} e^{-ik_{33}x_3} + g_4^{+} e^{-k_{34}x_3},$$

$$z_4 = p_3^{+} e^{ik_{43}x_4} + p_4^{+} e^{k_{44}x_4} \tag{12.55a}$$

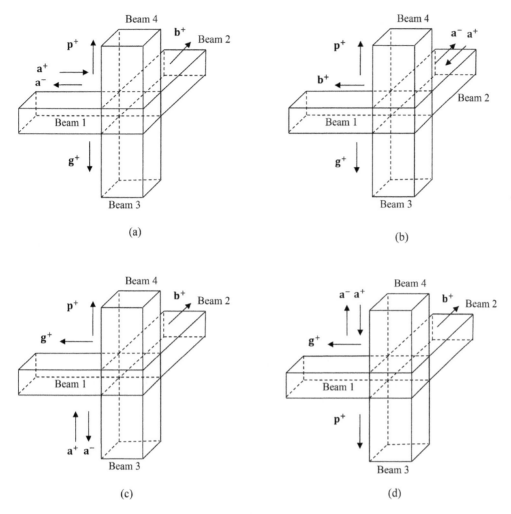

Figure 12.10 Wave transmission and reflection at the K joint of Figure 12.6 with waves (a) incident from Beam 1, (b) incident from Beam 2, (c) incident from Beam 3, and (d) incident from Beam 4. (Courtesy: (Mei and Sha 2015)).

and the bending slopes by the Euler–Bernoulli bending vibration theory are

$$\varphi_{y1} = -ik_{11}a_1^+ e^{-ik_{11}x_1} - k_{12}a_2^+ e^{-k_{12}x_1} + ik_{11}a_1^- e^{ik_{11}x_1} + k_{12}a_2^- e^{k_{12}x_1},$$

$$\varphi_{y2} = -ik_{21}b_1^+ e^{-ik_{21}x_2} - k_{22}b_2^+ e^{-k_{22}x_2},$$

$$\varphi_{y3} = -ik_{31}g_1^+ e^{-ik_{31}x_3} - k_{32}g_2^+ e^{-k_{32}x_3},$$

$$\varphi_{y4} = ik_{41}p_1^+ e^{ik_{41}x_4} + k_{42}p_2^+ e^{k_{42}x_4}$$

$$\varphi_{z1} = -ik_{13}a_3^+ e^{-ik_{13}x_1} - k_{14}a_4^+ e^{-k_{14}x_1} + ik_{13}a_3^- e^{ik_{13}x_1} + k_{14}a_4^- e^{k_{14}x_1},$$

$$\varphi_{z2} = -ik_{23}b_3^+ e^{-ik_{23}x_2} - k_{24}b_4^+ e^{-k_{24}x_2},$$

$$\varphi_{z3} = -ik_{33}g_3^+ e^{-ik_{33}x_3} - k_{34}g_4^+ e^{-k_{34}x_3},$$

$$\varphi_{z4} = ik_{43}p_3^+ e^{ik_{43}x_4} + k_{44}p_4^+ e^{k_{44}x_4} \tag{12.55b}$$

and the bending slopes by the Timoshenko bending vibration theory are

$$\varphi_{y1} = -iP_{1y}a_1^+ e^{-ik_{11}x_1} - N_{1y}a_2^+ e^{-k_{12}x_1} + iP_{1y}a_1^- e^{ik_{11}x_1} + N_{1y}a_2^- e^{k_{12}x_1},$$

$$\varphi_{y2} = -iP_{2y}b_1^+ e^{-ik_{21}x_2} - N_{2y}b_2^+ e^{-k_{22}x_2},$$

$$\varphi_{y3} = -iP_{3y}g_1^+ e^{-ik_{31}x_3} - N_{3y}g_2^+ e^{-k_{32}x_3},$$

$$\varphi_{y4} = iP_{4y}p_1^- e^{ik_{41}x_4} + N_{4y}p_2^- e^{k_{42}x_4}$$

$$\varphi_{z1} = -iP_{1z}a_3^+ e^{-ik_{13}x_1} - N_{1z}a_4^+ e^{-k_{14}x_1} + iP_{1z}a_3^- e^{ik_{13}x_1} + N_{1z}a_4^- e^{k_{14}x_1},$$

$$\varphi_{z2} = -iP_{2z}b_3^+ e^{-ik_{23}x_2} - N_{2z}b_4^+ e^{-k_{24}x_2},$$

$$\varphi_{z3} = -iP_{3z}g_3^+ e^{-ik_{33}x_3} - N_{3z}g_4^+ e^{-k_{34}x_3},$$

$$\varphi_{z4} = iP_{4z}p_3^- e^{ik_{43}x_4} + N_{4z}p_4^- e^{k_{44}x_4} \tag{12.55c}$$

where x_n denotes the position along the beam axis of the nth beam ($n = 1,2,3,4$), k_{nj} is the magnitude of the jth wavenumber (among them subscripts 1 and 2 are related to bending waves in the x–y plane, 3 and 4 are related to bending waves in the x–z plane, and 5 and 6 for longitudinal and torsional waves, respectively) of the nth beam member. P_{ny} and N_{ny} are the coefficients relating bending slope φ_y to bending deflection y; P_{nz} and N_{nz} are the coefficients relating bending slope φ_z to bending deflection z of the nth beam, these coefficients are defined in Eq. (12.5).

Substituting Eqs. (12.55a) and (12.55b) into a set of continuity equations from Eq. (12.43), and choosing the origin at the joint, a set of six scalar equations of continuity is obtained based on the Euler–Bernoulli bending vibration theory. Similarly, substituting Eqs. (12.55a) and (12.55c) into a set of continuity equations from Eq. (12.43), and choosing the origin at the joint, a set of six scalar equations of continuity is obtained based on the Timoshenko bending vibration theory. This set of six scalar equations of continuity, either by the Euler–Bernoulli or Timoshenko bending vibration theory, can be placed in a vector equation, in terms of wave components $\mathbf{a}^+$, $\mathbf{a}^-$, and $\mathbf{b}^+$, which are defined in Eq. (12.54). A vector equation from this group is named **K_Beam1**$_{C12}$, because it is a vector equation on the continuity between Beam 1 and Beam 2 at a spatial K joint corresponding to incident waves from Beam 1.

Substituting Eqs. (12.55a) and (12.55b) into a set of continuity equations from Eq. (12.44), and choosing the origin at the joint, a set of six scalar equations of continuity is obtained based on the Euler–Bernoulli bending vibration theory. Similarly, substituting Eqs. (12.55a) and (12.55c) into a set of continuity equations from Eq. (12.44), and choosing the origin at the joint, a set of six scalar equations of continuity is obtained based on the Timoshenko bending vibration theory. This set of six scalar equations of continuity, either by the Euler–Bernoulli or Timoshenko bending vibration theory, can be placed in a vector equation, in terms of wave components $\mathbf{a}^+$, $\mathbf{a}^-$, and $\mathbf{g}^+$, which are defined in Eq. (12.54). A vector equation from this group is named **K_Beam1**$_{C13}$, because it is a vector equation on the continuity between Beam 1 and Beam 3 at a spatial K joint corresponding to incident waves from Beam 1.

Substituting Eqs. (12.55a) and (12.55b) into a set of continuity equations from Eq. (12.45), and choosing the origin at the joint, a set of six scalar equations of continuity is obtained based on the Euler–Bernoulli bending vibration theory. Similarly, substituting Eqs. (12.55a) and (12.55c) into a set of continuity equations from Eq. (12.45), and choosing the origin at the joint, a set of six scalar equations of continuity is obtained based on the Timoshenko bending vibration theory. This set of six scalar equations of continuity, either by the Euler–Bernoulli or Timoshenko bending vibration theory, can be placed in a vector equation, in terms of wave components $\mathbf{a}^+$, $\mathbf{a}^-$, and $\mathbf{p}^+$, which are defined in Eq. (12.54). A vector equation from this group is named **K_Beam1**$_{C14}$, because it is a vector equation on the continuity between Beam 1 and Beam 4 at a spatial K joint corresponding to incident waves from Beam 1.

Eliminating the joint related parameters from Eq. (12.40) using Eqs. (12.49a) and (12.49b), substituting Eqs. (12.55a) and (12.55b) into Eqs. (12.7a), (12.7c), and (12.40) that is free of joint parameters, and choosing the origin at the joint, a set of six scalar equations of equilibrium is obtained based on the Euler–Bernoulli bending vibration theory. Similarly, eliminating the joint related parameters from Eq. (12.40) using Eqs. (12.49a) and (12.49b), substituting Eqs. (12.55a) and (12.55c) into Eqs. (12.7b), (12.7c), and (12.40) that is free of joint parameters, and choosing the origin at the joint, a set of six scalar equations of equilibrium based on the Timoshenko bending vibration theory. These six scalar equations of equilibrium can be placed in a vector equation in terms of $\mathbf{a}^+$, $\mathbf{a}^-$, $\mathbf{b}^+$, $\mathbf{g}^+$, and $\mathbf{p}^+$, which are defined in Eq. (12.54). A vector equation from this group is named $\mathbf{K\_Beam1}_E$, because it is a vector equation on equilibrium at a spatial K joint corresponding to incident waves from Beam 1.

The reflection and transmission matrices, $\mathbf{r}_{11}$, $\mathbf{t}_{12}$, $\mathbf{t}_{13}$, and $\mathbf{t}_{14}$, are found from Equation (12.50) and the four vector equations of continuity and equilibrium, namely, $\mathbf{K\_Beam1}_{C12}$, $\mathbf{K\_Beam1}_{C13}$, $\mathbf{K\_Beam1}_{C14}$, and $\mathbf{K\_Beam1}_E$.

Next, solve for $\mathbf{r}_{22}$, $\mathbf{t}_{21}$, $\mathbf{t}_{23}$, and $\mathbf{t}_{24}$, that is, the reflection and transmission matrices at the K joint of Figure 12.10b, corresponding to incident waves from Beam 2.

Following the local coordinate systems of Figure 12.6, the longitudinal, torsional, and bending deflections of the beam elements are

$$u_1 = b_5^+ e^{ik_{15}x_1},$$

$$u_2 = a_5^+ e^{ik_{25}x_2} + a_5^- e^{-ik_{25}x_2},$$

$$u_3 = g_5^+ e^{-ik_{35}x_3},$$

$$u_4 = p_5^+ e^{ik_{45}x_4}$$

$$\theta_1 = b_6^+ e^{ik_{16}x_1},$$

$$\theta_2 = a_6^+ e^{ik_{26}x_2} + a_6^- e^{-ik_{26}x_2},$$

$$\theta_3 = g_6^+ e^{-ik_{36}x_3},$$

$$\theta_4 = p_6^+ e^{ik_{46}x_4}$$

$$y_1 = b_1^+ e^{ik_{11}x_1} + b_2^+ e^{k_{12}x_1},$$

$$y_2 = a_1^+ e^{ik_{21}x_2} + a_2^+ e^{k_{22}x_2} + a_1^- e^{-ik_{21}x_2} + a_2^- e^{-k_{22}x_2},$$

$$y_3 = g_1^+ e^{-ik_{31}x_3} + g_2^+ e^{-k_{32}x_3},$$

$$y_4 = p_1^+ e^{ik_{41}x_4} + p_2^+ e^{k_{42}x_4}$$

$$z_1 = b_3^+ e^{ik_{13}x_1} + b_4^+ e^{k_{14}x_1},$$

$$z_2 = a_3^+ e^{ik_{23}x_2} + a_4^+ e^{k_{24}x_2} + a_3^- e^{-ik_{23}x_2} + a_4^- e^{-k_{24}x_2},$$

$$z_3 = g_3^+ e^{-ik_{33}x_3} + g_4^+ e^{-k_{34}x_3},$$

$$z_4 = p_3^+ e^{ik_{43}x_4} + p_4^+ e^{k_{44}x_4} \tag{12.56a}$$

and the bending slopes by the Euler–Bernoulli bending vibration theory are

$$\varphi_{y1} = ik_{11}b_1^+ e^{ik_{11}x_1} + k_{12}b_2^+ e^{k_{12}x_1},$$

$$\varphi_{y2} = ik_{21}a_1^+ e^{ik_{21}x_2} + k_{22}a_2^+ e^{k_{22}x_2} - ik_{21}a_1^- e^{-ik_{21}x_2} - k_{22}a_2^- e^{-k_{22}x_2},$$

$$\varphi_{y3} = -ik_{31}g_1^+ e^{-ik_{31}x_3} - k_{32}g_2^+ e^{-k_{32}x_3},$$

$$\varphi_{y4} = ik_{41}p_1^+ e^{ik_{41}x_4} + k_{42}p_2^+ e^{k_{42}x_4}$$

$$\varphi_{z1} = ik_{13}b_3^+ e^{ik_{13}x_1} + k_{14}b_4^+ e^{k_{14}x_1},$$

$$\varphi_{z2} = ik_{23}a_3^+ e^{ik_{23}x_2} + k_{24}a_4^+ e^{k_{24}x_2} - ik_{23}a_3^- e^{-ik_{23}x_2} - k_{24}a_4^- e^{-k_{24}x_2},$$

$$\varphi_{z3} = -ik_{33}g_3^+ e^{-ik_{33}x_3} - k_{34}g_4^+ e^{-k_{34}x_3},$$

$$\varphi_{z4} = ik_{43}p_3^+ e^{ik_{43}x_4} + k_{44}p_4^+ e^{k_{44}x_4} \qquad (12.56b)$$

and the bending slopes by the Timoshenko bending vibration theory are

$$\varphi_{y1} = iP_{1y}b_1^+ e^{ik_{11}x_1} + N_{1y}b_2^+ e^{k_{12}x_1},$$

$$\varphi_{y2} = iP_{2y}a_1^+ e^{ik_{21}x_2} + N_{2y}a_2^+ e^{k_{22}x_2} - iP_{2y}a_1^- e^{-ik_{21}x_2} - N_{2y}a_2^- e^{-k_{22}x_2},$$

$$\varphi_{y3} = -iP_{3y}g_1^+ e^{-ik_{31}x_3} - N_{3y}g_2^+ e^{-k_{32}x_3},$$

$$\varphi_{y4} = iP_{4y}p_1^+ e^{ik_{41}x_4} + N_{4y}p_2^+ e^{k_{42}x_4}$$

$$\varphi_{z1} = iP_{1z}b_3^+ e^{ik_{13}x_1} + N_{1z}b_4^+ e^{k_{14}x_1},$$

$$\varphi_{z2} = iP_{2z}a_3^+ e^{ik_{23}x_2} + N_{2z}a_4^+ e^{k_{24}x_2} - iP_{2z}a_3^- e^{-ik_{23}x_2} - N_{2z}a_4^- e^{-k_{24}x_2},$$

$$\varphi_{z3} = -iP_{3z}g_3^+ e^{-ik_{33}x_3} - N_{3z}g_4^+ e^{-k_{34}x_3},$$

$$\varphi_{z4} = iP_{4z}p_3^+ e^{ik_{43}x_4} + N_{4z}p_4^+ e^{k_{44}x_4} \qquad (12.56c)$$

Substituting Eqs. (12.56a) and (12.56b) into a set of continuity equations from Eq. (12.43), and choosing the origin at the joint, a set of six scalar equations of continuity is obtained based on the Euler–Bernoulli bending vibration theory. Similarly, substituting Eqs. (12.56a) and (12.56c) into a set of continuity equations from Eq. (12.43), and choosing the origin at the joint, a set of six scalar equations of continuity is obtained based on the Timoshenko bending vibration theory. This set of six scalar equations of continuity, either by the Euler–Bernoulli or Timoshenko bending vibration theory, can be placed in a vector equation, in terms of wave components $\mathbf{a}^+$, $\mathbf{a}^-$, and $\mathbf{b}^+$, which are defined in Eq. (12.54). A vector equation from this group is named $\mathbf{K\_Beam2}_{C21}$, because it is a vector equation on the continuity between Beam 2 and Beam 1 at a spatial K joint corresponding to incident waves from Beam 2.

Substituting Eqs. (12.56a) and (12.56b) into a set of continuity equations from Eq. (12.46), and choosing the origin at the joint, a set of six scalar equations of continuity is obtained based on the Euler–Bernoulli bending vibration theory. Similarly, substituting Eqs. (12.56a) and (12.56c) into a set of continuity equations from Eq. (12.46), and choosing the origin at the joint, a set of six scalar equations of continuity is obtained based on the Timoshenko bending vibration theory. This set of six scalar equations of continuity, either by the Euler–Bernoulli or Timoshenko bending vibration theory, can be placed in a vector equation, in terms of wave components $\mathbf{a}^+$, $\mathbf{a}^-$, and $\mathbf{g}^+$, which are defined in Eq. (12.54). A vector equation from this group is named $\mathbf{K\_Beam2}_{C23}$, because it is a vector equation on the continuity between Beam 2 and Beam 3 at a spatial K joint corresponding to incident waves from Beam 2.

Substituting Eqs. (12.56a) and (12.56b) into a set of continuity equations from Eq. (12.47), and choosing the origin at the joint, a set of six scalar equations of continuity is obtained based on the Euler–Bernoulli bending vibration theory. Similarly, substituting Eqs. (12.56a) and (12.56c) into a set of continuity equations from Eq. (12.47), and choosing the origin at the joint, a set of six scalar equations of continuity is obtained based on the Timoshenko bending vibration theory. This set of six scalar equations of continuity, either by the Euler–Bernoulli or Timoshenko bending vibration theory, can be placed in a vector equation, in terms of wave components $\mathbf{a}^+$, $\mathbf{a}^-$, and $\mathbf{p}^+$, which are defined in Eq. (12.54). A vector equation from this group is named **K_Beam2**$_{C24}$, because it is a vector equation on the continuity between Beam 2 and Beam 4 at a spatial K joint corresponding to incident waves from Beam 2.

Eliminating the joint related parameters from Eq. (12.40) using Eqs. (12.49a) and (12.49b), substituting Eqs. (12.56a) and (12.56b) into Eqs. (12.7a), (12.7c), and (12.40) that is free of joint parameters, and choosing the origin at the joint, a set of six scalar equations of equilibrium is obtained based on the Euler–Bernoulli bending vibration theory. Similarly, eliminating the joint related parameters from Eq. (12.40) using Eqs. (12.49a) and (12.49b), substituting Eqs. (12.56a) and (12.56c) into Eqs. (12.7b), (12.7c), and (12.40) that is free of joint parameters, and choosing the origin at the joint, a set of six scalar equations of equilibrium based on the Timoshenko bending vibration theory. These six scalar equations of equilibrium can be placed in a vector equation in terms of $\mathbf{a}^+$, $\mathbf{a}^-$, $\mathbf{b}^+$, $\mathbf{g}^+$, and $\mathbf{p}^+$, which are defined in Eq. (12.54). A vector equation from this group is named **K_Beam2**$_E$, because it is a vector equation on equilibrium at a spatial K joint corresponding to incident waves from Beam 2.

The reflection and transmission matrices, $\mathbf{r}_{22}$, $\mathbf{t}_{21}$, $\mathbf{t}_{23}$, and $\mathbf{t}_{24}$, are found from Eq. (12.51) and the four vector equations of continuity and equilibrium, namely, **K_Beam2**$_{C21}$, **K_Beam2**$_{C23}$, **K_Beam2**$_{C24}$, and **K_Beam2**$_E$.

Then, solve for $\mathbf{r}_{33}$, $\mathbf{t}_{31}$, $\mathbf{t}_{32}$, and $\mathbf{t}_{34}$, that is, the reflection and transmission matrices at the K joint of Figure 12.10c, corresponding to incident waves from Beam 3.

Following the local coordinate systems of Figure 12.6, the longitudinal, torsional, and bending deflections of the beam elements are

$$u_1 = g_5^+ e^{ik_{15}x_1},$$

$$u_2 = b_5^+ e^{-ik_{25}x_2},$$

$$u_3 = a_5^+ e^{ik_{35}x_3} + a_5^- e^{-ik_{35}x_3},$$

$$u_4 = p_5^+ e^{ik_{45}x_4};$$

$$\theta_1 = g_6^+ e^{ik_{16}x_1},$$

$$\theta_2 = b_6^+ e^{-ik_{26}x_2},$$

$$\theta_3 = a_6^+ e^{ik_{36}x_3} + a_6^- e^{-ik_{36}x_3},$$

$$\theta_4 = p_6^+ e^{+ik_{46}x_4};$$

$$y_1 = g_1^+ e^{ik_{11}x_1} + g_2^+ e^{k_{12}x_1},$$

$$y_2 = b_1^+ e^{-ik_{21}x_2} + b_2^+ e^{-k_{22}x_2},$$

$$y_3 = a_1^+ e^{ik_{31}x_3} + a_2^+ e^{k_{32}x_3} + a_1^- e^{-ik_{31}x_3} + a_2^- e^{-k_{32}x_3},$$

$$y_4 = p_1^+ e^{ik_{41}x_4} + p_2^+ e^{k_{42}x_4}$$

$$z_1 = g_3^+ e^{ik_{13}x_1} + g_4^+ e^{k_{14}x_1},$$

$$z_2 = b_3^+ e^{-ik_{23}x_2} + b_4^+ e^{-k_{24}x_2},$$

$$z_3 = a_3^+ e^{ik_{33}x_3} + a_4^+ e^{k_{34}x_3} + a_3^- e^{-ik_{33}x_3} + a_4^- e^{-k_{34}x_3},$$

$$z_4 = p_3^+ e^{ik_{43}x_4} + p_4^+ e^{k_{44}x_4} \tag{12.57a}$$

and the bending slopes by the Euler–Bernoulli bending vibration theory are

$$\varphi_{y1} = ik_{11}g_1^+ e^{ik_{11}x_1} + k_{12}g_2^+ e^{k_{12}x_1},$$

$$\varphi_{y2} = -ik_{21}b_1^+ e^{-ik_{21}x_2} - k_{22}b_2^+ e^{-k_{22}x_2},$$

$$\varphi_{y3} = ik_{31}a_1^+ e^{ik_{31}x_3} + k_{32}a_2^+ e^{k_{32}x_3} - ik_{31}a_1^- e^{-ik_{31}x_3} - k_{32}a_2^- e^{-k_{32}x_3},$$

$$\varphi_{y4} = ik_{41}p_1^+ e^{ik_{41}x_4} + k_{42}p_2^+ e^{k_{42}x_4}$$

$$\varphi_{z1} = ik_{13}g_3^+ e^{ik_{13}x_1} + k_{14}g_4^+ e^{k_{14}x_1},$$

$$\varphi_{z2} = -ik_{23}b_3^+ e^{-ik_{23}x_2} - k_{24}b_4^+ e^{-k_{24}x_2},$$

$$\varphi_{z3} = ik_{33}a_3^+ e^{ik_{33}x_3} + k_{34}a_4^+ e^{k_{34}x_3} - ik_{33}a_3^- e^{-ik_{33}x_3} - k_{34}a_4^- e^{-k_{34}x_3},$$

$$\varphi_{z4} = ik_{43}p_3^+ e^{ik_{43}x_4} + k_{44}p_4^+ e^{k_{44}x_4} \tag{12.57b}$$

and the bending slopes by the Timoshenko bending vibration theory are

$$\varphi_{y1} = iP_{1y}g_1^+ e^{ik_{11}x_1} + N_{1y}g_2^+ e^{k_{12}x_1},$$

$$\varphi_{y2} = -iP_{2y}b_1^+ e^{-ik_{21}x_2} - N_{2y}b_2^+ e^{-k_{22}x_2},$$

$$\varphi_{y3} = iP_{3y}a_1^+ e^{ik_{31}x_3} + N_{3y}a_2^+ e^{k_{32}x_3} - iP_{3y}a_1^- e^{-ik_{31}x_3} - N_{3y}a_2^- e^{-k_{32}x_3},$$

$$\varphi_{y4} = iP_{4y}p_1^+ e^{ik_{41}x_4} + N_{4y}p_2^+ e^{k_{42}x_4}$$

$$\varphi_{z1} = iP_{1z}g_3^+ e^{ik_{13}x_1} + N_{1z}g_4^+ e^{k_{14}x_1},$$

$$\varphi_{z2} = -iP_{2z}b_3^+ e^{-ik_{23}x_2} - N_{2z}b_4^+ e^{-k_{24}x_2},$$

$$\varphi_{z3} = iP_{3z}a_3^+ e^{ik_{33}x_3} + N_{3z}a_4^+ e^{k_{34}x_3} - iP_{3z}a_3^- e^{-ik_{33}x_3} - N_{3z}a_4^- e^{-k_{34}x_3},$$

$$\varphi_{z4} = iP_{4z}p_3^+ e^{ik_{43}x_4} + N_{4z}p_4^+ e^{k_{44}x_4} \tag{12.57c}$$

Substituting Eqs. (12.57a) and (12.57b) into a set of continuity equations from Eq. (12.44), and choosing the origin at the joint, a set of six scalar equations of continuity is obtained based on the Euler–Bernoulli bending vibration theory. Similarly, substituting Eqs. (12.57a) and (12.57c) into a set of continuity equations from Eq. (12.44), and choosing the origin at the joint, a set of six scalar equations of continuity is obtained based on the Timoshenko bending vibration theory. This set of six scalar equations of continuity, either by the Euler–Bernoulli or Timoshenko bending vibration theory, can be placed in a vector equation, in terms of wave components $\mathbf{a}^+$, $\mathbf{a}^-$, and $\mathbf{g}^+$, which are defined in Eq. (12.54). A vector equation from this group is named **K_Beam3**$_{C31}$, because it is a vector equation on the continuity between Beam 3 and Beam 1 at a spatial K joint corresponding to incident waves from Beam 3.

Substituting Eqs. (12.57a) and (12.57b) into a set of continuity equations from Eq. (12.46), and choosing the origin at the joint, a set of six scalar equations of continuity is obtained based on the Euler–Bernoulli bending vibration theory. Similarly, substituting Eqs. (12.57a) and (12.57c) into a set of continuity equations from Eq. (12.46), and choosing the origin at the joint, a set of six scalar equations of continuity is obtained based on the Timoshenko bending vibration theory. This set of six scalar equations of continuity, either by the Euler–Bernoulli or Timoshenko bending vibration theory, can be placed in a vector equation, in terms of wave components $\mathbf{a}^+$, $\mathbf{a}^-$, and $\mathbf{b}^+$, which are defined in Eq. (12.54). A vector equation from this group is named $\mathbf{K\_Beam3}_{C32}$, because it is a vector equation on the continuity between Beam 3 and Beam 2 at a spatial K joint corresponding to incident waves from Beam 3.

Substituting Eqs. (12.57a) and (12.57b) into a set of continuity equations from Eq. (12.48), and choosing the origin at the joint, a set of six scalar equations of continuity is obtained based on the Euler–Bernoulli bending vibration theory. Similarly, substituting Eqs. (12.57a) and (12.57c) into a set of continuity equations from Eq. (12.48), and choosing the origin at the joint, a set of six scalar equations of continuity is obtained based on the Timoshenko bending vibration theory. This set of six scalar equations of continuity, either by the Euler–Bernoulli or Timoshenko bending vibration theory, can be placed in a vector equation, in terms of wave components $\mathbf{a}^+$, $\mathbf{a}^-$, and $\mathbf{p}^+$, which are defined in Eq. (12.54). A vector equation from this group is named $\mathbf{K\_Beam3}_{C34}$, because it is a vector equation on the continuity between Beam 3 and Beam 4 at a spatial K joint corresponding to incident waves from Beam 3.

Eliminating the joint related parameters from Eq. (12.40) using Eqs. (12.49a) and (12.49b), substituting Eqs. (12.57a) and (12.57b) into Eqs. (12.7a), (12.7c), and (12.40) that is free of joint parameters, and choosing the origin at the joint, a set of six scalar equations of equilibrium is obtained based on the Euler–Bernoulli bending vibration theory. Similarly, eliminating the joint related parameters from Eq. (12.40) using Eqs. (12.49a) and (12.49b), substituting Eqs. (12.57a) and (12.57c) into Eqs. (12.7b), (12.7c), and (12.40) that is free of joint parameters, and choosing the origin at the joint, a set of six scalar equations of equilibrium based on the Timoshenko bending vibration theory. These six scalar equations of equilibrium can be placed in a vector equation in terms of $\mathbf{a}^+$, $\mathbf{a}^-$, $\mathbf{b}^+$, $\mathbf{g}^+$, and $\mathbf{p}^+$, which are defined in Eq. (12.54). A vector equation from this group is named $\mathbf{K\_Beam3}_E$, because it is a vector equation on equilibrium at a spatial K joint corresponding to incident waves from Beam 3.

The reflection and transmission matrices, $\mathbf{r}_{33}$, $\mathbf{t}_{31}$, $\mathbf{t}_{32}$, and $\mathbf{t}_{34}$, are found from Eq. (12.52) and the four vector equations of continuity and equilibrium, namely, $\mathbf{K\_Beam3}_{C31}$, $\mathbf{K\_Beam3}_{C32}$, $\mathbf{K\_Beam3}_{C34}$, and $\mathbf{K\_Beam3}_E$.

Last, solve for $\mathbf{r}_{44}$, $\mathbf{t}_{41}$, $\mathbf{t}_{42}$, and $\mathbf{t}_{43}$, that is, the reflection and transmission matrices at the K joint of Figure 12.10d, corresponding to incident waves from Beam 4.

Following the local coordinate systems of Figure 12.6, the longitudinal, torsional, and bending deflections of the beam elements are

$$u_1 = g_5^+ e^{ik_{15}x_1},$$

$$u_2 = b_5^+ e^{-ik_{25}x_2},$$

$$u_3 = p_5^+ e^{-ik_{35}x_3},$$

$$u_4 = a_5^+ e^{-ik_{45}x_4} + a_5^- e^{ik_{45}x_4}$$

$$\theta_1 = g_6^+ e^{ik_{16}x_1},$$

$$\theta_2 = b_6^+ e^{-ik_{26}x_2},$$

$$\theta_3 = p_6^+ e^{-ik_{36}x_3},$$

$$\theta_4 = a_6^+ e^{-ik_{46}x_4} + a_6^- e^{ik_{46}x_4}$$

$$y_1 = g_1^+ e^{ik_{11}x_1} + g_2^+ e^{k_{12}x_1},$$

$$y_2 = b_1^+ e^{-ik_{21}x_2} + b_2^+ e^{-k_{22}x_2},$$

$$y_3 = p_1^+ e^{-ik_{31}x_3} + p_2^+ e^{-k_{32}x_3},$$

$$y_4 = a_1^+ e^{-ik_{41}x_4} + a_2^+ e^{-k_{42}x_4} + a_1^- e^{ik_{41}x_4} + a_2^- e^{k_{42}x_4}$$

$$z_1 = g_3^+ e^{ik_{13}x_1} + g_4^+ e^{k_{14}x_1},$$

$$z_2 = b_3^+ e^{-ik_{23}x_2} + b_4^+ e^{-k_{24}x_2},$$

$$z_3 = p_3^+ e^{-ik_{33}x_3} + g_4^+ e^{-k_{34}x_3},$$

$$z_4 = a_3^+ e^{-ik_{43}x_4} + a_4^+ e^{-k_{44}x_4} + a_3^- e^{ik_{43}x_4} + a_4^- e^{k_{44}x_4} \qquad (12.58a)$$

and the bending slopes by the Euler–Bernoulli bending vibration theory are

$$\varphi_{y1} = ik_{11}g_1^+ e^{ik_{11}x_1} + k_{12}g_2^+ e^{k_{12}x_1},$$

$$\varphi_{y2} = -ik_{21}b_1^+ e^{-ik_{21}x_2} - k_{22}b_2^+ e^{-k_{22}x_2},$$

$$\varphi_{y3} = -ik_{31}p_1^+ e^{-ik_{31}x_3} - k_{32}p_2^+ e^{-k_{32}x_3},$$

$$\varphi_{y4} = -ik_{41}a_1^+ e^{-ik_{41}x_4} - k_{42}a_2^+ e^{-k_{42}x_4} + ik_{41}a_1^- e^{ik_{41}x_4} + k_{42}a_2^- e^{k_{42}x_4}$$

$$\varphi_{z1} = ik_{13}g_3^+ e^{ik_{13}x_1} + k_{14}g_4^+ e^{k_{14}x_1},$$

$$\varphi_{z2} = -ik_{23}b_3^+ e^{-ik_{23}x_2} - k_{24}b_4^+ e^{-k_{24}x_2},$$

$$\varphi_{z3} = -ik_{33}p_3^+ e^{-ik_{33}x_3} - k_{34}p_4^+ e^{-k_{34}x_3},$$

$$\varphi_{z4} = -ik_{43}a_3^+ e^{-ik_{43}x_4} - k_{44}a_4^+ e^{-k_{44}x_4} + ik_{43}a_3^- e^{ik_{43}x_4} + k_{44}a_4^- e^{k_{44}x_4} \qquad (12.58b)$$

and the bending slopes by the Timoshenko bending vibration theory are

$$\varphi_{y1} = iP_{1y}g_1^+ e^{ik_{11}x_1} + N_{1y}g_2^+ e^{k_{12}x_1},$$

$$\varphi_{y2} = -iP_{2y}b_1^+ e^{-ik_{21}x_2} - N_{2y}b_2^+ e^{-k_{22}x_2},$$

$$\varphi_{y3} = -iP_{3y}p_1^+ e^{-ik_{31}x_3} - N_{3y}p_2^+ e^{-k_{32}x_3},$$

$$\varphi_{y4} = -iP_{4y}a_1^+ e^{-ik_{41}x_4} - N_{4y}a_2^+ e^{-k_{42}x_4} + iP_{4y}a_1^- e^{ik_{41}x_4} + N_{4y}a_2^- e^{k_{42}x_4}$$

$$\varphi_{z1} = iP_{1z}g_3^+ e^{ik_{13}x_1} + N_{1z}g_4^+ e^{k_{14}x_1},$$

$$\varphi_{z2} = -iP_{2z}b_3^+ e^{-ik_{23}x_2} - N_{2z}b_4^+ e^{-k_{24}x_2},$$

$$\varphi_{z3} = -iP_{3z}p_3^+ e^{-ik_{33}x_3} - N_{3z}p_4^+ e^{-k_{34}x_3},$$

$$\varphi_{z4} = -iP_{4z}a_3^+ e^{-ik_{43}x_4} - N_{4z}a_4^+ e^{-k_{44}x_4} + iP_{4z}a_3^- e^{ik_{43}x_4} + N_{4z}a_4^- e^{k_{44}x_4} \qquad (12.58c)$$

Substituting Eqs. (12.58a) and (12.58b) into a set of continuity equations from Eq. (12.45), and choosing the origin at the joint, a set of six scalar equations of continuity is obtained based on the Euler–Bernoulli bending vibration theory. Similarly, substituting Eqs. (12.58a) and (12.58c) into a set of continuity equations from Eq. (12.45), and choosing the origin at the joint, a set of six scalar equations of continuity is obtained based on the Timoshenko bending vibration theory. This set of six scalar equations of continuity, either by the Euler–Bernoulli or Timoshenko bending vibration theory, can be placed in a vector equation, in terms of wave components $\mathbf{a}^+$, $\mathbf{a}^-$, and $\mathbf{g}^+$, which are defined in Eq. (12.54). A vector equation from this group is named **K_Beam4**$_{C41}$, because it is a vector equation on the continuity between Beam 4 and Beam 1 at a spatial K joint corresponding to incident waves from Beam 4.

Substituting Eqs. (12.58a) and (12.58b) into a set of continuity equations from Eq. (12.47), and choosing the origin at the joint, a set of six scalar equations of continuity is obtained based on the Euler–Bernoulli bending vibration theory. Similarly, substituting Eqs. (12.58a) and (12.58c) into a set of continuity equations from Eq. (12.47), and choosing the origin at the joint, a set of six scalar equations of continuity is obtained based on the Timoshenko bending vibration theory. This set of six scalar equations of continuity, either by the Euler–Bernoulli or Timoshenko bending vibration theory, can be placed in a vector equation, in terms of wave components $\mathbf{a}^+$, $\mathbf{a}^-$, and $\mathbf{b}^+$, which are defined in Eq. (12.54). A vector equation from this group is named $\mathbf{K\_Beam4}_{C42}$, because it is a vector equation on the continuity between Beam 4 and Beam 2 at a spatial K joint corresponding to incident waves from Beam 4.

Substituting Eqs. (12.58a) and (12.58b) into a set of continuity equations from Eq. (12.48), and choosing the origin at the joint, a set of six scalar equations of continuity is obtained based on the Euler–Bernoulli bending vibration theory. Similarly, substituting Eqs. (12.58a) and (12.58c) into a set of continuity equations from Eq. (12.48), and choosing the origin at the joint, a set of six scalar equations of continuity is obtained based on the Timoshenko bending vibration theory. This set of six scalar equations of continuity, either by the Euler–Bernoulli or Timoshenko bending vibration theory, can be placed in a vector equation, in terms of wave components $\mathbf{a}^+$, $\mathbf{a}^-$, and $\mathbf{p}^+$, which are defined in Eq. (12.54). A vector equation from this group is named $\mathbf{K\_Beam4}_{C43}$, because it is a vector equation on the continuity between Beam 4 and Beam 3 at a spatial K joint corresponding to incident waves from Beam 4.

Eliminating the joint related parameters from Eq. (12.40) using Eqs. (12.49a) and (12.49b), substituting Eqs. (12.58a) and (12.58b) into Eqs. (12.7a), (12.7c), and (12.40) that is free of joint parameters, and choosing the origin at the joint, a set of six scalar equations of equilibrium is obtained based on the Euler–Bernoulli bending vibration theory. Similarly, eliminating the joint related parameters from Eq. (12.40) using Eqs. (12.49a) and (12.49b), substituting Eqs. (12.58a) and (12.58c) into Eqs. (12.7b), (12.7c), and (12.40) that is free of joint parameters, and choosing the origin at the joint, a set of six scalar equations of equilibrium based on the Timoshenko bending vibration theory. These six scalar equations of equilibrium can be placed in a vector equation in terms of $\mathbf{a}^+$, $\mathbf{a}^-$, $\mathbf{b}^+$, $\mathbf{g}^+$, and $\mathbf{p}^+$, which are defined in Eq. (12.54). A vector equation from this group is named $\mathbf{K\_Beam4}_E$, because it is a vector equation on equilibrium at a spatial K joint corresponding to incident waves from Beam 4.

The reflection and transmission matrices, $\mathbf{r}_{44}$, $\mathbf{t}_{41}$, $\mathbf{t}_{42}$, and $\mathbf{t}_{43}$, are found from Eq. (12.53) and the four vector equations of continuity and equilibrium, namely, $\mathbf{K\_Beam4}_{C41}$, $\mathbf{K\_Beam4}_{C42}$, $\mathbf{K\_Beam4}_{C43}$, and $\mathbf{K\_Beam4}_E$.

With the derived matrices of wave propagation along a uniform waveguide, wave reflection at a boundary, wave reflection and transmission at a structural joint, as well as the availability of relationships between the injected waves and an external excitation, both free and forced vibrations in built up space frames can be obtained analytically. It involves a systematic assembling process as described in Section 12.4. Numerical examples of wave vibration analysis of built up space frames can be found in the references of the author.

References

Cremer L., Heckl M., Ungar E. E. *Structure-Borne Sound*, Springer-Verlag, Berlin (1987).

Mei C., Sha H. An Exact Analytical Approach for Vibrations in Built-up Space Frames, *ASME Journal of Vibration and Acoustics*, 137(3), 031005-1-12 (2015).

Mei C. Experimental Validation of Wave Vibration Analysis of Complex Vibrations in a Two-Story Metallic Space Frame based on the Advanced Timoshenko Bending Theory, *ASME Journal of Vibration and Acoustics*, 138(2), 021003-1-20 (2016).

Homework Project

Wave Analysis of Free Vibrations in a Y-shaped Space Frame

A spatial Y-shaped frame, as shown in the figure below, was built by welding three uniform steel beam elements together orthogonally. These beam elements are made of cold-rolled steel, whose material and geometrical properties are: the mass density is $\rho = 7664.2\ kg/m^3$, Young's modulus is $E = 198.87\ GPa$, Poisson's ratio is $\upsilon = 0.30$, and shear modulus is $G = \dfrac{E}{2(1+\nu)}$.

The width and thickness of Beam 1 and Beam 2, the two horizontal beam elements, are 25.4 *mm* and 12.7 *mm* (or 1.0 *in* and 0.5 *in*), respectively. Beam 3, the vertical beam element, is of a square cross section whose side length is 25.4 *mm* (or 1.0 *in*). The lengths of the three beam elements are $L_1 = 930\ mm$, $L_2 = 930\ mm$, and $L_3 = 890\ mm$.

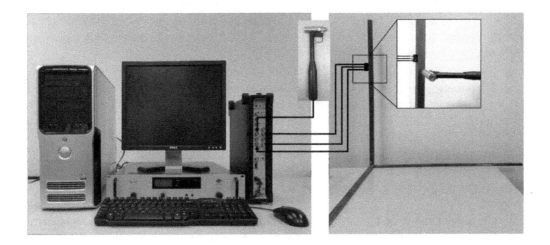

The natural frequencies of the Y-shaped space frame were measured in the laboratory. In the experiment, the Y-shaped space frame was laid on a large piece of foam, and all three boundaries were free. A Brüel & Kjær impact hammer 8202 was used to excite the space frame and a Brüel & Kjær triaxial accelerometer 4535 was used to measure the accelerations. A four-channel Pulse data acquisition system from Brüel & Kjaer was used to collect and analyze the measured data. The measured natural frequencies are listed in the table below.

Mode Sequence	1	2	3	4	5	6	7	8	9	10	11
Natural frequency (*Hz*)	15.7	17.67	28.07	71.4	76.7	106.6	129.2	134.3	160.8	208.0	214.7

Perform free vibration analysis of the Y-shaped space frame based on the elementary vibration theories using the wave analysis approach to space frame structures introduced in this chapter, plot the dB magnitude of the characteristic polynomial, read the natural frequencies from the plot, compare the analytical results with the tabulated experimental results, and identify the potential causes for discrepancy.

13

Passive Wave Vibration Control

Vibrations in structures can be controlled by adding damping. Vibrations can also be controlled by creating structural discontinuities to alter the vibrational characteristics of a structure. When the natural frequencies of a structure are shifted outside the frequency band of an external excitation, no resonant vibration will occur. Consequently, structural vibrations are suppressed.

Because it is not possible to analyze in detail all the different types of structural discontinuities that occur in practice, some example discontinuities are considered. These include discontinuities caused by a change in cross section or material, a spring (and viscous damper) attachment, a single degree-of-freedom (DOF) spring-mass (and viscous damper) attachment, and a two DOF spring-mass (and viscous damper) attachment. The analysis in this chapter focuses on bending vibrations, based on both the Euler–Bernoulli and Timoshenko bending vibration theories. Structures that undergo other types of vibrations with other types of discontinuities can be analyzed following a similar procedure.

Figure 13.1 shows the sign convention adopted for the analysis that follows.

Listed in Table 13.1 are governing equations of motion (EOM) for free bending vibrations of a uniform beam, wavenumbers, solutions of spatial responses to harmonic external excitations, and expressions of internal resistant shear force and bending moment, based on the Euler–Bernoulli and Timoshenko bending vibration theories. The definitions of parameters in Table 13.1 are: x is the position along the beam axis, t is the time, $y(x,t)$ is the bending deflection of the centerline of the beam, $\psi(x,t)$ is the slope due to bending, $\partial y(x,t)/\partial x$ is the slope of the centerline of the beam, and $\partial y(x,t)/\partial x - \psi(x,t)$ is the shear angle. A is the cross-sectional area, I is the area moment of inertia of the cross section, E is the Young's modulus, and ρ is the volume mass density. ω is the circular frequency and k is the wavenumber. G and κ are the shear modulus and shear coefficient, respectively.

In this chapter, all expressions based on the Timoshenko bending vibration theory are for frequency $\omega < \omega_c$, which is overwhelmingly the most common situation in audio frequency applications. When $\omega > \omega_c$, k_2 needs to be replaced by ik_2, in order to reflect the wave mode transition at the cut-off frequency ω_c predicted by the Timoshenko bending vibration theory.

13.1 Change in Cross Section or Material

Figure 13.2 shows that two beams with coincided centerlines but different geometric and/or material properties are joined. Incident waves from one beam give rise to reflected and transmitted waves at the structural discontinuity caused by a change in geometric and/or material property. Incident waves $\mathbf{a}^+$ and transmitted and reflected waves $\mathbf{b}^+$ and $\mathbf{a}^-$ are related by the transmission and reflection matrices $\mathbf{t}$ and $\mathbf{r}$,

$$\mathbf{b}^+ = \mathbf{t}\mathbf{a}^+, \quad \mathbf{a}^- = \mathbf{r}\mathbf{a}^+ \tag{13.1}$$

where the bending wave components are

$$\mathbf{a}^+ = \begin{bmatrix} a_1^+ \\ a_2^+ \end{bmatrix}, \quad \mathbf{a}^- = \begin{bmatrix} a_1^- \\ a_2^- \end{bmatrix}, \quad \mathbf{b}^+ = \begin{bmatrix} b_1^+ \\ b_2^+ \end{bmatrix}, \quad \mathbf{b}^- = \begin{bmatrix} b_1^- \\ b_2^- \end{bmatrix} \tag{13.2}$$

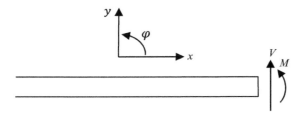

Figure 13.1 Definition of positive internal resistant shear force and bending moment.

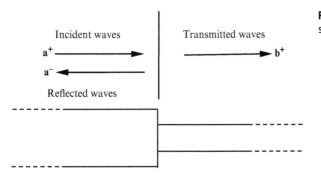

Figure 13.2 Wave reflection and transmission at a change in section and/or material.

Table 13.1 Summary on bending wave vibrations based on the Euler–Bernoulli and Timoshenko bending vibration theories.

Euler–Bernoulli bending vibration theory

EOM for free vibration

$$EI\frac{\partial^4 y(x,t)}{\partial x^4} + \rho A\frac{\partial^2 y(x,t)}{\partial t^2} = 0$$

Wavenumbers

$$k_1 = k_2 = k = \sqrt[4]{\rho A\omega^2 / EI}$$

Solutions

$$y(x) = a_1^+ e^{-ik_1 x} + a_2^+ e^{-k_2 x} + a_1^- e^{ik_1 x} + a_2^- e^{k_2 x}$$

$$\psi(x,t) = \frac{\partial y(x,t)}{\partial x}$$

where i is the imaginary unit

Internal resistant shear force and bending moment

$$V(x,t) = -EI\frac{\partial^3 y(x,t)}{\partial x^3}$$

$$M(x,t) = EI\frac{\partial \psi(x,t)}{\partial x} = EI\frac{\partial^2 y(x,t)}{\partial x^2}$$

Timoshenko bending vibration theory

EOM for free vibration

$$GA\kappa\left[\frac{\partial \psi(x,t)}{\partial x} - \frac{\partial^2 y(x,t)}{\partial x^2}\right] + \rho A\frac{\partial^2 y(x,t)}{\partial t^2} = 0$$

$$EI\frac{\partial^2 \psi(x,t)}{\partial x^2} + GA\kappa\left[\frac{\partial y(x,t)}{\partial x} - \psi(x,t)\right] - \rho I\frac{\partial^2 \psi(x,t)}{\partial t^2} = 0$$

Wavenumbers

$$k_1 = \left[\frac{1}{2}\left(\frac{C_r}{C_b}\right)^2 \omega^2 + \sqrt{\frac{\omega^2}{C_b^2} + \frac{1}{4}\left(\frac{C_r}{C_b}\right)^4 \omega^4}\right]^{\frac{1}{2}},$$

$$k_2 = \left|\left[\frac{1}{2}\left(\frac{C_r}{C_b}\right)^2 \omega^2 - \sqrt{\frac{\omega^2}{C_b^2} + \frac{1}{4}\left(\frac{C_r}{C_b}\right)^4 \omega^4}\right]\right|^{\frac{1}{2}}$$

Table 13.1 (Continued)

Timoshenko bending vibration theory

Solutions
$$y(x) = a_1^+ e^{-ik_1 x} + a_2^+ e^{-k_2 x} + a_1^- e^{ik_1 x} + a_2^- e^{k_2 x}$$

$$\psi(x) = -iPa_1^+ e^{-ik_1 x} - Na_2^+ e^{-k_2 x} + iPa_1^- e^{ik_1 x} + Na_2^- e^{k_2 x}$$

where i is the imaginary unit

Internal resistant shear force and bending moment
$$V(x,t) = GA\kappa \left[\frac{\partial y(x,t)}{\partial x} - \psi(x,t) \right], \quad M(x,t) = EI \frac{\partial \psi(x,t)}{\partial x}$$

$$C_b = \sqrt{\frac{EI}{\rho A}} \; , \; C_r = \sqrt{\frac{\rho I}{\rho A}} = \sqrt{\frac{I}{A}} \; , \; C_s = \sqrt{\frac{GA\kappa}{\rho A}} \; , \; P = k_1 \left(1 - \frac{\omega^2}{k_1^2 C_s^2} \right), \text{ and } N = k_2 \left(1 + \frac{\omega^2}{k_2^2 C_s^2} \right)$$

Cut-off frequency predicted by the Timoshenko bending vibration theory: $\omega_c = \dfrac{C_s}{C_r}$

The bending deflection $y(x,t)$, bending slope $\psi(x,t)$, bending moment $M(x,t)$, and shear force $V(x,t)$ are continuous at the junction, from which the reflection and transmission matrices at the discontinuity can be derived. Denoting physical parameters on the left and right side of the junction using subscripts L and R respectively, the continuities on deflections and internal resistant forces/moments are

$$y_L = y_R, \; \psi_L = \psi_R \tag{13.3a}$$

$$M_L = M_R, \; V_L = V_R \tag{13.3b}$$

13.1.1 Wave Reflection and Transmission at a Step Change by Euler–Bernoulli Bending Vibration Theory

From Table 13.1 and Figure 13.2, according to the Euler–Bernoulli bending vibration theory, the deflections are

$$
\begin{aligned}
y_L &= a_1^+ e^{-ik_{L1} x} + a_2^+ e^{-k_{L2} x} + a_1^- e^{ik_{L1} x} + a_2^- e^{k_{L2} x}, \\
y_R &= b_1^+ e^{-ik_{R1} x} + b_2^+ e^{-k_{R2} x}, \\
\psi_L &= \frac{\partial y_L}{\partial x} = -ik_{L1} a_1^+ e^{-ik_{L1} x} - k_{L2} a_2^+ e^{-k_{L2} x} + ik_{L1} a_1^- e^{ik_{L1} x} + k_{L2} a_2^- e^{k_{L2} x}, \\
\psi_R &= \frac{\partial y_R}{\partial x} = -ik_{R1} b_1^+ e^{-ik_{R1} x} - k_{R2} b_2^+ e^{-k_{R2} x}
\end{aligned}
\tag{13.4}
$$

Substituting Eq. (13.4) into the internal resistant shear force and bending moment expressions in Table 13.1 gives

$$
\begin{aligned}
M_L &= (EI)_L \frac{\partial^2 y_L}{\partial x^2} = (EI)_L \left(-k_{L1}^2 a_1^+ e^{-ik_{L1} x} + k_{L2}^2 a_2^+ e^{-k_{L2} x} - k_{L1}^2 a_1^- e^{ik_{L1} x} + k_{L2}^2 a_2^- e^{k_{L2} x} \right), \\
M_R &= (EI)_R \frac{\partial^2 y_R}{\partial x^2} = (EI)_R \left(-k_{R1}^2 b_1^+ e^{-ik_{R1} x} + k_{R2}^2 b_2^+ e^{-k_{R2} x} \right), \\
V_L &= -(EI)_L \frac{\partial^3 y_L}{\partial x^3} = (EI)_L \left(-ik_{L1}^3 a_1^+ e^{-ik_{L1} x} + k_{L2}^3 a_2^+ e^{-k_{L2} x} + ik_{L1}^3 a_1^- e^{ik_{L1} x} - k_{L2}^3 a_2^- e^{k_{L2} x} \right), \\
V_R &= -(EI)_R \frac{\partial^3 y_R}{\partial x^3} = (EI)_R \left(-ik_{R1}^3 b_1^+ e^{-ik_{R1} x} + k_{R2}^3 b_2^+ e^{-k_{R2} x} \right)
\end{aligned}
\tag{13.5}
$$

Choosing the origin at the discontinuity, from Eqs. (13.3a), (13.3b), and (13.5),

$$a_1^+ + a_2^+ + a_1^- + a_2^- = b_1^+ + b_2^+,$$
$$-ik_{L1}a_1^+ - k_{L2}a_2^+ + ik_{L1}a_1^- + k_{L2}a_2^- = -ik_{R1}b_1^+ - k_{R2}b_2^+$$

(13.6)

and

$$(EI)_L\left(-k_{L1}^2 a_1^+ + k_{L2}^2 a_2^+ - k_{L1}^2 a_1^- + k_{L2}^2 a_2^-\right) = (EI)_R\left(-k_{R1}^2 b_1^+ + k_{R2}^2 b_2^+\right),$$
$$(EI)_L\left(-ik_{L1}^3 a_1^+ + k_{L2}^3 a_2^+ + ik_{L1}^3 a_1^- - k_{L2}^3 a_2^-\right) = (EI)_R\left(-ik_{R1}^3 b_1^+ + k_{R2}^3 b_2^+\right)$$

(13.7)

Putting Eqs. (13.6) and (13.7) into matrix form gives

$$\begin{bmatrix} 1 & 1 \\ -ik_{L1} & -k_{L2} \end{bmatrix}\mathbf{a}^+ + \begin{bmatrix} 1 & 1 \\ ik_{L1} & k_{L2} \end{bmatrix}\mathbf{a}^- = \begin{bmatrix} 1 & 1 \\ -ik_{R1} & -k_{R2} \end{bmatrix}\mathbf{b}^+$$

(13.8)

$$\begin{bmatrix} -k_{L1}^2 & k_{L2}^2 \\ -ik_{L1}^3 & k_{L2}^3 \end{bmatrix}\mathbf{a}^+ + \begin{bmatrix} -k_{L1}^2 & k_{L2}^2 \\ ik_{L1}^3 & -k_{L2}^3 \end{bmatrix}\mathbf{a}^- = \begin{bmatrix} -k_{R1}^2\beta_{RL} & k_{R2}^2\beta_{RL} \\ -ik_{R1}^3\beta_{RL} & k_{R2}^3\beta_{RL} \end{bmatrix}\mathbf{b}^+$$

(13.9)

where $\beta_{RL} = (EI)_R / (EI)_L$ and the wave vectors are defined in Eq. (13.2).

From Eqs. (13.1), (13.8), and (13.9),

$$\mathbf{c}_1\mathbf{r}_{LL} + \mathbf{c}_2\mathbf{t}_{LR} = \mathbf{c}_3, \quad \mathbf{d}_1\mathbf{r}_{LL} + \mathbf{d}_2\mathbf{t}_{LR} = \mathbf{d}_3$$

(13.10a)

where $\mathbf{r}_{LL}$ and $\mathbf{t}_{LR}$ are the reflection and transmission matrices corresponding to incident waves from the left side of the discontinuity, and the coefficient matrices are

$$\mathbf{c}_1 = \begin{bmatrix} 1 & 1 \\ ik_{L1} & k_{L2} \end{bmatrix}, \quad \mathbf{c}_2 = \begin{bmatrix} -1 & -1 \\ ik_{R1} & k_{R2} \end{bmatrix}, \quad \mathbf{c}_3 = \begin{bmatrix} -1 & -1 \\ ik_{L1} & k_{L2} \end{bmatrix}$$

$$\mathbf{d}_1 = \begin{bmatrix} -k_{L1}^2 & k_{L2}^2 \\ ik_{L1}^3 & -k_{L2}^3 \end{bmatrix}, \quad \mathbf{d}_2 = \begin{bmatrix} k_{R1}^2\beta_{RL} & -k_{R2}^2\beta_{RL} \\ ik_{R1}^3\beta_{RL} & -k_{R2}^3\beta_{RL} \end{bmatrix}, \quad \mathbf{d}_3 = \begin{bmatrix} k_{L1}^2 & -k_{L2}^2 \\ ik_{L1}^3 & -k_{L2}^3 \end{bmatrix}$$

(13.10b)

Solving Eqs. (13.10a) gives

$$\mathbf{t}_{LR} = (\mathbf{c}_2 - \mathbf{c}_1\mathbf{d}_1^{-1}\mathbf{d}_2)^{-1}(\mathbf{c}_3 - \mathbf{c}_1\mathbf{d}_1^{-1}\mathbf{d}_3), \text{ and } \mathbf{r}_{LL} = (\mathbf{c}_1 - \mathbf{c}_2\mathbf{d}_2^{-1}\mathbf{d}_1)^{-1}(\mathbf{c}_3 - \mathbf{c}_2\mathbf{d}_2^{-1}\mathbf{d}_3)$$

(13.11)

There are various forms of solutions to the reflection and transmission matrices $\mathbf{r}_{LL}$ and $\mathbf{t}_{LR}$. Attention needs to be paid to the difference between solving a scalar and a vector/matrix form of systems of equations, in particular, the associative property of multiplication that applies to scalars does not hold for vectors/matrices.

The reflection and transmission matrices corresponding to incident waves from the right side of the discontinuity, namely $\mathbf{r}_{RR}$ and $\mathbf{t}_{RL}$, can be obtained by following a similar procedure.

Note that by the Euler–Bernoulli bending vibration theory, $k_{L1} = k_{L2}$ and $k_{R1} = k_{R2}$.

13.1.2 Wave Reflection and Transmission at a Step Change by Timoshenko Bending Vibration Theory

From Table 13.1 and Figure 13.2, according to the Timoshenko bending vibration theory, the deflections are

$$y_L = a_1^+ e^{-ik_{L1}x} + a_2^+ e^{-k_{L2}x} + a_1^- e^{ik_{L1}x} + a_2^- e^{k_{L2}x}$$
$$y_R = b_1^+ e^{-ik_{R1}x} + b_2^+ e^{-k_{R2}x}$$
$$\psi_L = -iP_L a_1^+ e^{-ik_{L1}x} - N_L a_2^+ e^{-k_{L2}x} + iP_L a_1^- e^{ik_{L1}x} + N_L a_2^- e^{k_{L2}x}$$
$$\psi_R = -iP_R b_1^+ e^{-ik_{R1}x} - N_R b_2^+ e^{-k_{R2}x}$$

(13.12)

Substituting Eq. (13.12) into the internal resistant shear force and bending moment expressions in Table 13.1 gives

$$M_L = (EI)_L \frac{\partial \psi_L}{\partial x} = (EI)_L \left(-P_L k_{L1} a_1^+ e^{-ik_{L1}x} + N_L k_{L2} a_2^+ e^{-k_{L2}x} - P_L k_{L1} a_1^- e^{ik_{L1}x} + N_L k_{L2} a_2^- e^{k_{L2}x} \right),$$

$$M_R = (EI)_R \frac{\partial \psi_R}{\partial x} = (EI)_R \left(-P_R k_{R1} b_1^+ e^{-ik_{R1}x} + N_R k_{R2} b_2^+ e^{-k_{R2}x} \right),$$

$$V_L = (GA\kappa)_L \left[\frac{\partial y_L}{\partial x} - \psi_L \right] = (GA\kappa)_L \left(-ik_{L1} a_1^+ e^{-ik_{L1}x} - k_{L2} a_2^+ e^{-k_{L2}x} + ik_{L1} a_1^- e^{ik_{L1}x} + k_{L2} a_2^- e^{k_{L2}x} \right) \quad (13.13)$$
$$- (GA\kappa)_L \left(-iP_L a_1^+ e^{-ik_{L1}x} - N_L a_2^+ e^{-k_{L2}x} + iP_L a_1^- e^{ik_{L1}x} + N_L a_2^- e^{k_{L2}x} \right),$$

$$V_R = (GA\kappa)_R \left[\frac{\partial y_R}{\partial x} - \psi_R \right] = (GA\kappa)_R \left(-ik_{R1} b_1^+ e^{-ik_{R1}x} - k_{R2} b_2^+ e^{-k_{R2}x} \right) - (GA\kappa)_R \left(-iP_R b_1^+ e^{-ik_{R1}x} - N_R b_2^+ e^{-k_{R2}x} \right)$$

Choosing the origin at the discontinuity, from Eqs. (13.3a) (13.3b), and (13.13),

$$a_1^+ + a_2^+ + a_1^- + a_2^- = b_1^+ + b_2^+,$$
$$-iP_L a_1^+ - N_L a_2^+ + iP_L a_1^- + N_L a_2^- = -iP_R b_1^+ - N_R b_2^+ \quad (13.14)$$

and

$$(EI)_L \left(-P_L k_{L1} a_1^+ + N_L k_{L2} a_2^+ - P_L k_{L1} a_1^- + N_L k_{L2} a_2^- \right) = (EI)_R \left(-P_R k_{R1} b_1^+ + N_R k_{R2} b_2^+ \right),$$
$$(GA\kappa)_L \left[(-ik_{L1} + iP_L) a_1^+ + (-k_{L2} + N_L) a_2^+ + (ik_{L1} - iP_L) a_1^- + (k_{L2} - N_L) a_2^- \right] \quad (13.15)$$
$$= (GA\kappa)_R \left[(-ik_{R1} + iP_R) b_1^+ + (-k_{R2} + N_R) b_2^+ \right]$$

Putting Eqs. (13.14) and (13.15) into matrix forms,

$$\begin{bmatrix} 1 & 1 \\ -iP_L & -N_L \end{bmatrix} \mathbf{a}^+ + \begin{bmatrix} 1 & 1 \\ iP_L & N_L \end{bmatrix} \mathbf{a}^- = \begin{bmatrix} 1 & 1 \\ -iP_R & -N_R \end{bmatrix} \mathbf{b}^+ \quad (13.16)$$

$$\begin{bmatrix} -P_L k_{L1} & N_L k_{L2} \\ -ik_{L1} + iP_L & -k_{L2} + N_L \end{bmatrix} \mathbf{a}^+ + \begin{bmatrix} -P_L k_{L1} & N_L k_{L2} \\ ik_{L1} - iP_L & k_{L2} - N_L \end{bmatrix} \mathbf{a}^- = \begin{bmatrix} -P_R k_{R1} \beta_{RL} & N_R k_{R2} \beta_{RL} \\ (-ik_{R1} + iP_R) \gamma_{RL} & (-k_{R2} + N_R) \gamma_{RL} \end{bmatrix} \mathbf{b}^+ \quad (13.17)$$

where $\beta_{RL} = (EI)_R / (EI)_L$, $\gamma_{RL} = (GA\kappa)_R / (GA\kappa)_L$, and the wave vectors are defined in Eq. (13.2).
From Eqs. (13.1), (13.16), and (13.17),

$$\mathbf{c}_1 \mathbf{r}_{LL} + \mathbf{c}_2 \mathbf{t}_{LR} = \mathbf{c}_3, \quad \mathbf{d}_1 \mathbf{r}_{LL} + \mathbf{d}_2 \mathbf{t}_{LR} = \mathbf{d}_3 \quad (13.18a)$$

where $\mathbf{r}_{LL}$ and $\mathbf{t}_{LR}$ are the reflection and transmission matrices corresponding to incident waves from the left side of the discontinuity. The coefficient matrices are

$$\mathbf{c}_1 = \begin{bmatrix} 1 & 1 \\ iP_L & N_L \end{bmatrix}, \mathbf{c}_2 = \begin{bmatrix} -1 & -1 \\ iP_R & N_R \end{bmatrix}, \mathbf{c}_3 = \begin{bmatrix} -1 & -1 \\ iP_L & N_L \end{bmatrix},$$
$$\mathbf{d}_1 = \begin{bmatrix} -P_L k_{L1} & N_L k_{L2} \\ ik_{L1} - iP_L & k_{L2} - N_L \end{bmatrix}, \mathbf{d}_2 = \begin{bmatrix} P_R k_{R1} \beta_{RL} & -N_R k_{R2} \beta_{RL} \\ (ik_{R1} - iP_R) \gamma_{RL} & (k_{R2} - N_R) \gamma_{RL} \end{bmatrix}, \mathbf{d}_3 = \begin{bmatrix} P_L k_{L1} & -N_L k_{L2} \\ ik_{L1} - iP_L & k_{L2} - N_L \end{bmatrix} \quad (13.18b)$$

$\mathbf{r}_{LL}$ and $\mathbf{t}_{LR}$ can be solved from Eq. (13.18a), which is identical to Eq. (13.10a). As a result, the solutions are given by the same equations obtained in Eq. (13.11), although the coefficient matrices are different.

The reflection and transmission matrices corresponding to incident waves from the right side of the discontinuity, namely $\mathbf{r}_{RR}$ and $\mathbf{t}_{RL}$, can be obtained by following a similar procedure.

13.2 Point Attachment

Figure 13.3 shows a discontinuity caused by a point support to a beam. The point support exerts both translational and rotational constraints on the beam. These are described by translational and rotational dynamic stiffnesses $\overline{K_T}$ and $\overline{K_R}$. If there exists coupling between the translation and rotation, in that the support inducing a moment in

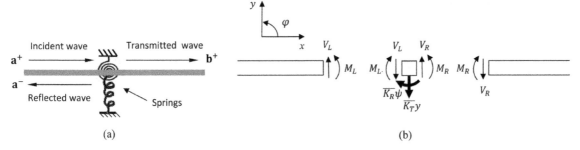

Figure 13.3 (a) Wave reflection and transmission at a point support and (b) free body diagram.

response to a translation and a force in response to a rotation, transfer dynamic stiffnesses $\overline{K_{TR}}$ and $\overline{K_{RT}}$ may be introduced.

At the point support, a set of incident waves $\mathbf{a}^+$ gives rise to transmitted and reflected waves $\mathbf{b}^+$ and $\mathbf{a}^-$, which are related to the incident waves by the transmission and reflection matrices $\mathbf{t}$ and $\mathbf{r}$, as described in Eq. (13.1).

The bending deflection $y(x,t)$ and bending slope $\psi(x,t)$ are continuous at the point support,

$$y_L = y_R = y, \; \psi_L = \psi_R = \psi \tag{13.19}$$

where subscripts L and R denote physical parameters on the left and right side of the point support, respectively.

Following the sign convention defined in Figure 13.1 and from the free body diagram of Figure 13.3, the equilibrium conditions at the point support are

$$V_R - V_L = \overline{K_T}y + \overline{K_{TR}}\psi, \; M_R - M_L = \overline{K_R}\psi + \overline{K_{RT}}y \tag{13.20}$$

Continuity in deflections at a point support described by Eq. (13.19) is identical to the continuity in deflections at a change of cross section described by Eq. (13.3a) in Section 13.1. As a result, all related derivations in Section 13.1 apply. In particular, the vector equations for continuity in deflections are the same as Eqs. (13.8) and (13.16) based on the Euler–Bernoulli and Timoshenko bending vibration theories, respectively. Furthermore, from Figures 13.2 and 13.3, the expressions for deflections and internal resistant forces/moments on the left and right side of both types of discontinuities are the same.

13.2.1 Wave Reflection and Transmission at a Point Attachment by Euler–Bernoulli Bending Vibration Theory

Choosing the origin at the point where the spring is attached, from Eqs. (13.5) and (13.20),

$$\begin{aligned}
&(EI)_R\left(-ik_{R1}^3 b_1^+ + k_{R2}^3 b_2^+\right) - (EI)_L\left(-ik_{L1}^3 a_1^+ + k_{L2}^3 a_2^+ + ik_{L1}^3 a_1^- - k_{L2}^3 a_2^-\right) = \overline{K_T}\left(b_1^+ + b_2^+\right) + \overline{K_{TR}}\left(-ik_{R1}b_1^+ - k_{R2}b_2^+\right), \\
&(EI)_R\left(-k_{R1}^2 b_1^+ + k_{R2}^2 b_2^+\right) - (EI)_L\left(-k_{L1}^2 a_1^+ + k_{L2}^2 a_2^+ - k_{L1}^2 a_1^- + k_{L2}^2 a_2^-\right) \\
&= \overline{K_R}\left(-ik_{R1}b_1^+ - k_{R2}b_2^+\right) + \overline{K_{RT}}\left(b_1^+ + b_2^+\right)
\end{aligned} \tag{13.21}$$

Note that in Eq. (13.21), the spring resistant forces are written in terms of y_R and ψ_R, they can also be expressed in terms of y_L and ψ_L because of continuity in the bending deflection $y(x,t)$ and bending slope $\psi(x,t)$ at the point where the spring is attached, as described in Eq. (13.19).

Putting Eq. (13.21) into matrix form,

$$\begin{aligned}
&\begin{bmatrix} -k_{L1}^2(EI)_L & k_{L2}^2(EI)_L \\ -ik_{L1}^3(EI)_L & k_{L2}^3(EI)_L \end{bmatrix}\mathbf{a}^+ + \begin{bmatrix} -k_{L1}^2(EI)_L & k_{L2}^2(EI)_L \\ ik_{L1}^3(EI)_L & -k_{L2}^3(EI)_L \end{bmatrix}\mathbf{a}^- = \\
&\begin{bmatrix} -k_{R1}^2(EI)_R + ik_{R1}\overline{K_R} - \overline{K_{RT}} & k_{R2}^2(EI)_R + k_{R2}\overline{K_R} - \overline{K_{RT}} \\ -ik_{R1}^3(EI)_R + ik_{R1}\overline{K_{TR}} - \overline{K_T} & k_{R2}^3(EI)_R + k_{R2}\overline{K_{TR}} - \overline{K_T} \end{bmatrix}\mathbf{b}^+
\end{aligned} \tag{13.22}$$

where the wave vectors are defined in Eq. (13.2).

From Eqs. (13.1), (13.8), and (13.22),

$$\mathbf{c}_1\mathbf{r}_{LL} + \mathbf{c}_2\mathbf{t}_{LR} = \mathbf{c}_3, \quad \mathbf{d}_1\mathbf{r}_{LL} + \mathbf{d}_2\mathbf{t}_{LR} = \mathbf{d}_3 \tag{13.23a}$$

where $\mathbf{r}_{LL}$ and $\mathbf{t}_{LR}$ are the reflection and transmission matrices corresponding to incident waves from the left side of the discontinuity, and the coefficient matrices are

$$\mathbf{c}_1 = \begin{bmatrix} 1 & 1 \\ ik_{L1} & k_{L2} \end{bmatrix}, \ \mathbf{c}_2 = \begin{bmatrix} -1 & -1 \\ ik_{R1} & k_{R2} \end{bmatrix}, \ \mathbf{c}_3 = \begin{bmatrix} -1 & -1 \\ ik_{L1} & k_{L2} \end{bmatrix},$$

$$\mathbf{d}_1 = \begin{bmatrix} -k_{L1}^2(EI)_L & k_{L2}^2(EI)_L \\ ik_{L1}^3(EI)_L & -k_{L2}^3(EI)_L \end{bmatrix},$$

$$\mathbf{d}_2 = \begin{bmatrix} k_{R1}^2(EI)_R - ik_{R1}\overline{K_R} + \overline{K_{RT}} & -k_{R2}^2(EI)_R - k_{R2}\overline{K_R} + \overline{K_{RT}} \\ ik_{R1}^3(EI)_R - ik_{R1}\overline{K_{TR}} + \overline{K_T} & -k_{R2}^3(EI)_R - k_{R2}\overline{K_{TR}} + \overline{K_T} \end{bmatrix}, \tag{13.23b}$$

$$\mathbf{d}_3 = \begin{bmatrix} k_{L1}^2(EI)_L & -k_{L2}^2(EI)_L \\ ik_{L1}^3(EI)_L & -k_{L2}^3(EI)_L \end{bmatrix}$$

$\mathbf{r}_{LL}$ and $\mathbf{t}_{LR}$ can be solved from Eq. (13.23a), which is identical to Eq. (13.10a). As a result, the solutions are given by the same equations obtained in Eq. (13.11), although the coefficient matrices are different.

Equations (13.23a) and (13.23b) can be simplified, because by the Euler–Bernoulli bending vibration theory, $k_{L1} = k_{L2}$ and $k_{R1} = k_{R2}$. In addition, under the situation where the geometrical and material properties on both sides of the point support are the same, Eqs. (13.23a) and (13.23b) can be further simplified, because $k_{L1} = k_{L2} = k_{R1} = k_{R2}$ and $(EI)_L = (EI)_R$.

The reflection and transmission matrices corresponding to incident waves from the right side of the discontinuity, namely $\mathbf{r}_{RR}$ and $\mathbf{t}_{RL}$, can be obtained by following a similar procedure.

13.2.2 Wave Reflection and Transmission at a Point Attachment by Timoshenko Bending Vibration Theory

Choosing the origin at the point where the spring is attached, from Eqs. (13.13) and (13.20),

$$(GA\kappa)_R\left(-ik_{R1}b_1^+ - k_{R2}b_2^+\right) - (GA\kappa)_R\left(-iP_Rb_1^+ - N_Rb_2^+\right) - (GA\kappa)_L\left(-ik_{L1}a_1^+ - k_{L2}a_2^+ + ik_{L1}a_1^- + k_{L2}a_2^-\right)$$
$$+ (GA\kappa)_L\left(-iP_La_1^+ - N_La_2^+ + iP_La_1^- + N_La_2^-\right)$$
$$= \overline{K_T}\left(b_1^+ + b_2^+\right) + \overline{K_{TR}}\left(-iP_Rb_1^+ - N_Rb_2^+\right),$$
$$(EI)_R\left(-P_Rk_{R1}b_1^+ + N_Rk_{R2}b_2^+\right) - (EI)_L\left(-P_Lk_{L1}a_1^+ + N_Lk_{L2}a_2^+ - P_Lk_{L1}a_1^- + N_Lk_{L2}a_2^-\right)$$
$$= \overline{K_R}\left(-iP_Rb_1^+ - N_Rb_2^+\right) + \overline{K_{RT}}\left(b_1^+ + b_2^+\right) \tag{13.24}$$

Note that in Eq. (13.24), the spring resistant forces are written in terms of y_R and ψ_R, they can also be expressed in terms of y_L and ψ_L because of continuity in bending deflection $y(x,t)$ and bending slope $\psi(x,t)$ at the point where the spring is attached, as described in Eq. (13.19).

Putting Eq. (13.24) into matrix form,

$$\begin{bmatrix} -P_Lk_{L1}(EI)_L & N_Lk_{L2}(EI)_L \\ (-ik_{L1}+iP_L)(GA\kappa)_L & (-k_{L2}+N_L)(GA\kappa)_L \end{bmatrix}\mathbf{a}^+ + \begin{bmatrix} -P_Lk_{L1}(EI)_L & N_Lk_{L2}(EI)_L \\ (ik_{L1}-iP_L)(GA\kappa)_L & (k_{L2}-N_L)(GA\kappa)_L \end{bmatrix}\mathbf{a}^-$$
$$= \begin{bmatrix} -P_Rk_{R1}(EI)_R + iP_R\overline{K_R} - \overline{K_{RT}} & N_Rk_{R2}(EI)_R + N_R\overline{K_R} - \overline{K_{RT}} \\ (-ik_{R1}+iP_R)(GA\kappa)_R + iP_R\overline{K_{TR}} - \overline{K_T} & (-k_{R2}+N_R)(GA\kappa)_R + N_R\overline{K_{TR}} - \overline{K_T} \end{bmatrix}\mathbf{b}^+ \tag{13.25}$$

where the wave vectors are defined in Eq. (13.2).

From Eqs. (13.1), (13.16), and (13.25),

$$\mathbf{c}_1 \mathbf{r}_{LL} + \mathbf{c}_2 \mathbf{t}_{LR} = \mathbf{c}_3, \quad \mathbf{d}_1 \mathbf{r}_{LL} + \mathbf{d}_2 \mathbf{t}_{LR} = \mathbf{d}_3 \tag{13.26a}$$

where $\mathbf{r}_{LL}$ and $\mathbf{t}_{LR}$ are the reflection and transmission matrices corresponding to incident waves from the left side of the discontinuity, and the coefficient matrices are

$$\mathbf{c}_1 = \begin{bmatrix} 1 & 1 \\ iP_L & N_L \end{bmatrix}, \quad \mathbf{c}_2 = \begin{bmatrix} -1 & -1 \\ iP_R & N_R \end{bmatrix}, \quad \mathbf{c}_3 = \begin{bmatrix} -1 & -1 \\ iP_L & N_L \end{bmatrix},$$

$$\mathbf{d}_1 = \begin{bmatrix} -P_L k_{L1}(EI)_L & N_L k_{L2}(EI)_L \\ (ik_{L1} - iP_L)(GA\kappa)_L & (k_{L2} - N_L)(GA\kappa)_L \end{bmatrix},$$

$$\mathbf{d}_2 = \begin{bmatrix} P_R k_{R1}(EI)_R - iP_R \overline{K_R} + \overline{K_{RT}} & -N_R k_{R2}(EI)_R - N_R \overline{K_R} + \overline{K_{RT}} \\ (ik_{R1} - iP_R)(GA\kappa)_R - iP_R \overline{K_{TR}} + \overline{K_T} & (k_{R2} - N_R)(GA\kappa)_R - N_R \overline{K_{TR}} + \overline{K_T} \end{bmatrix},$$

$$\mathbf{d}_3 = \begin{bmatrix} P_L k_{L1}(EI)_L & -N_L k_{L2}(EI)_L \\ (ik_{L1} - iP_L)(GA\kappa)_L & (k_{L2} - N_L)(GA\kappa)_L \end{bmatrix} \tag{13.26b}$$

$\mathbf{r}_{LL}$ and $\mathbf{t}_{LR}$ can be solved from Eq. (13.26a), which is identical to Eq. (13.10a). As a result, the solutions are given by the same equations obtained in Eq. (13.11), although the coefficient matrices are different.

The reflection and transmission matrices corresponding to incident waves from the right side of the discontinuity, namely $\mathbf{r}_{RR}$ and $\mathbf{t}_{RL}$, can be obtained by following a similar procedure.

13.3 Beam with a Single Degree of Freedom Attachment

When a single DOF spring-mass system is attached to a beam, the spring-mass system applies an external force to the beam at the point where the spring-mass system is attached. This external force is special because it is a function of the deflection of the beam at the point where it is attached. The functional relationship is derived from the free body diagram of the combined system.

Figure 13.4 shows a beam carrying a single DOF spring-mass system and the free body diagram. The rigid mass block has mass m and the linear elastic spring has spring constant K_s. The transverse deflections of the center of gravity of the rigid mass block and the point on the beam where the spring-mass is attached, measured from the equilibrium positions, are y_m and y_1, respectively. Assuming that the transverse deflection of the rigid mass block is greater than that of point on the beam where the spring-mass system is attached, the equation of motion of the single DOF spring-mass system is obtained from the free body diagram by applying Newton's second law,

$$-F_s = m\ddot{y}_m \tag{13.27a}$$

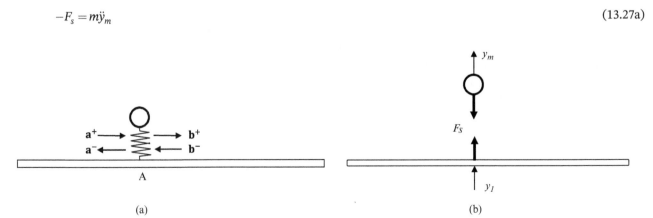

(a)   (b)

Figure 13.4 (a) A beam carrying a single DOF spring-mass system and (b) free body diagram.

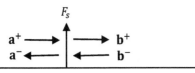

Figure 13.5 Waves generated by an external force.

where

$$F_s = K_s(y_m - y_1) \tag{13.27b}$$

For time harmonic motion of circular frequency ω, $\ddot{y}_m = -\omega^2 y_m$. Substituting this expression and Eq. (13.27b) into (13.27a),

$$y_m = \frac{K_s}{K_s - \omega^2 m} y_1 \tag{13.28}$$

Substituting Eq. (13.28) into Eq. (13.27b),

$$F_s = \beta_0 y_1 \tag{13.29}$$

where $\beta_0 = \dfrac{\omega^2 M K_s}{K_s - \omega^2 M}$.

An externally applied force injects vibration waves into a continuous beam. The relationships between the injected waves and an externally applied point transverse force F_s, as shown in Figure 13.5, are found in Chapters 3 and 9 by considering the continuity and equilibrium conditions at the point where the external force is applied,

$$\mathbf{b}^+ - \mathbf{a}^+ = \mathbf{q}, \ \mathbf{b}^- - \mathbf{a}^- = -\mathbf{q} \tag{13.30}$$

where

$$\mathbf{q} = \begin{bmatrix} -i \\ -1 \end{bmatrix} \frac{F_s}{4EIk^3} \ \text{(by the Euler – Bernoulli bending vibration theory)} \tag{13.31a}$$

$$\mathbf{q} = \begin{bmatrix} -iN \\ -P \end{bmatrix} \frac{F_s}{2GA\kappa(Nk_1 - Pk_2)} \ \text{(by the Timoshenko bending vibration theory)} \tag{13.31b}$$

F_s in Eqs. (13.31a) and (13.31b) is related to y_1 by Eq. (13.29). From the wave vibration standpoint, deflection y_1 can be expressed in terms of the related wave components,

$$y_1 = \begin{bmatrix} 1 & 1 \end{bmatrix} \mathbf{a}^+ + \begin{bmatrix} 1 & 1 \end{bmatrix} \mathbf{a}^- \tag{13.32}$$

Note that in Eq. (13.32), y_1 can also be expressed in terms of $\mathbf{b}^+$ and $\mathbf{b}^-$ because of the continuity in deflections on both sides of the point where the spring is attached.

From Eqs. (13.29) to (13.32), the matrix equations that describe the relationships of the vibration waves at the point where the spring is attached are

$$\begin{bmatrix} -1 + i\beta_0/\alpha_1 & i\beta_0/\alpha_1 \\ \beta_0/\alpha_1 & -1 + \beta_0/\alpha_1 \end{bmatrix} \mathbf{a}^+ + \begin{bmatrix} i\beta_0/\alpha_1 & i\beta_0/\alpha_1 \\ \beta_0/\alpha_1 & \beta_0/\alpha_1 \end{bmatrix} \mathbf{a}^- + \begin{bmatrix} 1 & 0 \\ 0 & 1 \end{bmatrix} \mathbf{b}^+ = 0,$$

$$\begin{bmatrix} i\beta_0/\alpha_1 & i\beta_0/\alpha_1 \\ \beta_0/\alpha_1 & \beta_0/\alpha_1 \end{bmatrix} \mathbf{a}^+ + \begin{bmatrix} 1 + i\beta_0/\alpha_1 & i\beta_0/\alpha_1 \\ \beta_0/\alpha_1 & 1 + \beta_0/\alpha_1 \end{bmatrix} \mathbf{a}^- - \begin{bmatrix} 1 & 0 \\ 0 & 1 \end{bmatrix} \mathbf{b}^- = 0; \tag{13.33a}$$

$$\left(\text{by the Euler – Bernoulli bending vibration theory} \right)$$

$$\begin{bmatrix} -1+iN\beta_0/\alpha_2 & iN\beta_0/\alpha_2 \\ P\beta_0/\alpha_2 & -1+P\beta_0/\alpha_2 \end{bmatrix}\mathbf{a}^+ + \begin{bmatrix} iN\beta_0/\alpha_2 & iN\beta_0/\alpha_2 \\ P\beta_0/\alpha_2 & P\beta_0/\alpha_2 \end{bmatrix}\mathbf{a}^- + \begin{bmatrix} 1 & 0 \\ 0 & 1 \end{bmatrix}\mathbf{b}^+ = 0,$$

$$\begin{bmatrix} iN\beta_0/\alpha_2 & iN\beta_0/\alpha_2 \\ P\beta_0/\alpha_2 & P\beta_0/\alpha_2 \end{bmatrix}\mathbf{a}^+ + \begin{bmatrix} 1+iN\beta_0/\alpha_2 & iN\beta_0/\alpha_2 \\ P\beta_0/\alpha_2 & 1+P\beta_0/\alpha_2 \end{bmatrix}\mathbf{a}^- - \begin{bmatrix} 1 & 0 \\ 0 & 1 \end{bmatrix}\mathbf{b}^- = 0 \tag{13.33b}$$

$\left(\text{by the Timoshenko bending vibration theory}\right)$

where $\beta_0 = \dfrac{\omega^2 M K_s}{K_s - \omega^2 M}$, $\alpha_1 = 4EIk_1^3 = 4EIk_2^3$, and $\alpha_2 = 2GA\kappa(Nk_1 - Pk_2)$.

13.4 Beam with a Two Degrees of Freedom Attachment

Figure 13.6 shows a beam carrying a two DOF spring-mass system and the free body diagram. The mass of the rigid mass block is m. The mass moment of inertia of the rigid mass block about its center of gravity G is I_G. The two DOF spring-mass system is connected to the beam at Points A to B. The two linear elastic springs are with spring constants K_{sA} and K_{sB}, and the distances from Points A and B and the center of gravity G are D_A and D_B, respectively. The transverse deflections of the center of gravity of the rigid mass block G and Points A and B on the beam where the springs are attached are y_m, y_A, and y_B, respectively. The angular rotation of the rigid mass block is α_m. Assuming that the transverse deflections of the rigid mass block are greater than those of the two points on the beam where the springs are attached, the equations of motion of the rigid mass block are obtained from the free body diagram by applying Newton's second law for translational and rotational motion,

$$-F_{sA} - F_{sB} = m\ddot{y}_m, \quad F_{sA}D_A - F_{sB}D_B = I_G\ddot{\alpha}_m \tag{13.34}$$

where

$$F_{sA} = K_{sA}(y_m - y_A - D_A\alpha_m), \quad F_{sB} = K_{sB}(y_m - y_B + D_B\alpha_m) \tag{13.35}$$

For time harmonic motion of the spring-mass system with frequency ω, $\ddot{y}_m = -\omega^2 y_m$ and $\ddot{\alpha}_m = -\omega^2 \alpha_m$. Substituting these expressions and Eq. (13.35) into Eq. (13.34) and putting the equations into matrix form,

$$\begin{bmatrix} K_{sA} + K_{sB} - \omega^2 m & -K_{sA}D_A + K_{sB}D_B \\ K_{sA}D_A - K_{sB}D_B & -K_{sA}D_A^2 - K_{sB}D_B^2 + \omega^2 I_G \end{bmatrix}\begin{bmatrix} y_m \\ \alpha_m \end{bmatrix} = \begin{bmatrix} K_{sA} & K_{sB} \\ K_{sA}D_A & -K_{sB}D_B \end{bmatrix}\begin{bmatrix} y_A \\ y_B \end{bmatrix} \tag{13.36}$$

From Eq. (13.36), the deflection and angular rotation of the discrete rigid body mass y_m and α_m are solved in terms of y_A and y_B, which are the deflections of the two points on the beam where the two springs are attached,

$$\begin{bmatrix} y_m \\ \alpha_m \end{bmatrix} = \begin{bmatrix} K_{sA} + K_{sB} - \omega^2 m & -K_{sA}D_A + K_{sB}D_B \\ K_{sA}D_A - K_{sB}D_B & -K_{sA}D_A^2 - K_{sB}D_B^2 + \omega^2 I_G \end{bmatrix}^{-1}\begin{bmatrix} K_{sA} & K_{sB} \\ K_{sA}D_A & -K_{sB}D_B \end{bmatrix}\begin{bmatrix} y_A \\ y_B \end{bmatrix} \tag{13.37}$$

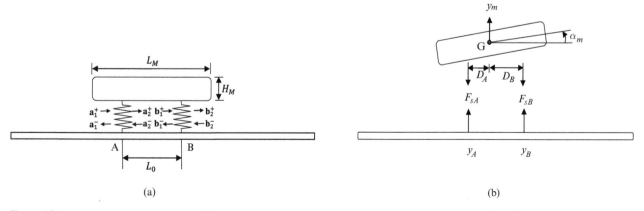

(a)   (b)

Figure 13.6 (a) A beam carrying a two DOF spring-mass system and (b) free body diagram (Courtesy: (Mei 2018)).

Equation (13.37) can be written in a simpler form,

$$\begin{bmatrix} y_m \\ \alpha_m \end{bmatrix} = \begin{bmatrix} \gamma_{11} & \gamma_{12} \\ \gamma_{21} & \gamma_{22} \end{bmatrix} \begin{bmatrix} y_A \\ y_B \end{bmatrix} \tag{13.38}$$

where γ_{nj} $(n, j = 1, 2)$ denotes the corresponding elements of the coefficient matrix of Eq. (13.37).

In scalar form, Eq. (13.38) is

$$y_m = \gamma_{11} y_A + \gamma_{12} y_B, \ \alpha_m = \gamma_{21} y_A + \gamma_{22} y_B \tag{13.39}$$

Substituting Eq. (13.39) into Eq. (13.35),

$$F_{sA} = \beta_{11} y_A + \beta_{12} y_B, \ F_{sB} = \beta_{21} y_A + \beta_{22} y_B \tag{13.40}$$

where

$$\begin{aligned} \beta_{11} &= K_{sA} \left(\gamma_{11} - D_A \gamma_{21} - 1 \right), \ \beta_{12} = K_{sA} \left(\gamma_{12} - D_A \gamma_{22} \right) \\ \beta_{21} &= K_{sB} \left(\gamma_{11} + D_B \gamma_{21} \right), \ \beta_{22} = K_{sB} \left(\gamma_{12} - 1 + D_B \gamma_{22} \right) \end{aligned} \tag{13.41}$$

Equation (13.40) describes the relationships between the spring forces and deflections of the two points on the beam where the two springs are attached.

Applying the relationships between the injected waves and an externally applied transverse force described in Eqs. (13.30) and (13.31) to the waves generated by spring forces F_{sA} and F_{sB} on a beam carrying a two DOF spring-mass system shown in Figure 13.6a, the relationships between the waves and forces at Points A and B are

$$\mathbf{a}_2^+ - \mathbf{a}_1^+ = \mathbf{q}_1, \ \mathbf{a}_2^- - \mathbf{a}_1^- = -\mathbf{q}_1 \tag{13.42}$$

$$\mathbf{b}_2^+ - \mathbf{b}_1^+ = \mathbf{q}_2, \ \mathbf{b}_2^- - \mathbf{b}_1^- = -\mathbf{q}_2 \tag{13.43}$$

where

$$\mathbf{q}_1 = \begin{bmatrix} -i \\ -1 \end{bmatrix} \frac{F_{sA}}{4EIk^3}, \ \mathbf{q}_2 = \begin{bmatrix} -i \\ -1 \end{bmatrix} \frac{F_{sB}}{4EIk^3} \ (\text{by the Euler} - \text{Bernoulli bending vibration theory}) \tag{13.44a}$$

$$\mathbf{q}_1 = \begin{bmatrix} -iN \\ -P \end{bmatrix} \frac{F_{sA}}{2GA\kappa \left(Nk_1 - Pk_2 \right)}, \ \mathbf{q}_2 = \begin{bmatrix} -iN \\ -P \end{bmatrix} \frac{F_{sB}}{2GA\kappa \left(Nk_1 - Pk_2 \right)} \ (\text{by the Timoshenko bending vibration theory}) \tag{13.44b}$$

F_{sA} and F_{sB} are related to y_A and y_B, the deflections of the two pints where the springs are attached, by Eq. (13.40). From the wave vibration standpoint, deflections y_A and y_B can be expressed in terms of their related wave components,

$$y_A = \begin{bmatrix} 1 & 1 \end{bmatrix} \mathbf{a}_2^+ + \begin{bmatrix} 1 & 1 \end{bmatrix} \mathbf{a}_2^-, \ y_B = \begin{bmatrix} 1 & 1 \end{bmatrix} \mathbf{b}_1^+ + \begin{bmatrix} 1 & 1 \end{bmatrix} \mathbf{b}_1^- \tag{13.45}$$

Because of the continuity in deflections on both sides of each spring attached point, y_A and y_B in Eq. (13.45) can also be expressed in other forms such as in terms of $\mathbf{a}_1^\pm$ and $\mathbf{b}_2^\pm$; or $\mathbf{a}_1^\pm$ and $\mathbf{b}_1^\pm$; or $\mathbf{a}_2^\pm$ and $\mathbf{b}_2^\pm$.

From Eqs. (13.40) to (13.45), the coupled relationships of vibration waves at discontinuities caused by the two spring attachments are (Mei 2018 and 2021)

$$\begin{bmatrix} \mathbf{A}_{11} & \vdots & \mathbf{A}_{12} \\ \cdots & \vdots & \cdots \\ \mathbf{A}_{21} & \vdots & \mathbf{A}_{22} \end{bmatrix} \begin{bmatrix} \mathbf{a}_1^+ \\ \mathbf{a}_1^- \\ \mathbf{a}_2^+ \\ \mathbf{a}_2^- \end{bmatrix} + \begin{bmatrix} \mathbf{B}_{11} & \vdots & \mathbf{B}_{12} \\ \cdots & \vdots & \cdots \\ \mathbf{B}_{21} & \vdots & \mathbf{B}_{22} \end{bmatrix} \begin{bmatrix} \mathbf{b}_1^+ \\ \mathbf{b}_1^- \\ \mathbf{b}_2^+ \\ \mathbf{b}_2^- \end{bmatrix} = 0 \tag{13.46}$$

where

$$\begin{aligned} \mathbf{A}_{11} &= -\mathbf{B}_{22} = \mathbf{D}(4), \ \mathbf{A}_{12} = -\mathbf{D}(4) + \beta_{11} \mathbf{O}(4), \ \mathbf{A}_{21} = \mathbf{B}_{12} = 0; \\ \mathbf{A}_{22} &= \beta_{21} \mathbf{O}(4), \ \mathbf{B}_{11} = \beta_{12} \mathbf{O}(4), \ \mathbf{B}_{21} = \mathbf{D}(4) + \beta_{22} \mathbf{O}(4) \end{aligned} \tag{13.47}$$

In which $\mathbf{O}(4)$ denotes a constant square matrix of size 4 by 4 with each element being unity,

$$\mathbf{O}(4) = \begin{bmatrix} 1 & 1 & 1 & 1 \\ 1 & 1 & 1 & 1 \\ 1 & 1 & 1 & 1 \\ 1 & 1 & 1 & 1 \end{bmatrix} \tag{13.48}$$

and $\mathbf{D}(4)$ is a square diagonal matrix of size 4 by 4 whose elements are vibration theory dependent,

$$\mathbf{D}(4) = \begin{bmatrix} i\alpha_1 & & & \\ & -\alpha_1 & & \\ & & -i\alpha_1 & \\ & & & \alpha_1 \end{bmatrix} \left(\text{by the Euler} - \text{Bernoulli bending vibration theory}\right) \tag{13.49a}$$

$$\mathbf{D}(4) = \begin{bmatrix} i\alpha_2/N & & & \\ & -\alpha_2/P & & \\ & & -i\alpha_2/N & \\ & & & \alpha_2/P \end{bmatrix} \left(\text{by the Timoshenko bending vibration theory}\right) \tag{13.49b}$$

13.5 Vibration Analysis of a Beam with Intermediate Discontinuities

From the wave point of view, vibrations propagate along uniform structural elements, are reflected at boundaries, and are reflected and transmitted at each intermediate structural discontinuity. In addition, an external excitation, if it exists, has the effect of injecting waves into a continuous structure. Assembling these wave relationships, both free and forced vibration analysis of a structure containing various structural discontinuities are obtained systematically.

Figure 13.7 shows a uniform beam CD of length L that is subjected to an external excitation at Point G and comprises two intermediate structural discontinuities at Points A and B.

The beam is divided into four uniform waveguides by the boundaries, intermediate discontinuities, and external excitation. Vibrations propagate along the uniform waveguides CG, GA, AB, and BD. The four pairs of propagation relationships along the uniform beam elements are

$$\begin{aligned} \mathbf{g}_1^+ &= \mathbf{f}(L_{11})\mathbf{c}^+, \quad \mathbf{c}^- = \mathbf{f}(L_{11})\mathbf{g}_1^- \\ \mathbf{a}_1^+ &= \mathbf{f}(L_{12})\mathbf{g}_2^+, \quad \mathbf{g}_2^- = \mathbf{f}(L_{12})\mathbf{a}_1^- \\ \mathbf{b}_1^+ &= \mathbf{f}(L_0)\mathbf{a}_2^+, \quad \mathbf{a}_2^- = \mathbf{f}(L_0)\mathbf{b}_1^- \\ \mathbf{d}^+ &= \mathbf{f}(L - L_1 - L_0)\mathbf{b}_2^+, \quad \mathbf{b}_2^- = \mathbf{f}(L - L_1 - L_0)\mathbf{d}^- \end{aligned} \tag{13.50}$$

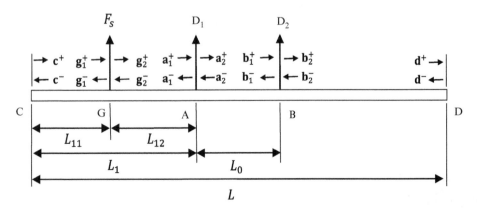

Figure 13.7 Waves in a beam with intermediate structural discontinuities.

where the bending propagation matrix for a distance x is $\mathbf{f}(x) = \begin{bmatrix} e^{-ik_1 x} & 0 \\ 0 & e^{-k_2 x} \end{bmatrix}$.

At Boundaries C and D, incoming waves are reflected. The incident and reflected vibration waves are related by the reflection matrices $\mathbf{r}_C$ and $\mathbf{r}_D$,

$$\mathbf{c}^+ = \mathbf{r}_C \mathbf{c}^-, \ \mathbf{d}^- = \mathbf{r}_D \mathbf{d}^+ \tag{13.51}$$

At Point G where the external force is applied,

$$\mathbf{g}_2^+ - \mathbf{g}_1^+ = \mathbf{q}, \ \mathbf{g}_2^- - \mathbf{g}_1^- = -\mathbf{q} \tag{13.52}$$

where load vector $\mathbf{q}$ by the Euler–Bernoulli and Timoshenko bending vibration theories are obtained in Chapters 3 and 9, respectively. They are given in Eqs. (13.31a) and (13.31b).

Wave relationships at each of the intermediate discontinuities are dependent upon the type of structural discontinuity. For example, when a discontinuity is caused by a geometric/material change or a point attachment,

$$\mathbf{a}_1^- = \mathbf{r}_{LL}\mathbf{a}_1^+ + \mathbf{t}_{RL}\mathbf{a}_2^-, \ \mathbf{a}_2^+ = \mathbf{r}_{RR}\mathbf{a}_2^- + \mathbf{t}_{LR}\mathbf{a}_1^+ \tag{13.53}$$

where the transmission and reflection matrices are obtained in Sections 13.1 and 13.2.

When a discontinuity is caused by the attachment of a single DOF spring-mass system, waves at the discontinuity are related by Eqs. (13.33a) and (13.33b). When a discontinuity is caused by the attachment of a two DOF spring-mass system, waves at the discontinuities are related by Eqs. (13.46) to (13.49).

Assembling these propagation, reflection, transmission, and force generated wave relationships into matrix form,

$$\mathbf{A}_f \mathbf{z}_f = \mathbf{F} \tag{13.54}$$

where $\mathbf{A}_f$ is a square coefficient matrix, $\mathbf{z}_f$ is a wave vector, and $\mathbf{F}$ is a vector related to external excitations.

From Eq. (13.54), the wave components are

$$\mathbf{z}_f = \mathbf{A}_f^{-1}\mathbf{F} \tag{13.55}$$

With the availability of wave components, the responses at any point on the beam can be obtained. For example, the response of a point between Boundary C and external force/moment applied Point G that is distance x away from Point C is

$$y = \begin{bmatrix} 1 & 1 \end{bmatrix} \mathbf{f}(x)\mathbf{c}^+ + \begin{bmatrix} 1 & 1 \end{bmatrix} \mathbf{f}(-x)\mathbf{c}^- \tag{13.56a}$$

or equivalently is

$$y = \begin{bmatrix} 1 & 1 \end{bmatrix} \mathbf{f}(L_{11} - x)\mathbf{g}_1^- + \begin{bmatrix} 1 & 1 \end{bmatrix} \mathbf{f}(-(L_{11} - x))\mathbf{g}_1^+ \tag{13.56b}$$

The natural frequencies of the beam can be found by setting the determinant of $\mathbf{A}_f$ to zero, which gives a polynomial equation with circular frequency ω being the variable. Solving for the roots of the characteristic polynomial equation gives the natural frequencies of the beam that comprises various types of discontinuities.

13.6 Numerical Examples

Bending vibrations of a uniform steel beam, consisting of various types of discontinuities, are analyzed based on the Timoshenko bending vibration theory.

The material and geometrical properties of the example steel beam are: the Young's modulus E is 198.87 GN/m^2, shear modulus G is 77.5 GN/m^2, Poisson's ratio ν is 0.30, and mass density ρ is 7664.5 kg/m^3. The shear coefficient κ is related to the Poisson's ratio ν by $\kappa = 10(1+\nu)/(12+11\nu)$. The cross section of the beam is a rectangular shape whose width and thickness are 25.4 mm and 12.7 mm (or 1.0 in and 0.5 in), respectively.

The length of the beam L is 2.0 m, and both boundaries of the beam are free. The discontinuity occurs at $0.55L$, the point external force is applied at $0.22L$, and the response is observed at $0.64L$, all distances are measured from the left boundary of the beam.

Both free and forced vibrations are obtained for the example beam with a single discontinuity component of various types. These include discontinuity from a step change, discontinuities from mounting a two DOF spring-mass system, discontinuity from mounting a single DOF spring-mass system, and discontinuity from a spring/damper attachment.

a) Discontinuity from a step change

Assuming the steel beam undergoes a geometrical step change in the thickness, as shown in Figure 13.2, without change in materials. Figures 13.8 and 13.9 show the overlaid magnitudes of the characteristic polynomials and the overlaid receptance frequency responses of the beam before and after the step changes in its thickness, corresponding to thickness ratios (right/left) of 0.7 and 0.5, respectively. The natural frequencies are the x values of the local minima in the magnitude plots of the characteristic polynomial, or those of the local maxima in the receptance frequency response. A discontinuity caused by a thickness change modifies the vibration characteristics of a beam by altering its natural frequencies.

MATLAB scripts for free and forced wave vibration analysis of a stepped Timoshenko beam are available in Section 13.7.

b) Discontinuities from mounting a two DOF spring-mass system

Consider the uniform steel beam carrying a two DOF spring-mass system, as shown in Figure 13.6. It is assumed that the right side spring is mounted at $0.55L$, measured from the left boundary, the spacing between the left and right springs is $L_0 = 0.20L$, and the center of gravity G of the mass block is of equal distance to the two springs, namely, $D_A = D_B = 0.10L$. The spring constants of the two springs are assumed equal, that is, $K_{sA} = K_{sB} = K_s$. The length L_m and height H_m of the rigid mass block are $L_m = 1.20L_0$ and $H_m = 0.30L_0$. The mass moment of inertia of the rigid mass block about its center of gravity is $I_G = m\ (L_m^2 + H_m^2)$.

The natural frequencies of the two DOF spring-mass system by itself can be found as a special situation by setting y_A and y_B on the right side of Eq. (13.36) to zero. Physically, this is equivalent to having the two DOF spring-mass system attached to a fixed unmoving frame,

$$\begin{bmatrix} K_{sA} + K_{sB} - \omega^2 m & -K_{sA}D_A + K_{sB}D_B \\ K_{sA}D_A - K_{sB}D_B & -K_{sA}D_A^2 - K_{sB}D_B^2 + \omega^2 I_G \end{bmatrix} \begin{bmatrix} y_m \\ \alpha_m \end{bmatrix} = 0 \tag{13.57}$$

Setting the determinant of the coefficient matrix of Eq. (13.57) to zero gives a quadratic polynomial in terms of ω^2. The two positive roots of this characteristic polynomial are the two natural frequencies of the two DOF spring-mass system. The two natural frequencies can also be obtained numerically by plotting the magnitude of the characteristic polynomial of the two DOF spring-mass system, as shown in Figure 13.10, and finding the x values of the local minima.

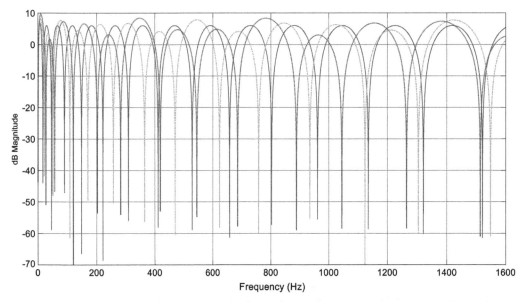

Figure 13.8 Magnitudes of the characteristic polynomials of the beam: without (___), and with a step change of thickness ratio 0.7 (_._._) and 0.5 (...), respectively.

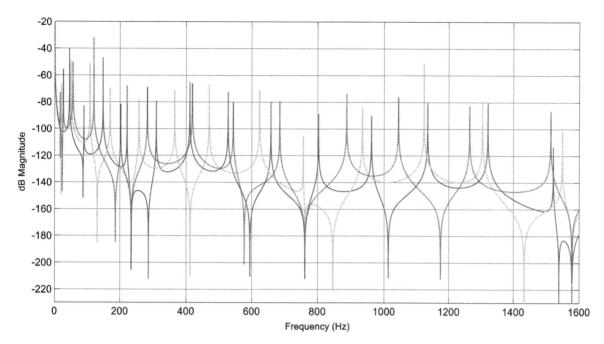

Figure 13.9 Receptance frequency response of the beam: without (___) and with a step change of thickness ratio 0.7 (_._._) and 0.5 (...), respectively.

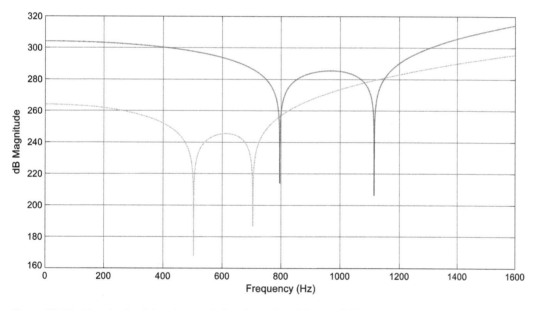

Figure 13.10 Magnitude of the characteristic polynomial of the two DOF spring-mass system of $m = 2\,kg$ and $K_s = 10^7\,N/m$ (_._._), and $m = 8\,kg$ and $K_s = 10^8\,N/m$ (...), respectively.

Figures 13.11 and 13.12 present the overlaid magnitudes of the characteristic polynomials and the overlaid receptance frequency responses of the uniform beam before and after the attachment of a two DOF discrete system, corresponding to $m = 2\,kg$ and $K_s = 10^7\,N/m$, and $m = 8\,kg$ and $K_s = 10^8\,N/m$, respectively. The two natural frequencies of the two DOF spring-mass system are marked on Figures 13.11 and 13.12, the characteristic and receptance response plots of the beam, using vertical lines. The large magnitudes in Figures 13.10 and 13.11 are caused by the close to singular or badly scaled coefficient matrices.

The attached two DOF spring-mass system changes the vibration characteristics of the beam by altering its natural frequencies, which correspond to the x values of the local minima in the magnitude of the characteristic polynomial or those of the local maxima of the receptance frequency response.

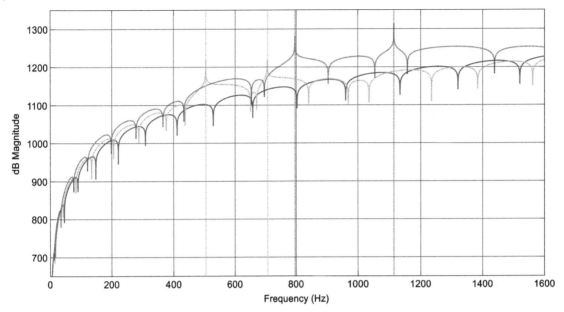

Figure 13.11 Magnitudes of the characteristic polynomials of the uniform beam: without (__), and with a two DOF attachment of $m = 2\,kg$ and $K_s = 10^7\,N/m$ (_._._), and $m = 8\,kg$ and $K_s = 10^8\,N/m$ (…), respectively.

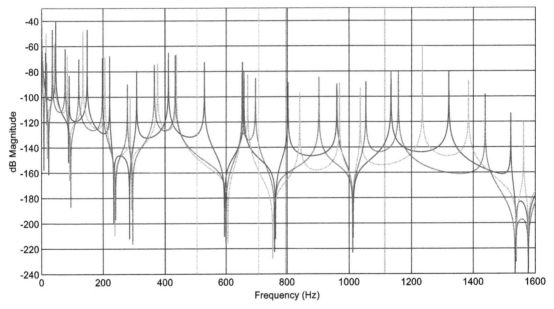

Figure 13.12 Receptance frequency response of the uniform beam: without (__), and with a two DOF attachment of $m = 2\,kg$ and $K_s = 10^7\,N/m$ (_._._) and $m = 8\,kg$ and $K_s = 10^8\,N/m$ (…), respectively.

c) Discontinuity from mounting a single DOF spring-mass system

Consider the uniform steel beam carrying a single DOF spring-mass system, as shown in Figure 13.4. The natural frequency of the single DOF spring-mass system by itself can be found as a special situation of Eqs. (13.27a) and (13.27b) by first setting y_1 in Eq. (13.27b) to zero. Physically, this is equivalent to having the single DOF spring-mass system attached to a fixed unmoving frame. Substituting Eq. (13.27b) into (13.27a) and recalling that $\ddot{y}_m = -\omega^2 y_m$ for time harmonic motion,

$$\left(m\omega^2 - K_s\right)y_m = 0 \tag{13.58}$$

Setting the coefficient of y_m in Eq. (13.58) to zero gives the well-known natural frequency formula of a single DOF spring-mass system, namely, $\omega_n = \sqrt{K_s/m}$ in rad/s.

Figures 13.13 and 13.14 present the overlaid magnitudes of the characteristic polynomials and the overlaid receptance frequency responses of the uniform beam before and after the attachment of a single DOF discrete system, corresponding to $m = 2\,kg$ and $K_s = 10^7\,N/m$, and $m = 8\,kg$ and $K_s = 10^8\,N/m$, respectively.

The natural frequency of the corresponding single DOF spring-mass system is marked on Figures 13.13 and 13.14 using vertical lines.

The attached single DOF spring-mass system changes the vibration characteristics of the beam by altering its natural frequencies, which correspond to the x values of the local minima in the magnitudes of the characteristic polynomials, or those of the local maxima of the receptance frequency responses.

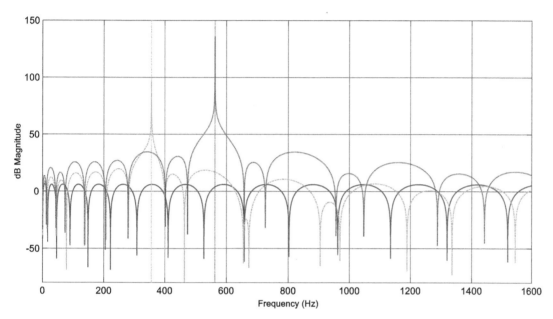

Figure 13.13 Magnitudes of the characteristic polynomials of the uniform beam: without (__), and with a single DOF attachment of $m = 2\,kg$ and $K_s = 10^7\,N/m$ (_._._), and $m = 8\,kg$ and $K_s = 10^8\,N/m$ (...), respectively.

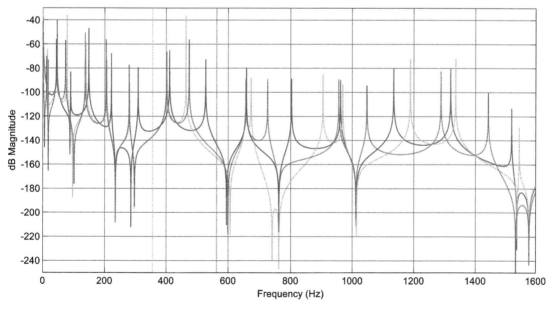

Figure 13.14 Receptance frequency response of the uniform beam: without (__), and with a single DOF attachment of $m = 2\,kg$ and $K_s = 10^7\,N/m$ (_._._), and $m = 8\,kg$ and $K_s = 10^8\,N/m$ (...), respectively.

d) Discontinuity from a spring/damper attachment

Consider the uniform steel beam attached to a spring, as shown in Figure 13.3. Recall that the viscous damping effect, if it exists, can be taken into account by adding a frequency dependent imaginary term to the spring stiffness, that is, by introducing a dynamic spring stiffness. For example, for the beam attached to a translational spring of stiffness $\overline{K_T}$, $i\omega\overline{C_T}$ term can be added to the translational stiffness $\overline{K_T}$ when viscous damping exists. The spring stiffness becomes complex, which is $\overline{K_T} + i\omega\overline{C_T}$. If the beam is attached to a viscous damper instead of a spring, the spring stiffness is pure imaginary and becomes $i\omega\overline{C_T}$. In the studies that follow, the effects of a spring only, a damper only, and a spring and damper combination are examined.

Figures 13.15 and 13.16 present the overlaid magnitudes of the characteristic polynomials and the overlaid receptance frequency responses of the uniform beam before and after the attachment of a translational spring, corresponding to spring stiffness $K_s = 10^7\,N/m$ and $K_s = 10^8\,N/m$, respectively. The spring attachment changes the vibration characteristics of the

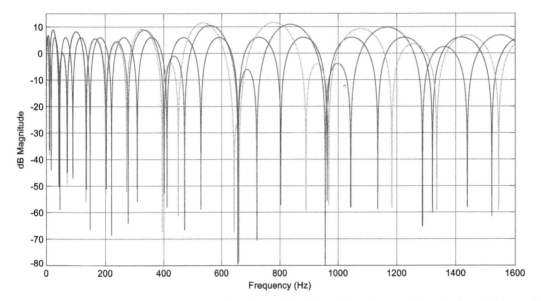

Figure 13.15 Magnitudes of the characteristic polynomials of the uniform beam: without (___), and with a spring attachment of $K_s = 10^7\,N/m$ (_._._) and $K_s = 10^8\,N/m$ (...), respectively.

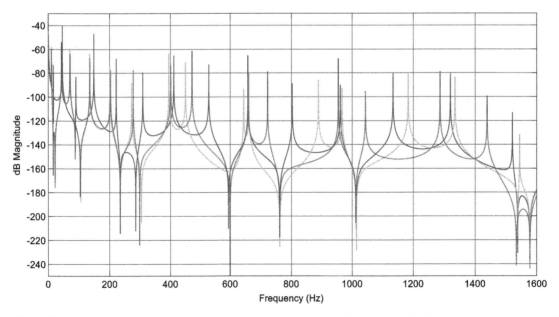

Figure 13.16 Receptance frequency response of the uniform beam: without (___), and with a spring attachment of $K_s = 10^7\,N/m$ (_._._) and $K_s = 10^8\,N/m$ (...), respectively.

beam by altering its natural frequencies, which correspond to the *x* values of the local minima of the magnitudes of the characteristic polynomials or those of the local maxima of the receptance frequency responses.

Figures 13.17 and 13.18 present the overlaid magnitudes of the characteristic polynomials and the overlaid receptance frequency responses of the uniform beam before and after the attachment of a viscous damper whose damping constant is $C = 10^3 \, Ns/m$. The damper attachment significantly reduces the sharpness of resonance at the natural frequencies, therefore, suppressing the vibrations in the beam.

Figures 13.19 and 13.20 present the overlaid magnitudes of the characteristic polynomials and the overlaid receptance frequency responses of the uniform beam, after the attachment of a spring with stiffness $K_s = 10^7 \, N/m$, and after the attachment of a spring and viscous damper combination, with spring of stiffness $K_s = 10^7 \, N/m$ and viscous damping

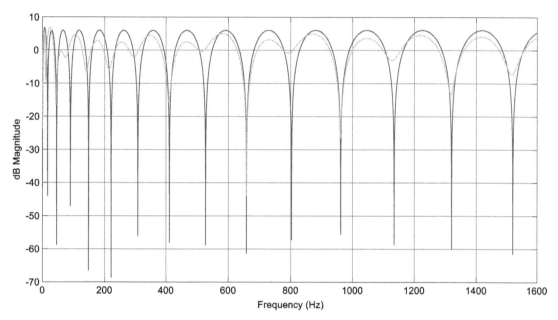

Figure 13.17 Magnitudes of the characteristic polynomials of the uniform beam: without (___), and with a damper attachment of $C = 1000 \, Ns/m$ (_._._).

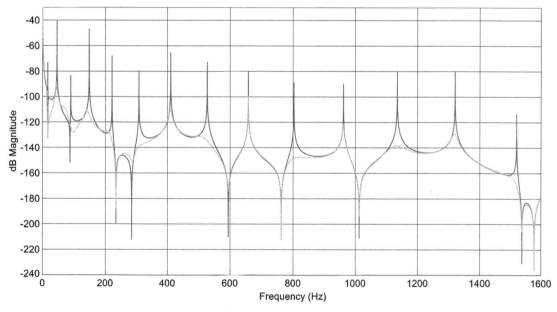

Figure 13.18 Receptance frequency responses of the uniform beam: without (___), and with a damper attachment of $C = 1000 \, Ns/m$ (_._._).

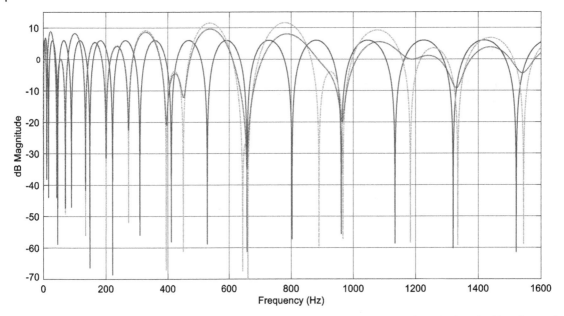

Figure 13.19 Magnitudes of the characteristic polynomials of the uniform beam: without (___), and with only a spring attachment of $K_s = 10^7\, N/m$ (_._._.), and with a combined spring-damper attachment of $K_s = 10^7\, N/m$ and $C = 1000\, Ns/m$ (...), respectively.

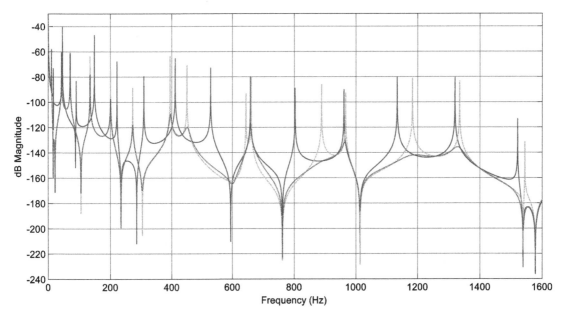

Figure 13.20 Receptance frequency responses of the uniform beam: without (___), and with only a spring attachment of $K_s = 10^7\, N/m$ (_._._.), and with a combined spring-damper attachment of $K_s = 10^7\, N/m$ and $C = 1000\, Ns/m$ (...), respectively.

constant $C = 10^3\, Ns/m$. As observed earlier, the spring alone changes the natural frequencies of the beam. The addition of a damper to the spring attachment reduces the sharpness of resonances at the natural frequencies altered by the attached spring. Although the combined effect is only demonstrated under this spring-damper attachment scenario, the same phenomenon applies to all previous studies involving springs.

In conclusion, passive vibration control can be achieved by creating structural discontinuities to change the vibration characteristics of a structure and by adding damping to suppress resonant vibrations in the structure.

Periodical structures are known to form passing and stopping bands to broad band frequency vibrations (Mead 1996). A periodical structure can be constructed, for example, by mounting or attaching multiple identical discontinuity components on a uniform beam. Figure 13.21 illustrates such an example structure supported by multiple evenly spaced translational springs.

Consider the example uniform steel beam that is supported by various numbers of evenly spaced identical translational springs whose stiffness is $K_s = 10^7 \, N/m$.

Figure 13.22 shows the overlaid magnitudes of characteristic polynomials of the uniform beam before and after carrying five, ten, and fifteen evenly spaced spring attachments. Magnified responses of individual magnitude plots of the characteristic polynomials of Figure 13.22 are presented in Figure 13.23. MATLAB scripts for free wave vibration analysis of a uniform beam with periodical spring attachments are available in Section 13.7.

Recall that on the magnitude plot of the characteristic polynomial of a beam, the x values of the local minima are natural frequencies of the beam. Figures 13.22 and 13.23 show that the local minima of the uniform beam are clustered after being supported by evenly spaced springs. When the material and geometrical properties of the system remain unchanged, the location and length of the clustered bands, as well as gaps between the adjacent clustered groups, are dependent on the number of discrete spring attachments.

This confirms the forming of passing and stopping bands of a periodical structure. The clustered local minima correspond to vibration passing bands because they correspond to the natural frequencies, and hence the resonant responses of the structure. The gaps between two adjacent groups of clusters are vibration stopping bands.

The filtering characteristics of a periodical structure consisting of evenly spaced spring attachments is localized in the lower frequency range. As the number of discrete spring attachments increases, the periodical design is seen to push the natural frequencies farther away from zero frequency and form a wider range of vibration stopping band in this region. Such a design is useful in suppressing low frequency mechanical vibrations. A common application is in railway/railroad track design: the steel rails are held down to periodical sleepers/ties using resilient fastenings, often with rubber pads to provide an additional damping effect.

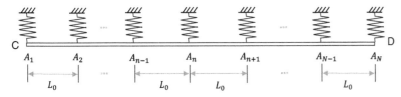

Figure 13.21 A uniform beam supported by multiple evenly spaced translational springs.

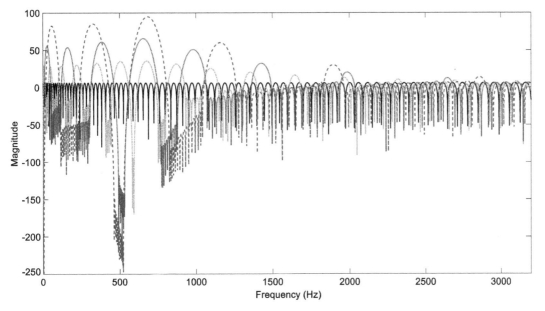

Figure 13.22 Magnitudes of the characteristic polynomials of the uniform beam: without (___), and with five (-·-·), ten (...), and fifteen (---) spring attachments, respectively.

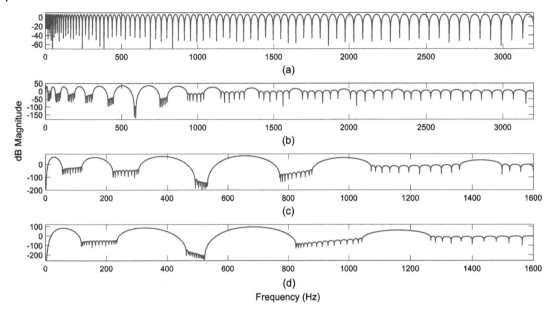

Figure 13.23 Magnitude of the characteristic polynomial of the uniform beam with various numbers of spring attachments: (a) zero, (b) five, (c) ten, and (d) fifteen.

In Chapter 5, the reflection and transmission matrices at a general intermediate spring support that exerts translational, bending, and torsional constraints on a uniform composite beam of coupled bending-torsional vibration are derived. The reflection and transmission are functions of the stiffnesses of the springs. This indicates that the same passive vibration control strategy, that is, altering the vibration characteristics of the structure by creating structural discontinuities and suppressing vibration resonances by adding damping, is applicable to structures that undergo both simple and coupled vibration motions.

13.7 MATLAB Scripts

<u>Script 1:</u> Scripts for free and forced wave vibration analysis of a stepped Timoshenko beam

clc; clear all; close all;

L=2 %Length
b=2.54/100 %Width
h=0.5*2.54/100 %Thickness

%Steel beam
E=198.87*10^9;
p=7664.5
v=0.3;
K=10*(1+v)/(12+11*v)
G=77.5*10^9;

A=b*h;
I=b*h^3/12;
EA=E*A;
EI=E*I;
KAG=K*A*G
GAk=G*A*K

```
%Left side

CbL=sqrt(E*I/p/A);
CsL=sqrt(G*K*A/p/A);
CrL=sqrt(p*I/p/A);
wcL=CsL/CrL
fcL=wcL/(2*pi)

% Right/Left=Sigma---the thickness ratio
% Assuming the width and the materials are the same on both sides

%Right side

StepChange = input('Enter 1 for with step change, enter any other number for uniform beam: ')

if StepChange == 1
   ThicknessRatio=input('Thickness ratio (right/left): ')
   else
   ThicknessRatio=1
end

SigmaL=ThicknessRatio

SigmaR=1/SigmaL

BetaL=SigmaL^3
GamaL=SigmaL

BetaR=SigmaR^3
GamaR=SigmaR

CbR=SigmaR*CbL;
CsR=CsL;
CrR=SigmaR*CrL;
wcR=CsR/CrR
fcR=wcR/(2*pi)

L1 = 0.55*L    %Location of step change
L2 = L-L1;

L11=0.40*L1    %Disturbance force applied point
L12=L1-L11;
xd=L11;

%Sensor to view responses at L1+0.20*L2
xsensor=0.20*L2;

%Frequency range in Hz
f1 = 0;   %Starting frequency in Hz
f2 = 1600; %Ending frequency in Hz
stepsize=0.1

freq = f1:stepsize:f2;
```

```
for n = 1:length(freq)

w = 2*pi*freq(n);

kL1=abs(sqrt(0.5*((1/CsL)^2+(CrL/CbL)^2)*w^2+sqrt(w^2/CbL^2+0.25*((1/CsL)^2-(CrL/CbL)^2)^2*w^4)));
kL2=abs(sqrt(0.5*((1/CsL)^2+(CrL/CbL)^2)*w^2-sqrt(w^2/CbL^2+0.25*((1/CsL)^2-(CrL/CbL)^2)^2*w^4)));

kR1=abs(sqrt(0.5*((1/CsR)^2+(CrR/CbR)^2)*w^2+sqrt(w^2/CbR^2+0.25*((1/CsR)^2-(CrR/CbR)^2)^2*w^4)));
kR2=abs(sqrt(0.5*((1/CsR)^2+(CrR/CbR)^2)*w^2-sqrt(w^2/CbR^2+0.25*((1/CsR)^2-(CrR/CbR)^2)^2*w^4)));

%%%%%%%%%%%%%%%%
%When w<wc, k2=k2
%When w>wc, k2=ik2
%%%%%%%%%%%%%%%%%

if w<wcL
kL2=kL2;
else
kL2=i*kL2;
end

PL=(-w^2+CsL^2*kL1^2)/(kL1*CsL^2);
NL=(w^2+CsL^2*kL2^2)/(kL2*CsL^2);

if w<wcR
kR2=kR2;
else
kR2=i*kR2;
end

PR=(-w^2+CsR^2*kR1^2)/(kR1*CsR^2);
NR=(w^2+CsR^2*kR2^2)/(kR2*CsR^2);

%Left side boundary
rclampedLEFT=[(-i*NL+PL)/(i*NL+PL),-2*i*NL/(i*NL+PL);-2*PL/(i*NL+PL),-(-i*NL+PL)/(i*NL+PL)];
rsimpLEFT=[-1,0;0,-1];
rfreeLEFT=[(kL1*PL*kL2-kL1*PL*NL-i*kL2*NL*kL1+i*kL2*NL*PL)/(-kL1*PL*kL2+kL1*PL*NL-
i*kL2*NL*kL1+i*kL2*NL*PL),2*(-kL2+NL)/(-kL1*PL*kL2+kL1*PL*NL-i*kL2*NL*kL1+i*kL2*NL*PL)*kL2*NL;2*i*(-
kL1+PL)/(-kL1*PL*kL2+kL1*PL*NL-i*kL2*NL*kL1+i*kL2*NL*PL)*kL1*PL,(-kL1*PL*kL2+kL1*PL*NL+i*kL2*NL*kL1-i*kL2*NL*PL)/
(-kL1*PL*kL2+kL1*PL*NL-i*kL2*NL*kL1+i*kL2*NL*PL)];

%Right side boundary
rclampedRIGHT=[(-i*NR+PR)/(i*NR+PR),-2*i*NR/(i*NR+PR);-2*PR/(i*NR+PR),-(-i*NR+PR)/(i*NR+PR)];
rsimpRIGHT=[-1,0;0,-1];
rfreeRIGHT=[(kR1*PR*kR2-kR1*PR*NR-i*kR2*NR*kR1+i*kR2*NR*PR)/(-kR1*PR*kR2+kR1*PR*NR-
i*kR2*NR*kR1+i*kR2*NR*PR),2*(-kR2+NR)/(-kR1*PR*kR2+kR1*PR*NR-i*kR2*NR*kR1+i*kR2*NR*PR)*kR2*NR;
2*i*(-kR1+PR)/(-kR1*PR*kR2+kR1*PR*NR-i*kR2*NR*kR1+i*kR2*NR*PR)*kR1*PR,(-kR1*PR*
kR2+kR1*PR*NR+i*kR2*NR*kR1-i*kR2*NR*PR)/(-kR1*PR*kR2+kR1*PR*NR-i*kR2*NR*kR1+i*kR2*NR*PR)];

%Boundary conditions
rc=rfreeLEFT;
rd=rfreeRIGHT;
```

```
fL1 = diag([exp(-i*kL1*L1) exp(-kL2*L1)]);
fL2 = diag([exp(-i*kR1*L2) exp(-kR2*L2)]);

fL11=diag([exp(-i*kL1*L11) exp(-kL2*L11)]);
fL12=diag([exp(-i*kL1*L12) exp(-kL2*L12)]);

fxsensor=diag([exp(-i*kR1*xsensor) exp(-kR2*xsensor)]);

%Forming the characteristic matrix

%Force generated waves

%(g12-)-(g11-)=-qforce+mforce
%(g12+)-(g11+)=qforce+mforce
ShearForce=1;
Moment=0;

qforce=(ShearForce/(2*KAG*(kL2*PL-kL1*NL)))*[i*NL;PL];
mforce=(Moment/(2*EI*(kL1*PL+kL2*NL)))*[-1;1];

%Building up the characteristic matrix

BigA=zeros(24,24);

%Boundary reflections
RefC=[-eye(2) rc];
RefD=[rd -eye(2)];

BigB=[zeros(2,1);zeros(2,1);qforce+mforce;qforce+mforce;zeros(2,1);zeros(2,1);zeros(2,1);zeros(2,1);zeros(2,1);zeros(2,1);
zeros(2,1);zeros(2,1)];
%(g2-)-(g1-)=-qforce+mforce+fforce)
%(g2+)-(g1+)=qforce+mforce-fforce)

ExternalForce1=[zeros(2) eye(2) zeros(2) -eye(2)];
ExternalForce2=[-eye(2) zeros(2) eye(2) zeros(2)];

%Propagations along L1 and L2
PropL11=[fL11 zeros(2) -eye(2) zeros(2); zeros(2) -eye(2) zeros(2) fL11];
PropL12=[fL12 zeros(2) -eye(2) zeros(2); zeros(2) -eye(2) zeros(2) fL12];
PropL2=[fL2 zeros(2) -eye(2) zeros(2); zeros(2) -eye(2) zeros(2) fL2];

if StepChange==1

%Solve for the transmission and reflection matrices at change of section

%From left to right

c1=[PL*kL1,-NL*kL2;i*(-PL+kL1),-NL+kL2];
c2=[-BetaL*PR*kR1,BetaL*NR*kR2;i*GamaL*(-PR+kR1),GamaL*(-NR+kR2)];
c3=[-PL*kL1,NL*kL2;i*(-PL+kL1),-NL+kL2];

d1=[1,1;i*PL,NL];
d2=[-1,-1;i*PR,NR];
d3=[-1,-1;i*PL,NL];
```

```
t12=inv(c2-c1*inv(d1)*d2)*(c3-c1*inv(d1)*d3);
r11=inv(c1-c2*inv(d2)*d1)*(c3-c2*inv(d2)*d3);

%From right to left

c1=[PR*kR1,-NR*kR2;i*(-PR+kR1),-NR+kR2];
c2=[-BetaR*PL*kL1,BetaR*NL*kL2;i*GamaR*(-PL+kL1),GamaR*(-NL+kL2)];
c3=[-PR*kR1,NR*kR2;i*(-PR+kR1),-NR+kR2];

d1=[1,1;i*PR,NR];
d2=[-1,-1;i*PL,NL];
d3=[-1,-1;i*PR,NR];

t21=inv(c2-c1*inv(d1)*d2)*(c3-c1*inv(d1)*d3);
r22=inv(c1-c2*inv(d2)*d1)*(c3-c2*inv(d2)*d3);

else  %No step change (that is, a uniform beam)

t12=eye(2);
t21=eye(2);
r11=zeros(2);
r22=zeros(2);

end

Discontinuity=[r11,-eye(2),zeros(2),t21;t12,zeros(2),-eye(2) r22];

BigA(21:24,13:20)=Discontinuity;

BigA(1:2,1:4)=RefC;
BigA(3:4,21:24)=RefD;
BigA(5:6,5:12)=ExternalForce1;
BigA(7:8,5:12)=ExternalForce2;

BigA(9:12,1:8)=PropL11;
BigA(13:16,9:16)=PropL12;
BigA(17:20,17:24)=PropL2;

ForcedResponse=BigA\BigB;

% c1+ c2+ c1- c2- g11+ g12+ g11- g12- g21+ g22+ g21- g22- [a1+ a2+ a1- a2- b1+ b2+ b1- b2-] d1+ d2+ d1- d2-

cplus=ForcedResponse(1:2);
cminus=ForcedResponse(3:4);

g1plus=ForcedResponse(5:6);
g1minus=ForcedResponse(7:8);
g2plus=ForcedResponse(9:10);
g2minus=ForcedResponse(11:12);

aplus=ForcedResponse(13:14);
aminus=ForcedResponse(15:16);
```

```
bplus=ForcedResponse(17:18);
bminus=ForcedResponse(19:20);

dplus=ForcedResponse(21:22);
dminus=ForcedResponse(23:24);

Ysensor(n)=[1 1]*fxsensor*bplus+[1 1]*inv(fxsensor)*bminus;

result(n)=det(BigA);

end

figure(1);
plot(freq,abs(result),'b','LineWidth',2);
xlabel('Frequency (Hz)')
ylabel('Magnitude')
grid on
hold on;

figure(2);
plot(freq,20*log10(abs(result)),'b','LineWidth',2);
xlabel('Frequency (Hz)')
ylabel('dB Magnitude')
grid on
hold on;

figure(3), plot(freq,20*log10(abs(Ysensor)),'b','LineWidth',2); hold on
grid on
xlabel('Frequency (Hz)')
ylabel('dB Magnitude')
```

Script 2: Scripts for free wave vibration analysis of a uniform beam with periodical spring attachments

```
close all, clear all, clc

%For spring attachment
KRbar=0; KRTbar=0; KTRbar=0;
KTbar=10^7

L=8      %The length of the beam

Number = input('Enter number of intermediate spring attachment (excluding springs at both boundaries): ')
BeamType=input('Enter 1 for Euler-Bernoulli Beam, enter any other number for Timoshenko Beam: ')

L0=L/(Number+1)

b=2.54/100 %Width
h=0.5*2.54/100 %Thickness

%Steel beam
E=198.87*10^9;
p=7664.5
```

```
v=0.3;
K=10*(1+v)/(12+11*v)
G=77.5*10^9;

A=b*h;
I=b*h^3/12;
EA=E*A;
EI=E*I;
GAk=G*A*K

Cb=sqrt(E*I/p/A);
Cs=sqrt(G*K*A/p/A);
Cr=sqrt(p*I/p/A);

%Cutoff frequency
wc=Cs/Cr
fc=wc/(2*pi)

f1=0 % Starting frequency in Hz
f2=3200 % Ending frequency in Hz
stepsize=0.1

freq=f1:stepsize:f2;

for n=1:length(freq)

w = 2*pi*freq(n);

if BeamType==1   % The Euler–Bernoulli model
k = sqrt(sqrt(p*A*w^2/(E*I)));
k1 = k;
k2 = k;

rclamped = [ -i -(1+i);-(1-i) i];
rsimp = [-1 0; 0 -1];
rfree = [ -i (1+i);(1-i) i];

AminusCoeff=[i*EI*k1^3+KTbar,-EI*k2^3+KTbar;-
EI*k1^2+i*k1*KRbar,EI*k2^2+k2*KRbar];
AplusCoeff=[i*EI*k1^3-KTbar,-EI*k2^3-KTbar;EI*k1^2+i*k1*KRbar,-EI*k2^2+k2*KRbar];
rSpring=inv(AminusCoeff)*(AplusCoeff);

%For Intermediate Spring attachment

c1=[1,1;i,1];
c2=[-1,-1;i,1];
c3=[-1,-1;i,1];

d1= [-1,1;i,-1];
d2= [1-i*KRbar/(E*I*k),-1-KRbar/(E*I*k);KTbar/(E*I*k^3)+i, KTbar/(E*I*k^3)-1];
d3= [1,-1;i,-1];

else  % The Timoshenko bending model
```

```
k1=abs(sqrt(0.5*((1/Cs)^2+(Cr/Cb)^2)*w^2+sqrt(w^2/Cb^2+0.25*((1/Cs)^2-(Cr/Cb)^2)^2*w^4)));
k2=abs(sqrt(0.5*((1/Cs)^2+(Cr/Cb)^2)*w^2-sqrt(w^2/Cb^2+0.25*((1/Cs)^2-(Cr/Cb)^2)^2*w^4)));

%%%%%%%%%%%%%%%%%
%When w<wc, k2=k2
%When w>wc, k2=ik2
%%%%%%%%%%%%%%%%%%

if w<wc
k2=k2;
else
k2=i*k2;
end

P=(-(w^2-Cs^2*k1^2)/(k1*Cs^2));
N=((w^2+Cs^2*k2^2)/(k2*Cs^2));

rclamped=[(-i*N+P)/(i*N+P),-2*i*N/(i*N+P);-2*P/(i*N+P),-(-i*N+P)/(i*N+P)];
rsimp=[-1,0;0,-1];
rfree=[(k1*P*k2-k1*P*N-i*k2*N*k1+i*k2*N*P)/(-k1*P*k2+k1*P*N-i*k2*N*k1+i*k2*N*P),2*(-k2+N)/
(-k1*P*k2+k1*P*N-i*k2*N*k1+i*k2*N*P)*k2*N;2*i*(-k1+P)/(-k1*P*k2+k1*P*N-i*k2*N*k1+i*k2*N*P)*k1*P,(-
k1*P*k2+k1*P*N+i*k2*N*k1-i*k2*N*P)/(-k1*P*k2+k1*P*N-i*k2*N*k1+i*k2*N*P)];

KT=KTbar/GAk;
KR=KRbar/EI;
KRT=KRTbar/GAk;
KTR=KTRbar/EI;
AminusCoeff =[- KT + P*1i - k1*1i - KTR*P*1i, N - KT - k2 - KTR*N;
P*k1 - KRT - KR*P*1i, - KRT - N*k2 - KR*N];
AplusCoeff =[- KT - P*1i + k1*1i + KTR*P*1i, k2 - N - KT + KTR*N;
P*k1 - KRT + KR*P*1i,  KR*N - N*k2 - KRT];
rSpring=-inv(AminusCoeff)*(AplusCoeff);

% For Intermediate Spring attachment

c1=[1,1;i*P,N];
c2=[-1,-1;i*P,N];
c3=[-1,-1;i*P,N];

d1= [-P*k1,N*k2;i*(k1-P), k2-N];
d2= [P*k1-i*P*KRbar/(E*I), -N*k2-N*KRbar/(E*I); i*(k1-P)+KTbar/GAk, k2-N+KTbar/GAk];
d3= [P*k1,-N*k2;i*(k1-P), k2-N];

end

t=inv(c2-c1*inv(d1)*d2)*(c3-c1*inv(d1)*d3);
r=inv(c1-c2*inv(d2)*d1)*(c3-c2*inv(d2)*d3);

Discontinuity=[r,-eye(2),zeros(2),t;t,zeros(2),-eye(2) r];

ra=rSpring;
rb=rSpring;
```

```
fL0=[exp(-i*k1*L0),0;0,exp(-k2*L0)];

%Building up the characteristic matrix

BigA=zeros(8*(Number+1));

if Number==0 %For no intermediate spring discontinuity

BigA(1:4,1:8)=[eye(2),-ra,zeros(2),zeros(2);zeros(2),zeros(2),rb,-eye(2)];
BigA(5:8,1:8)=[fL0,zeros(2),-eye(2),zeros(2);zeros(2),-eye(2),zeros(2),fL0];

else %For the existence of intermediate spring discontinuity

%Boundary related matrix components
BigA(1:4,1:8)=[eye(2),-ra,zeros(2),zeros(2);zeros(2),zeros(2),rb,-eye(2)];

%The very left side bay (the 1st bay)
BigA(5:8,1:12)=[fL0,zeros(2),zeros(2),zeros(2),-eye(2),zeros(2);zeros(2),-eye(2),zeros(2),zeros(2),zeros(2),fL0];

%The very right side bay (the (n+1)th bay)
BigA((4*(Number+1)+1):(4*(Number+1)+4),5:8)=[-eye(2),zeros(2);zeros(2),fL0];
BigA((4*(Number+1)+1):(4*(Number+1)+4),(8*(Number+1)-3):(8*(Number+1)))=[fL0,zeros(2);zeros(2),-eye(2)];

%General intermediate bays (2nd to nth)
if Number>=2
for m=2:Number
BigA((4*m+1):(4*m+4),(4*(2*m-1)+1):(4*(2*m-1)+8))=[fL0,zeros(2),-eye(2),zeros(2);zeros(2),-eye(2),zeros(2),fL0];
end
else
end

%R and T at each discontinuity
if Number>=1
for m=1:Number
BigA((4*(Number+m+1)+1):(4*(Number+m+1)+4),(8*m+1):(8*m+8))=Discontinuity;
end
else
end

end

result(n)=det(BigA);

end

figure(1),plot(freq,abs(result),'b','LineWidth',2),hold on
xlabel('Frequency (Hz)'),
ylabel('Magnitude')

figure(2),
plot(freq,20*log10(abs(result)),'b','LineWidth',2),hold on
xlabel('Frequency (Hz)'),
ylabel('dB Magnitude')
```

References

Mead D.J. Wave Propagation in Continuous Periodic Structures: Research Contributions from Southampton, *Journal of Sound and Vibration*, 190(3), 495–524, (1996).

Mei C. A Wave based Analytical Solution to Free Vibrations in a Combined Euler–Bernoulli Beam/ Frame and a Two Degree-of-Freedom Spring-Mass System, *ASME Journal of Vibration and Acoustics*, 140(6) 061001–061001 to 8, (2018).

Mei C. Free and Forced Wave Vibration Analysis of a Timoshenko Beam/Frame Carrying a Two Degrees of Freedom Spring-Mass System, *ASME Journal of Vibration and Acoustics*, 143, 061008–1 to 10, (2021).

Homework Project

Free and Forced Wave Vibration Analysis of a Uniform Beam with an Intermediate Spring Attachment

Write MATLAB scripts to verify both Figures 13.15 and 13.16, which are the overlaid magnitudes of the characteristic polynomials and the overlaid receptance frequency responses of the uniform beam before and after the attachment of a translational spring, as illustrated in the figure below. Two different stiffness values are assumed for the spring, which are $K_s = 10^7$ N/m and $K_s = 10^8$ N/m, respectively. The response plots of Figures 13.15 and 13.16 are obtained using the Timoshenko bending vibration theory.

The material and geometrical properties of the example steel beam are: the Young's modulus E is 198.87 GN/m^2, shear modulus G is 77.5 GN/m^2, Poisson's ratio ν is 0.30, and mass density ρ is 7664.5 kg/m^3. The shear coefficient κ is related to the Poisson's ratio ν by $\kappa = 10(1+\nu)/(12+11\nu)$. The cross section of the beam is a rectangular shape whose width and thickness are 25.4 mm and 12.7 mm (or 1.0 in and 0.5 in), respectively.

The length of the beam L is 2.0 m, and both boundaries are free. The spring is attached at 0.55 L, the external force is applied at 0.22 L, and the response is observed at 0.64 L, all measured from the left boundary of the beam.

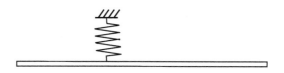

14

Active Wave Vibration Control

Active wave vibration control is particularly useful for one-dimensional structures or built-up structures that comprise one-dimensional structural elements, where a finite number of waves with given directions of propagation exist. The control of vibration waves can be feedforward, where a disturbance wave is detected and a control force is applied somewhere downstream to produce a destructive wave to cancel out the incoming wave or to absorb the associated vibration energy. The control of vibration waves can also be feedback because vibration waves are reflected and transmitted at a discontinuity.

A feedback controller can be designed to minimize the transmitted and/or reflected vibrations in a structure (Brennan 1994). In this chapter, collocated feedback wave control is studied, where the point control force and the measurement sensor are at the same location (Mace and Jones 1996), as illustrated in Figure 14.1. In the frequency domain, the wave control force $F(x,\omega)$ is related to the deflection measured at the same location by the feedback wave controller $\bar{H}(x,\omega)$, where ω denotes the circular frequency and x is the location of the controller. For longitudinal and bending vibration control, with deflections denoted using $u(x,\omega)$ and $y(x,\omega)$, respectively, the wave control forces in the frequency domain are

$$F(x,\omega) = -\bar{H}(x,\omega)u(x,\omega) \tag{14.1}$$

$$F(x,\omega) = -\bar{H}(x,\omega)y(x,\omega) \tag{14.2}$$

From Figure 14.1, an active control force, applied either on the span or at a free boundary of a structure, creates a discontinuity at the point where the control force is applied. A vibration wave a^+ incident upon such an "active discontinuity" located on the span, as shown in Figure 14.1a, gives rise to reflected wave a^- and transmitted wave b^+; while a vibration wave a^+ incident upon such an "active discontinuity" located at a free boundary, as shown in Figure 14.1b, gives rise to reflected wave a^-. Consequently, collocated feedback control is dynamically identical to a structure with an attached spring, either on the span or at a free boundary, of dynamic spring stiffness $\bar{H}(x,\omega)$ that is frequency dependent and complex in general.

Feedback wave controllers for suppressing longitudinal and bending vibration waves are designed, based on the Elementary longitudinal vibration theory and the Euler–Bernoulli bending vibration theory, respectively.

14.1 Wave Control of Longitudinal Vibrations

Figure 14.2 shows the sign convention adopted in wave control of longitudinal vibration that follows.

14.1.1 Feedback Longitudinal Wave Control on the Span

First, take a look at the reflection and transmission of longitudinal waves, with a point spring and viscous damper attachment on the span of a uniform beam, as shown in Figure 14.3. The transmitted and reflected longitudinal vibration waves are related to the incident wave by the transmission and reflection coefficients t and r,

Mechanical Wave Vibrations: Analysis and Control, First Edition. Chunhui Mei.
© 2023 Chunhui Mei. Published 2023 by John Wiley & Sons Ltd.
Companion Website: www.wiley.com/go/Mei/MechanicalWaveVibrations

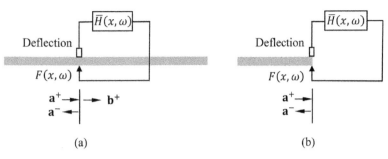

Figure 14.1 Collocated feedback wave control (a) on the span and (b) at a free boundary.

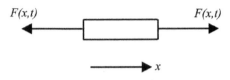

Figure 14.2 Definition of positive internal resistant force along the axial direction.

$$b^+ = ta^+, \ a^- = ra^+ \tag{14.3}$$

The reflection and transmission coefficients can be found by considering the continuity and equilibrium conditions at the discontinuity. Following the sign convention defined in Figure 14.2 and from Figure 14.3, the continuity and equilibrium conditions are

$$u_-(x,t) = u_+(x,t)$$

$$F_+(x,t) - F_-(x,t) + \bar{F}(x,t) = 0 \tag{14.4}$$

where $F(x,t)$ is the internal resistant axial force, and $\bar{F}(x,t)$ is the external force applied to the beam by the attached spring and viscous damper system at the point discontinuity. Subscripts $-$ and $+$ denote physical parameters on the left and right side of the point where the spring and viscous damper are attached.

$\bar{F}(x,t)$ is related to the spring stiffness $\overline{K_T}$ and viscous damping constant $\overline{C_T}$ by

$$\bar{F}(x,t) = -\overline{K_T}u(x,t) - \overline{C_T}\dot{u}(x,t) \tag{14.5a}$$

In the frequency domain, $i\omega$ is a differentiator, Eq. (14.5a) becomes

$$\bar{F}(x,\omega) = -(\overline{K_T} + i\omega\overline{C_T})u(x,\omega) \tag{14.5b}$$

From the wave components shown in Figure 14.3a, the axial deflections are

$$u_- = a^+ e^{-ik_i x} + a^- e^{ik_i x}$$

$$u_+ = b^+ e^{-ik_i x} \tag{14.6a}$$

and according to the Elementary longitudinal vibration theory and following the sign convention defined in Figure 14.2, the internal resistant axial force and axial deflection are related by

$$F_\pm(x) = EA \frac{\partial u_\pm(x)}{\partial x} \tag{14.6b}$$

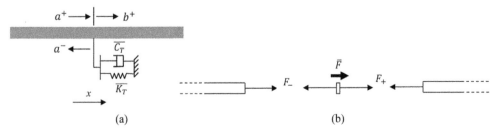

Figure 14.3 (a) Reflection and transmission of longitudinal waves at a point spring and viscous damper attachment on the span and (b) free body diagram.

In Eqs. (14.6a) and (14.6b), i is the imaginary unit, E is the Young's modulus, A is the cross-sectional area of the beam. The wavenumber for longitudinal vibration is $k_l = \sqrt{\rho/E}\omega$, where ρ is the volume mass density of the beam.

Choosing the origin at the point where the spring and viscous damper system is attached, from Eqs. (14.4), (14.6a), and (14.6b), the continuity and equilibrium conditions become

$$a^+ + a^- = b^+$$

$$EA(-ik_l)b^+ - EA(-ik_l a^+ + ik_l a^-) - (\overline{K_T} + i\omega\overline{C_T})b^+ = 0 \tag{14.7}$$

From Eqs. (14.3) and (14.7), the reflection and transmission coefficients are

$$r = \frac{-\overline{K_T} - i\omega\overline{C_T}}{i2EAk_l + \overline{K_T} + i\omega\overline{C_T}} = \frac{-\overline{K_T} - i\omega\overline{C_T}}{i2\sqrt{\rho E}A\omega + \overline{K_T} + i\omega\overline{C_T}} \tag{14.8a}$$

$$t = \frac{i2EAk_l}{i2EAk_l + \overline{K_T} + i\omega\overline{C_T}} = \frac{i2\sqrt{\rho E}A\omega}{i2\sqrt{\rho E}A\omega + \overline{K_T} + i\omega\overline{C_T}} \tag{14.8b}$$

Comparing Eqs. (14.1) with (14.5b), the feedback longitudinal wave controller $\bar{H}(\omega)$ is related to the parameters of the spring and viscous damper attachment as follows

$$\bar{H}(\omega) = \overline{K_T} + i\omega\overline{C_T} \tag{14.9}$$

where variable x in $\bar{H}(x,\omega)$ is dropped for simplicity.

From Eqs. (14.8a), (14.8b), and (14.9), the reflection and transmission coefficients in terms of the feedback wave controller are

$$r = \frac{-\bar{H}(\omega)}{i2\sqrt{\rho E}A\omega + \bar{H}(\omega)} \tag{14.10a}$$

$$t = \frac{i2\sqrt{\rho E}A\omega}{i2\sqrt{\rho E}A\omega + \bar{H}(\omega)} \tag{14.10b}$$

Equations (14.10a) and (14.10b) can be rewritten in a simpler form,

$$r = \frac{iH(\omega)}{1 - iH(\omega)} \tag{14.11a}$$

$$t = \frac{1}{1 - iH(\omega)} \tag{14.11b}$$

where

$$H(\omega) = \bar{H}(\omega)/(2EAk_l) = \bar{H}(\omega)/\left(2\sqrt{\rho E}A\omega\right) \tag{14.12}$$

Because the transmission and reflection coefficients are functions of the feedback controller, feedback wave control can be designed to absorb the transmitted and reflected vibration energy, that is, to minimize

$$E_{rt} = |r|^2 + |t|^2 \tag{14.13}$$

A feedback controller is complex in general, which can be expressed either in the form of

$$H_{PD}(\omega) = C_1 + iC_2 \tag{14.14a}$$

or

$$H_D(\omega) = iC \tag{14.14b}$$

where the control gains C_1, C_2, and C are of real values, and i is the imaginary unit. Subscripts P and D in the controllers stand for proportional and derivative, respectively.

From Eqs. (14.1), (14.12), (14.14a), and (14.14b), the control forces in the frequency domain are

$$F(x,\omega) = -2\sqrt{\rho E} A\omega C_1 u(x,\omega) - 2\sqrt{\rho E} A C_2 (i\omega) u(x,\omega) \tag{14.15a}$$

$$F(x,\omega) = -2\sqrt{\rho E} A C(i\omega) u(x,\omega) \tag{14.15b}$$

Because $i\omega$ is a differentiator in the frequency domain, the control forces of Eqs. (14.15a) and (14.15b) in the time domain are

$$F(x,t) = -2\sqrt{\rho E} A\omega C_1 u(x,t) - 2\sqrt{\rho E} A C_2 \dot{u}(x,t) \tag{14.16a}$$

$$F(x,t) = -2\sqrt{\rho E} A C\dot{u}(x,t) \tag{14.16b}$$

The control forces obtained in Eqs. (14.16a) and (14.16b) are related to the deflections by proportional and derivative (PD) and derivative (D) functional relationships, respectively. For this reason, the controllers in the form of Eqs. (14.14a) and (14.14b) are called PD and D controller, respectively. The D feedback controller is sometimes called velocity feedback controller because the first-order derivative of deflection is velocity.

To design an optimal energy absorbing PD controller, substituting Eq. (14.14a) into Eqs. (14.11a) and (14.11b)

$$r = \frac{i(C_1 + iC_2)}{1 - i(C_1 + iC_2)} = \frac{-C_2 + iC_1}{1 + C_2 - iC_1} \tag{14.17a}$$

$$t = \frac{1}{1 - i(C_1 + iC_2)} = \frac{1}{1 + C_2 - iC_1} \tag{14.17b}$$

From Eqs. (14.17a), (14.17b) and (14.13),

$$E_{rt} = \frac{C_1^2 + C_2^2 + 1}{C_1^2 + (C_2 + 1)^2} \tag{14.18}$$

To minimize E_{rt}, first finding partial derivatives of E_{rt} to variables C_1 and C_2, respectively, then setting these two equations to zero, the optimal energy absorbing PD control gains are

$$C_1 = 0, \text{ and } C_2 = 1 \tag{14.19}$$

From Eqs. (14.14a), (14.14b), and (14.19), the PD controller is a D controller because of $C_1 = 0$. Substituting Eq. (14.19) into Eq. (14.18), the total vibration energy after control is $E_{rt} = 0.5$. This indicates that the on span controller absorbs 50% of the incoming longitudinal vibration energy.

Similarly, to design an optimal D controller, substituting Eq. (14.14b) into Eqs. (14.11a) and (14.11b)

$$r = \frac{-C}{C + 1} \tag{14.20a}$$

$$t = \frac{1}{C + 1} \tag{14.20b}$$

From Eqs. (14.20a), (14.20b), and (14.13),

$$E_{rt} = \frac{C^2 + 1}{(C + 1)^2} \tag{14.21}$$

To minimize E_{rt}, finding derivative of E_{rt} to variable C and setting the equation to zero, the optimal D control gain is

$$C = 1 \tag{14.22}$$

Substituting Eq. (14.22) into Eq. (14.21), the total vibration energy after control is $E_{rt} = 0.5$. This indicates that the on span controller absorbs 50% of the incoming longitudinal vibration energy.

The two controllers perform the same, because they are identical. The optimal energy absorbing PD controller obtained in Eq. (14.19) is a D controller with control gain $C_2 = 1$. This is the same as the control gain of the D controller obtained in Eq. (14.22).

From Eqs. (14.16a) and (14.16b), the control force in the time domain is (Mei 2002)

$$F(t) = -2\sqrt{\rho E} A \dot{u}(t) \tag{14.23}$$

Equation (14.23) shows that the actively designed wave controller can be implemented passively by using a viscous damper whose damping constant is $2\sqrt{\rho E} A$, which is the control gain of the feedback controller.

14.1.2 Feedback Longitudinal Wave Control at a Free Boundary

Figure 14.4 shows the reflection of longitudinal waves with a point spring and viscous damper attachment at a free boundary. At the boundary, incident wave a^+ and reflected wave a^- are related by the reflection coefficient r

$$a^- = ra^+ \tag{14.24}$$

Following the sign convention defined in Figure 14.2 and from the free body diagram of Figure 14.4, the equilibrium condition at the boundary is

$$-F(x) + \bar{F}(x) = 0 \tag{14.25}$$

where $\bar{F}(x)$ denotes the external force from the attached spring and viscous damper system and is given by Eq. (14.5b). The internal resistant axial force is $F(x) = EA\dfrac{\partial u(x)}{\partial x}$ and the axial deflection, from the wave components shown in Figure 14.4a, is $u(x) = a^+ e^{-ik_l x} + a^- e^{ik_l x}$. Substituting these expressions into Eq. (14.25) and choosing the origin at the boundary gives,

$$-EA\left(-ik_l a^+ + ik_l a^-\right) - \left(\overline{K_T} + i\omega\overline{C_T}\right)\left(a^+ + a^-\right) = 0 \tag{14.26}$$

From Eqs. (14.24) and (14.26), the reflection coefficient at the spring and viscous damper attached boundary is

$$r = \frac{ik_l EA - \left(\overline{K_T} + i\omega\overline{C_T}\right)}{ik_l EA + \left(\overline{K_T} + i\omega\overline{C_T}\right)} = \frac{i\sqrt{\rho E} A\omega - \left(\overline{K_T} + i\omega\overline{C_T}\right)}{i\sqrt{\rho E} A\omega + \left(\overline{K_T} + i\omega\overline{C_T}\right)} \tag{14.27}$$

Equation (14.27) can also be obtained from Eq. (2.21a) in Chapter 2, which is the reflection coefficient at a spring attached boundary, by replacing spring stiffness $\overline{K_T}$ using dynamic spring stiffness $\overline{K_T} + i\omega\overline{C_T}$.

From Eqs. (14.1) and Eq. (14.5b), Eq. (14.27) can be rewritten as

$$r = \frac{1 + iH(\omega)}{1 - iH(\omega)} \tag{14.28}$$

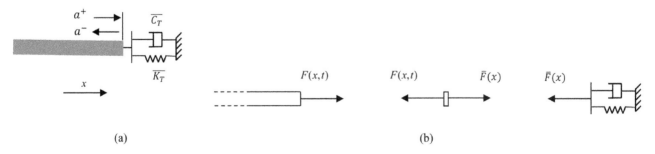

(a) (b)

Figure 14.4 (a) Reflection and transmission of longitudinal waves at a point spring and viscous damper attachment at the boundary and (b) free body diagram.

where

$$H(\omega) = \bar{H}(\omega) / (EAk_l) = \bar{H}(\omega) / (\sqrt{\rho E} A\omega) \tag{14.29}$$

Because the reflection coefficient is a function of the feedback controller, feedback wave control can be designed to absorb the reflected vibration energy, that is, to minimize

$$E_r = |r|^2 \tag{14.30}$$

The feedback controllers are complex in general and can be expressed in the forms of Eqs. (14.14a) and (14.14b).
From Eqs. (14.1), (14.14a), (14.14b), and (14.29),

$$F(\omega) = -\sqrt{\rho E} A\omega C_1 u(\omega) - \sqrt{\rho E} A C_2(i\omega)u(\omega) \tag{14.31a}$$

$$F(\omega) = -\sqrt{\rho E} A C(i\omega)u(\omega) \tag{14.31b}$$

Because in the frequency domain $i\omega$ is a differentiator, the control forces in the time domain are

$$F(t) = -\sqrt{\rho E} A\omega C_1 u(t) - \sqrt{\rho E} A C_2 \dot{u}(t) \tag{14.32a}$$

$$F(t) = -\sqrt{\rho E} A C \dot{u}(t) \tag{14.32b}$$

To design an optimal energy absorbing PD controller, substituting Eq. (14.14a) into Eq. (14.28)

$$r = \frac{1 + i(C_1 + iC_2)}{1 - i(C_1 + iC_2)} = \frac{-C_2 + 1 + iC_1}{C_2 + 1 - iC_1} \tag{14.33}$$

From Eqs. (14.30) and (14.33), the reflected vibration energy E_r is

$$E_r = \frac{C_1^2 + (C_2 - 1)^2}{C_1^2 + (C_2 + 1)^2} \tag{14.34}$$

To minimize E_r, finding partial derivatives of E_r to variables C_1 and C_2, respectively, and setting these two equations to zero, the optimal energy absorbing PD control gains are

$$C_1 = 0, \text{ and } C_2 = 1 \tag{14.35}$$

The optimal PD controller becomes a D controller, which is similar to the design of feedback wave control on the span.
Substituting Eq. (14.35) into Eq. (14.34), the reflected vibration energy after control is $E_r = 0$. This indicates that the boundary controller absorbs 100% of the incoming longitudinal vibration energy.
Similarly, to design of an optimal D controller, substituting Eq. (14.14b) into Eq. (14.28)

$$r = \frac{1 - C}{1 + C} \tag{14.36}$$

From Eqs. (14.30) and (14.36),

$$E_r = \frac{(C - 1)^2}{(C + 1)^2} \tag{14.37}$$

To minimize E_r, finding the derivative of E_r to variable C and setting the equation to zero, the optimal energy absorbing D control gain is

$$C = 1 \tag{14.38}$$

Substituting Eq. (14.38) into Eq. (14.37), the reflected vibration energy after control is $E_r = 0$. This indicates that the boundary controller absorbs 100% of the incoming longitudinal vibration energy.
Both controllers perform the same, because from Eqs. (14.14a), (14.14b), (14.35) and (14.38), the two controllers are identical D controllers.

From Eqs. (14.32a) and (14.32b), the control force in the time domain is (Mei 2002)

$$F(t) = -\sqrt{\rho E}\, A \dot{u}(t) \tag{14.39}$$

Equation (14.39) shows that the actively designed wave controller can be implemented passively by attaching at the boundary a viscous damper of damping constant $\sqrt{\rho E}\, A$, which is the control gain of the feedback controller.

MATLAB scripts for feedback wave control of longitudinal vibrations are available in Section 14.4.

14.2 Wave Control of Bending Vibrations

14.2.1 Feedback Bending Wave Control on the Span

Figure 14.5 shows a uniform beam with an intermediate translational spring and viscous damper attachment, whose spring stiffness and damping constant are K_T and C_T, respectively.

As illustrated in Figure 14.5a, incident bending waves $\mathbf{a}^+$ are transmitted and reflected at the point where the spring and viscous damper system is attached. The transmitted and reflected bending waves $\mathbf{b}^+$ and $\mathbf{a}^-$ are related to the incident waves $\mathbf{a}^+$ by the transmission and reflection matrices $\mathbf{t}$ and $\mathbf{r}$

$$\mathbf{b}^+ = \mathbf{t}\mathbf{a}^+, \ \mathbf{a}^- = \mathbf{r}\mathbf{a}^+ \tag{14.40a}$$

where

$$\mathbf{a}^+ = \begin{bmatrix} a_1^+ \\ a_2^+ \end{bmatrix}, \ \mathbf{a}^- = \begin{bmatrix} a_1^- \\ a_2^- \end{bmatrix}, \ \mathbf{b}^+ = \begin{bmatrix} b_1^+ \\ b_2^+ \end{bmatrix} \tag{14.40b}$$

Superscripts $+$ and $-$ and subscripts 1 and 2 in $a_{1,2}^{\pm}$ and $b_{1,2}^+$ denote positive- and negative-going propagating and decaying bending waves.

Equations (14.40a) and (14.40b) can be written in terms of components t_{mn} and r_{mn} (where $m, n = 1,2$) of the 2 by 2 transmission and reflection matrices $\mathbf{t}$ and $\mathbf{r}$,

$$\begin{bmatrix} b_1^+ \\ b_2^+ \end{bmatrix} = \begin{bmatrix} t_{11} & t_{12} \\ t_{21} & t_{22} \end{bmatrix} \begin{bmatrix} a_1^+ \\ a_2^+ \end{bmatrix}, \ \begin{bmatrix} a_1^- \\ a_2^- \end{bmatrix} = \begin{bmatrix} r_{11} & r_{12} \\ r_{21} & r_{22} \end{bmatrix} \begin{bmatrix} a_1^+ \\ a_2^+ \end{bmatrix} \tag{14.41a}$$

In scalar form, Eq. (14.41a) is

$$\begin{aligned} b_1^+ &= t_{11}a_1^+ + t_{12}a_2^+ \\ b_2^+ &= t_{21}a_1^+ + t_{22}a_2^+ \\ a_1^- &= r_{11}a_1^+ + r_{12}a_2^+ \\ a_2^- &= r_{21}a_1^+ + r_{22}a_2^+ \end{aligned} \tag{14.41b}$$

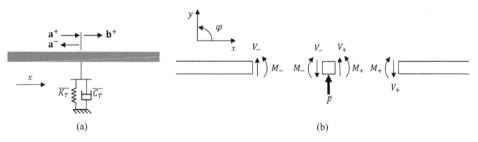

(a) (b)

Figure 14.5 (a) Reflection and transmission of bending waves at a point spring and viscous damper attachment on the span and (b) free body diagram.

From Eq. (14.41b), both the incident propagating bending wave component a_1^+ and the decaying bending wave component a_2^+ contribute to the transmitted propagating bending wave component b_1^+ and decaying bending wave component b_2^+, as well as to the reflected propagating bending wave component a_1^- and decaying bending wave component a_2^-. The existence of multiple wave components in bending vibration appears to complicate the design of wave feedback controllers.

However, the decaying waves are nearfield waves. If the point where the spring and viscous damper system is attached is at a sufficient distance away from its adjacent structural discontinuities, the incident waves upon the point where the spring and damper system is attached comprise predominantly the propagating wave component. This allows the transmitted and reflected propagating waves in Eq. (14.41b) to be approximated,

$$b_1^+ \approx t_{11} a_1^+, \; a_1^- \approx r_{11} a_1^+ \tag{14.42}$$

In addition, under the situation where there is no discontinuity nearby, both the transmitted and reflected nearfield wave components b_2^+ and a_2^- will die out with distance. This allows the remaining two equations in Eq. (14.41b) to be neglected in the control design.

With the above consideration, feedback bending wave control design becomes similar to feedback longitudinal wave control design. Feedback bending wave control can be designed to optimally absorb vibration energy carried by the reflected and transmitted propagating wave components, that is, to minimize

$$E_{rt} = \left| r_{11} \right|^2 + \left| t_{11} \right|^2 \tag{14.43}$$

The reflection and transmission matrices can be obtained by considering the continuity and equilibrium conditions at the discontinuity.

Continuity in the bending deflection y and bending slope ψ gives

$$y_-(x,t) = y_+(x,t), \; \psi_-(x,t) = \psi_+(x,t) \tag{14.44}$$

From the free body diagram of Figure 14.5, the equilibrium conditions at the discontinuity are

$$\begin{aligned} V_+(x,t) - V_-(x,t) + \bar{F}(x,t) &= 0 \\ M_+(x,t) - M_-(x,t) &= 0 \end{aligned} \tag{14.45}$$

where $M(x,t)$ and $V(x,t)$ are the internal resistant bending moment and shear force, and $\bar{F}(x,t)$ is the external force applied to the beam by the attached spring and viscous damper at the point discontinuity.

In Eqs. (14.44) and (14.45), subscripts $-$ and $+$ denote physical parameters on the left and right side of the point where the spring and viscous damper system is attached.

$\bar{F}(x,t)$ is related to the spring stiffness $\overline{K_T}$ and viscous damping constant $\overline{C_T}$ by

$$\bar{F}(x,t) = -\overline{K_T} y(x,t) - \overline{C_T} \dot{y}(x,t) \tag{14.46a}$$

In the frequency domain, $i\omega$ is a differentiator, Eq. (14.46a) becomes

$$\bar{F}(x,\omega) = -\left(\overline{K_T} + i\omega \overline{C_T} \right) y(x,\omega) \tag{14.46b}$$

Comparing Eqs. (14.2) with (14.46b), the feedback bending wave controller $\bar{H}(\omega)$ is related to the parameters of the spring and viscous damper attachment by the same expression obtained in Eq. (14.9).

The transmission and reflection matrices at a spring attachment have been derived in Chapter 13, following the sign convention defined in Figure 14.6. Recall that the viscous damping effect can be taken into account by introducing complex and frequency dependent dynamic spring stiffness. For a uniform beam of the same material and cross-sectional dimension on both sides of the point where the spring and viscous damper system is attached, the transmission and reflection matrices, by the Euler–Bernoulli bending vibration theory, are

$$\begin{aligned} \mathbf{t} &= \frac{1}{4k_b^3 EI - (1+i)\left(\overline{K_T} + i\omega \overline{C_T} \right)} \begin{bmatrix} 4k_b^3 EI - \left(\overline{K_T} + i\omega \overline{C_T} \right) & i\left(\overline{K_T} + i\omega \overline{C_T} \right) \\ \overline{K_T} + i\omega \overline{C_T} & 4k_b^3 EI - i\left(\overline{K_T} + i\omega \overline{C_T} \right) \end{bmatrix}, \\ \mathbf{r} &= \frac{\overline{K_T} + i\omega \overline{C_T}}{4k_b^3 EI - (1+i)\left(\overline{K_T} + i\omega \overline{C_T} \right)} \begin{bmatrix} i & i \\ 1 & 1 \end{bmatrix} \end{aligned} \tag{14.47}$$

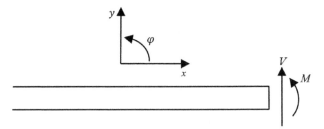

Figure 14.6 Definition of positive internal resistant shear force and bending moment.

where bending wavenumber $k_b = \sqrt[4]{\rho A \omega^2 / EI}$, and I is the area moment of inertia of the cross section.

From Eq. (14.9), Eq. (14.47) can be written as

$$\mathbf{t} = \frac{1}{4k_b^3 EI - (1+i)\bar{H}(\omega)}\begin{bmatrix} 4k_b^3 EI - \bar{H}(\omega) & i\bar{H}(\omega) \\ \bar{H}(\omega) & 4k_b^3 EI - i\bar{H}(\omega) \end{bmatrix},$$

$$\mathbf{r} = \frac{\bar{H}(\omega)}{4k_b^3 EI - (1+i)\bar{H}(\omega)}\begin{bmatrix} i & i \\ 1 & 1 \end{bmatrix}$$

(14.48)

Equation (14.48) can be rewritten in a simpler form

$$\mathbf{t} = \frac{1}{4 - (1+i)H(\omega)}\begin{bmatrix} 4 - H(\omega) & iH(\omega) \\ H(\omega) & 4 - iH(\omega) \end{bmatrix},$$

$$\mathbf{r} = \frac{H(\omega)}{4 - (1+i)H(\omega)}\begin{bmatrix} i & i \\ 1 & 1 \end{bmatrix}$$

(14.49)

where

$$H(\omega) = \bar{H}(\omega) / (k_b^3 EI) = \bar{H}(\omega) / \left(\omega \sqrt[4]{(\rho A)^3 EI \omega^2} \right)$$

(14.50)

From Eqs. (14.41a) and (14.49),

$$t_{11} = \frac{4 - H(\omega)}{4 - (1+i)H(\omega)}$$

$$r_{11} = \frac{iH(\omega)}{4 - (1+i)H(\omega)}$$

(14.51)

The controllers can be of two types, a PD and D controller, as described in Eqs. (14.14a) and (14.14b).

From Eqs. (14.2), (14.14a), (14.14b), and (14.50),

$$F(x,\omega) = -\omega \sqrt[4]{(\rho A)^3 EI \omega^2} C_1 y(x,\omega) - \sqrt[4]{(\rho A)^3 EI \omega^2} C_2 (i\omega) y(x,\omega)$$

(14.52a)

$$F(x,\omega) = -\sqrt[4]{(\rho A)^3 EI \omega^2} C(i\omega) y(x,\omega)$$

(14.52b)

Recall that in the frequency domain $i\omega$ is a differentiator, the control forces in the time domain are

$$F(x,t) = -\omega \sqrt[4]{(\rho A)^3 EI \omega^2} C_1 y(x,t) - \sqrt[4]{(\rho A)^3 EI \omega^2} C_2 \dot{y}(x,t)$$

(14.53a)

$$F(x,t) = -\sqrt[4]{(\rho A)^3 EI \omega^2} C\dot{y}(x,t)$$

(14.53b)

To design an optimal energy absorbing PD controller, first substituting Eq. (14.14a) into Eqs. (14.51), then minimizing E_{rt} of Eq. (14.43) by finding partial derivatives of E_{rt} to variables C_1 and C_2 and setting these two equations to zero, the optimal energy absorbing PD control gains are (Mace and Jones 1996)

$$C_1 = 2, \text{ and } C_2 = 2$$

(14.54)

From Eqs. (14.14a), (14.43), (14.51), and (14.54), the vibration energy after PD control is $E_{rt} = 0.5$. This indicates that the controller absorbs 50% of the bending vibration energy carried by the reflected and transmitted propagating waves.

To design an optimal D controller, first substituting Eq. (14.14b) into Eq. (14.51), then minimizing E_{rt} of Eq. (14.43) by finding the derivative of E_{rt} to variable C and setting the equation to zero, the optimal D control gain is (Mei 1998)

$$C = 2\sqrt{2} \tag{14.55}$$

From Eqs. (14.14b), (14.43), (14.51), and (14.55), the vibration energy after D control, is $E_{rt} = 2 - \sqrt{2} \approx 0.5858$. This indicates that the controller absorbs 41.42% of the bending vibration energy carried by the reflected and transmitted propagating waves.

The optimal energy absorbing PD controller performs better than the optimal D controller for bending vibration control.

Substituting the PD and D control gains of Eqs. (14.54) and (14.55) into Eqs. (14.53a) and (14.53b), the control forces in the time domain are

$$F(x,t) = -2\omega\sqrt[4]{(\rho A)^3 EI\omega^2}\,y(x,t) - 2\sqrt[4]{(\rho A)^3 EI\omega^2}\,\dot{y}(x,t) \tag{14.56a}$$

$$F(x,t) = -2\sqrt{2}\sqrt[4]{(\rho A)^3 EI\omega^2}\,\dot{y}(x,t) \tag{14.56b}$$

From Eqs. (14.56a) and (14.56b), the control gains are frequency dependent, hence non-causal. This indicates that in real-time implementation, approximations to the ideal controllers need to be found.

One way is to implement the controller using a digital finite impulse response (FIR) filter, in which a causal controller is found by truncating the non-causal part of the ideal controller. The choice of the FIR filter length is often a compromise between the accuracy of the controller and the calculation time for the control output. The other way to implement the controller is to tune the controller to be optimal at some specific frequency. Assuming that it is desired for the controllers to have optimal performance at a frequency ω_d, the control forces of Eqs. (14.56a) and (14.56b) become

$$F(x,t) = -2\sqrt[4]{(\rho A)^3 EI}\,\omega_d^{3/2}y(x,t) - 2\sqrt[4]{(\rho A)^3 EI}\,\omega_d^{1/2}\dot{y}(x,t) \tag{14.57a}$$

$$F(x,t) = -2\sqrt{2}\sqrt[4]{(\rho A)^3 EI}\,\omega_d^{1/2}\dot{y}(x,t) \tag{14.57b}$$

The control gains are now constant, hence causal.

14.2.2 Feedback Bending Wave Control at a Free Boundary

Figure 14.7 shows a boundary with a translational spring and viscous damper attachment. The stiffness and damping constants are $\overline{K_T}$ and $\overline{C_T}$, respectively.

As illustrated in Figure 14.7a, incident bending waves are reflected at the boundary where the spring and viscous damper system is attached. The reflected bending waves $\mathbf{a}^-$ are related to the incident waves $\mathbf{a}^+$ by the reflection matrix $\mathbf{r}$

$$\mathbf{a}^- = \mathbf{r}\mathbf{a}^+ \tag{14.58a}$$

where

$$\mathbf{a}^+ = \begin{bmatrix} a_1^+ \\ a_2^+ \end{bmatrix}, \text{ and } \mathbf{a}^- = \begin{bmatrix} a_1^- \\ a_2^- \end{bmatrix} \tag{14.58b}$$

Superscripts $+$ and $-$ and subscripts 1 and 2 in $a_{1,2}^{\pm}$ denote positive- and negative-going propagating and decaying bending waves.

Equations (14.58a) and (14.58b) can be written in terms of components r_{mn} (where m, $n = 1, 2$) of the 2 by 2 reflection matrix $\mathbf{r}$,

$$\begin{bmatrix} a_1^- \\ a_2^- \end{bmatrix} = \begin{bmatrix} r_{11} & r_{12} \\ r_{21} & r_{22} \end{bmatrix} \begin{bmatrix} a_1^+ \\ a_2^+ \end{bmatrix} \tag{14.59}$$

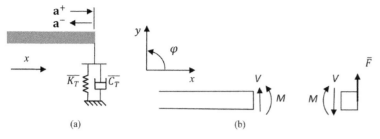

Figure 14.7 (a) Boundary with spring attachments and (b) free body diagram.

In scalar form, Eq. (14.59) is

$$a_1^- = r_{11}a_1^+ + r_{12}a_2^+$$
$$a_2^- = r_{21}a_1^+ + r_{22}a_2^+$$

(14.60)

Assuming that there is no other structural discontinuity nearby the boundary, waves incident upon the boundary where the spring and damper system is attached comprise predominantly the propagating wave component. This allows the reflected propagating waves in Eq. (14.60) to be approximated,

$$a_1^- \approx r_{11}a_1^+$$

(14.61)

Furthermore, under the situation where there is no discontinuity nearby, the reflected nearfield wave components a_2^- will die out with distance. This allows the other equation in Eq. (14.60) to be neglected in the control design.

With the above consideration, feedback bending wave control design becomes similar to feedback longitudinal wave control design. To optimally absorb the reflected bending vibration energy carried by the propagating wave component is to minimize

$$E_r = |r_{11}|^2$$

(14.62)

The reflection matrix can be obtained by considering the equilibrium conditions at the boundary. From the free body diagram of Figure 14.7,

$$-V(x,t) + \bar{F}(x,t) = 0, \quad -M(x,t) = 0$$

(14.63)

$\bar{F}(x,t)$, the external force applied to the beam by the attached spring and viscous damper at the free boundary, is related to the stiffness and viscous damping constant by the same expressions of Eqs. (14.46a) and (14.46b).

The reflection matrix at a spring attached boundary has been derived in Chapter 3, following the sign convention defined in Figure 14.6. Recall that viscous damping effect can be taken into account by introducing a complex and frequency dependent dynamic spring stiffness. The reflection matrix, by the Euler–Bernoulli bending vibration theory, is

$$\mathbf{r} = \begin{bmatrix} i + \left(\overline{K_T} + i\omega\overline{C_T}\right)/\left(k_b^3 EI\right) & -1 + \left(\overline{K_T} + i\omega\overline{C_T}\right)/\left(k_b^3 EI\right) \\ -1 & 1 \end{bmatrix}^{-1} \begin{bmatrix} i - \left(\overline{K_T} + i\omega\overline{C_T}\right)/\left(k_b^3 EI\right) & -1 - \left(\overline{K_T} + i\omega\overline{C_T}\right)/\left(k_b^3 EI\right) \\ 1 & -1 \end{bmatrix}$$

(14.64)

where the bending wavenumber is $k_b = \sqrt[4]{\rho A \omega^2 / EI}$.

Comparing Eqs. (14.2) with (14.46b), the feedback bending wave controller $\bar{H}(\omega)$ is related to the parameters of the spring and viscous damper attachment by the same expression obtained in Eq. (14.9). As a result, Eq. (14.64) can be written as

$$\mathbf{r} = \begin{bmatrix} i + \bar{H}(\omega)/(k_b^3 EI) & -1 + \bar{H}(\omega)/(k_b^3 EI) \\ -1 & 1 \end{bmatrix}^{-1} \begin{bmatrix} i - \bar{H}(\omega)/(k_b^3 EI) & -1 - \bar{H}(\omega)/(k_b^3 EI) \\ 1 & -1 \end{bmatrix}$$

(14.65)

Equation (14.65) can be rewritten in a simpler form,

$$\mathbf{r} = \begin{bmatrix} i+H(\omega) & -1+H(\omega) \\ -1 & 1 \end{bmatrix}^{-1} \begin{bmatrix} i-H(\omega) & -1-H(\omega) \\ 1 & -1 \end{bmatrix} \qquad (14.66)$$

where $H(\omega) = \bar{H}(\omega)/(k_b^3 EI) = \bar{H}(\omega)/\left(\omega\sqrt[4]{(\rho A)^3 EI\omega^2}\right)$, as described in Eq. (14.50).

Solving Eq. (14.66) gives

$$\mathbf{r} = \frac{1}{2H(\omega)-1+i} \begin{bmatrix} -2H(\omega)+1+i & -2 \\ 2i & -2H(\omega)-1-i \end{bmatrix} \qquad (14.67)$$

From Eqs. (14.41a) and (14.67),

$$r_{11} = \frac{-2H(\omega)+1+i}{2H(\omega)-1+i} \qquad (14.68)$$

The controllers can be of two types, namely, a PD and D controller, as described in Eqs. (14.14a) and (14.14b). The control forces, in the frequency and time domain, are given by Eqs. (14.52) and (14.53), respectively.

To design an optimal energy absorbing PD controller, first substituting Eq. (14.14a) into Eqs. (14.68), then minimizing E_r of Eq. (14.62) by finding partial derivatives of E_r to variables C_1 and C_2 and setting these two equations to zero, the optimal energy absorbing PD control gains are

$$C_1 = \frac{1}{2}, \text{ and } C_2 = \frac{1}{2} \qquad (14.69)$$

From Eqs. (14.14a), (14.62), (14.68), and (14.69), the vibration energy after PD control is $E_r = 0$. This indicates that the controller absorbs 100% of the bending vibration energy carried by the reflected propagating waves.

To design an optimal D controller, first substituting Eq. (14.14b) into Eqs. (14.68), then minimizing E_r of Eq. (14.62) by finding the derivative of E_r to variable C and setting the equation to zero, the optimal D control gain is

$$C = \frac{\sqrt{2}}{2} \qquad (14.70)$$

From Eqs. (14.14b), (14.62), (14.68), and (14.70), the vibration energy after D control is $E_r = 0.1716$. This indicates that the controller absorbs 82.84% of the bending vibration energy carried by the reflected propagating waves.

Both types of controllers perform exceptionally well, with the optimal energy absorbing PD controller perform better than the optimal D controller for bending vibration control.

Substituting the PD and D control gains of Eqs. (14.69) and (14.70) into Eqs. (14.53), the control forces in the time domain are

$$F(x,t) = -\frac{1}{2}\omega\sqrt[4]{(\rho A)^3 EI\omega^2}\, y(x,t) - \frac{1}{2}\sqrt[4]{(\rho A)^3 EI\omega^2}\, \dot{y}(x,t) \qquad (14.71a)$$

$$F(x,t) = -\frac{\sqrt{2}}{2}\sqrt[4]{(\rho A)^3 EI\omega^2}\, \dot{y}(x,t) \qquad (14.71b)$$

From Eqs. (14.71a) and (14.71b), the control gains are frequency dependent, hence non-causal. Causal approximations to the ideal controllers need to be found for real-time implementation, which can be obtained by FIR filter or tuning, as mentioned in Section 14.2.1.

Assuming that it is desired for the controllers to have optimal performance at a frequency ω_d, the control forces of Eqs. (14.71a) and (14.71b) become

$$F(x,t) = -\frac{1}{2}\sqrt[4]{(\rho A)^3 EI}\,\omega_d^{3/2} y(x,t) - \frac{1}{2}\sqrt[4]{(\rho A)^3 EI}\,\omega_d^{1/2} \dot{y}(x,t) \qquad (14.72a)$$

$$F(x,t) = -\frac{\sqrt{2}}{2}\sqrt[4]{(\rho A)^3 EI}\,\omega_d^{1/2} \dot{y}(x,t) \qquad (14.72b)$$

The control gains become constant and causal.

Table 14.1 lists the controllers and their corresponding energy absorption.

Table 14.1 Controller and the absorbed vibration energy.

	On Span Control		At Boundary Control	
	Control Force $F(t)$	Energy Absorption	Control Force $F(t)$	Energy Absorption
Axial Vibration	$-2\sqrt{\rho E A}\,\dot{u}(t)$	50%	$-\sqrt{\rho E A}\,\dot{u}(t)$	100%
Bending Vibration	$-2\omega\sqrt[4]{(\rho A)^3 EI\omega^2}\,y(x,t) - 2\sqrt[4]{(\rho A)^3 EI\omega^2}\,\dot{y}(x,t)$	50%	$-\frac{1}{2}\omega\sqrt[4]{(\rho A)^3 EI\omega^2}\,y(x,t) - \frac{1}{2}\sqrt[4]{(\rho A)^3 EI\omega^2}\,\dot{y}(x,t)$	100%
	$-2\sqrt{2}\sqrt[4]{(\rho A)^3 EI\omega^2}\,\dot{y}(x,t)$	41.42%	$-\frac{\sqrt{2}}{2}\sqrt[4]{(\rho A)^3 EI\omega^2}\,\dot{y}(x,t)$	82.84%

14.3 Numerical Examples

Longitudinal and bending vibrations in a uniform steel beam are controlled using the designed collocated feedback wave controllers, on the span or at a free boundary.

The material and geometrical properties of the example steel beam are: the Young's modulus and mass density are 198.87 GN/m^2 and 7664.5 kg/m^3, respectively. The cross section of the beam is of rectangular shape whose width and thickness are 25.4 mm and 12.7 mm (or 1.0 in and 0.5 in), respectively. The length of the beam is $L = 2.0\,m$, and both boundaries of the beam are free.

The external excitation force is applied at 0.22 L. The on span and the boundary control forces are applied at 0.55 L and L, respectively. The deflection is observed at 0.64 L. All distances are measured from the left boundary of the beam.

Figure 14.8 shows the receptance frequency response (displacement/force) of longitudinal vibrations before control, after on span D control, and after boundary D control. The receptance frequency responses show that the feedback wave controllers are very efficient in suppressing longitudinal vibrations in the beam. The feedback wave controller at the boundary performs better than the on span feedback wave controller. This observation agrees with the predicted energy absorption of the two types of controllers, which are 100% and 50%, respectively, as shown in Table 14.1.

Figure 14.9 shows the receptance frequency response of bending vibrations before control, after ideal on span PD control, and after ideal on span D control. The receptance frequency responses show that the feedback wave controllers are quite efficient in suppressing bending vibrations in the beam. The ideal on span feedback PD wave controller performs better

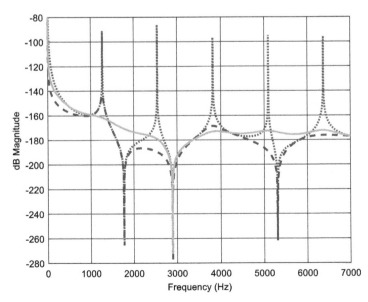

Figure 14.8 Receptance frequency response of longitudinal vibrations before control (...), after on span D control (---), and after boundary D control (__).

than the ideal on span feedback D wave controller. This observation agrees with the predicted energy absorption of the two types of controllers, which are 50% and 41.42%, respectively, as shown in Table 14.1.

Figure 14.10 shows the receptance frequency response of bending vibrations before control, after ideal boundary PD control, and after ideal boundary D control. The feedback wave controllers are very efficient in suppressing bending vibrations in the beam. The ideal boundary feedback PD wave controller performs better than the ideal boundary feedback D wave controller. This observation agrees with the predicted energy absorption of the two types of controllers, which are 100% and 82.84%, respectively, as shown in Table 14.1.

Figures 14.11 to 14.14 compare the performance of the ideal and the tuned controllers. Figures 14.11 and 14.12 are the receptance frequency response of bending vibrations before control, after ideal on span PD and D control, and after tuned on span PD and D control, respectively. Figures 14.13 and 14.14 are the receptance frequency response of bending vibrations before control, after ideal boundary PD and D control, and after tuned boundary PD and D control, respectively. In all cases, the controller is tuned to be optimal at 100 Hz, that is, $\omega_d = 200\pi\ rad/s$.

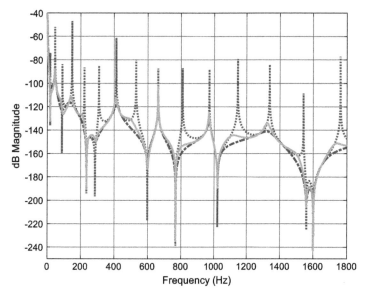

Figure 14.9 Receptance frequency response of bending vibrations before control (…), after ideal on span PD control (---), and after ideal on span D control (__).

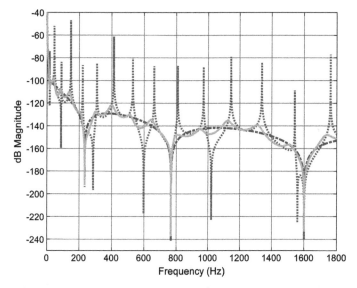

Figure 14.10 Receptance frequency response of bending vibrations before control (…), after ideal boundary PD control (---), and after ideal boundary D control (__).

Figure 14.11 Receptance frequency response of bending vibrations before control (...), after ideal on span PD control (---), and after tuned on span PD control (__).

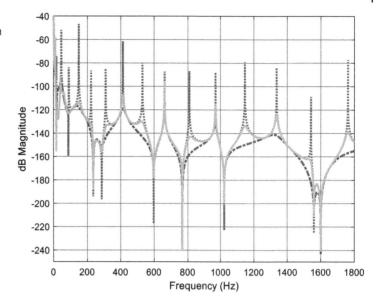

Figure 14.12 Receptance frequency response of bending vibrations before control (...), after ideal on span D control (---), and after tuned on span D control (__).

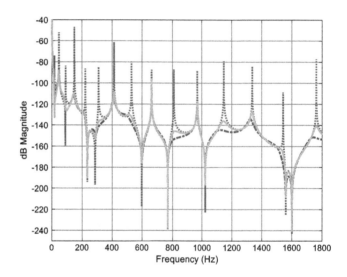

Figure 14.13 Receptance frequency response of bending vibrations before control (...), after ideal boundary PD control (---), and after tuned boundary PD control (__).

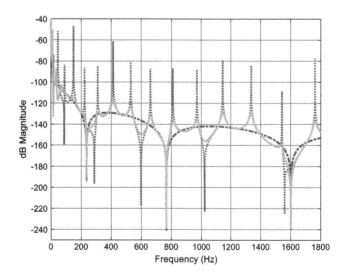

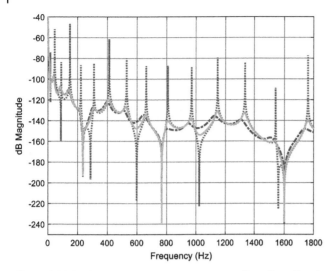

Figure 14.14 Receptance frequency response of bending vibrations before control (...), after ideal boundary D control (---), and after tuned boundary D control (___).

The receptance frequency responses of Figures 14.11 to 14.14 show that the performance of the tuned feedback wave controller is as efficient as the ideal controller in the vicinity of the tuned frequency. Away from the tuned region, the control efficiency somewhat deteriorates. To obtain the desired bending vibration control performance, one may identify a targeted frequency region and set the tuned frequency ω_d accordingly.

Although the feedback wave controllers for longitudinal and bending vibration control are designed based on the elementary longitudinal vibration theory and the Euler–Bernoulli bending vibration theory, the same wave control strategy applies to the suppression of vibrations modeled using advanced vibration theories (Mei 2009).

14.4 MATLAB Script

<u>Script</u>: Feedback wave control of longitudinal vibrations

```
close all, clear all, clc

%Controller on the span
H=i

%Controller at a free boundary
endH=i*sqrt(2)/2

L=2 %Length
b=2.54/100 %Width
h=0.5*2.54/100 %Thickness

%Steel beam
E=198.87*10^9;
p=7664.5
A=b*h;

L1 = 0.55*L
L2 = L-L1;
```

```
L11=0.40*L1        %Disturbance force applied point
L12=L1-L11;
xd=L11;

%Sensors to view responses
x1sensor=0.20*L11;
x2sensor=0.20*L12;         %at  L11+0.20*L12
x3sensor=0.20*L2;      %at  L1+0.20*L2

%Frequency range in Hz
f1 = 0;    %Starting frequency in Hz
f2 = 7000;         %Ending  frequency in Hz
stepsize=0.1

Nub = input('Enter 1 for on span control, enter any other number for no span control: ')
endNub=input('Enter 1 for boundary control, enter any other number for no end con-
trol: ')

freq = f1:stepsize:f2;

for n = 1:length(freq)

w = 2*pi*freq(n);

k=w*sqrt(p/E);

%Force generated waves

F=1;
q=(F/(2*E*A*k))*i;

rfixed = -1;
rfree = 1;

if Nub==1
r=i*H/(1-i*H);   % On span D control
t=1/(1-i*H);
else
t=1;   %No control
r=0;
end

endr=(i-endH)/(i+endH);

Discontinuity=[r,-1,0,t;t,0,-1 r];

rc=rfree;

if endNub==1
    rd=endr;     % At boundary D control
else
rd=rfree;   % No boundary control
end
```

```
fL1 = exp(-i*k*L1);
fL2 = exp(-i*k*L2);

fL11=exp(-i*k*L11);
fL12=exp(-i*k*L12);

fx1sensor=exp(-i*k*x1sensor);
fx2sensor=exp(-i*k*x2sensor);
fx3sensor=exp(-i*k*x3sensor);

%Building up the characteristic matrix

BigA=zeros(12,12);

%Boundary reflections
RefC=[-1 rc];
RefD=[rd -1];

BigB=[0;0;q;q;0;0;0;0;0;0;0;0];
%(g2-)-(g1-)=q
%(g2+)-(g1+)=-q
ExternalForce1=[0 -1 0 1];
ExternalForce2=[1 0 -1 0];

%c+ c- g1+ g1- g2+ g2- (force G) a+ a- b+ b- d+ d- (2 by 2 matrix form)
%Propagations along L1 and L2
PropL11=[fL11 0 -1 0; 0 -1 0 fL11];
PropL12=[fL12 0 -1 0; 0 -1 0 fL12];
PropL2=[fL2 0 -1 0; 0 -1 0 fL2];


BigA(11:12,7:10)=Discontinuity;

BigA(1,1:2)=RefC;
BigA(2,11:12)=RefD;
BigA(3,3:6)=ExternalForce1;
BigA(4,3:6)=ExternalForce2;

BigA(5:6,1:4)=PropL11;
BigA(7:8,5:8)=PropL12;
BigA(9:10,9:12)=PropL2;

ForcedResponse=BigA\BigB;

cplus=ForcedResponse(1);
cminus=ForcedResponse(2);

g1plus=ForcedResponse(3);
g1minus=ForcedResponse(4);
g2plus=ForcedResponse(5);
g2minus=ForcedResponse(6);

aplus=ForcedResponse(7);
```

```
aminus=ForcedResponse(8);

bplus=ForcedResponse(9);
bminus=ForcedResponse(10);

dplus=ForcedResponse(11);
dminus=ForcedResponse(12);

Y1sensor(n)=fx1sensor*cplus+inv(fx1sensor)*cminus;
Y2sensor(n)=fx2sensor*g2plus+inv(fx2sensor)*g2minus;
Y3sensor(n)=fx3sensor*bplus+inv(fx3sensor)*bminus;

result(n)=det(BigA);

end

figure(1),
plot(freq,20*log10(abs(Y1sensor)),'g','LineWidth',2); hold on
grid on
xlabel('Frequency (Hz)')
ylabel('dB Magnitude')

figure(2),
plot(freq,20*log10(abs(Y2sensor)),'g','LineWidth',2); hold on
grid on
xlabel('Frequency (Hz)')
ylabel('dB Magnitude')

figure(3),
plot(freq,20*log10(abs(Y3sensor)),'g','LineWidth',2); hold on
grid on
xlabel('Frequency (Hz)')
ylabel('dB Magnitude')
```

References

Brennan M.J. *Active Control of Waves on One-Dimensional Structure*, University of Southampton, United Kingdom, Ph.D. Thesis, (1994).

Mace B.R., Jones R.W. Feedback Control of Flexural Waves in Beams, *Journal of Structural Control*, 3(1–2), 89–98, (1996).

Mei C. *Hybrid Active Vibration Control of a Distributed Structure*, University of Auckland, New Zealand, Ph.D. Thesis, (1998).

Mei C. The Analysis and Control of Longitudinal Vibrations from Wave Viewpoint, *ASME Journal of Vibration and Acoustics*, 124 645–649, (2002).

Mei C. Hybrid Wave/Mode Active Vibration Control of Bending Vibrations in Beams based on Advanced Timoshenko Theory, *Journal of Sound and Vibration*, 322(1–2), 29–38, (2009).

Homework Project

Feedback Wave Control of Bending Vibrations

As illustrated in the figures below, a collocated feedback bending wave controller on the span of a beam is dynamically identical to an attached spring and viscous damper on the span,

$$\bar{H}(\omega) = \overline{K_T} + i\omega \overline{C_T}$$

The optimal energy absorbing PD and D controllers are

$$H_{PD}(\omega) = 2 + 2i, \; H_D(\omega) = 2\sqrt{2}i$$

where $H(\omega) = \bar{H}(\omega) / \left(\omega \sqrt[4]{(\rho A)^3 EI\omega^2} \right)$. The transmission and reflection relationships at the point where the controller is applied are

$$\mathbf{t} = \frac{1}{4 - (1+i)H(\omega)} \begin{bmatrix} 4 - H(\omega) & iH(\omega) \\ H(\omega) & 4 - iH(\omega) \end{bmatrix}, \mathbf{r} = \frac{H(\omega)}{4 - (1+i)H(\omega)} \begin{bmatrix} i & i \\ 1 & 1 \end{bmatrix}$$

Write MATLAB scripts to verify Figure 14.9, which are the receptance frequency responses of bending vibrations in the steel beam before control, after ideal on span PD control, and after ideal on span D control.

The material and geometrical properties of the steel beam are: the Young's modulus is 198.87 GN/m^2 and volume mass density is 7664.5 kg/m^3. The cross-section of the beam is a rectangular shape whose width and thickness are 25.4 mm and 12.7 mm (or 1.0 in and 0.5 in), respectively. The length of the beam is $L = 2.0\,m$, and both boundaries are free.

The external excitation force is applied at 0.22 L. The on span control force is applied at 0.55 L. The deflection is observed at 0.64 L. All distances are measured from the left boundary of the beam.

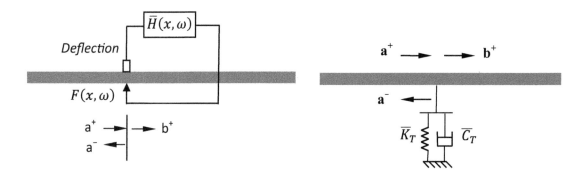

Index

Mechanical Wave Vibrations: Analysis and Control, First Edition. Chunhui Mei.
© 2023 Chunhui Mei. Published 2023 by John Wiley & Sons Ltd.
Companion Website: www.wiley.com/go/Mei/MechanicalWaveVibrations